Biobased Products from Food Sector Waste

Teresa Cecchi • Carla De Carolis

Biobased Products from Food Sector Waste

Bioplastics, Biocomposites, and Biocascading

Springer

Teresa Cecchi
ITT Montani
Fermo, Italy

Carla De Carolis
DIISM Department
UNIVPM - University Polytechnic of Marche
Ancona, Italy

ISBN 978-3-030-63435-3 ISBN 978-3-030-63436-0 (eBook)
https://doi.org/10.1007/978-3-030-63436-0

This Springer imprint is published by the registered company Springer Nature Switzerland AG
The registered company address is: Gewerbestrasse 11, 6330 Cham, Switzerland

Contents

Part I Introduction

The original version of this book was revised. The correction to this book can be found at https://doi.org/10.1007/978-3-030-63436-0_14

Part I
Introduction

Chapter 1
Food Waste in the Sustainable Development Framework

Teresa Cecchi

We live in an age of very fast and incomparable planetary change, referred to as the 'Great Acceleration'. It has brought many benefits to human society, such as an overall rise in our health, wealth, food, and security, even if the distribution of these benefits is unequal and we assist to a declining state of the Earth's natural systems. Our ever-increasing consumption is driving what can be considered a new geological epoch, the Anthropocene because for the 1st time in the Earth's history a single species – *Homo sapiens* – has such a powerful impact on the planet. This Chapter deals with the importance of using food waste as feedstock for biorefineries.

1.1 The Triple Bottom Line: "People, Planet, and Profit"

The world is facing important sustainability issues and opportunities. Sustainable development is a challenge, and society has to respond to it in the decades to come. It seems to be a dreadful task. Myriads of necessities have evolved in modern society: food, travel, housing, leisure are only examples of needs influenced by 'culture', differences between individuals and nations. The Information and Communication Technology (ICT) is shaping the society at large as steam power, electricity, and automobility did in the past.

The concept of sustainability is related to the enlightening etymology of this word. It comes from the Latin *sustĭnēre* that means "hold up, support, bear, endure". Anyhow the exact description of a sustainable future is not straightforward.

In this context, 1972 is a cornerstone year. At the UNO Conference on the Human Environment held in Stockholm, Sweden, the concept of sustainability had yet to really take off but, nevertheless, the whole international community agreed to the crucial notion that both development and the environment are not separate issues, and they could be managed in a mutually beneficial way. Environmental concerns were introduced in the political realm. The summit declaration contained the following 26 principles concerning the environment and development.

T. Cecchi, C. De Carolis, *Biobased products from food sector waste*,
https://doi.org/10.1007/978-3-030-63436-0_1

1. "Human rights must be asserted, apartheid and colonialism condemned"
2. "Natural resources must be safeguarded"
3. "The Earth's capacity to produce renewable resources must be maintained"
4. "Wildlife must be safeguarded"
5. "Non-renewable resources must be shared and not exhausted"
6. "Pollution must not exceed the environment's capacity to clean itself"
7. "Damaging oceanic pollution must be prevented"
8. "Development is needed to improve the environment"
9. "Developing countries therefore need assistance"
10. "Developing countries need reasonable prices for exports to carry out environmental management"
11. "Environment policy must not hamper development"
12. "Developing countries need money to develop environmental safeguards"
13. "Integrated development planning is needed"
14. "Rational planning should resolve conflicts between environment and development"
15. "Human settlements must be planned to eliminate environmental problems"
16. "Governments should plan their own appropriate population policies"
17. "National institutions must plan development of states' natural resources"
18. "Science and technology must be used to improve the environment"
19. "Environmental education is essential"
20. "Environmental research must be promoted, particularly in developing countries"
21. "States may exploit their resources as they wish but must not endanger others"
22. "Compensation is due to states thus endangered"
23. "Each nation must establish its own standards"
24. "There must be cooperation on international issues"
25. "International organizations should help to improve the environment"
26. "Weapons of mass destruction must be eliminated"

Sustainable development is a concept that was defined by the Brundtland Commission of the United Nations in 1987 (Brundtland Commission 1987) (Robert et al. 2005). The Brundtland Commission correctly argues: "the environment is where we live; and development is what we all do in attempting to improve our lot within that abode. The two are inseparable." According to the Brundtland Commission, sustainable development meets the needs of the actual generation without impairing the capacity of future generations to meet their own needs within a framework of intergenerational equity. In 1992, the term "sustainability" was referred to explicitly during the United Nations Conference on Environment and Development held in Rio de Janeiro, in which major world leaders acknowledged sustainable development as the major challenge. The Rio Summit marked the first international attempt to recognize the necessity for a more sustainable pattern of development. That summit was attended by over 100 Heads of State and representatives from 178 national governments and civil society which adopted an agenda (literally "things to do") for the 21st century, called Agenda 21, which "recognized each nation's right to pursue social and economic progress and assigned to States the responsibility of adopting a model of sustainable development" (United Nations 1992). Agenda 21 is the first action plan for making sure the world will change into a more just, secure, and wealthy habitat where all peoples had access to the natural resources they needed to thrive.

Agenda 21 is divided into 40 chapters, grouped into 4 sections:

Section I: Social and Economic Dimensions.
It is aimed at combating poverty, especially in developing countries, changing consumption patterns, and promoting health.
Section II: Conservation and Management of Resources for Development.
It is aimed at protecting atmosphere, forests, fragile environments, and biodiversity and at controlling of pollution and radioactive wastes.
Section III: Strengthening the Role of Major Groups.
It is aimed at strengthening the roles of youth, women, indigenous peoples, NGOs, local authorities, business and industry, and workers.
Section IV: Means of Implementation.
It is aimed at promoting science, technology transfer, education, institutions, and financial mechanisms.

Notwithstanding the Agenda 21 action plan, unprecedented and interconnected sustainability challenges now face the future development of the world, spanning food security, climate change, pollution, biodiversity loss, poverty, energy, and an over-dependence on non-renewable resources. These challenges are easily predicted to intensify when taking into account that the world population is expected to strongly increase, reaching ca. 11 billion people in 2100 (United Nations, Department of Economic and Social Affairs 2019).

The total energy need in the world is expected to increase by 48% between 2012 to 2040, with estimates of 860 quadrillion kilojoules (KJ) in 2040 (EIA 2016).

Moreover, according to the Food and Agriculture Organization (FAO) there is an annual growth rate of worldwide consumption of all agricultural products of 1.1% per year (Alexandratos and Bruinsma 2012), which translates into an additional 70 million ha of cultivated land required by 2050. Unfortunately, most of the projected lands for expansion in cultivation are in developing countries in Africa, which are often characterized by water shortages. Furthermore, urbanization and agriculture would compete for land availability (Bren d'Amour et al. 2017).

Additionally, the Earth is suffering a big climate change due to an increased global greenhouse gas emissions (GHG). In 2019, total GHG emissions reached a new high of 59.1 gigatonnes of carbon dioxide (CO_2) equivalent. The brief dip in CO_2 emissions caused by the COVID-19 pandemic may provide an opportunity for recovery and for the world's progress towards the Paris Agreement goals of keeping global warming to well below 2 °C compared to pre-industrial levels (UNEP and UNEP DTU Partnership 2021).

Environmental pollution is an additional problem. Air pollution is responsible for the deaths of 3 million people annually, and only one-tenth of the population lives in a city that complies with the WHO air quality standards (World Health Organization 2016).

The ever-increasing volume of solid wastes is an added major concern. The World Bank has estimated that cities will produce 2.2 billion tonnes of solid waste by 2025, with concomitant increases in waste management costs to $375.5 billion (Hassan et al. 2019a).

Most human activities make use of water and produce wastewater. The quantity of wastewater produced and its overall pollution load are unremittingly growing worldwide. Over 80% of the world's wastewater is discharged into the environment each year without treatment (WWAP 2017).

In addition, biodiversity is threatened. The astonishing decline in wildlife populations– a 60% fall in just over 40 years – is a grim memento and the decisive meter of our negative influence on Nature that is underpinned by biodiversity (Grooten and Almond 2018).

Only a holistic approach can be effective in identifying and proposing suitable solutions that tackle these challenges. Strategies that harness renewable resources to maintain ecological sustainability while maintaining economic growth are needed.

In Rio, two decades after the first Rio summit, the "Rio+20" Conference took place; the Leaders from 180 nations attending the United Nations Conference on Sustainable Development reaffirmed their commitment to Agenda 21 in their outcome document called "The Future We Want".

To overcome these unprecedented environmental challenges, during the last UNO Sustainable Development Summit (2015) Agenda 2030 was put forth (United Nations). It re-asserts all of the goals set by Agenda 21 as the basis for sustainable development.

Figure 1.1 illustrates a total of 17 Sustainable Development Goals (SDGS) (detailed by 169 targets) that have been agreed on by 193-member states of the United Nations. People, planet, prosperity, peace, profit, and partnership are the focus of interest of the Agenda 2030 (United Nations).

From Fig. 1.1 it is clear that there is growing attention towards a more integrated vision fully including three pillars of the multifaceted concept 'sustainability' linked to:

1. People and human aspects, such as equity, quality of work conditions, education, health, and respect for human rights. Fair business practices toward labor and the community in which the corporation conducts its business are crucial for coordinating stakeholders' interests and not only shareholders' ones. Aside from the moral trait of being "good" to society, disregarding social responsibility can impair the performance of the business because there are economic costs linked to overlooking social responsibility. Employee relations and fair wages must also be considered. Quantifying these aspects is relatively new, problematic, and fraught with potential subjectivity problems. The Global Reporting Initiative (an

Fig. 1.1 Agenda 2030's sustainable development goals. Reprinted with permission from SGD Permissions. https://www.un.org/sustainabledevelopment/ "The content of this publication has not been approved by the United Nations and does not reflect the views of the United Nations or its officials or Member States"

independent international organization that has pioneered sustainability) has developed guidelines to enable corporations and non-governmental organizations (NGOs) to comparably report on the social impact of a business.

2. Planet and physical aspects, such as preventing depletion of essential resources, their wise management, environmental protection and biodiversity. In this context, the quantitation and reduction of the ecological footprint of a business are crucial. This can be achieved by carefully managing the consumption of energy and non-renewables and reducing manufacturing waste and its toxicity before disposing of it in a safe and legal manner. However, this "cradle to grave" approach to production is not the smartest one since it would be even better if the process residuals are recycled or upcycled in other processes in a "cradle to cradle" approach with the help of digital strategies and opportunities. In a healthy sales stream, a life cycle assessment of products helps decision-makers to know what the product's true environmental cost is from the growth and harvesting of raw materials to manufacture, to supply, to eventual disposal by the enduser. It would be equitable for the business which sells a problematic product with toxic waste to endure part of the cost of its ultimate disposal. Ecologically destructive practices, such as endangering depletions of resources, must be avoided. In the long run, environmentally sound business is not less profitable than hazardous processed. For the quantitation of environmental aspects, the Global Reporting Initiative, nonprofit sustainability organizations, such as CERES, and the Institute 4 Sustainability, among others, provide companies with rigorous environmental metrics that are standardized better than social ones.
3. Profit and economic aspects, such as sound businesses, wise management, and distribution of economic resources also in developing countries, and equity in

trading. Profit is the goal shared by all businesses. Real economic benefits enjoyed by the host society and its economic environment at large must be considered. Within the sustainability framework, profit should be considered as the economic benefit shared by all stakeholders non only the company's stakeholders.

In this perspective, many apparently disconnected topics (such as energy and safety, waste reduction, and economic development) are strictly interconnected within three boundary conditions: social, environmental, and financial.

It follows that "Planet People and Profit" has become a refrain known as the Triple Bottom Line (TBL) framework that enables organizations to take a longer-term outlook and thus evaluate the future consequences of decisions.

Business writer John Elkington claims to have coined the phrase (Elkington 1998). In traditional business accounting, the very bottom line on a statement of revenue and expenses refers to profit (or loss, if the case). Corporations are nowadays requested to perform a full societal costbenefit analysis, thereby taking into account two more "bottom lines" regarding, respectively, social and environmental concerns.

This accounting framework with financial, social, and environmental details has become the dominant approach to public sector full cost accounting. Similar UNO standards apply to natural capital and human capital measurement to assist in measurements required by TBL.

The TBL approach has also been extended to a quadruple bottom line (QBL) approach to encompass a fourth pillar concerning a future-oriented approach (future generations, intergenerational equity, etc.) (Waite 2013). An integrated balance sheet was also proposed to provide a more holistic look into a company's performance (Robert Sroufe 2018), (Eccles and Krzus 2012).

If in the early stages of this process, the demand for sustainably manufactured products was putting pressure on the industries, now corporations are aware of the fact that sustainable products and processes are an economic opportunity. The concept of materials coming from natural biomass with environmental advantages of being biodegradable and/or biobased is becoming very attractive to the industry and to the consumers.

TBL companies can find financially profitable niches that were missed when money alone was the driving factor; ecotourism, fair trade and B corporations (benefit corporations), and social enterprises are common examples of these niches.

The major concern about the triple bottom line is that the three separate accounts cannot easily be added up because the planet and people accounts can not be measured in the same terms as cash profits. This has led to TBL being augmented with cost-benefit analysis.

Sustainable development should recognize the well-being of human systems that is supported by a healthy, natural environment in which future generations have an equal claim on our planet's resources.

1.2 Decoupling of Production from Fossil Feedstock and the Breakthrough of Renewable Resources

The twentieth century saw a colossal growth in chemical manufacturing which fed the world development mostly in an unsustainable way. Energy-intensive and inefficient processes reliant on fossil fuels resulted in unacceptable levels of pollution. The approach to Nature involved exploitation and control over it, and this came at the cost of hazardous operations resulting in a number of well-publicized disasters. The inadequate disposal of huge amounts of waste posed many concerns to public health, and product safety issues led to an exponential growth in chemicals legislation. It's a shame how chemistry has been demonized during the last decades, but it is true that some products typical of the developed world became an emblem of waste, pollution, and ecotoxicity. This is the case of plastics, whose attractive qualities lead us, around the world, to a voracious appetite and over-consumption of disposable plastic goods.

Continued dependence on fossil fuel energy reserves is unmaintainable because world crude oil reserves are limited and dwindling; moreover, they are associated with GHG emissions responsible for major climate change. Fossil energy and raw materials for chemicals are rapidly depleting. There is a serious need to change the industry and human civilization to sustainable habits for assuring a constant improvement of life quality. This is well perceived by consumers who ask for environmentally-sustainable products. Increased restrictions for the use of products with high "carbon footprint" are continuously put forth, industries start focusing on the development of new biobased feedstocks and are now very attentive to the reduction of GHG emission and to the sequestration of CO_2 to comply with legislation aimed at curbing GHG presence in the atmosphere. The concerns over the mounting energy demand and price, as well as environmental pollution from fossil fuel, have prompted research into renewable energy. There is a worldwide trend to produce alternative, renewable biobased fuels and chemicals to those derived from petroleum. By this scenario, the hunt for a sustainable alternative feedstock for fuels and materials for our society continues to grow.

The European Commission (EC) set mandatory targets for an overall share of 20% renewable energy and a 10% share of renewable energy in transport by 2020 (van Dam and Junginger 2011). The European Green Deal aims at improving the well-being of people by (i) making Europe climate-neutral by 2050; (ii) protecting human life, animals, and plants, (iii) cutting pollution (iv) helping companies become world leaders in clean products and technologies (European Commission 2019).

In order to decouple production from fossil feedstock and non-renewable resources, the concept of circular economy was launched in 2010 to preserve physical stocks; it is gaining momentum, and it is seen more than from a research viewpoint (Kiser 2016), shifting our approach to seeing raw materials as assets to be preserved, rather than continually consumed. The linear extract-process-consume-dispose concept is replaced by a restorative and regenerative industrial system by intention and design in order to maintain the value of resources in the economy as long as practically possible. For those familiar with the gross domestic product

(GDP), creating wealth by making things last is the opposite of what economists teach; hence governments and regulators should adopt policy levers, including taxation, to promote a circular economy in the industry. Anyhow, the circular economy idea of turning goods that are at the end of their service life into resources is easily predicted to stimulate the development of green technologies and services, thereby creating new green jobs.

A concrete framework aimed at closing the loop of product life cycles while creating businesses and jobs for people needs the participation of researchers and policymakers, but also involvement and incentives to businesses and people overall.

Manufacturers should also consider the costs and risks associated with the waste generated. Industrial symbiosis by producing innovative and useful products in one industry via utilizing co- or byproducts from another industry needs (i) the development of quality patterns for secondary raw materials to gain confidence in the market; (ii) revision of regulations concerning identification of bio- or waste-based products in the market; (iii) economic incentives for companies to launch and maintain sustainable products in the market, and support recovery and recycling systems (Lieder and Rashid 2016).

This life-cycle thinking is underpinned by (i) the use of renewable energy, (ii) the replacement of toxic chemicals with bio-benign building blocks via the green chemistry approach (Kharisov et al. 2019) to promote human health and environmental protection (Sheldon 2016), (iii) the reuse via recycling or upcycling, (iv) the biodegradability and return to the biosphere.

In this context, the approach to waste needs to be deeply revised also from a linguistic point of view.

According to the Waste Framework Directive (European Parliament 2008) "waste means any substance or object which the holder discards or intends or is required to discard" and food waste is considered a "bio-waste". Certain specified waste can cease to be waste when it has undergone a recovery, and the item can be used for a specific purpose with clear market demand. The end-of-waste status requires that the item "fulfills the technical requirements for the specific purposes and meets the existing legislation and standards applicable to products"; moreover, its use should not lead to overall adverse environmental or human health impacts.

A by-product is an item that is not intentionally produced during the production process of other products. Byproducts can be produced as a residual of, or incidental to, the production process. According to article 5 of the Waste Framework Directive (European Parliament 2008) an item resulting from a production activity, the primary aim of which is not its production, may be regarded as being a by-product only if:

(a) "further use of the substance or object is certain";
(b) "the substance or object can be used directly without any further processing other than normal industrial practice";
(c) "the substance or object is produced as an integral part of a production process; and"
(d) "further use is lawful, i.e. the substance or object fulfils all relevant product, environmental and health protection requirements for the specific use and will not lead to overall adverse environmental or human health impacts."

Along with byproducts, co-products are often referred to as valuable secondary goods generated during a production run together with other valuable materials.

The management of process waste is usually presented diagrammatically in the form of a pyramid since there is a hierarchy of actions from the least favorable to the most favorable (European Parliament 2008) to extract the maximum practical benefits from production processes and to dispose the minimum amount of waste within the circular economy framework.

The following waste hierarchy, illustrated in Fig. 1.2, shall apply as a priority order in waste prevention and management legislation and policy: (i) prevention of waste production via superior eco-design of materials, products, systems, and business models; (ii) preparing for reuse via cleaning or repairing discarded products; (iii) recycling or up-cycling via material and chemical sorting processes; (iv) other recovery strategies, e.g., energy recovery; (v) disposal (Kiser 2016), (European Parliament 2008).

Environmental protection principles, technical, economic, and social viability, as well as protection of resources and of human health, are the underpinning pillars of this legislation.

In 2012, the European Commission (EC) put forth the European bioeconomy strategy (Hassan et al. 2019a). It relies on the production of biomass and the utilization of lignocellulosic wastes for their conversion into bio-energy, as well as the production of novel biobased value-added products. At the EU level, the bioeconomy had an annual turnover of 2.3 trillion EURO, and a total employment of 18.5 million people in 2019 (Hassan et al. 2019a). Cultivating the worldwide bioeconomy ethos as the pathway for achieving SDGs is crucial.

The project "Biomass Futures" (2010–2012) assessed the role of bioenergy in meeting Europe's renewable energy targets established by the 2009 Renewable Energy Directive (The European Parliament and the Council of the European Union 2009) for 2020 and provided outlooks to 2030 and 2050. The project estimated the future availability of lignocellulosic biomass. Agricultural wastes are recognized as the largest reservoir of cost-effective feedstocks. On the converse, forestry residues are very expensive because of the lack of commercial harvesting, and many competing uses are possible. The modeling of the biomass supply chain provides data for decision-makers and other stakeholders (Panoutsou et al. 2013).

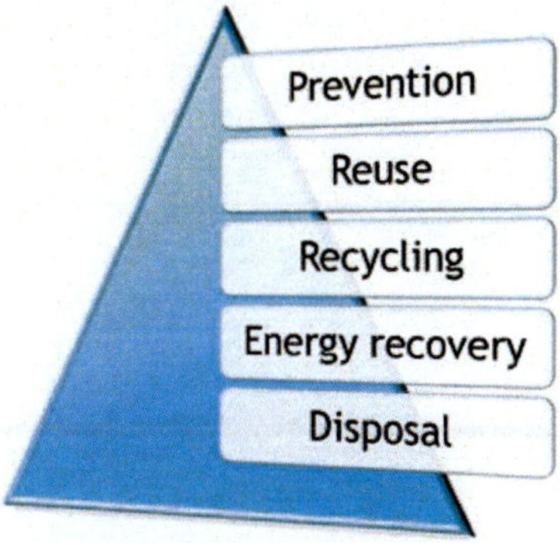

Fig. 1.2 The Waste Hierarchy (European Parliament 2008)

Decoupling of production from fossil feedstock and non-renewable resources necessarily passes through the biorefinery concept (David et al. 2017). According to the FAO, biorefineries are a pillar of bioeconomic strategy in different countries and regions, including the EU (Hassan et al. 2019a). In the biorefinery biomass is sustainably processed into a range of marketable products (food, feed, chemicals, and materials) and energy (fuels, power and/or heat) (International Energy Agency (IEA) 2009). Compared to fossilbased refineries, biorefineries are an embryonic business, with a portfolio of different biomass feedstocks and a variety of outputs.

There are 67 lignocellulosic biorefineries around the world (only about one-third operating at commercial scale) with biofuel output, while additional advanced biorefineries are under development (Hassan et al. 2019b).

The S2Biom project has projected that a total of 476 million tons of biomass need to be secured to match the demand for biobased products by 2030 (Hassan et al. 2019a). The twin challenge of recalcitrant lignin breakdown and conversion into viable products is a hot research topic with almost 200 annual patent filings and intense publication activity (Toivanen and Novotny 2017).

Biorefineries can be categorized according to a number of different ways. The easiest classification parameter is the raw material used as feedstock.

As illustrated in Fig. 1.3, we are now familiar with four generations of biorefineries, according to the raw material selection.

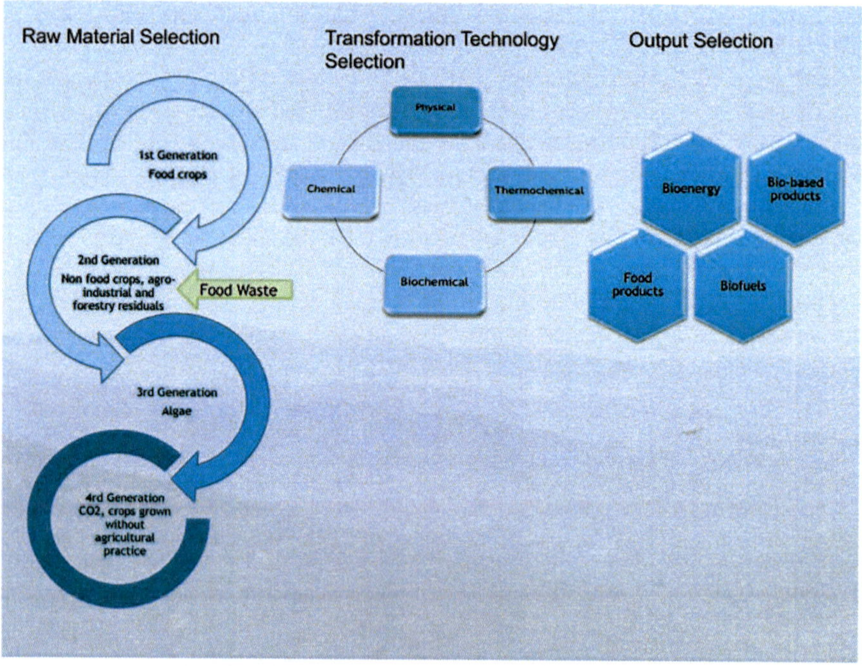

Fig. 1.3 Biorefinery classification according to the raw material selection, transformation technologies, and possible outputs

The first generation of biorefinery relied on food crops, with the main product being biofuel. In recent years, the "food versus fuel" issue for biofuel production using edible food crops and the competition in land use have arisen as a direct consequence of incentivizing energy and oil crops at the expense of food crops (Mohr and Raman 2013). However, the separation of plant biomass intended for the biorefinery from that which may be used in the food chain is a key aspect of future sustainability. For this reason the biorefinery concept has experienced a surge in popularity but also a vocal opposition to the hypothesis of diverting food-grade land and crops for non-food purposes.

The second generation of biorefinery was based on non-food crops (grass, and lignocellulosic wastes, short-rotation woody crops), agricultural and forestry residues (e.g., vegetable leaves, forest thinning, sawdust, sugarcane bagasse, rice husk and bran, corn stover, wheat straw and bran), and agroindustrial or food wastes (e.g., potatoes peel, and fruit peel and stone, spent coffee grounds, apple or tomato pomace, soybean oil cake, spent coffee grounds and so on) (Sadh et al. 2018) (Amaducci et al. 2017) The idea of using the unavoidable wastes arising from biomass processing, farming and food production as the feedstock for the biorefinery is gaining momentum.

The global production of some of these humble wastes is significant; for example, 70–140 thousand tons of potato peels are available worldwide (Wu 2016), and over 4.5 million tons of brewer's spent grain are produced in the USA (Buffington 2014).

Second-generation feedstocks might be recalcitrant and complex lignocellulosic materials that contain variable levels of cellulose, tough substrates, such as hemicellulose and lignin, and other composites. It can be observed that low lignin content and high digestibility render herbaceous biomass crops particularly suitable for second-generation biofuel production (Amaducci et al. 2017). The improvement of technologies to release fermentable sugars represents the major challenge for their efficient and scalable exploitation. In this regard, the utilization of food waste as chemical feedstock is viewed as an interesting opportunity.

The third generation of biorefinery relied on non-food marine biomass, spanning from green, red, and brown macroalgae (seaweed) to microalgae endowed with precious photosynthetic pigments. Seaweeds include *Ulva lactuca, Gracilaria vermiculophylla,* and *Saccharina latissimi*. Seaweeds are currently used in the production of food, feed, and nutritional supplements. They do not require either arable land or freshwater resources to grow, and their ash content can reach up to 60%, while the cellulose content is generally low (Hassan et al. 2019a). Microalgae, such as *Schiochytrium sp., Botryococcus braunii, Nitzschia, Hantzschia,* and *Neochloris oleoabundans,* are generally richer in lipid content compared with carbohydrate.

Algal species are rich sources of oils, and therefore attention has focused on their use for biodiesel production (Mata et al. 2010) and of a number of products of high added-value, such as docosahexaenoic acid (Trivedi et al. 2015). However, for the majority of algal species, a variable efficacy of conversion technologies, high production cost compared with the use of lignocellulosic biomass, and technical difficulties in the scale-up of cultivation operations represent major challenges. The major impediment to any substantial production of algal fuels is the insufficiency of concentrated carbon dioxide they use in their photosynthetic pathways; the development of an ability to almost fully recycle the phosphorous and nitrogen nutrients necessary for algae culture is challenging. These are serious constraints to the commercially-viable exploitation of algae (Chisti 2013).

The fourth generation of feedstocks capitalizes on carbon dioxide, the main pollutant generated through industrial processes, with evident environmental and economic benefits (Mata et al. 2010). Crops spontaneously grown in marginal lands without agricultural activities, such as Jatropha, Castor, and Karanja, could also be categorized under fourth-generation feedstocks (Moncada et al. 2016).

As regards the scale, while high added-value products, such as fine chemicals can be profitably produced in small biorefineries, for biofuel production, the scale must necessarily be large; the output itself is very sensitive to scale changes (Moncada et al. 2016).

Valorization of transformation technologies is still a formidable hurdle facing the expansion of the nascent biorefinery industry, and productive integration of individual biorefinery conversion methods remains at a relatively early stage that will be discussed in Chap. 4.

By this scenario, the use of food waste-based feedstock, illustrated in Chap. 2, can pave the way to new and negative cost approaches to produce a greater variety of higher value end innovative products that will be discussed in Chap. 5, 6, and 7, along with more traditional outputs, which are the focus of interest of Chap. 3.

The current ongoing research in the hot area of biorefineries is therefore focused on developing an advanced model which can utilize a wide range of feedstocks, including food waste.

Bibliography

Alexandratos N, Bruinsma J (2012) World agriculture towards 2030/2050: the 2012 revision

Amaducci S, Facciotto G, Bergante S et al (2017) Biomass production and energy balance of herbaceous and woody crops on marginal soils in the Po Valley. GCB Bioenergy 9:31–45. https://doi.org/10.1111/gcbb.12341

Bren d'Amour C, Reitsma F, Baiocchi G et al (2017) Future urban land expansion and implications for global croplands. Proc Natl Acad Sci U S A 114:8939–8944. https://doi.org/10.1073/pnas.1606036114

Brundtland Commission (1987) Report of the world commission on environment and development: our common future towards sustainable development 2. Part II Common Challenges Population and Human Resources 4

Buffington J (2014) The economic potential of Brewer's spent grain (BSG) as a biomass feedstock. Adv Chem Eng Sci 04:308–318. https://doi.org/10.4236/aces.2014.43034

Chisti Y (2013) Constraints to commercialization of algal fuels. J Biotechnol 167:201–214. https://doi.org/10.1016/J.JBIOTEC.2013.07.020

David C, Gutiérrez B, Lorena D et al (2017) A comprehensive review on the implementation of the biorefinery concept in biodiesel production plants. 15:691–703. https://doi.org/10.18331/BRJ2017.4.3.6

Eccles RG, Krzus MP (eds) (2012) One report. Wiley, Hoboken

EIA (2016) International Energy Outlook 2016: U.S. Energy Information Administration (EIA), Washington DC

Elkington J (1998) Cannibals with forks : the triple bottom line of 21st century business. New Society Publishers

European Parliament (2008) Directive 2008/98/EC of the European Parliament and of the Council of 19 November 2008 on waste and repealing certain Directives. EU, Brussels. Consolidated text available at https://eur-lex.europa.eu/legal-content/EN/TXT/?uri=CELEX%3A02008L0098-20180705. Accessed 8th Mar 2021

European Commission (2019) Communication from the commission to the European parliament, the European Council, The Council, The European Economic and Social Committee and The Committee of The Regions, The European Green Deal. EU, Brussels, pp 1–24. https://eur-lex.europa.eu/legal-content/EN/TXT/?uri=CELEX%3A52019DC0640. Accessed 7th Mar 2021

Grooten M, Almond REA (2018) Living planet report. 2018:41

Hassan SS, Williams GA, Jaiswal AK (2019a) Moving towards the second generation of lignocellulosic biorefineries in the EU: drivers, challenges, and opportunities. Renew Sust Energ Rev:590–599

Hassan SS, Williams GA, Jaiswal AK (2019b) Lignocellulosic biorefineries in Europe: current state and prospects. Trends Biotechnol 37:231–234. https://doi.org/10.1016/j.tibtech.2018.07.002

International Energy Agency (IEA) (2009) Task 42 Biorefinery: Bio based chemicals value added products from Biorefineries, pp 1–16

Kharisov BI, Kharissova OV, Máximo C et al (2019) Greener synthesis of chemical compounds and materials. https://doi.org/10.1098/rsos.191378

Kiser B (2016) Circular economy: getting the circulation going. Nature 531:443–446. https://doi.org/10.1038/531443a

Lieder M, Rashid A (2016) Towards circular economy implementation: a comprehensive review in context of manufacturing industry. J Clean Prod 115:36–51. https://doi.org/10.1016/J.JCLEPRO.2015.12.042

Mata TM, Martins AA, Caetano NS (2010) Microalgae for biodiesel production and other applications: a review. Renew Sust Energ Rev 14:217–232. https://doi.org/10.1016/j.rser.2009.07.020

Mohr A, Raman S (2013) Lessons from first generation biofuels and implications for the sustainability appraisal of second generation biofuels. Energy Policy 63:114–122. https://doi.org/10.1016/J.ENPOL.2013.08.033

Moncada B J, Aristizábal M V, Cardona A CA (2016) Design strategies for sustainable biorefineries. Biochem Eng J 116:122–134. https://doi.org/10.1016/j.bej.2016.06.009

Panoutsou C, Bauen A, Böttcher H et al (2013) Biomass futures: an integrated approach for estimating the future contribution of biomass value chains to the European energy system and inform future policy formation. Biofuels Bioprod Biorefin 7:106–114. https://doi.org/10.1002/bbb.1367

Robert KW, Parris TM, Leiserowitz AA (2005) What is sustainable development? Goals, indicators, values, and practice. Environ Sci Policy Sustain Dev 47:8–21. https://doi.org/10.1080/00139157.2005.10524444

Sadh PK, Duhan S, Duhan JS (2018) Agro-industrial wastes and their utilization using solid state fermentation: a review. Bioresour Bioprocess 5:1. https://doi.org/10.1186/s40643-017-0187-z

Sheldon RA (2016) Green chemistry and resource efficiency: towards a green economy. Green Chem 18:3180–3183. https://doi.org/10.1039/C6GC90040B

Sroufe R (2018) Integrated management: how sustainability creates value for any business. Emerald Publishing

The European Parliament and the Council of the European Union (2009) Renewable energy directive 2009/28/EC OF. Off J Eur Union 1:32–38

Toivanen H, Novotny M (2017) The emergence of patent races in lignocellulosic biofuels, 2002–2015. Renew Sust Energ Rev 77:318–326. https://doi.org/10.1016/J.RSER.2017.03.089

Trivedi J, Aila M, Bangwal DP et al (2015) Algae based biorefinery—how to make sense? Renew Sust Energ Rev 47:295–307. https://doi.org/10.1016/j.rser.2015.03.052

UNEP, UNEP DTU Partnership (2021) Emissions Gap Report 2020. https://unepdtu.org/publications/emissions-gap-report-2020/
United Nations (1992) United Nations Sustainable Development, pp 254–266
United Nations (2015) Transforming our world: the 2030 agenda for sustainable development. https://sdgs.un.org/2030agenda
United Nations, Department of Economic and Social Affairs PD (2019) World population prospects. Highlights
van Dam J, Junginger M (2011) Striving to further harmonization of sustainability criteria for bioenergy in Europe: recommendations from a stakeholder questionnaire. Energy Policy 39:4051–4066
Waite M (2013) SURF framework for a sustainable economy. J Manag Sustain 3:25–40. https://doi.org/10.5539/jms.v3n4p25
World Health Organization (2016) Ambient air pollution: a global assessment of exposure and burden of disease. World Health Organanization, pp 1–131
Wu D (2016) Recycle technology for Potato Peel waste processing: a review. Procedia Environ Sci 31:103–107. https://doi.org/10.1016/j.proenv.2016.02.014
WWAP (2017) The United Nations world water development report 201. Wastewater -The Untapped Resource

Chapter 2
Food Processing Industries, Food Waste Classification and Handling, Target Compounds

Teresa Cecchi

One of the major moral challenges for modern society is the fact that nearly one in ten people in the world is exposed to severe levels of food insecurity, and an estimated 2 billion people worldwide cannot rely on safe, nutritious and sufficient food (FAO 2020a), whereas industrialized countries are dealing with over-consumption of food, food-related diseases and increasing food waste production; this represents an environmental, economic, and ethical challenge for the modern society.

From a historical perspective, concerns related to food waste (FW) were brought to the scientific community since the 90s (Kroyer 1995).

Figure 2.1 illustrates the food volumes of all commodity groups produced in their primary form, in the regions of the world studied in the last decade.

A decade ago, as per the reports of FAO, about one-third of the food produced in the world for human consumption every year—approximately 1.3 billion tonnes—got lost or wasted during agricultural production (e.g., mechanical damage), post-harvest handling, storage and transportation (e.g., spillage and degradation during storage, damage during transport or non-appropriate transport systems), processing (e.g., degradation during industrial or contamination processing), distribution at the market system (e.g., improper packaging or istorage), and consumption at the household level (Gustavsson et al. 2011). It is rewarding to observe that, nowadays, lost or wasted food represents only 14% of the food produced globally, according to the newly-released FAO 2019 State of Food and Agriculture (SOFA) report. The wastage decrease is significant, and it is a first step towards the end goal of zero hunger, as indicated by the Agenda 2030 Sustainable Development Goal 2. The zero hunger goal can be achieved via Target 12.3 of the Sustainable Development Goals (SDGs), which aims at halving, by 2030, per capita global food waste and reduce food losses (Gustafson, 2019).

The huge generation of FW has been harshly impacting the environment, especially due to the emission of GHG.

Both food loss (FL) and FW are used in scientific literature to identify potential foods that are discharged, lost, degraded, or contaminated, and that will not be used as food. Unfortunately, the definitions of FL and FW often overlap.

T. Cecchi, C. De Carolis, *Biobased products from food sector waste*,
https://doi.org/10.1007/978-3-030-63436-0_2

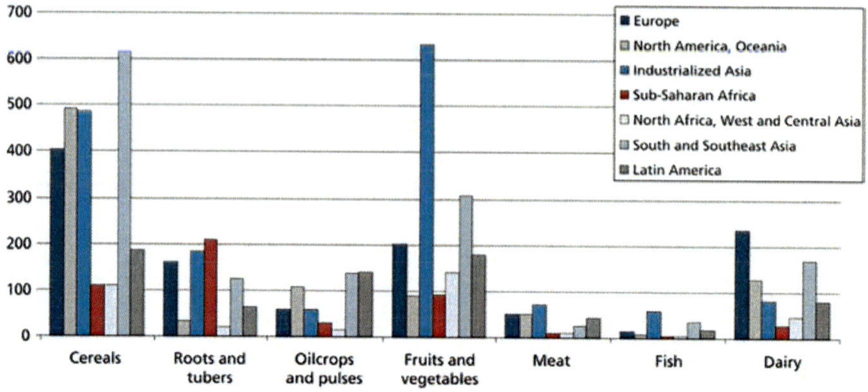

Fig. 2.1 Production volumes of each commodity group, per region (million tonnes) (Gustavsson et al. 2011). (Source: Food and Agriculture Organization of the United Nations, [2011], [Jenny Gustavsson, Christel Cederberg, Ulf Sonesson, Robert van Otterdijk, Alexandre Meybeck], [Global food losses and food waste: extent, causes and prevention], [http://www.fao.org/fileadmin/user_upload/suistainability/pdf/Global_Food_Losses_and_Food_Waste.pdf]. Reproduced with permission)

According to the Food and Agriculture Organisation of the United Nations (FAO) "*food losses refer to the decrease in edible food mass throughout the part of the supply chain that specifically leads to edible food for human consumption. Food losses take place at production, post-harvest and processing stages in the food supply chain*" (Gustavsson et al. 2011; Parfitt et al. 2010).

Food waste is the food loss occurring at the end of the food chain, at the retail and final consumption stages, and its generation is related to retailers' and consumers' behavior (Gustavsson et al. 2011; Parfitt et al. 2010).

In the following, we will refer to "food waste" to indicate both food losses and food waste. The fact that these substances are production process undesirable left over defines them as "wastes" (Commission Regulations 442/1975/EEC; 689/1991/EEC).

FW can be classified into avoidable FW (edible) and unavoidable FW (non-edible). The avoidable FW generation (due to leftover food, improper storage, unnecessary purchase) can be decreased by taking preventive measures at each level from its production to consumption. For unavoidable FW (such as shells, stones, bones, peels), there is an extreme necessity to have proper waste management and reuse practices. Unfortunately, the word "waste" does not account for the potentiality of re-utilizing it. The term "food byproducts" would be more appropriate to notify that "food wastes" are precious substrates for the extraction of functional compounds and the synthesis of marketable products.

Figure 2.2 illustrates the per capita food loss and waste. Per capita waste by consumers was between 95 and 115 kg a year in Europe and North America, while consumers in developing countries each throw away only 6–11 kg a year.

In developing countries, 40% of losses occured in the supply chain: improved infrastructures, better transportation, and the expansion of the packaging industry could help to reduce the amount of food loss and waste. On the converse, in industrialized countries, FW at retail and consumer levels due to quality standards that

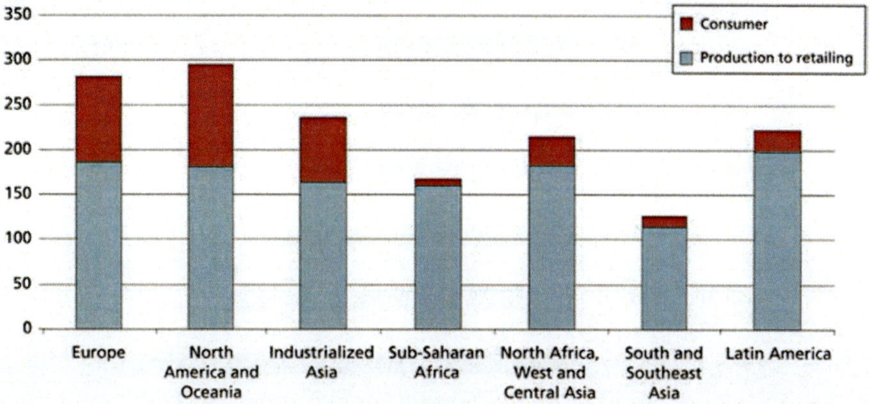

Fig. 2.2 Per capita food losses and waste (kg/year), at consumption and pre-consumptions stages, in different regions (Gustavsson et al. 2011). (Source: Food and Agriculture Organization of the United Nations, [2011], [Jenny Gustavsson, Christel Cederberg, Ulf Sonesson, Robert van Otterdijk, Alexandre Meybeck], [Global food losses and food waste: extent, causes and prevention], [http://www.fao.org/fileadmin/user_upload/suistainability/pdf/Global_Food_Losses_and_Food_Waste.pdf]. Reproduced with permission)

over-emphasize appearance and consumption habits at the household level were impressive. FW at the consumer level in developed countries is almost as high as the total net food production in sub-Saharan Africa: this is an unethical and paradoxical absurdity (Gustavsson et al. 2011). The graph of the seven commodity groups below shows the percentage food losses and waste of the edible parts of food products that were produced for human consumption.

In 2016, plant-based food, such as fruits, vegetables, roots, and tubers had the highest wastage rates of any food (ca. 45%), followed by fish (35%), cereal (30%), oilseeds, meat, and dairy products (20%) as graphically depicted in Fig. 2.3.

Figure 2.4 depicts the percentages of FW derived from a specific food industry sector (Baiano 2014). It is surprising to observe that circa one-fourth of the global waste comes from the drink industry.

Due to the notorious problem of land competition between food and biomass feedstock dedicated crops, residues represent a highly eligible biorefineries feedstock for reducing the burden on virgin raw materials. The first step for applying the industrial symbiosis concept in the food chain sector is their identification, quantification, and characterization followed by the study of the regional availability of FW producers and of potential users. This is a crucial starting point for deciding their best market positioning.

As regards the valorization of FW, it is easier to focus on the recovery of functional compounds derived from byproducts from the early stages of the food processing sector since these sources are abundant and concentrated in production or processing locations and less susceptible to deterioration compared to the wastes, produced across a broad range of retailers and households, that need to be subsequently accumulated. This collection stage complicates their valorization as sources for valuable components since the biological stability might be dramatically reduced due to the growth of pathogens.

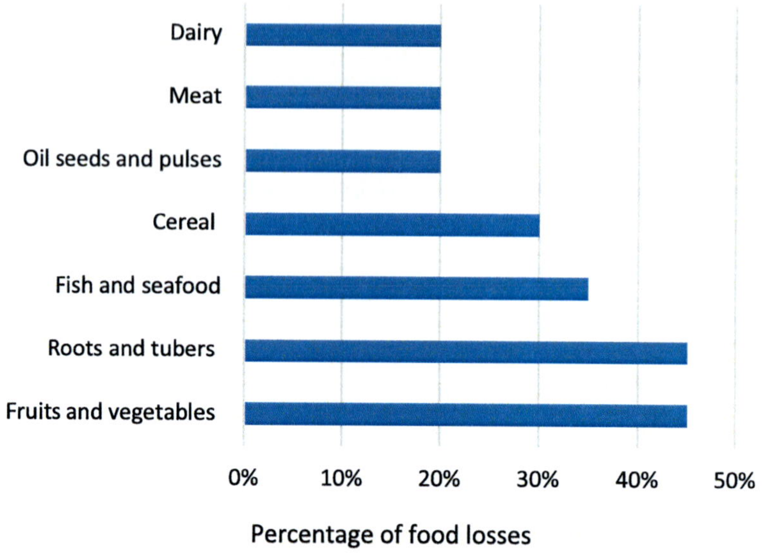

Fig. 2.3 Percentage of food losses and waste of the seven commodity groups (FAO 2016)

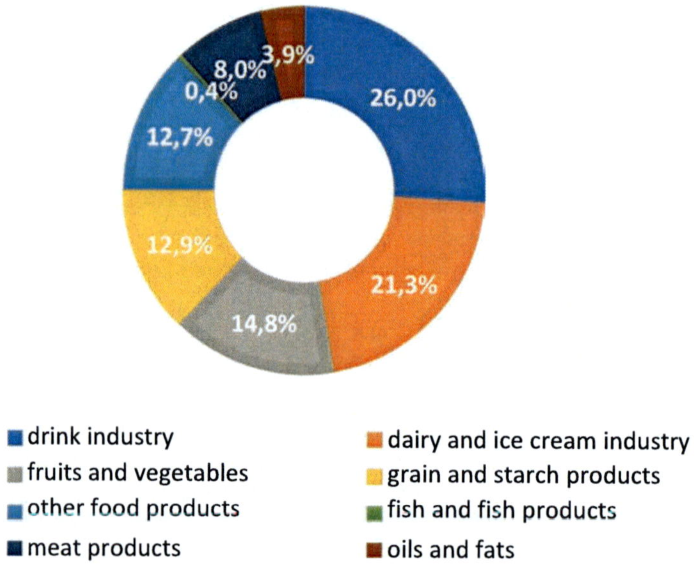

Fig. 2.4 Food waste generation; percentages of FW derived from a specific food industry sector (Baiano 2014).

Functional foods meet strengthening consumer demand for foods that enhance health and wellbeing providing a specific health benefit over and above their nutritional value. Modern science has unraveled the secrets of many FW components and established a sound basis for their nutritional and functional value.

In parallel, technology developments and cost-effectiveness have been the main drivers for a range of functional isolates. Generally speaking, the potential of the valorization of target compounds in specific FW has to be supported by evidence-based information on their valuable content promoting health. In parallel, the time between waste generation and valorization activities has a direct impact on the final concentration and quality of bioactive compounds. Isolation, purification, and recovery procedures to improve and standardize target compounds profile are needed and will be dealt with in Chap. 4.

In the following, we will detail the main FW sources and the high-value target compounds that can be obtained from them.

2.1 Fruits and Vegetables

Fruits and vegetables are high energy food items; soluble carbohydrates (glucose, fructose), vitamins, minerals, fibers, polyphenols and other bioactive compounds make their nutritive profile very rich (Schieber 2017). Fruit and vegetable spoilage is very common, and it is often indicated by discoloration, biochemical reactions, and oxidation, thermal treatments, rotting, microbial proliferation, and over-ripening. Wastes from the fruit and vegetable sectors mainly consist of hydrocarbons and relatively small amounts of proteins and fat, with an impressive moisture content of 80–90%. The wastewaters contain large amounts of suspended solids (SS), that result in high biochemical (BOD) and chemical oxygen demand (COD); among dissolved compounds, pesticides, herbicides, and cleaning chemicals can be found (Kosseva 2011). In general, fruits and vegetable byproducts can be profitably used as animal feedstuff (Angulo et al. 2012). Common disposal methods, such as composting, land filling, and incineration, give rise to environmental concerns because of toxic and greenhouse gas emissions, microbial proliferation, and landfill leachate (Dessie et al. 2018). Since wastes from the fruit and vegetable sectors are well renowned as one of the best waste source of polyphenols, dietary fiber, pigments, enzymes (cellulase, amylase, protease, phytase, etc.), fragrances and flavors (vanillin), essential oils, biopesticides, plant growth regulators, and other precious bioactive compounds (Balasundram et al. 2006; Sharma et al. 2020) their valorization is mandatory.

Fruit and vegetable waste can also be a cheap source of starch, cellulose and/or hemicelluloses; they can be hydrolyzed to soluble sugars and further fermented to produce biohydrogen, bioethanol, and biogas (Díaz et al. 2017). Microbial processing of fruits and vegetable waste has paved the way for their valorization as fermented beverages (fenny, vineger), single-cell proteins (*Saccharomyces* sp., *Candida utilis*, *Endomycopsis fibuligera* and *Pichia burtonii*), single-cell oils (Sharma et al. 2020). Acidogenic fermentation of fruit and vegetable wastes, via the bifidus pathway by *Bifidobacterium,* resulted in lactic acid production (Wu et al. 2015).

Fruit and vegetable wastes were also used to produce succinic acid: firstly, they were hydrolyzed; the fungal hydrolysis by *Aspergillus niger* and *Rhizopus oryzae* produced glucose and fructose; *Actinobacillus succinogenes* used fruit and vegetable wastes hydrolysate as the sole feedstock to produce succinic acid with high yield (Dessie et al. 2018).

In Chap. 6, the importance of this platform chemical, which has considerable potential economics and environmental meaning for the bioplastics sector, will be detailed.

Many different kinds of fruit and vegetable waste were exploited as biosorbents for heavy metal removal (Sud et al. 2008). Compared to conventional treatment methods, biosorption advantages includes: low cost, high efficiency, minimization of sludge, and good metal recovery. Biosorption is discussed in Chap. 5 (Sect. 5.8).

Tables 2.1 and 2.2 detail, for fruit and vegetable wastes, respectively, the specific source and the target molecules which can be upcycled.

In the following, we will detail a classical gamut of source byproducts and associated target compounds.

Table 2.1 Fruit waste: specific source and target molecules for recovery

Source	Target compound	References
Apple pomace	Pectin, polyphenols, fiber, lactic acid, citric acid, aroma compounds, pectinases	Galanakis (2012), Ravindran and Jaiswal (2016), Wolfe and Liu (2003), Huber and Rupasinghe (2009), Henríquez et al. (2010), Schieber (2017), Kiran et al. (2014), Dhillon et al. (2011, 2012), Liu et al. (2011), and Vorobiev and Lebovka (2010)
Apple skin	Phenols	Galanakis (2012) and Huber and Rupasinghe (2009)
Apple, black currant, chokeberry, pear, cherry and carrot pomace	Dietary fiber	Nawirska and Kwaśniewska (2005)
Chokeberry and apple pomace	Hemicellulose and pectin as sorbents for heavy metals	Nawirska and Kwaśniewska (2005)
Apple, golden rod and artichoke	Polyphenol	Peschel et al. (2006)
Berries residuals	Phytochemicals, polyphenols, pectin, dietary fiber	Górecka et al. (2010), Rohm et al. (2015), Paes et al. (2014), and Kryževičiute et al. (2016)
Black currant seeds	Essential fatty acids, tocopherols, phytosterols	Bakowska-Barczak et al. (2009)
Lingonberry (*Vaccinium vitis-idaea L.*) Pomace	Linolenic and linoleic fatty acids	Kitrytė et al. (2020)
Peach pomace	Pectin	Galanakis (2012)
Peach	Antioxidants	Plazzotta et al. (2020)

(continued)

Table 2.1 (continued)

Source	Target compound	References
Jackfruit waste (peel).	Pectin	Naik et al. (2020)
Apricot kernel	Protein isolate	Galanakis (2012) and Ravindran and Jaiswal (2016)
Kalahari melon seed	Phytosterol	Ravindran and Jaiswal (2016)
Grape pomace	Dietary fiber	Galanakis (2012)
Grape skin	Phenols, fragrance compounds, anthocyanins	Galanakis (2012) and Ghafoor et al. (2010)
Grape processing industry	Ethanol, dietary fibre, grape seed oil, pomace oil, oleanolic acid, catechin, epicatechin, and proanthocyanidins (flavanols), gallic acid (phenolic acid), resveratrol (stilbene), enocyanin (anthocyanin), procyanidins, tartates, malates, citric acid, single cell protein, short-chain fatty acids from vinasse	Schieber (2017), Galanakis (2017), Makris (2018), Luo et al. (2019), Barba et al. (2016), Santamaría et al. (2002), Löf et al. (2011), Spigno et al. (2015), Dahmoune et al. (2013), Vergara-Salinas et al. (2015) Pedras et al. (2017), Duba et al. (2015), Drosou et al. (2015), Casazza et al. (2010), Dang et al. (2014), González-Centeno et al. (2015), Bonfigli et al. (2017), Caldas et al. (2018), Boussetta et al. (2009), Liu et al. (2011), Mouratoglou et al. (2016), Patsea et al. (2017), and Bosiljkov et al. (2017)
Wine lees	Calcium tartrate, enocyanin	Galanakis (2012)
Pomace of six fruits (*Fragaria ananassa* (strawberry), *Prunus cerasus* (sourcherry), *Ribes nigrum* (black currant), *Ribes rubrum* (red currant), *Rubus fruticosus* (blackberry) and *Rubus idaeus* (raspberry)	Natural antioxidants, preservatives, antimicrobial compounds	Krisch et al. (2009)
Fresh cut fruit peels, seeds, and unused flesh	Bioactive compounds, (phenolic compounds, carotenoids, vitamins) endowed with antioxidant anti-microbial activity	Ayala-Zavala et al. (2010)
Cold hardy mandarin peel	Narirutin	Galanakis (2012)
Orange peel and other byproducts	Hesperidin, apocarotenoid, limonene, cellulose, fiber	Galanakis (2012), Ravindran and Jaiswal (2016), Bicu and Mustata (2011), and de Moraes Crizel et al. (2013)
Lemon by-product	Pectin	Galanakis (2012)

(continued)

Table 2.1 (continued)

Source	Target compound	References
Citrus by-product	Essential oil (limonene), phenolics, pectin, antioxidants, ethanol, organic acids (citric, malic, malonic, and oxalic acids), and flavonoids, fibers, amino acids and proteins, minerals, lipids, vitamins, carotenoids, xanthan gum	Matharu et al. (2016), Fernández-López et al. (2004), Scordino et al. (2007), Satari and Karimi (2018), Farhat et al. (2011), Ferhat et al. (2006), Yang et al. (2009), Fishman et al. (1999), and Liu et al. (2006)
Clementine	Antioxidants and natural preservatives polyphenols and carotenoids, flavonoids and anthocyanins, proanthocyanidins ascorbic acid	Pfukwa et al. (2019)
Rejected and processed kiwifruits	Soluble and insoluble dietary fiber	Galanakis (2012)
Banana waste	Cyanidin-3-rutinoside Lignin, pectin, cellulose, hemicellulose, phenolic compounds (prodelphinidins, flavonol glycosides), procyanidins, flavanols	Galanakis (2012) Schieber (2017), Kiran et al. (2014), and Maneerat et al. (2017)
Mango peels and byproducts	Starch, fiber, sterols, tocopherols, tannins, polyphenols, flavonols, xanthones, anthocyanins, alkylresorcinols, unsaponifiable matter, essential amino acids, antioxidant, carotenoids, vitamins, enzymes, dietary fibers, pectin	Schieber (2017), Abdalla et al. (2007), Ajila et al. (2010), and Berardini et al. (2005).
Passion fruit peels	Pectin	Canteri et al. (2010) and Seixas et al. (2014)
Papaya processing wastes	Pectin production Bioactive compounds such as benzyl isothiocyanate	Boonrod et al. (2006) and Hall et al. (2018)
Pineapple	Pectin, phenolic compounds, carotenoids and bromelain, dietary fibre	Schieber (2017), Kiran et al. (2014), Upadhyay et al. (2012), da Silva et al. (2010), and Martins et al. (2014)
Brazilian tropical fruits	Total phenols, anthocyanins, yellow flavonoids, flavonoids, resveratrol, coumarin, other bioactives	Da Silva et al. (2014)
Watermelon fruit rinds	Pectin	Prakash Maran et al. (2014)

Table 2.2 Vegetables waste: specific source and target molecules for recovery

Source	Target compound	References
Various vegetable and fruits water extracts	Soluble sugars and organic acids for yeast cultivation without nutrient addition	Stabnikova et al. (2005)
Tomatoes byproducts	Carotenoids (lycopene, β-carotene) and pectin	Schieber (2017), Kiran et al. (2014), Grabowska et al. (2019), Riggi and Avola (2008), Zuorro et al. (2011), Strati and Oreopoulou (2011), Ishida and Chapman (2009), MacHmudah et al. (2012), Baysal et al. (2000), Sarkar and Kaul (2014), and Kehili et al. (2017).
Tomato pomace	Lycopene	Galanakis (2012) and Ravindran and Jaiswal (2016)
Tomato skin	Carotenoids, fibers	Galanakis (2012), Ravindran and Jaiswal (2016), Urbonaviciene et al. (2012), and García Herrera et al. (2010)
Carrot peel	α-carotene, β-carotene, phenols	Galanakis (2012) and Schieber (2017)
Onion	Pectin, fructans, phenolic compounds and dietary fiber myricetin, quercetin, quercetin, luteolin and kaempferol	Kiran et al. (2014), Corell et al. (2018), Nile et al. (2017), Choi et al. (2015), and Zill-E-Huma et al. (2011)
Leeks	Myricetin, quercetin, luteolin and kaempferol Pectins	Corell et al. (2018) Christiaens et al. (2015)
Cauliflower floret curd	Pectin	Galanakis (2012)
Cauliflower	Dietary fiber with high water absorption index, suitable for functional food production	Stojceska et al. (2008)
Cabbage (white)	Sulforaphane Pectin	Tanongkankit et al. (2013) Westereng et al. (2008)
Cabbage (red)	Di-acylated cyanidin Anthocyanins	Valencia-Arredondo et al. (2020) and Ravanfar et al. (2018)
Cabbage (black)	Flavonoids	Romani et al. (2003)
Cabbage, watermelon husk and peach peels	Carotene production, using blakeslea trispora, a heterothallic fungus	Papaioannou and Liakopoulou-Kyriakides (2012)
Verza	Sulforaphane (a biologically active phytochemical)	Sivakumar et al. (2007)

(continued)

Table 2.2 (continued)

Source	Target compound	References
Broccoli	Bioactive compounds (glucosinolates, phenolic acids, and flavonoids) and nutrients (vitamin c, minerals, and trace elements) with antioxidant activity	Dominguez-Perles et al. (2011) and Aires et al. (2017)
Chicory roots, citrus peel, cauliflower florets and leaves, endive, and sugar beet pulps	Pectic oligosaccharides and pectin using proteases and cellulases enzymes	Zykwinska et al. (2008)
Chicory	Antioxidants	Lante et al. (2011)
	Fructan and sugars for the production of 5-hydroxymethylfurfural	Stökle and Kruse (2019)
	Hydrochar	Stökle et al. (2019)
Horse chestnut (from gardening activities)	Quercetin and kaempferol glycosides, saponins of oleane type, flavonoids used in pharmacy and cosmetic industries	Kapusta et al. (2007)
Asparagus	Phenols (flavonoids and hydroxycinnamic acids) and saponins useful as functional ingredients	Fuentes-Alventosa et al. (2013)
Carob pulp aqueous extracts	Carbon source for the production of the biocontrol agent *Pantoea agglomerans* PBC-1, usable as biopesticide	Manso et al. (2010)
Turnip greens	Organic acids, fatty acids, and tocopherols.	Chihoub et al. (2019)
	Phenolic compounds	
Field herbs	Tannin	Kardel et al. (2013)
Thistle	Antioxidant and Anti-aging Flavonolignans	Drouet et al. (2019)
	Cynaropicrin and Cnicin	Mizuno and Usuki (2018)
	Silymarin	Saleh et al. (2017)
	Total flavonoids and total polyphenols	Lee et al. (2017)
	Silybin A and silybin B	Çelik and Gürü (2015)
	Organic acids, fatty acids, and tocopherols.	Chihoub et al. (2019)
	Phenolic compounds with antioxidant, antibacterial	
Swiss chard	Polyphenols	Santos et al. (2014)

(continued)

Table 2.2 (continued)

Source	Target compound	References
Lettuce	Bioaerogel-like materials antioxidants	Plazzotta et al. (2018a)
	Functional flour or natural adsorbents	Plazzotta et al. (2018b)
	Bitter and gelling compounds, poly-phenols	Plazzotta et al. (2017)
	Cellulose nanofibers for biocomposite	Zhang et al. (2019)
	Polyphenols	Santos et al. (2014)
	Antioxidants	Lafarga et al. (2019)
Rocket salad	Phenolic and glucosinolate	Solana et al. (2016)
Radicchio	Bioplastics with antioxidant capacity	Perotto et al. (2018)
	Polyphenols	Cefola et al. (2016)
Spinach	Polyphenols	Santos et al. (2014)
	Protein complexes associated with oxygenic photosynthesis from thylakoid membranes.	Korotych et al. (2019)
	Lutein and chlorophyll	Derrien et al. (2018)
	Flavonoids	Singh et al. (2018)
	Chlorophylls	Leite et al. (2018)
	Carotenoids and polyphenols	Jaime et al. (2015)
Garlic	Organosulfur compounds	Ferioli et al. (2020)
	Polyphenols	Ciric et al. (2020)
	Polyphenols	Noda et al. (2019)
Artichokes	Inositols and caffeoylquinic acids	Mena-García et al. (2020)
	Inulin-Type Fructans	Zeaiter et al. (2019)
	Pentacyclic Triterpenes	Dai et al. (2019)
	Caffeoylquinic acids and flavone glycosides (antioxidants)	Pagano et al. (2018)
	Phenolic acids and flavonoids	Stumpf et al. (2020)
	Cynaropicrin and Cnicin	Conidi et al. (2014)
	Inulin and functional protein (from the tuber Jerusalem Artichoke)	Mizuno and Usuki (2018) Maumela et al. (2020)
Fennel	Anisole, Estragole, Fenchone, and Limonene	Eyvazkhani et al. (2020)
	Sterols, phenols from seed oil	Bettaieb Rebey et al. (2019)
Cucumbers	Antioxidant polysaccharide	Chen et al. (2019b)
Beans	Tannins, flavonoids, cardiac glycosides, anthocyanins, terpenoids, carotenoids, ascorbic acid and reducing compounds	Nawaz et al. (2020)
	Quercetin	Aghajanian et al. (2020)
	Antioxidant with anticancer activity	Yang et al. (2019)

(continued)

Table 2.2 (continued)

Source	Target compound	References
Green beans	Flavonoids	Gbashi et al. (2017)
	Antioxidant phenolic acids and flavonoids	Aires et al. (2017)
	Pectin	Christiaens et al. (2015)
Legumes	Proteins, lipids, fatty acids, vitamins, minerals and phenolic compounds	Kiran et al. (2014) and Parate and Talib (2015)
Pea peel waste	Carbon source to produce cellulase enzymes	Verma et al. (2011)
Pea pod, broad bean pod and okara	Dietary fiber, potassium, iron Fats rich in linoleic and oleic acid.	Mateos-Aparicio et al. (2010)
Soybean wastewater	Albumin, isoflavone aglycone	Galanakis (2012) and Ravindran and Jaiswal (2016)
Soybean vinasse	L(+)-lactic acid	Karp et al. (2011)
Eggplant	Anthocyanins (nasunin) polyphenols (5-O-caffeoylquinic acid) and steroidal glycoalkaloids (alpha-solasonine and alpha-solamargine)	Mauro et al. (2020)
	Phenolics, pectin, pullulan for starch replacement and edible coatings	Kazemi et al. (2019)
Sweet Peppers	Polyphenols, carotenoids	Nath et al. (2015)
	Capsinoids	Oğuzkan (2019)
	Capsaicin and dihydrocapsaicin	Martinez-Sena et al. (2017)
Celery	Flavonoids polyphenols	He et al. (2016)
	Luteolin and apigenin (bioactive flavonoids)	Han and Row (2011) and Zhang et al. (2011)
Pumpkin	Pumpkin polysaccharide	Chen et al. (2020)
	Vanillic acid	Mitić et al. (2020)
	Linoleic and oleic acids, beta-carotene and tocopherols, phytosterols and phenolic compounds	Cuco et al. (2019)
	Polyphenols and fatty acids	Ferreira et al. (2019)
	Polyphenols and fatty acids	Massa et al. (2019)
	Carotenoids	Lima et al. (2019)
	Beta-carotene, tocopherols and phytosterol	Cuco et al. (2019)
	Squalene	Hataminia et al. (2018)
	Antioxidants	Kulczyński et al. (2020)
Zucchini	Antioxidants	Lafarga et al. (2019)
	Flavonols and flavones for cosmeceutical valorization	Piccolella et al. (2019)
Mushroom waste	Ergosterol and vitamin d2	Papoutsis et al. (2020)

(continued)

Table 2.2 (continued)

Source	Target compound	References
Coffee	Antioxidants, vitamins, enzymes, cellulose, starch, lipids, proteins, pigments, citric acid, gibberellic acid, ethanol, dyes, and dietary fibres (cellulose, hemicelluloses, lignin, pectin, gums)	Murthy and Madhava Naidu (2012) and Kiran et al. (2014)
Spent coffee grounds	Lipids	Ahangari and Sargolzaei (2013)
Spent filter coffee	Polyphenols	Pavlović et al. (2013)
Tea	Caffeine, polyphenols, triacontanol, and saponins	Sui et al. (2019)
Filter-tea by-product (yarrow–Rose Hip Mixtures)	Chlorophylls and Carotenoids	Pavlić et al. (2016)
Cocoa husks	Pectin	Mollea et al. (2008)

2.1.1 *Apple*

Apple juice industry produces a million pounds a year of wastes with high costs of disposal and threat to the environment (Wolfe and Liu 2003). Up to 25% of the processed apples turn into by-product (Shalini and Gupta 2010), which is largely available during harvesting season. This FW refers to a variable mixture of peels, pomace and seed, with a high content of water, simple sugars (glucose, fructose, and sucrose) and insoluble carbohydrates, such as hemicellulose, cellulose and lignin, and pectin; it has a great biotechnological potential.

Apple pomace was used as a carbon source in the propionic-acetic fermentation via wild strain *Propionibacterium freudenreichii* T82 bacteria. After 120 hours of cultivation, propionic acid biosynthesis reached its maximum (1.771 g/L) while the content of the acetic acid reached the level of 7.049 g/L (Piwowarek et al. 2016).

High concentrations of phenolic compounds in apple peels may support the prevention of chronic diseases, such as cardiovascular diseases and cancer. Besides, several microorganisms can use apple residues as a substrate for growth (Kosseva 2011). This by-product can be used as fuel or cattle feed, but it is also a source of food ingredients, pectin, fibers; moreover, it can be used in biotransformation (Shalini and Gupta 2010).

Valuable food ingredients could be made using the waste peels, dried, and ground to a powder, without large losses of phytochemicals. Blanching treatments for 10 s and freeze-drying represented the best processing condition to keep the total phenolic and flavonoid contents similar to those of the fresh apple peels. The phenolic, flavonoid, and anthocyanin contents in 100 g of dried peels were, respectively, 3342 ± 12 mg gallic acid equivalents, 2299 ± 52 mg catechin equivalents, and 169.7 ± 1.6 mg cyanidin 3-glucoside equivalents. The apple peel powder had a total antioxidant activity of

220 mg vitamin C equivalents/g: it also had a strong antiproliferative effect on HepG 2 liver cancer cells. Apple peel powder may be used in a range of healthy food products to add phytochemicals and promote the antioxidant activity of foods (Wolfe and Liu 2003).

Huber and Rupasinghe (2009) demonstrated that apple peel extracts are natural antioxidants and effective inhibitors of oxidation of polyunsaturated fatty. Indeed, lipid oxidation is a major concern in the food industry, impacting both the food quality and health of consumers, and new natural food antioxidants would be highly eligible candidates to solve the problem.

Henríquez et al. (2010) optimized a pilot scale extraction procedure using a double drum-dryer as drying technology and maneuvered temperature and extraction time in order to maximize the retention of phenolic compounds and dietary fiber.

2.1.2 *Berries*

High amounts of wastes from currant, raspberries, and blackberries released during juices, jams, and jellies preparations manufacturing are extremely rich in antioxidants, phenolic acids, flavonoids, polyphenols, and fibers, which are eligible to be recovered for different industrial purposes.

Bakowska-Barczak et al. (2009) focused on residues of seeds from five black currant (*Ribes nigrum L.*) cultivars. Canadian black currant seed oil can be profitably upcycled for the recovery of essential fatty acids, tocopherols, and phytosterols. Extraction of polyphenols from the seed residues remaining after the oil extraction even allowed the recovery of additional valuable phenolic antioxidants.

Pap et al. (2004) studied waste production prevention and minimization in berry juice processing. Less process waste and effluent resulted from the eco-innovative practices. Supercritical fluid extraction with natural carbon dioxide (CO_2) was deemed the most environmentally friendly solvent-free extraction method for the recovery of value-added compounds.

The amount of fibers in berries residuals makes them as replacement of flour in cookies at the level 25 and 50% (Górecka et al. 2010). The added raspberry pomace in a dried form resulted in a functional product with high fiber content and with a palatable taste whose organoleptic characteristics were positively accepted by consumers.

2.1.3 *Citrus Fruits*

In contrast with other types of fruits, citrus fruits processing results in large amounts of waste materials, such as peels and seeds, usually discarded during juice manufacturing. These residues are a precious source of a gamut of compounds, including soluble sugars, fiber, organic acids, amino acids and proteins, minerals, lipids, flavonoids and vitamins (Fernández-López et al. 2004).

Different concentrations of lemon albedo and orange dietary fiber powder (including that obtained from the orange juice wastewater) were added to cooked and dry-cured sausages to increase their dietary fiber content (Fernández-López et al. 2004).

Viuda-Martos et al. (2011) studied the microbiological properties of the orange juice wastewater and their physico-chemical characteristics, including pH, phenolic compounds and antioxidant activity, organic acids, sugars, soluble solids, and color. Narirutin and hesperidin were the most concentrated phenolic compounds, whereas the main sugar and organic acid were, respectively, glucose and succinic acid. De Moraes Crizel et al. (2013) suggested that orange juice fibers byproducts can be used as a fat replacer in ice cream because they have an ideal ratio between soluble and insoluble fiber and high water and oil retention capacity. Their use as a substitute of fat in ice cream led to a 70% reduction of fat without causing significant changes in attributes such as color, odor, and texture while providing a high content of phenolic compounds and carotenoids.

The residue of pigmented orange pulp wash (Scordino et al. 2007) was stabilized by removal of enzymes and microorganisms via ultra-filtration; 80% of the water was eliminated by a reverse osmosis treatment to obtain a purified transparent liquid of slight amber color that is a sugar concentrate (28 Brix) containing about 250 g/L of sugars (glucose, fructose, and sucrose), 9 g/L of citric acid and 1 g/L of pectin. It could be easily used as a natural sweetener in the food and beverage industries.

Citrus fruits byproducts were also effectively used as cattle feed, especially for ruminants. Bampidis and Robinson (2006) studied the physical and nutrient composition, digestibility, and effects on ruminants (weight and lactating production) according to the size of the rations of these feeds to support growth and lactation in ruminants.

Citrus waste was also used for xanthan gum production (see Sect. 5.1.4) using the plant pathogen using *Xanthomonas campestris* (Murad et al. 2019).

Cellulose extraction from orange peels is a promising upcycling of these byproducts. Bicu and Mustata (2011) studied sodium sulfite and sodium metabisulfite as pulping reagents and optimized the main process parameters (reagent dosage and reaction time), to maximize cellulose yield. Recovered cellulose was characterized by good purity, whiteness, and water retention, low crystallinities, and moderate molecular weight. It can be used as filler, water-absorbent, or as raw material to produce cellulose derivatives.

2.1.4 Exotic Fruits

The exotic fruits processing industry also produces a large percentage of waste byproducts very rich in bioactive compounds, phenols, carotenoids, vitamins, fibers that are precious as antioxidants, antimicrobials, flavoring, colorants, and texturizer additives (Ayala-Zavala et al. 2010).

New uses for these types of waste are continuously envisioned because fruit byproducts presented higher bioactive compound contents than their respective fruit pulps.

Passion fruit peels, treated with nitric acid, are a major source of pectin, a dietary fiber that is a gelling agent and stabilizer (Canteri et al. 2010) due to its rheological properties.

Pineapple stem, a typical pineapple processing waste, was extensively used for bromelain extraction (Upadhyay et al. 2012). The waste obtained during this process is a nutraceutical resource of antioxidants, antimicrobials, and inhibitors against 15-lipoxygenase and advanced glycation end products. The anti fungal activity is probably due to high amounts of benzoic acid present in the waste. The bioactivities in all the performed assays indicate that pineapple stem waste could be utilized as an economical source of preventive or therapeutic phytochemicals that have a promising usage in functional food based industries.

Mango "waste mining" seems to be one of the most promising FW upcycling. Abdalla et al. (2007) studied mango seed kernel extract and oil. Their combination had synergistic antioxidant potency and could be used as natural antioxidant and antimicrobial in different kinds of food. Phenolic compounds, unsaponifiable matter, and crude protein rich in all essential amino acids are typical of this waste. Mango peels powders rich in bioactive compounds, such as antioxidant polyphenols, carotenoids, vitamins, enzymes, and dietary fibers, were incorporated into macaroni (up to 7.5%) to enhance its nutritional quality without affecting its cooking, textural and sensory properties (Ajila et al. 2010). Polyphenolics were extracted from Mango peel while pectin was recovered by alcohol precipitation (Berardini et al. 2005).

Virgin coconut oil byproducts, namely coconut skim milk and insoluble protein, can be used to obtain coconut protein powder, with good emulsifying properties that can be used in emulsified foods (Naik et al. 2012).

Besides, coconut and peanuts shell powder could be upcycled as filler in natural rubber (Sareena et al. 2012).

Papaya processing wastes served as a substrate for yeast (*Saccharomyces cerevisiae*) growth as well as a feed supplement for shrimp (Kang et al. 2010) but it is also used for pectin production (Boonrod et al. 2006).

Ueno et al. (2003) proposed to use pineapple and grapes processing wastes as substrate and enzyme for successful lactic acid production.

Da Silva et al. (2014) quantified the levels of total phenols, anthocyanins, yellow flavonoids, flavonoids, resveratrol, coumarin, and other bioactives in pulps and byproducts of twelve typical tropical fruits from Brazil (pineapple, acerola, cashew apple, guava, soursop, papaya, mango, passion fruit, surinam cherry, sapodilla, and tamarind). They are interesting functional food ingredients.

2.1.5 *Tomatoes*

Tomato is the second most important global vegetable crop next to potato, with a yearly production of 100 million tons fresh fruit in 144 countries (Kalogeropoulos et al. 2012). Tomatoes are rich in bioactive and valuable phytochemicals (Dumas et al. 2003), such as carotenoids, vitamin C and E and various phenolic compounds. The strong correlations between carotenoid consumption and a reduced risk of cancer and coronary and cardiovascular diseases, probably because of the carotenoids antioxidant activity, made tomato processing byproducts very interesting as a potential and inexpensive source of these bioactive compounds (Urbonaviciene et al. 2012). Actually a huge amount of carotenoids are discarded along with the peels during the processing of tomatoes into pastes or sauces. Lycopene is the most abundant and the most antioxidant carotenoid in tomatoes, but β-carotene is the major dietary precursor of vitamin A.

Kalogeropoulos et al. (2012) pointed out that antioxidants from tomatoes byproducts could be used for the formulation of functional foods or serve as additives in food systems to extend their shelf-life.

Ćetković et al. (2012) confirmed the importance of tomato wastes as a potential source of phenolic antioxidants and anticancer agents, identifying and quantifying some individual phenolic compounds, including phenolic acids and flavonoids and their antioxidant activity.

For the commercial exploitation of these wastes as nutraceutical resources, it is necessary to consider both environmental and extraction profitability.

The majority of the studies published in the literature focus on two target carotenoids: lycopene and β-carotene (Grabowska et al. 2019). Riggi and Avola (2008) studied the dependence of carotenoids content in tomato wastes on their source and the month in which they were produced. The highest lycopene and β-carotene contents were recorded in summer due to the higher proportion of red fruits that is statistically correlated with the lycopene content. Zuorro et al. (2011) indicated that the recovery of lycopene could be greatly enhanced by using cellulolytic and pectinolytic enzymes. Thermal processing of tomatoes (for example, the blanching process) increases carotenoid bioavailability in tomato peels. It was also found that the best source for purified lycopene was tomato peel with lycopene and β-carotene concentrations higher in the blanched tomatoes than in the non-blanched tomatoes (Urbonaviciene et al. 2012).

As regards the extraction of carotenoids, it is clear that the use of a food-friendly and nontoxic solvent is mandatory. Ethyl lactate gave the highest carotenoid yield, compared to other organic solvents, after optimizing the extraction parameters (time, temperature and steps) (Strati and Oreopoulou 2011; Ishida and Chapman 2009). MacHmudah et al. 2012) discussed the use of supercritical carbon dioxide to extract lycopene from tomato peel by-product containing tomato seeds which exerted a synergistic effect on extraction yield. Baysal et al. (2000) assessed various parameters for lycopene extraction by supercritical carbon dioxide such as temperature, pressure, time, CO_2 flow rate, and co-solvent addition.

García Herrera et al. (2010) demonstrated that peels are mainly composed of carbohydrates, with an average value of 80% of total dietary fiber; insoluble fiber is the major component. Tomatoes peel fiber can be used as a food ingredient of new functional foods.

2.1.6 *Pulses*

Pulses represent an excellent source of high-quality dietary proteins, dietary fiber, micronutrients, and phytochemicals.

Pulses are prone to serious damage due to insect pests which cause serious losses; fumigation is the current way to prevent crop damage, but the use of an anti-metabolic protein from wild pulses, involved in plant defense mechanism, such as arcelin, is promising (Karuppiah et al. 2018).

Horse gram is an underutilized pulse crop that has been used as a food item for millennia. It grows in a wide assortment of adverse climatic conditions. Besides its importance for human nutrition, its non-nutritive bioactive substances, such as phytic acid, phenolic acid, fiber, enzymatic/proteinase inhibitors, flavonoids, and high molecular tannins, have significant metabolic and/or physiological effects and might be responsible for its use in ethnic and traditional medicine as a potential therapeutic agent to treat kidney stones, urinary diseases, respiratory infections etc. The use of antioxidants from legumes decreases the risk of intestinal diseases, diabetes, coronary heart disease, and dental caries; hence the seed coat fractions and other waste can be profitably used (Prasad and Singh 2015).

Tannins, flavonoids, cardiac glycosides, anthocyanins, terpenoids, carotenoids, ascorbic acid, and reducing compounds (Nawaz et al. 2020), antioxidants with anti-cancer activity (Yang et al. 2019), and quercetin (Aghajanian et al. 2020) are important bioactives that can be recovered from beans waste. Flavonoids (Gbashi et al. 2017), antioxidant phenolic acids (Aires et al. 2017), and pectin (Christiaens et al. 2015) are specific target compounds that can be recovered from green beans waste.

Peel pea waste was used as a carbon source for cellulase production under solid-state cultivation by *Tricoderma reesei*. Cellulase (Sect. 5.7.2) is a crucial enzyme for lignocellulosic biomass waste bioconversion (Verma et al. 2011). Dietary fiber, potassium, iron, and fats rich in linoleic and oleic acid can be recovered from pea pod and broad bean pod (Mateos-Aparicio et al. 2010).

2.2 Winemaking Industries

According to FAO statistics (FAO 2020b), grapes are one of the largest and highest value crops grown worldwide in order to produce wine. In the Mediterranean area, Italy, France and Spain are the three main wine producers. The annual production worldwide can reach almost 80 million tons, but 20% of processed grapes remain as a by-product.

The waste from the winemaking process mainly consists of pomace, which is the solid waste that remains after the grape pressing and the fermentation processes, and lees, which are clarification sediment.

Grape pomace consists of the seedless fraction (pulp, skin and stem) and the grape seed itself. Wine lees mainly contain ethanol, tartaric acid, and yeast cells with phenolic compounds (Kalli et al. 2018). The traditional disposal of large amounts of pomace in the environment or its incineration may be detrimental because the high polyphenols content decreases the pH and increases the resistance to biodegradation.

The winemaking byproducts generated after grape exploitation constitute a very cheap source of high-value ingredients, including hydrocolloids, dietary fibers, lipids, proteins, and natural antioxidants, which can be used as dietary supplements and nutraceuticals in functional foods, or in the production of phytochemicals, medical remedies, antimicrobial components, and cosmetics (Yu and Ahmedna 2013). The winery waste biomass contains an unparalleled amount and assortment of polyphenols, a broad class of secondary plant metabolites. Target compound content obviously depends on the grape variety, pedo-climatological conditions, and agricultural conditions. In red skins the polyphenol content is usually higher than for corresponding white skins. Specific polyphenols are contained in stems and seeds, and their apportionment is significant. Given the importance of polyphenols as target compounds in wine and fruit industry byproducts Fig. 2.5 illustrates the principal polyphenolic subclasses.

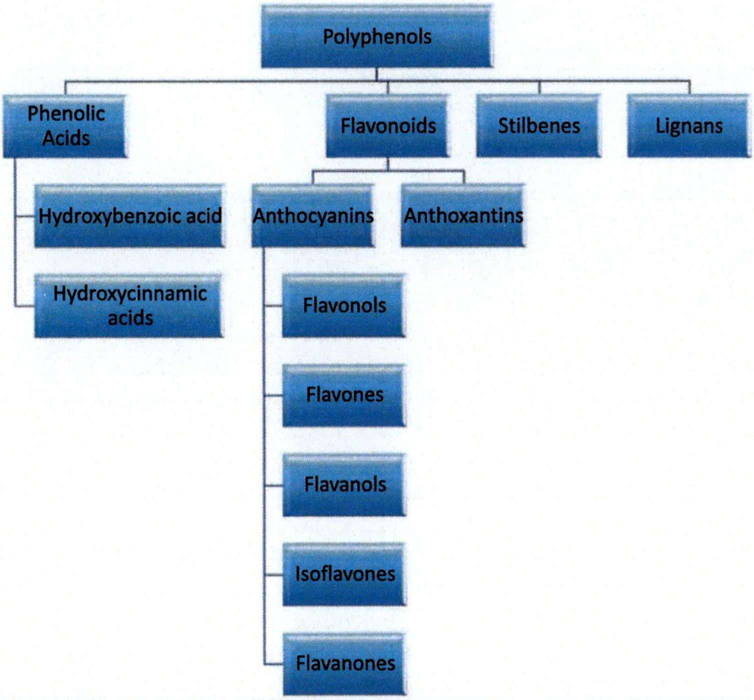

Fig. 2.5 Principal polyphenolic subclasses

Growing evidence suggests that polyphenolic molecules are endowed with a gamut of beneficial bioactivities and have the potential to exert antioxidant, free radical scavenging, anti-inflammatory, antimicrobial, cardioprotective, anticancer, anti-aging, metal chelating, antithrombotic, and antidepressant activity (Rabiei et al. 2017). These multifunctional compounds are capable of interfering with the progression of diabetes, cardiovascular, carcinogenic, neurodegenerative diseases and inhibiting allergic reactions.

The effective recovery of polyphenols from vinification waste is the focus of interest of many laboratory-scale research studies; the use of organic solvents is unsuitable in terms of safety and poses environmental concerns. Green, nontoxic solvents and eco-compatible technologies represent a hot research field with respect to health-related and environmental risks. In this context, an aqueous bio-ethanol mixture is a green and effective solution for extraction of polyphenols from pomace (skins and seeds), from seeds, and for the extraction of trans-resveratrol from grape stems. Glycerol is a good bio-solvent competitor. Acidity was an important parameter to maximize the extraction yield; lactic acid was successfully used for flavonoid extraction, while acetic acid was shown to perform better for flavanol recovery. Along with conventional green solvent extraction, additional specific techniques, such as pressurized-liquid extraction, microwave-assisted extraction, ultrasound-assisted extraction, supercritical fluid extraction, accelerated solvent extraction, and subcritical water, and deep eutectic solvent extraction (that will be dealt with in Chap. 4) were successfully used (Makris 2018; Aliakbarian et al. 2012; Kalli et al. 2018).

Novel applications of winery by-product will be detailed in the following.

2.2.1 *Winery By-Product for the Food Industry*

First of all, within a circular economy approach, grape pomace finds application in the wine industry itself. It was exploited as a fining agent for red wines to remove tannins (Gil et al. 2018), and to improve the antioxidant potential of red wines, as indicated by total phenolic and flavonoid content (Jara-Palacios et al. 2016). Dehydrated waste grape skins were able to release significant amounts of pigments, bioactive compounds, and aromas into white wines (Pedroza et al. 2011), and positive sensory repercussion from grape seed flour usage was also reported (Rosales Soto et al. 2012). Fermentation of grape pomace resulted in hydrolytic enzymes such as xylanase, and exo-polygalacturonase, and cellulase commonly used in the clarification processes in wine cellars and juices industries as well as in paper and pulp industries (Walia et al. 2017; Díaz et al. 2012).

Grape waste was also used as a functional ingredient to produce cheese (Marchiani et al. 2016), ice creams (Sagdic et al. 2012), and fermented milk (Frumento et al. 2013) with improved antioxidant activity. Cereal bars, pancakes, and noodles were enriched with phenols from Merlot grape seed flour to provide these food products with high antioxidant activity (Rosales Soto et al. 2012), and, similarly, grape pomace was used in bakery products and cookies for the same reason (Acun and Gül 2014).

Grape byproducts are a source of natural antioxidants that can replace synthetic antioxidants such as butylated hydroxyanisole and butylated hydroxytoluene, studied for potential toxicities, including hazard for carcinogenicity and genotoxicity (Kalli et al. 2018).

As polyphenols delay lipid peroxidation, grape pomace and its extracts have the potential to protect foods from oxidative damage. In this context, grape byproducts were used for lipid oxidation protection in yoghurt and salad dressings (Tseng and Zhao 2013), cooked beef, and pork patties (Rojas and Brewer 2007; Lorenzo et al. 2014). lamb meat (Guerra-Rivas et al. 2016), pork burgers (Garrido et al. 2011), minced fish muscle during frozen storage (Sánchez-Alonso et al. 2007), cooked and uncooked chicken meat (Selani et al. 2011). Extracts of grape pomace were incorporated into chitosan edible films providing antioxidant properties and promising shelf-life extension (Ferreira et al. 2014).

Along with the antioxidant activity, different components of grape pomace show antimicrobial activity against food spoilage microorganisms and pathogens, limiting the color deterioration, acidification, slime forming bacteria, and gas production. The most active compounds were reported to be gallic acid, hydroxybenzoic, and vanillic acid (Tesaki et al. 1999; Kalli et al. 2018).

In the fruit processing industry, white grape skin extract improved the antioxidant activity and color stability of a model fruit juice and retained the stability of probiotic strains, during storage (Shah et al. 2010).

Grape byproducts provided protection from potentially hazardous substances, such as acrylamide, classified as "a probably carcinogenic agent to humans", in heat-processed foods: polyphenols from skin extracts and grape seeds blocked its formation during frying (Xu et al. 2015). Grape seeds proved to be able to reduce residuals of nitrite after ripening of dry-cured sausages thereby inhibiting the formation of N-nitrosodimethylamine, identified as a mutagenic and a carcinogenic agent in all species examined (Li et al. 2013; Wang et al. 2015).

Grape pomace is a potential substrate for the production of microbial proteins (single-cell protein) with a high concentration of essential amino acids like lysine and valine for livestock feed or human nutrition (Kalli et al. 2018).

Tartaric acid from lees may serve as a natural preservative, as an emulsifier in the bread-making industry, as an acidifier in the winemaking industry, and as an ingredient of many food products and beverages. Similarly, enocyanin, known as E163 in the food sector, is a natural pigment derived from the anthocyanins in grape skin, approved by EFSA (European Food Safety Authority) as food dyes and in pharmaceutical products (Smith and Hong-Shum 2011).

Target compounds that can be extracted from winery waste are outlined in Table 2.1.

2.2.2 Winery By-Product for Cosmetics and Pharmaceutical Industry

Natural formulations are the focus of interest of the anti-aging market to avoid health risks of allergies from long-term exposure to synthetic preservatives, colorants, aromas, and stabilizers.

Anti-ageing properties of polyphenols from grape pomace make them a cost-effective source for the cosmetic industry also because they are endowed with antibacterial and antifungal properties that are crucial in cosmetic formulations (Baydar et al. 2006). Special care during encapsulation and delivery is needed to prevent the degradation of bioactive compounds (Davidov-Pardo and McClements 2015).

Many dietary supplements include grape seed, grape extract, and red wine powder for therapeutic purposes due to their antioxidant, hypoglycemic and hypolipidemic effects (Weseler and Bast 2017; Sun et al. 2016). The addition of grape seed extract and freeze-dried skin pomace to white bread was reported to reduce the post-prandial glycemic response and cholesterol levels (Coe and Ryan 2016; Mildner-Szkudlarz and Bajerska 2013). The grape seed extract was shown to reduce *Campylobacter* induced pro-inflammatory cytokines secretion in human intestinal epithelial cells HT-29 in a dose-dependent manner (Silvan et al. 2017).

Grape pomace was also exploited as a drug delivery vehicle to create microemulsions (Kumar et al. 2014).

2.2.3 Winery By-Product for the Environment, Agriculture, and Animal Feeding

Removal of heavy metals from wastewater is a hot research topic; conventional methods such as chemical precipitation, ion exchange, membrane separation, reverse osmosis, and electro dialysis, are costly and not very effective. Plant wastes, including grape stalks, rice husk, black gram, waste tea, coffee, and walnut shell proved to be potent adsorbents for heavy metal removal (Tripathi and Rawat Ranjan 2015). Winery wastes adsorb heavy metals from aqueous solutions thanks to an ion-exchange mechanism with a significant release of Ca^{2+}, Mg^{2+}, K^+ and H^+ from grape stalks and uptake of Cu^{2+} and Ni^{2+}. In addition, grape stalk wastes can be reused for the decontamination of metal-containing effluents (Villaescusa et al. 2004).

Grape pomace can be composted at first and then can be used as a soil conditioner with a low content in phosphoric acids and nitrogen (Ferrer et al. 2001).

Grape seed can be used in animal feeding (Brenes et al. 2016) with beneficial effects on animal health (Moate et al. 2014; Sehm et al. 2011; Hao et al. 2015; Ebrahimzadeh et al. 2018).

2.2.4 Winery By-Product as Biorefinery Feedstock

Following the biorefinery idea, grape marc, pomace, vinasses, wine lees, and winery wastewater have been used as raw materials to produce lactic acid, biofuels including ethanol, carboxylic acids (butyric acid, acetic acid), butanol, acetone, enzymes, high-value chemicals and energy via enzymatic or chemical hydrolysis and anaerobic digestion and pyrolysis (Luguel et al. 2011). Factors that may affect the revenue generation of winery waste biorefinery include harvesting, transport, treatment, storage, and distribution (Zacharof 2017).

Anaerobic digestion of grape marc waste was performed under mesophilic conditions at a laboratory scale to produce methane with a coefficient of 252 ± 31 mL of CH_4 per gram of added volatile solids at a temperature of 0 °C, and pressure of 1 atm. The volatile fatty acids concentration increased, and the pH and alkalinity slightly declined over the experimental time. The biomethanization can be implemented for waste management, reducing the large volume of waste generated (Javier et al. 2019).

High-quality short-chain fatty acids can be recovered from the anaerobic fermentation of wine vinasse and waste activated sludge mixtures. The diversity and abundance of fermentative bacteria resulted in the rapid hydrolysis and acidification of bioavailable substrates for short-chain fatty acids generation. Moreover, the presence of wine vinasses enhanced the dewaterability of fermentation residues (Luo et al. 2019).

2.3 Cereals and Tubers

Globally, cereal grains obtained from seeds of *Gramineae* family (such as wheat, rice, barley, maize, sorghum, millet, oat, and rye) fill a significant proportion of the human diet. Wheat is the dominant crop supply in high-income countries, while in South Asia rice is more prevalent (Gustavsson et al. 2011). Cereals processing is a key point of supply chains for important products as bread, pasta, and sweets. Cereal processing produces various byproducts like bran and germ (Sharma et al. 2020). Quantifying and correctly qualifying byproducts are an important starting point for deciding their best market positioning. Traditionally, straw, husk, and dried leaves of crops like wheat, corn, rice, and barley are used for making thatching roofs, broom, hand fans, baskets, and other commodities. Husks is actually a thermal insulator in the structures of environmentally friendly buildings (Hỳsek et al. 2019) and can be used in construction materials like lightweight bricks (Kumar and Mor 2013). Cereals products often undergo pretreatments to produce semi-finished products for various agri-food manufacturing. The typical example is represented by the milling-industry for the production of flours. These semi-finished products are used, in turn, as raw materials for various food productions. Pretreatments are generally characterized by low production yields with significant production of byproducts. Their quantification seems, therefore, to be important for both economic and environmental reasons. In addition to cereals residues from the early steps of the manufacturing process, bakeries, confectioneries, and pasta factories generate additional very variegate cereal-based FW. The complexity of data collection does not enable satisfactory official data assessments. Table 2.3 details, for cereal and tubers waste, specific sources, and target molecules for recovery.

2.3.1 *Soft and Durum Wheat Processing*

In addition to flours suitable for the production of bread, pasta, and other products for the human diet, wheat milling provides a series of byproducts that are mainly used for feed preparation.

Processing soft and durum wheat aims to the maximum separation between the flour containing endosperm from the germ and outer layers (bran). Notably, the milling process is represented by a sequence of physical operations like cleaning, conditioning, breaking, sifting, and grinding.

This last step aims at breaking seeds, opening caryopsis, and detaching, as much as possible, endosperm from the outermost seed parts, leaving them as flat flakes. Further grinding processes aim to reduce particles size. The flour yield is about 75%; hence we have 25% of other byproducts such as bran (10%) which contains outer integument fragments and other seed parts almost entirely deprived by the floury nucleus and other cereal residues (15%), namely outer skins fragments and seed parts deprived of wheat bran (Riva 2013). The grinding of wheat produces byproducts, rich in high nutritional proteins, dietary fibers, and glucuronoarabinoxylans, that represent a form of hemicellulose found in the cell walls of grasses (Hollmann et al. 2009).

For production of refined flour, bran and germ are removed as they adversely affect the processing properties (Sharma et al. 2020). Wheat bran was used as a bread nutritional improver after an integrated biotechnological approach, combining xylanase treatment and lactic acid bacteria fermentation with a relevant decrease of the predicted glycemic index of the fortified bread (Pontonio et al. 2020). Water insoluble fibers (i.e., hemicelluloses) are able to regulate intestinal function and are used in supplement food products or ready meals (Rodríguez et al. 2006).

Wheat bran is an agroindustrial residual which represents an eligible feedstock in biorefinery processes. In line with the need to better utilize underutilized agricultural fractions, a sequential processing for extraction of starch, lipids, and proteins from wheat brans leaving a rest-material enriched in dietary fiber was

Table 2.3 Cereal, roots and tubers waste: specific source and target molecules for recovery

Source	Target compound	References
Wheat byproducts	Arabinoxylans	Galanakis (2012) and Hollmann et al. (2009)
Wheat straw	Hemicellulose	Galanakis (2012)
Wheat bran	Glucuronoarabinoxylans, fructans, nonstarch carbohydrates and oligosaccharides, phenolic compounds, lipid soluble vitamins and folic acid, and phytosterols, arabinose and xylose	Galanakis (2012), Ravindran and Jaiswal (2016), Schieber (2017), Sardari et al. (2019), and Bedő et al. (2019)
Straw	Lignin nanoparticles with (applications in the biomedical and environmental fields, such as tissue engineering or regeneration, artificial muscles, and strong underwater antifouling materials)	Lou et al. (2020) and Iravani and Varma (2020)
	Furfural	Brandenburg et al. (2018)
	Cellulose and Methylcellulose	Kumar and Kumar Walia (2017)
Rice bran	Albumin, globulin, hemicellulose b, insoluble dietary fiber, lipids, minerals, and antioxidants (vitamin e and oryzanol)	Galanakis (2012), Ravindran and Jaiswal (2016), Kiran et al. (2014), and Schieber (2017)
Rice husk	Silicon, silica	Kong et al. (2019) and Chen et al. (2019a)
Sesame husk	Insoluble dietary fiber	Ravindran and Jaiswal (2016)
Oat mill waste	β-Glucan, antioxidants	Galanakis (2012), Ravindran and Jaiswal (2016), and Kock et al. (2018).
Malt dust	Glucose, arabinose & galactose	Galanakis (2012),
Brewery's spent grains	Arabinoxylans, ferulic acid, fermentable sugars, bio-oil, bio-char, bioactive peptides (Connolly et al. 2019)	Galanakis (2012), Ravindran and Jaiswal (2016), Luft et al. (2019), López-Linares et al. (2019), Lorente et al. (2019), Connolly et al. (2019), and Ferrentino et al. (2019)
Brewery spent liquid	Brewery spent liquid as fermentation substrate for citric acid production or microalgae production.	Dhillon et al. (2012) and Amenorfenyo et al. (2019).
	Antioxidants	Barbosa-Pereira et al. (2013)
Potato peel	Phenols, antioxidants, flavonoids and flavonols	Galanakis (2012), Mohdaly et al. (2010), Sabeena Farvin et al. (2012), Kanatt et al. (2005), Okuno et al. (2002), and Amado et al. (2014)
Potatoes	Lysine, protein (patatin), steroidal alkaloids, cellulolytic enzymes, adsorption dyes, and biopolymer films	Kiran et al. (2014), Matharu et al. (2016), and Schieber (2017)
Sugar beet molasses	Organic acids	Galanakis (2012)
Sugar beet pulp	Phenols, antioxidants, flavonoid and flavonol	Mohdaly et al. (2010)
Cassava	Sugars	Ona et al. (2019)

designed and developed. After a mild water-based extraction of starch, supercritical fluid extraction of lipids, and alkaline extraction of proteins, the remaining material had a fiber content that on average corresponded to 73±3% (Sardari et al. 2019).

A microbial saccharification of wheat bran was the first step for bioethanol fermentation (Farkas et al. 2019). A biotechnological approach was also used, via an integrated biorefinery process, for arabinose and xylitol production from wheat bran; a dilute-sulphuric-acid fractionation process resulted in xylose-rich and arabinose-enriched hydrolysates, separately. Arabinose biopurification of the hydrolysates by *Ogataea zsoltii* gave pure arabinose solutions. Xylitol fermentation by *Ogataea zsoltii* on the xylose-rich hydrolysate under the best pH and aeration conditions resulted in a xylitol yield of 53% in the bioreactor (Bedő et al. 2019).

Bioconversion of wheat bran into high value-added products, such as ethanol, lactic acid, and inulinase, was optimized, and mathematical modeling of fermentations from acid-pretreated and non-detoxified wheat bran predicted the experimental production data. Ethanol and lactic acid yields were up to 30.44% and 66.68%, respectively, when the free *Scheffersomyces stipitis* (ATCC 58,784) and the immobilized *Lactobacillus casei* were used. The feasibility of the process was clearly demonstrated (Germec et al. 2019).

2.3.2 *ByProducts from Rice Processing (Oryza sativa L.)*

The processing needed to transform raw rice into edible rice aims at improving the quality of the edible product consisting of cleaning, husking, bleaching, and optical separation.

During the cleaning process impurities like straw, stones, soil, and foreign seeds are eliminated. Husking is the second step in which peeling machines remove husk (the outer shell that covers the grain) by rubbing; this way, brown or whole-rice is obtained. Husking of rice produces byproducts, which are rich in high nutritional proteins, dietary fibers, and glucuronoarabinoxylans (Hollmann et al. 2009).

The bleaching aims at removing chaff and other residues. The product obtained must be further separated by broken seeds with a separator machine. By using the optical separation, a further refinement of rice is carried out on the basis of seed color. Staining defects are often produced during processing operations.

Thus, the yield of refined rice is around 60–65%, while different byproducts could reach 40%.

Byproducts can vary due to a gamut of rice varieties, but a common classification can be summarized as follow:

- Rice husk (ca. 20%) is the main brown-beige by-product obtained by rice processing. It is the coating on a grain of rice, made of hard materials, including silica and lignin. It is available in large quantities continuously. Its average humidity is around 10%, and the gross energy (or higher heat value) of its dry matter can be up to 15–16 MJ/kg. It is much more resistant than wheat husk. Its low density ranges from 132 to 140 kg/m^3, and it is biologically very stable. The ashes are made almost entirely of silicon oxide and are around 17% (Riva 2013; Kumar et al. 2016).

- Rice bud and bran (ca. 10%). Rice germ essentially consists of embryos and small rice fragments. It is generated during the whitening process. Its humidity is around 12%; the gross energy of its dry matter is about 22.3 MJ/kg, while ashes represent 8.7%; Bran contains dietary fibers (β-glucan, pectin, and gum) various antioxidants, rice bran oil, and unsaponifiable components including tocotrienols, γ-oryzanol, and β-sitosterol; all these components may be healthy as regards the plasma lipid profile (Riva 2013; Watson et al. 2014).
- Rice broken grains (ca.10%) with smaller dimensions. They have a humidity value of 12%; the gross energy of its dry matter is about 18.6 MJ/kg, while ashes are around 0.8% (Riva 2013).

Rice husk is useful as pet feed fiber, fertilizer and it is also a substrate for vermicomposting technique (Kumar and Mor 2013) or in fermentation processes to adjust moisture, maintain the porosity of fermentative material for gaseous exchange (Sharma et al. 2020).

Recycling of carbonized rice husk was used for producing high purity silicon (the purity of the silicon can reach up to 99.9) by the combination of electric arc smelting and slag refining (Kong et al. 2019). A zinc silicate glass system (ZnO-SiO2) synthesized using the melt and quench method and derived from white rice husk ash was doped with various europium ion concentrations; the luminescent properties are useful for optical devices such as white light-emitting diodes (Khaidir et al. 2019). Rice-husk-derived porous silica (Kumar et al. 2016), modified with recycled copper from industrial wastewater and cerium, was found to be useful to remove mercury and NO from simulated flue gases (Chen et al. 2019a). Rice husk ash was also profitably mixed with some cementitious structural materials (Shimamoto and Suzuki 2019) or to produce silica fertilizer for rice plants (Sekifuji and Tateda 2019), steel fiber-reinforced concrete (Koushkbaghi et al. 2019), and for pavement base material (Poltue et al. 2019).

2.3.3 Potatoes

Potato is the largest crop worldwide since it is a quantitatively important energy source due to the high content of carbohydrates, especially starch. Processing of potatoes mainly for the production of chips or French fries generates peels or cull potatoes as byproducts. The use of byproducts from potatoes industry as cattle feed is very common (Nelson 2010). The upcycling of their peel waste would solve a potentially massive disposal problem. Potatoes processing byproducts are very interesting since they also contain vitamins, minerals, and phytochemicals, such as carotenoids and natural phenols. Sabeena Farvin et al. (2012) demonstrated that the ethanol extract of potato peel is a natural antioxidant for retarding lipid and protein oxidation in minced mackerel via comparison to a control with no added extracts. Radiation processing is very effective for sterilization but may impair the sensory quality of meat. Kanatt et al. (2005) exploited the antioxidant activity of potatoes peel to reduce lipid peroxidation of radiation-processed meat without affecting its flavor.

Starch can be recovered from wastewater produced during potato washing for chips, snacks, or fries, thereby avoiding starch sedimentation in drainage pipelines (Catarino et al. 2007).

Okuno et al. (2002) extracted antioxidants from waste powder derived from sweet-potatoes with orange flesh, using supercritical carbon dioxide extraction with no toxic solvent residue.

Among the several roots and tubers, cassava is very popular in South Asia and America; its peel can be enzymatically hydrolyzed to simple sugars as a useful feedstock from a cheap non-food biomass waste (Ona et al. 2019).

2.4 Breweries

Brewer's spent grain is the typical waste produced in large amounts by the brewing industry. Its high potential in obtaining fermentable sugars through enzymatic hydrolysis makes it an eligible source of sugars for bioprocesses. The optimization of operational variables such as mechanical stirring, temperature, enzyme and substrate concentrations, and ultrasound (pulse factor and amplitude) were analyzed. The best sugar yield (370.9 g /kg) was obtained using direct sonication that performed better than mechanical stirring and indirect sonication processes. Besides, direct sonication decreased the treatment time, confirming that it is a promising technology for improving the yields of fermentable sugars from brewer's spent grain (Luft et al. 2019).

A microwave-assisted hydrothermal pretreatment technology was similarly developed to recover fermentable sugars from brewer's spent grain without acid or alkali catalysts. These sugars were fermented, yielding 46 kg of butanol per ton of brewer's spent grain (López-Linares et al. 2019).

An integral valorization of brewers' spent grains within the biorefinery framework relied on a microwave-assisted, hydrothermal process catalyzed by Ni-Co/Al-Mg. Using a temperature of 250 °C and a pressure of 125 bar for 2 h, it is possible to convert the original material into (i) 8% high-energy bio-oil made up a complex mixture of phenols, ketones, aldehydes, carboxylic acids, and nitrogen compounds with a higher heating value of 26 MJ/kg; (ii) 35% bio-char (spent solid) with a higher heating value of 32 MJ/kg; (iii) a saccharide-rich (31%) aqueous solution (Lorente et al. 2019).

Brewer's spent grain was used to generate a range of bioactive peptides with antioxidant activity via direct enzymatic hydrolysis without the prior use of an alkali protein extraction step (Connolly et al. 2019).

Supercritical fluid extraction was successfully used to extract oils from brewer's spent grain without and with ethanol as co-solvent in percentages, and this procedure was compared to the Soxhlet using hexane as solvent. The fatty acid composition, the yield, the recovery, and the antioxidant activity were similar, but supercritical carbon dioxide is obviously a greener procedure with no health hazard (Ferrentino et al. 2019).

Alternative use of spent brewer's yeast, which is both biologically and nutritionally rich, exploits its angiotensin converting enzyme inhibitory activity (ACE-I), after autolysis and subsequent hydrolysis via an extract of *Cynara cardunculus.* Nanofiltration produced a fraction with <3 kDa, stable in the gastrointestinal system and in vitro, exhibiting the highest inhibitory activity. This ingredient can be useful to deal with hypertension (Amorim et al. 2019).

Brewer's spent grain and yeast are not the only byproducts of the brewing process. In fact, brewery industries release about 70% of the production water as wastewater into water bodies. Brewery effluents contain huge amounts of organic compounds such as sugars, soluble starch, ethanol, volatile fatty acids, and nitrogen and phosphorous compounds; its discharge raises environmental concerns, including excessive growth of undesirable microbes that causes loss of aquatic lifeforms. Physical wastewater treatment to reduce suspended solids through sedimentation by gravitational force, chemical treatments such as pH adjustment or coagulation/flocculation and final aerobic (activated sludge) or anaerobic digestion are usually applied for the sake of environmental safety and to comply with strict regulations. Membrane filtration is considered environmentally friendly and safe. Nanofiltration can subtract COD, Na^+, and Cl^- with an average removal rate of 100%, 55%, and 70%, respectively but fouling and high energy consumption are major challenges for this process. Microbial fuel cells (see Sect. 4.3.4) are able to convert organic pollutants into electricity, but this technology is still in its infancy. The microalgae wastewater treatment method is an emerging environmentally friendly biotechnological process to turn nutrient-rich wastewaters into useful biomass; this process is interesting for the CO_2 sequestration involved in biomass production. Microalgae such as *Chlorella* sp., *Scenedesmus dimorphus*, and *Scenedesmus obliquus* are actually able to absorb organic nutrients, and efficiently reduce the organic pollutants, but light and temperature are limiting factors of the process; removal of salt, odor and color is not efficient. The harvested algal biomass from the wastewater may be processed into plant fertilizers, animal feed, or biofuel production (Amenorfenyo et al. 2019).

Aspergillus niger NRRL-567 was used to ferment several food industry wastes, including brewery spent liquid, lactoserum, and starch industry water sludge to produced citric acid. The production was optimized by manipulating several parameters, such as temperature (25–35 °C), pH (3–5), presence of inducers, incubation time, and supplementation with different proportions of apple pomace ultrafiltration sludge (Dhillon et al. 2012).

Target compounds that can be extracted from brewery waste are outlined in Table 2.3.

2.5 Olive Oil and Edible Oil Industry

Virgin olive oil is valuable vegetable oil, extracted from olive fruits (*Olea europaea* L.) by mechanical and other physical methods since ancient times. This agroindustrial activity is of crucial economic significance for many Mediterranean countries (Cecchi et al. 2010a, 2011). Nowadays, it is well-known that regular dietary consumption of virgin olive oil manifests in health benefits associated with a Mediterranean diet (Cecchi and Alfei 2013).

Olive oil production results in the generation of large quantities of solid wastes (Lozano-Sánchez et al. 2011) and wastewaters (Federici et al. 2009; Cardinali et al. 2012; Bertin et al. 2011) rich in antioxidant phenolic compounds (Cecchi et al. 2010b) (phenolic acids and alcohols, secoiridoids, lignans, and flavones) with a great nutraceutical and pharmaceutical interest (Cecchi et al. 2011; Ghanbari et al. 2012). The traditional release of effluents from the oil processing industry into the soil and groundwater results in oily film formation on the aquatic surface that endangers aquatic animals and may block sewage and drains due to emulsification of organic matter (Okino-Delgado et al. 2017).

Waste valorization from olive industry processing regards the recovery of bioactive compounds from wastewaters and the reduction of water pollution problems. To exploit the potential of olive mill wastewaters, Federici et al. (2009) evaluated various challenging technologies for extraction of fine chemicals and for different biotechnological applications such as the production of important metabolites; technology costs are the major concern for these procedures. Again, antioxidants are the main target compounds in olive mill wastewater with many promising applications in foods, cosmetics, and pharmaceuticals. Among them, oleuropein and hydroxytyrosol (Bouaziz et al. 2008) are the most exhaustively studied due to their presence in all olive byproducts and their outstanding biological activities. Cardinali et al. used a membrane technology coupled to low-pressure gel filtration chromatography and focused on polyphenols recovery and structural characterization (Cardinali et al. 2012). For the same purpose, Bertin et al. (2011) applied a solid-phase extraction procedure.

A biotechnological degradation of the organic matter from oil industry effluents or kitchen waste oil with the concomitant production of high value-added bio-based microbial biosurfactants can be an attractive valorization alternative. For example, *Pseudomonas aeruginosa* synthesized biosurfactants such as rhamnolipids and glycolipids using lipase (Chen et al. 2018), (Henkel et al. 2012). Edible oil industry waste was reported to be a cheap source for bioactives such as tocopherols, sterols, and squalene which represents important target compounds of the pharmaceutical and cosmetics industry (Sherazi et al. 2016). Solid byproducts include olive mill pomace or olive press cake containing fragments of skin, pulp, pieces of kernels, and some oil. The oil industry annually produces about 350.9 million tonnes of the de-oiled cake; they are a good source of protein and they are commonly further utilized for preparing animal feed and fertilizer (Chang et al. 2018). Their water and methanol/water derived extracts are reported to have antioxidant and breast cancer antiproliferative activities due to the phenolic compounds (hydroxytyrosol in particular) and can be useful for nutraceutical applications, as food supplements (Ramos et al. 2013). Polar lipids able to protect against cardiovascular diseases were extracted, fractionated, and tested for their bioactivity; the presence of platelet-activating factor antagonists in the polar lipid extract from olive oil waste makes it a value-added materials for the functional food industry (Karantonis et al. 2008).

Accelerated solvent extraction of polyphenols from the solid residue of olive cake could be proposed as a simple and rapid extraction procedure (Suárez et al. 2009); the phenolic extract characterized by a strong antioxidant activity is a putative natural antioxidant also for the development of supplemented olive oil. Convective dehydration of olive cakes resulted in a dried olive cake that could be upcycled in many food and cosmetic applications or animal feed (Uribe et al. 2013).

Along with polyphenols and polar lipids, there are other important target compounds, including carbohydrates, proteins, fatty acids, polyalcohols, and pigments. A hydrothermal treatment and an autohydrolysis of the olive cake, gave a water-soluble fraction comprising monosaccharides, oligosaccharides, and mannitol (Fernández-Bolaños et al. 2004).

Application of olive cake to soil was proposed as a simple disposal strategy due to its high content in organic matter, able to adsorb triazine herbicides and to reduce desorption of the most hydrophobic terbuthylazine and prometryn (Delgado-Moreno et al. 2007). Actually, olive mill wastes could be used as a source of carbon for soil carbon sequestration strategy (Sánchez-Monedero et al. 2008).

Waste cooking oil is the residue of the deep fat frying process, and it is unsuitable for human consumption due to the formation of polar compounds like free short-chain fatty acids, mono- and di-glycerides, aldehydes, ketones, polymers, cyclic and aromatic compounds; cooking oil wastes were bioremediated using lipases extracted from orange wastes through transesterification reactions that decreased the toxicity of the waste cooking oil (Okino-Delgado et al. 2017).

Olive oil wastewaters were also used for xanthan gum production (see Sect. 5.1.4) using the plant pathogen using *Xanthomonas campestris* (Murad et al. 2019).

Target compounds that can be extracted from oil crops waste are outlined in Table 2.4.

Table 2.4 Oil crops waste: specific source and target molecules for recovery

Source	Target compound	References
Sunflower seed	Phytosterols	Galanakis (2012)
Sunflower	Proteins and chlorogenic acid Polyphenols and flavonoids	Albe Slabi et al. (2020) Rodríguez et al. (2019) Daraee et al. (2019)
Rapeseed meal	Isothiocyanate and Glucosinolates	Ishikawa et al. (2014)
Rapeseed oil waste	Phytosterols and tocopherols	Jafarian Asl et al. (2020)
Soybean seed	Phytosterols	Galanakis (2012)
Soybean oil waste	Phytosterols	Galanakis (2012)
Olive mill waste and wastewater, olive pomace	Phenols, pectin, substrate for biosurfactants (rhamnolipids and glycolipids), and biodiesel production, tocopherols, sterols, squalene, and single cell protein, phenolic acids (ferulic, cinnamic, gallic etc.) and alcohols, secoiridoids, lignans, and flavones, oleuropein and hydroxytyrosol, xanthan gum	Galanakis (2012), Ravindran and Jaiswal (2016), Kiran et al. (2014), Henkel et al. (2012), Ghanbari et al. (2012), Suárez et al. (2009), Bouaziz et al. (2008), Federici et al. (2009), Bouaziz et al. (2008), Murad et al. (2019), Cassano et al. (2013), Rahmanian et al. (2014), Crea (2002), Caballero et al. (2020), Di Mauro et al. (1999) and Ferri et al. (2011).
Olives processing byproducts	Phenolic compounds, polyphenols, carotenoids, phytosterols, squalene, and dietary fiber	Schieber (2017), Karantonis et al. (2008), Suárez et al. (2009), Cardinali et al. (2012) and Bertin et al. (2011)
Winter oil seed rape waste	Phytosterol	Ravindran and Jaiswal (2016)
Sesame cake	Antioxidants, phenolics, flavonoid and flavonol	Mohdaly et al. (2010) and Sarkis et al. (2014)

2.6 Meat Products

Globally, meat consumption has been continuously increasing, and the meat processing industry is a huge segment of the food industry chain system. Animal originated wastes are rich in proteins that are too valuable to be discharged into the environment.

Meat industry discards large quantities of slaughterhouse byproducts, such as blood, entrails, and some muscles (widely consumed in the past due to poverty needs) and non-edible parts such as skin, bones, deboning residues, tendons, skull, hooves, feathers, and detergent residues.

Nearly 40% and 20% of the slaughtered carcass are made of edible and inedible byproducts, respectively (Bhaskar et al. 2007). The former can obviously be used as human food or animal feed (Toldrá et al. 2012).

Due to microbiological hazard, the recovery of meat industry byproducts is bound by severe hygiene limitations. For example, the European Union promulgated legislative measures in order to avoid the spread of the bovine spongiform encephalopathy (BSE) (Regulation 999/2001and 853/2004). In this respect, combustion of bone byproducts (Deydier et al. 2005) is a safe procedure that produces residues rich in calcium (30.7%) and phosphate (56.3%), with significant levels of sodium (2.7%), potassium (2.5%) and magnesium (0.8%), and with no heavy metal content. Almost 90% of ash particles are smaller than 1 mm.

Anyhow the recovery of protein from meat byproducts is the focus of interest of many studies.

Selmane et al. investigated the operating conditions of the extraction, purification, and concentration of proteins from meat byproducts in order to maximize their recovery (ca. 50%) and to enhance their functional properties, namely gelling, emulsifying, and foaming properties if used as food ingredients (Selmane et al. 2008).

Rendering industries are the oldest recycling industries; they convert waste animal tissue into stable bone meal, hydrolyzed feather meal, blood meal, fish meal, and purified animal fats (lard and tallow) (Yaakob et al. 2019).

Poultry slaughterhouse waste mixed with molasses were fermented to eliminate pathogens and to produce lactic acid bacteria that can be used as probiotics (Ashayerizadeh et al. 2017). Chicken slaughterhouse wastewater was used for green algae (*Scenedesmus* sp.) biomass generation to be used as fish feed (Yaakob et al. 2019). Apart from food applications such as blood sausage, blood cake, blood pudding, and blood curd, blood can also be used as a fertilizer or a binder (Adhikari et al. 2018).

The use of a fungal protease for the enzymatic treatment of sheep visceral mass (including stomach, large and small intestines) resulted in a protein hydrolysate rich in some of the essential amino acids that can be used as a flavor enhancer, functional ingredient in foods of low protein quality (Bhaskar et al. 2007). Tahergorabi et al. used isoelectric solubilization/precipitation to recover functional proteins, useful for novel food preparations, from chicken meat processing byproducts. The authors recommended the addition of TiO_2 during the extraction process to increase the gel strength (Tahergorabi et al. 2012).

Krasnoshtanova developed methods for increasing the efficiency of enzymatic hydrolysis of meat byproducts. A kinetic models of the processes enabled the determination of the optimal conditions ensuring a degree of substrate conversion of at least 95%; the presence of alkyl hydroxy benzenes improved the efficiency of enzymatic processes. Increased temperatures (55–65 °C) made it superfluous to maintain aseptic conditions (Krasnoshtanova 2010).

Gomez-Guillen et al. focused on collagen and gelatin extraction from non-mammalian species (fish and poultry). They can be easily be upcycled as emulsifiers, foaming agents, and colloid stabilizers (Gomez-Guillen et al. 2011).

Cellulosic spent casing from frankfurter/sausage production was found to be an excellent source of cellulose that can be hydrolyzed into simple sugars or used as a substrate for the production of cellulase (Cumba and Bellmer 2005).

Target compounds that can be extracted from meat waste are outlined in Table 2.5.

Table 2.5 Animal food waste origin, specific source and target molecules for recovery

Origin	Source	Target compound	References
Meat	Chicken byproducts	Protein	Galanakis (2012), Ravindran and Jaiswal (2016), Tahergorabi et al. (2012) and Gomez-Guillen et al. (2011)
	Slaughterhouse byproducts	Protein, bone meal, hydrolyzed feather meal, blood meal, animal fats (lard and tallow); substrate for lactic acid bacteria and biomass generation as fish feed	Galanakis (2012), Ravindran and Jaiswal (2016), Selmane et al. (2008), Yaakob et al. (2019), Ashayerizadeh et al. (2017) and Yaakob et al. (2019).
	Bovine blood	Protein	Galanakis (2012)
	Beef lung	Protein	Galanakis (2012)
	Sheep visceral mass	Protein hydrolisate	Galanakis (2012) and Bhaskar et al. (2007)
	Bone byproducts	Calcium, phosphate, sodium, potassium, magnesium, with no heavy metal content	Deydier et al. (2005)
Fish and seafood	Fish leftovers (skin, head, bones)	Protein, lipids, gelatin, chitosan, glycosaminoglycans	Galanakis (2012), Ravindran and Jaiswal (2016), Gehring et al. (2011), Aspevik et al. (2017), Gomez-Guillen et al. (2011), Sousa et al. (2017), and Prameela et al. (2017)
	Shrimp and crab shells	Chitosan, chitin, chitinase, glycosaminoglycans; astaxanthin, calcium carbonate	Galanakis (2012), Ravindran and Jaiswal (2016), Kiran et al. (2014), Yan and Chen (2015), Prameela et al. (2017), Kumar et al. (2018), and Valcarcel et al. (2017)
	Lobster	Chitin	Yan and Chen (2015)
	Surimi wastewater	Proteins	Galanakis (2012)
	Sea products waste	Lipids and pigments from algae, and oils from fish-processing plant waste streams	Kerton et al. (2013)
	Tuna by-product hydrolysate	Antioxidant bioactive peptides.	Saidi et al. (2018)
	Fish cartilage	Chondroitin sulfate	Kim et al. (2012), Gargiulo et al. (2009), and Murado et al. (2010).

(continued)

Table 2.5 (continued)

Origin	Source	Target compound	References
Dairy products	Whey	Lactose, whey protein concentrate and isolate, β-lactoglobulin, α-lactalbumin, probiotics, antibacterials, immunoglobulins, whey peptides, whey permeate, fermentation products, single cell protein, and probiotics, xanthan gum, bioactive peptides.	Smithers (2008), Yadav et al. (2015), Cheirsilp and Radchabut (2011), Nguyen et al. (2003), Santos et al. (2019), Borad et al. (2019) Yadav et al. (2015), Barba et al. (2001), Pereira et al. (2002), Cheirsilp and Radchabut (2011), Minhalma et al. (2007), Nguyen et al. (2003), Jeantet et al. (2000), Rektor and Vatai (2004), Kelly et al. (2000), DeSilva et al. (2003), Borad et al. (2019), Alonso et al. (2010), Maragkoudakis et al. (2010), Koutinas et al. (2009), Murad et al. (2019), Patel and Murthy (2010), Sánchez-García et al. (2018), and Monti et al. (2018)
	Milk	Biodiesel, ethanol, whey protein, Lactose, baker's yeast, and minerals	Parashar et al. (2016), Matharu et al. (2016), and Kiran et al. (2014)
	Dairy wastes and wastewaters	Lactose, proteins, peptides, salts, fatty substances, and cleaning chemicals	Kosseva (2011).

2.7 Fishery

Fishery supplies around 20% of the world's food to humans. The utilizable mass of marine animal is unfortunately quite low: for example crabs only yield 40% meat, and in tuna fish only 75% of the fillet is available (Sharma et al. 2020).

Processing of fish and seafood results in appreciable amounts of wastes (i.e., 9–15% and 6–8% of marine catches in industrialized and developing countries, respectively) rich in proteins and lipids from fish skin, head frames, and viscera (Gehring et al. 2011). After industrial filleting, fish processing byproducts are *ca.* 60–70% of the total fish weight (Tahergorabi and Jaczynski 2014).

These residuals contain potentially bioactive compounds, which can be made use of as food or feed. Optimal utilization of proteins in such residual raw materials makes use of advanced technologies such as freezing/cooling, acid preservation, salting, rendering and protein hydrolysis (Aspevik et al. 2017).

Upcycling fish byproducts is a crucial practice for marine resources conservation. Different added-value products can be recovered, thereby avoiding the costs associated with the management of wastes.

Strict legislation regulates the upcycling of various animal-based residuals Fish skins constitute an important fraction of the huge amount of fish byproducts. Fish skins can be valorized through the extraction of gelatins. The gel-forming properties of gelatins are profitably used in food, photographic, cosmetic, and pharmaceutical applications. Gelatins are gummy when moist; hence they are natural biodegradable emulsifiers, foaming agents, and colloid stabilizers. Gelatins are also endowed with diverse antimicrobial, antioxidant or antihypertensive activities (Gomez-Guillen et al. 2011).

Extraction of gelatin from fish skin can be optimized manipulating the extraction parameters, such as reagents concentration, water consumption, and time of processing, while maintaining the highest extraction yield. Characterization of the chemical composition, rheology, structure, texture, and molecular weight, enable the introduction in the market of gelatins with unique properties for specific applications (Sousa et al. 2017).

A strictly pharmaceutical compound (i.e., chondroitin sulfate) used for osteoarthritis, can be recovered from scapular cartilage of Shortfin mako shark (Kim et al. 2012), waste of *Scyliorhinus canicula*, (Gargiulo et al. 2009), and skate cartilage (Murado et al. 2010).

Shells of aquatic animals such as shrimps, lobsters, crabs, prawns, fish scales, and endoskeletons harbor useful chemicals such as protein, chitin, and calcium carbonate. Crustaceans are a large, diverse arthropod taxon. Their processing results in important byproducts; seafood waste is chitinaceous in nature. Actually, the shell of crabs, lobsters, shrimps, and prawns all contain chitin from which chitosan can be extracted (Galanakis 2012; Ravindran and Jaiswal 2016; Kiran et al. 2014; Yan and Chen 2015; Prameela et al. 2017; Kumar et al. 2018). Chitin is insoluble in water and inert to most chemical agents; hence its common disposal (ocean dumping, incineration, landfilling) leads to environmental pollution, eutrophication, and biofouling. Its recovery would eradicate the environmental issue along with revenue generation (Yan and Chen 2015; Kumar et al. 2018). Dried shrimps and crabs contain about 50% of chitin on a dried basis (Yan and Chen 2015), and even if it can be used as an animal feed supplement, bait, or fertilizer, it is the main source of chitosan, an important biopolymer which exhibits antibacterial activity against *Enterococcus faecalis*, *Escherichia coli*, *Staphylococcus aureus*, and *Candida albicans* (Hussein et al. 2013). The importance of chitosan will be clearly detailed in Sect. 5.9. Seafood waste is also a potential source for functional and bioactive compounds: for example, glycosaminoglycans extracted from marine animal waste are better than those obtained from terrestrial organisms (Valcarcel et al. 2017); shrimp waste was used to produce nutraceutical compound such as astaxanthin (3,3′-dihydroxy-ß-carotene-4,4′-Dione), a xanthophyll carotenoid present in crustacean waste that was extracted through oxidative transformations of ingested β-carotene or zeaxanthin by feed microalgae (Prameela et al. 2017).

Target compounds that can be extracted from fish waste are outlined in Table 2.5.

2.8 Dairy Products

The dairy industry is an important food industry sector; dairy waste is derived from the processing industry, microbial spoilage of the dairy products, and inappropriate storage or handling. Dairy wastes and wastewaters are a significant source of lactose, proteins, peptides, salts, fatty substances, and cleaning chemicals used for maintaining proper hygiene during processing (Kosseva 2011). During the processing of raw milk, up to 3.0 L of wastewater are generated per liter of processed milk; the high content of the nutrients poses serious environmental concerns and demands wastewater conversion methods such as activated sludge, sequencing batch reactor, trickling filter, anaerobic sludge blanket, and aerated lagoons (Sharma et al. 2020).

Cheese whey was discovered about 3000 years ago when calves' stomachs were used as milk containers, and the naturally occurring enzyme chymosin (rennet) coagulated the milk and spawned the start of the cheese-making activity. The manufacturing of cheese generates 9 kg of whey per kg of cheese produced (Parashar et al. 2016).

After milk is coagulated during cheese production when pH is dropped to 4.6, 50% of the milk total solids remain in the whey; interestingly, all the lactose (ca. 5%, w/v) and all the typical whey protein (ca. 0.7% w/v) such as β-lactoglobulin, α-lactalbumin, glycomacropeptide, minor protein/peptide components including immunoglobulins, lactoferrin (a whey globular glycoprotein of the immune system, involved in the binding and transport of iron ions, with strong antibacterial, antiviral, antiparasitic, anticancer, and anti-allergic functions, with bone growth enhancement properties), lactoperoxidase (a whey enzyme that catalyzes the oxidation of substrates by hydrogen peroxide, endowed with natural antibacterial activity), serum albumin, lysozyme, and growth factors can be found in whey since they are not involved in curd formation. Whey also include 1.0% (w/v) salts, and 0.1%±0.8% (w/v) lactic acid (Lee et al. 2000). Whey proteins are the most important target compounds in whey because of their exceptional biological value (15% better than egg protein). This means that the body can well and quickly exploit this kind of proteins; for this reason, whey protein is the protein of choice for bodybuilders, athletes, and debilitated people. Whey proteins are a rich source of the essential and branched amino acids that represent metabolic regulators in protein, glucose and lipid metabolism, thereby playing a role in weight control (Smilowitz et al. 2005). Whey proteins are also a balanced source of the sulfur amino acids (methionine, cysteine), precursors to the potent intracellular antioxidant glutathione (Shoveller et al. 2005). By this scenario, nowadays, the bioactivity of whey is receiving considerable scientific scrutiny around the world even if whey medicinal applications during the seventeenth and eighteenth centuries were already well renowned (Smithers 2008).

Many recent studies established a sound basis for its nutritional and functional value and this underpinning knowledge paved the way to technology developments for its valorization. It was found that the bioactivity of whey also includes (i) physical fitness, (ii) satiety and weight control, (iii) cardiovascular health, (iv) anticancer activity on in vitro cell culture, in vivo animal studies, and some epidemiological investigations (v) growth factor activity and wound healing (vi) antimicrobial effects of lactoferrin for the management of infections and for enhancing the safety of meat (vii) infant nutrition (viii) potent bone growth enhancement properties of lactoferrin for the prevention and treatment of osteoporosis (ix) antimicrobial effects of lactoperoxidase again oral infection and halitosis (Smithers 2008).

Historically, the first management method of the dairy activity was whey disposal via its spraying onto fields, although the smell and salt content were major problems. Discharging the whey into rivers, lakes, sea, or municipal sewage system was also exploited; however environmental concerns progressively restricted and banned this disposal approach. Actually, whey emission into river systems is an environmental hazard due to its high organic content (Biochemical oxygen demand (BOD) >35,000 ppm; chemical oxygen demand (COD) >60,000 ppm) (Smithers 2008).

Transformation of whey from a nuisance into a valuable dairy raw material and valuable feedstock has been facilitated by environmental considerations and legislative restrictions on whey disposal, knowledge about target bioactive compounds in whey, and technology advances in whey-processing technologies (concentration, transformation, fractionation, and dehydration of whey, and modern biochemical techniques).

Whey was also used for xanthan gum production (see Sect. 5.1.4) using the plant pathogen using *Xanthomonas campestris* (Murad et al. 2019).

Whey can be simply concentrated and dried as whey powder, or it can be used as substrate for ethanol production; technological advancements have enhanced whey utilization up to 50% of the total produced whey to produce lactose, whey protein, whey permeate, but also fermentation products such as bioethanol, hydrogen, methane, whey peptides, and probiotics (Yadav et al. 2015). The recovery of protein and lactose is of utmost importance: no whey utilization strategy is feasible without suitable attention being paid to lactose, which represents more than 75% of whey solids. Whey proteins, containing all 20 amino acids, are widely used to prepare drinks and bars, which are the most popular dietary supplements for athletes; anyhow, whey-based ingredients successfully compete for a share of the burgeoning functional foods market. There are two types of whey protein useful for dietary supplements: whey concentrate and whey isolate. The latter undergoes more processing, which results in higher protein content with fewer carbohydrates, lactose, and fat.

The processes and techniques for the full valorization of target compounds from whey will be described in Chap. 4, they include microfiltration, ultrafiltration, nanofiltration, reverse osmosis, electrodialysis, and ion exchange, anyhow we will detail in the following whey target compounds.

Microfiltration of whey (to remove fat, bacteria, and residual casein micelles) is the first step of the process.

Whey protein concentrates (up to 80% protein) are then obtained by ultrafiltration, a membrane technique that lowers the fat and lactose content and limits protein denaturation (Barba et al. 2001). Pereira et al. valorized ovine cheese whey, by means of a thermo-calcic precipitation and microfiltration before ultrafiltration. The clarification step improved the next ultrafiltration treatments. The chemical composition of retentate powders was comparable to that of conventional ultrafiltration powders (Pereira et al. 2002).

The membrane ultrapermeate usually contains lactose, salts, nonprotein nitrogen, and water. If it is subsequently nanofiltrated, lactose can be recovered because it is retained by the nanomembrane, while salts and water constitute the permeate. The salt depleted lactose concentrate can be used as a raw material in the pharmaceutical, food, and paper industries; it is a lowcost carbon source for the biotechnological production of kefiran, a water-soluble exopolysaccharide that comprises equal amounts of glucose and galactose, by *Lactobacillus kefiranofaciens* (Cheirsilp and Radchabut 2011). The valorization of this dairy sugar may involve the use of crystallization, fermentation, hydrolysis, and transformation (into various valuable derivatives).

The salt-enriched permeate from nanofiltration, almost free from organic matter, could be reused in the process: nanofiltration, using the optimal operating conditions, could therefore minimize the wastewater environmental impact with a maximum water recovery of approximately 80% (Minhalma et al. 2007).

Nguyen et al. used nanofiltration to obtain a four-fold concentration of solids from cottage cheese whey, while removing about three-quarters of the sodium and potassium salts and some acids. The desalted nanoconcentrate could be considered an ingredient in dairy and other food products (Nguyen et al. 2003). Anyhow, other techniques such as electrodialysis and chromatography are also used for demineralisation of whey and separation of whey protein from the abundant lactose and residual fat (Jeantet et al. 2000; Rektor and Vatai 2004) to produce functional whey ingredients for athletes and debilitated people.

At variance with whey protein concentrate, whey protein isolate contains up to 90% protein without fat and lactose (Kelly et al. 2000); it can be obtained by ionic exchange chromatography that is able to selectivity adsorb and subsequently desorb proteins even if it might trigger their denaturation.

Given the high value of specific proteins, it would be highly desirable to fractionate whey proteins.

Procedures usually based on selective ion exchange have been developed for selective manufacture of β-lactoglobulin and α-lactalbumin isolates via precipitation of α-lactalbumin from sweet whey concentrated by ultrafiltration, under specified conditions of pH and temperature, leaving β-lactoglobulin in solution and unaffected by the pH/temperature treatment (Smithers 2008). Alternatively, fractionation can be obtained via liquid chromatography as well as continuous chromatographic separation technology and ultra-high pressure processing aiming at maintaining the functional and nutritional traits of the wide array of whey bioactive compounds (DeSilva et al. 2003).

Apart from the major whey proteins (β-lactoglobulin, α-lactalbumin, and glycomacropeptide) whey contains a number of minor proteins. Among them immunoglobulins represent the most abundant. Colostrum is a precious ingredient for the manufacture of immunoglobulin-enriched milk products that are gaining significance in the global nutraceutical market as dietary supplements. These immunoglobulin-rich isolates might be important because they contain an abundance of immune components that can be exploited to address the poor immune status of immuno-compromised people and enhance gut health. High-pressure processing enables the preservation of the colostrum immunoglobulins (Borad et al. 2019).

Whey peptides have additional and often varied bioactivity compared to that of the parent molecule (Herrera-Ponce et al. 2019). A number of these bioactive whey peptides, together with the parent protein source, have been the focus of interest of in vitro and in vivo studies to investigate their positive influence on cardiovascular, endocrine, and immune systems. Desolvation and internal gelation have been so far employed for bioactive peptides particulation (Yadav et al. 2015). Commercially, the most promising of these peptides is the potent antimicrobial lactoferricin and closely related peptides, derived by peptic or chymosin digestion of lactoferrin (Herrera-Ponce et al. 2019).

Engineering developments have led to a number of innovations and improvements in the drying of whey, whey concentrates, isolate, and permeates: spray drying, freeze-drying, and filtermat drying have different advantages and disadvantages (Písecký 2005), and also influence on the organoleptic outcomes (Vincenzetti et al. 2018).

Lactic acid is widely used in food and pharmaceutical sectors and recently for biopolymers production, as illustrated in Chap. 6. To obtain lactic acid from whey, Vasala et al. (2005) used *Lactobacillus salivarius,* an ideal microorganism for lactic acid fermentation in high-salt high lactose conditions, typical of whey. Yoghurt whey from expired products could be used for lactic acid production by *Lactobacillus casei* (Alonso et al. 2010).

Raw whey permeate can serve as a natural sanitizing agent in the washing of vegetables (Martin-Diana et al. 2006). The possibility to obtain lactic acid from whey fermentation led to the production of a whey-based food disinfectant. A scalable fermentation protocol to produce a disinfectant, with the highest lactic acid concentration and the lowest lactose levels, made use of a specific mesophilic-lactic acid bacteria starter mix over 120 h. Antibacterial activity against food pathogenic and spoilage strains was successfully tested. This fermented whey represents an effective and environmentally-safe alternative to the use of chlorine as a disinfectant for shredded lettuce on the basis of microbial quality and other quality indicators such as texture, color, and sensory perception, pH, O_2 and CO_2 determinations (Santos et al. 2019).

Whey permeate, obtained from ultrafiltration, containing lactose, salts, nonprotein nitrogen, and water, is a dairy effluent usually dealt with via land spreading or incorporation into animal feed. It was recently demonstrated that it could be hydrolyzed using enzymes to release fermentable sugars and then integrated into wheat fermentation as a co-substrate or to partially replace process water for ethanol production using *Saccharomyces cerevisiae* (Parashar et al. 2016).

Edible filamentous fungi such as *Aspergillus oryzae* and *Neurospora intermedia,* were profitably used to enzymatically convert dairy waste into high-quality biomass for animal feed or human consumption, and chemicals, such as ethanol and glycerol (Mahboubi et al. 2017).

Whey was also used for the production of probiotics. The whey by-product from Ricotta cheese manufacturing, called "Scotta" presented favorable characteristics as a potential growth substrate for the production of a probiotic fermented drink (Maragkoudakis et al. 2010). The manufacture of novel dairy starter cultures for better cheese ripening and extended shelf life, using whey as raw material, was also proposed. The developed technology involved biomass production from whey followed by thermal drying of cultures. Authors found that the most suitable culture in lactose and milk whey fermentation was kefir due to its technological properties (Koutinas et al. 2009).

Target compounds that can be extracted from dairy waste are outlined in Table 2.5.

2.9 Kitchen Waste

Kitchen waste from public catering rooms, restaurants, households, and canteens broadly consists of fruits, vegetables, staple food, shells, bones, pits, and meat (Zhao et al. 2017). It is compostable. It contains high organic matter and particularly carbohydrates (cellulose, hemicellulose, pectin and starch), protein, lignin, fats, organic acids, inorganic salts, and others (Chen et al. 2017). Traditional disposal of kitchen waste such as land-filling, incineration, and direct or indirect discharge into the environment is detrimental to the ecosystem and human health (Sharma et al. 2020). The high moisture content and a pool of nutrients facilitate the growth of pathogens, rotting, breeding of flies, emission of toxic volatiles, including ammonia and greenhouse gases.

Interestingly, kitchen waste could be used as a substrate to generate of a huge array of value-added products. Xanthan gum was produced via chemical and enzymatic treatments of kitchen waste (Li et al. 2017); volatile fatty acids were obtained using anaerobic fermentation, using activated sludge to inoculate FW (Wang et al. 2014); short-chain fatty acids were similarly obtained by co-fermentation of kitchen waste with waste activated sludge under alkaline conditions and the fermentation liquid was successfully used in microbial fuel cells (Chen et al. 2013) described in Sect. 4.3.4; kitchen waste was investigated as a nutrient resource for L(+)-lactic acid production by *Lactobacillus manihotivorans* LMG18011 (Ohkouchi and Inoue 2006), and for cellulase manufacture via *Aspergillus niger* in solid-state fermentation (Bansal et al. 2012); glucoamylase is another interesting target compound in kitchen waste valorization, it was produced by *Aspergillus niger* UV-60 under submerged fermentation (Wang et al. 2008).

Carbohydrate from kitchen waste after mild acid pretreatment of hydrochloric acid and glucoamylase gave fermentable sugars (93.25g/L glucose, 0.542g/L sucrose, 0.348g/L maltose, and 0.321g/L fructose) as the feedstock for bioethanol production (Hafid et al. 2015). Ethanol production from kitchen waste was studied using successive liquefaction, pre-saccharification, and simultaneous saccharification and fermentation (Nishimura et al. 2017), while biobutanol is the product of enzymatic hydrolysis and *Clostridium acetobutylicum* fermentation of kitchen waste (Chen et al. 2017); methane and hydrogen can also be obtained by anaerobic fermentation (Campuzano and González-Martínez 2016; Parthiba Karthikeyan et al. 2018).

Bacillus agaradhaerens C9 degrading enzymes (amylase, protease, lipase, cellulase, xylanase, and pectinase) promoted the hydrolysis of kitchen waste and strong alkaline fermentation conditions induced the production of an environment-friendly and biodegradable macromolecular bioflocculant, using unsterilized kitchen waste. The bioflocculant was characterized by a high flocculation efficiency, and it was applied to treat mineral mining processing wastewater comprising heavy metal ions, thereby enabling its recycling (Liu et al. 2019).

Bibliography

Abdalla AEM, Darwish SM, Ayad EHE, El-Hamahmy RM (2007) Egyptian mango by-product 1. Compositional quality of mango seed kernel. Food Chem 103:1134–1140. https://doi.org/10.1016/j.foodchem.2006.10.017

Acun S, Gül H (2014) Effects of grape pomace and grape seed flours on cookie quality. Qual Assur Saf Crop Foods 6:81–88. https://doi.org/10.3920/QAS2013.0264

Adhikari BB, Chae M, Bressler DC (2018) Utilization of slaughterhouse waste in value-added applications: recent advances in the development of wood adhesives. Polymers (Basel) 10:176. https://doi.org/10.3390/polym10020176

Aghajanian S, Kazemi S, Esmaeili S et al (2020) Sequential microwave-assisted extraction for isolation of quercetin from red kidney bean. Int J Eng Trans A Basics 33:12–17. https://doi.org/10.5829/ije.2020.33.01a.02

Ahangari B, Sargolzaei J (2013) Extraction of lipids from spent coffee grounds using organic solvents and supercritical carbon dioxide. J Food Process Preserv 37:1014–1021. https://doi.org/10.1111/j.1745-4549.2012.00757.x

Aires A, Carvalho R, Saavedra MJ (2017) Reuse potential of vegetable wastes (broccoli, green bean and tomato) for the recovery of antioxidant phenolic acids and flavonoids. Int J Food Sci Technol 52:98–107. https://doi.org/10.1111/ijfs.13256

Ajila CM, Aalami M, Leelavathi K, Rao UJSP (2010) Mango peel powder: a potential source of antioxidant and dietary fiber in macaroni preparations. Innov Food Sci Emerg Technol 11:219–224. https://doi.org/10.1016/j.ifset.2009.10.004

Albe Slabi S, Mathe C, Basselin M et al (2020) Multi-objective optimization of solid/liquid extraction of total sunflower proteins from cold press meal. Food Chem 317. https://doi.org/10.1016/j.foodchem.2020.126423

Aliakbarian B, Fathi A, Perego P, Dehghani F (2012) Extraction of antioxidants from winery wastes using subcritical water. J Supercrit Fluids 65:18–24. https://doi.org/10.1016/j.supflu.2012.02.022

Alonso S, Herrero M, Rendueles M, Díaz M (2010) Residual yoghurt whey for lactic acid production. Biomass Bioenergy 34:931–938. https://doi.org/10.1016/j.biombioe.2010.01.041

Amado IR, Franco D, Sánchez M et al (2014) Optimisation of antioxidant extraction from Solanum tuberosum potato peel waste by surface response methodology. Food Chem 165:290–299. https://doi.org/10.1016/j.foodchem.2014.05.103

Amenorfenyo DK, Huang X, Zhang Y et al (2019) Microalgae brewery wastewater treatment: potentials, benefits and the challenges. Int J Environ Res Public Health 16. https://doi.org/10.3390/ijerph16111910

Amorim M, Pinheiro H, Pintado M (2019) Valorization of spent brewer's yeast: optimization of hydrolysis process towards the generation of stable ACE-inhibitory peptides. LWT 111:77–84. https://doi.org/10.1016/j.lwt.2019.05.011

Angulo J, Mahecha L, Yepes SA et al (2012) Nutritional evaluation of fruit and vegetable waste as feedstuff for diets of lactating Holstein cows. J Environ Manag. https://doi.org/10.1016/j.jenvman.2011.06.050

Ashayerizadeh O, Dastar B, Samadi F et al (2017) Study on the chemical and microbial composition and probiotic characteristics of dominant lactic acid bacteria in fermented poultry slaughterhouse waste. Waste Manag 65:178–185. https://doi.org/10.1016/j.wasman.2017.04.017

Aspevik T, Oterhals Å, Rønning SB et al (2017) Valorization of proteins from co- and by-products from the fish and meat industry. Top Curr Chem 375:53

Ayala-Zavala JF, Rosas-Domínguez C, Vega-Vega V, González-Aguilar GA (2010) Antioxidant enrichment and antimicrobial protection of fresh-cut fruits using their own byproducts: looking for integral exploitation. J Food Sci 75. https://doi.org/10.1111/j.1750-3841.2010.01792.x

Baiano A (2014) Recovery of biomolecules from food wastes — a review. Molecules 19:14821–14842. https://doi.org/10.3390/molecules190914821

Bakowska-Barczak AM, Schieber A, Kolodziejczyk P (2009) Characterization of Canadian black currant (Ribes nigrum L.) seed oils and residues. J Agric Food Chem 57:11528–11536. https://doi.org/10.1021/jf902161k

Balasundram N, Sundram K, Samman S (2006) Phenolic compounds in plants and agri-industrial by-products: antioxidant activity, occurrence, and potential uses. Food Chem 99:191–203. https://doi.org/10.1016/j.foodchem.2005.07.042

Bampidis VA, Robinson PH (2006) Citrus by-products as ruminant feeds: a review. Anim Feed Sci Technol 128:175–217

Bansal N, Tewari R, Soni R, Soni SK (2012) Production of cellulases from Aspergillus niger NS-2 in solid state fermentation on agricultural and kitchen waste residues. Waste Manag 32:1341–1346. https://doi.org/10.1016/j.wasman.2012.03.006

Barba D, Beolchini F, Cifoni D, Vegliò F (2001) Whey protein concentrate production in a pilot scale two-stage diafiltration process. Sep Sci Technol 36:587–603. https://doi.org/10.1081/SS-100102948

Barba FJ, Zhu Z, Koubaa M et al (2016) Green alternative methods for the extraction of antioxidant bioactive compounds from winery wastes and by-products: a review. Trends Food Sci Technol 49:96–109

Barbosa-Pereira L, Pocheville A, Angulo I et al (2013) Fractionation and purification of bioactive compounds obtained from a brewery waste stream. Biomed Res Int 2013:11. https://doi.org/10.1155/2013/408491

Baydar NG, Sagdic O, Ozkan G, Cetin S (2006) Determination of antibacterial effects and total phenolic contents of grape (*Vitis vinifera* L.) seed extracts. Int J Food Sci Technol 41:799–804. https://doi.org/10.1111/j.1365-2621.2005.01095.x

Baysal T, Ersus S, Starmans DAJ (2000) Supercritical CO2 extraction of β-carotene and lycopene from tomato paste waste. J Agric Food Chem 48:5507–5511. https://doi.org/10.1021/jf000311t

Bedő S, Antal B, Rozbach M et al (2019) Optimised fractionation of wheat bran for arabinose biopurification and xylitol fermentation by Ogataea zsoltii within a biorefinery process. Ind Crops Prod 139. https://doi.org/10.1016/j.indcrop.2019.111504

Berardini N, Knödler M, Schieber A, Carle R (2005) Utilization of mango peels as a source of pectin and polyphenolics. Innov Food Sci Emerg Technol 6:442–452. https://doi.org/10.1016/j.ifset.2005.06.004

Bertin L, Ferri F, Scoma A et al (2011) Recovery of high added value natural polyphenols from actual olive mill wastewater through solid phase extraction. Chem Eng J 171:1287–1293. https://doi.org/10.1016/j.cej.2011.05.056

Bettaieb Rebey I, Bourgou S, Detry P et al (2019) Green extraction of fennel and anise edible oils using bio-based solvent and supercritical fluid: assessment of chemical composition, antioxidant property, and oxidative stability. Food Bioprocess Technol 12:1798–1807. https://doi.org/10.1007/s11947-019-02341-8

Bhaskar N, Modi VK, Govindaraju K et al (2007) Utilization of meat industry by products: protein hydrolysate from sheep visceral mass. Bioresour Technol 98:388–394. https://doi.org/10.1016/j.biortech.2005.12.017

Bicu I, Mustata F (2011) Cellulose extraction from orange peel using sulfite digestion reagents. Bioresour Technol 102:10013–10019. https://doi.org/10.1016/j.biortech.2011.08.041

Bonfigli M, Godoy E, Reinheimer MA, Scenna NJ (2017) Comparison between conventional and ultrasound-assisted techniques for extraction of anthocyanins from grape pomace. Experimental results and mathematical modeling. J Food Eng 207:56–72. https://doi.org/10.1016/j.jfoodeng.2017.03.011

Boonrod D, Reanma K, Niamsup H (2006) Extraction and Physicochemical characteristics of acid-soluble pectin from raw papaya (Carica papaya) peel. Chiang Mai J Sci 33:129–135

Borad SG, Singh AK, Kapila S et al (2019) Influence of unit operations on immunoglobulins and thermal stability of colostrum fractions. Int Dairy J 93:85–91. https://doi.org/10.1016/j.idairyj.2019.02.007

Bosiljkov T, Dujmić F, Cvjetko Bubalo M et al (2017) Natural deep eutectic solvents and ultrasound-assisted extraction: green approaches for extraction of wine lees anthocyanins. Food Bioprod Process 102:195–203. https://doi.org/10.1016/j.fbp.2016.12.005

Bouaziz M, Hammami H, Bouallagui Z et al (2008) Production of antioxidants from olive processing by-products. Electron J Environ Agric Food Chem 7:3231–3236

Boussetta N, Lanoisellé JL, Bedel-Cloutour C, Vorobiev E (2009) Extraction of soluble matter from grape pomace by high voltage electrical discharges for polyphenol recovery: effect of sulphur dioxide and thermal treatments. J Food Eng 95:192–198. https://doi.org/10.1016/j.jfoodeng.2009.04.030

Brandenburg J, Poppele I, Blomqvist J et al (2018) Bioethanol and lipid production from the enzymatic hydrolysate of wheat straw after furfural extraction. Appl Microbiol Biotechnol 102:6269–6277. https://doi.org/10.1007/s00253-018-9081-7

Brenes A, Viveros A, Chamorro S, Arija I (2016) Use of polyphenol-rich grape by-products in monogastric nutrition. A review. Anim Feed Sci Technol 211:1–17

Caballero AS, Romero-García JM, Castro E, Cardona CA (2020) Supercritical fluid extraction for enhancing polyphenolic compounds production from olive waste extracts. J Chem Technol Biotechnol 95:356–362. https://doi.org/10.1002/jctb.5907

Caldas TW, Mazza KEL, Teles ASC et al (2018) Phenolic compounds recovery from grape skin using conventional and non-conventional extraction methods. Ind Crops Prod 111:86–91. https://doi.org/10.1016/j.indcrop.2017.10.012

Campuzano R, González-Martínez S (2016) Characteristics of the organic fraction of municipal solid waste and methane production: a review. Waste Manag 54:3–12

Canteri MHG, Scheer AP, Wosiacki G et al (2010) A comparative study of pectin extracted from passion fruit rind flours. J Polym Environ 18:593–599. https://doi.org/10.1007/s10924-010-0206-z

Cardinali A, Pati S, Minervini F et al (2012) Verbascoside, isoverbascoside, and their derivatives recovered from olive mill wastewater as possible food antioxidants. J Agric Food Chem 60:1822–1829. https://doi.org/10.1021/jf204001p

Casazza AA, Aliakbarian B, Mantegna S et al (2010) Extraction of phenolics from Vitis vinifera wastes using non-conventional techniques. J Food Eng 100:50–55. https://doi.org/10.1016/j.jfoodeng.2010.03.026

Cassano A, Conidi C, Giorno L, Drioli E (2013) Fractionation of olive mill wastewaters by membrane separation techniques. J Hazard Mater 248–249:185–193. https://doi.org/10.1016/j.jhazmat.2013.01.006

Catarino J, Mendonça E, Picado A et al (2007) Getting value from wastewater: by-products recovery in a potato chips industry. J Clean Prod 15:927–931. https://doi.org/10.1016/j.jclepro.2005.12.003

Cecchi T, Alfei B (2013) Volatile profiles of Italian monovarietal extra virgin olive oils via HS-SPME-GC-MS: newly identified compounds, flavors molecular markers, and terpenic profile. Food Chem 141. https://doi.org/10.1016/j.foodchem.2013.05.090

Cecchi T, Passamonti P, Cecchi P (2010a) Study of the quality of extra virgin olive oil stored in PET bottles with or without an oxygen scavenger. Food Chem 120:730–735. https://doi.org/10.1016/J.FOODCHEM.2009.11.001

Cecchi T, Passamonti P, Cecchi P (2010b) Optimization of the measurement of Italian monocultivar extra virgin olive oil antioxidant power via the Briggs-Rauscher reaction. Food Anal Methods 3:1–6. https://doi.org/10.1007/s12161-009-9071-6

Cecchi T, Passamonti P, Alfei B, Cecchi P (2011) Monovarietal extra virgin olive oils from the Marche region, Italy: analytical and sensory characterization. Int J Food Prop 14. https://doi.org/10.1080/10942910903254811

Cefola M, Carbone V, Minasi P, Pace B (2016) Phenolic profiles and postharvest quality changes of fresh-cut radicchio (Cichorium intybus L.): nutrient value in fresh vs. stored leaves. J Food Compos Anal 51:76–84. https://doi.org/10.1016/j.jfca.2016.06.004

Çelik HT, Gürü M (2015) Extraction of oil and silybin compounds from milk thistle seeds using supercritical carbon dioxide. J Supercrit Fluids 100:105–109. https://doi.org/10.1016/j.supflu.2015.02.025

Ćetković G, Savatović S, Čanadanović-Brunet J et al (2012) Valorisation of phenolic composition, antioxidant and cell growth activities of tomato waste. Food Chem 133:938–945. https://doi.org/10.1016/j.foodchem.2012.02.007

Chang FC, Tsai MJ, Ko CH (2018) Agricultural waste derived fuel from oil meal and waste cooking oil. Environ Sci Pollut Res 25:5223–5230. https://doi.org/10.1007/s11356-017-9119-x

Cheirsilp B, Radchabut S (2011) Use of whey lactose from dairy industry for economical kefiran production by Lactobacillus kefiranofaciens in mixed cultures with yeasts. N Biotechnol 28:574–580. https://doi.org/10.1016/j.nbt.2011.01.009

Chen Y, Luo J, Yan Y, Feng L (2013) Enhanced production of short-chain fatty acid by co-fermentation of waste activated sludge and kitchen waste under alkaline conditions and its application to microbial fuel cells. Appl Energy 102:1197–1204. https://doi.org/10.1016/j.apenergy.2012.06.056

Chen H, Shen H, Su HF et al (2017) High-efficiency bioconversion of kitchen garbage to biobutanol using an enzymatic cocktail procedure. Bioresour Technol 245:1110–1121. https://doi.org/10.1016/j.biortech.2017.09.056

Chen C, Sun N, Li D et al (2018) Optimization and characterization of biosurfactant production from kitchen waste oil using Pseudomonas aeruginosa. Environ Sci Pollut Res 25:14934–14943. https://doi.org/10.1007/s11356-018-1691-1

Chen MY, Tsai YC, Tseng CF et al (2019a) Using rice-husk-derived porous silica modified with recycled cu from industrial wastewater and ce to remove Hg0 and NO from simulated flue gases. Aerosol Air Qual Res 19:2557–2567. https://doi.org/10.4209/aaqr.2019.09.0468

Chen S, Huang H, Huang G (2019b) Extraction, derivatization and antioxidant activity of cucumber polysaccharide. Int J Biol Macromol 140:1047–1053. https://doi.org/10.1016/j.ijbiomac.2019.08.203

Chen L, Long R, Huang G, Huang H (2020) Extraction and antioxidant activities in vivo of pumpkin polysaccharide. Ind Crops Prod 146. https://doi.org/10.1016/j.indcrop.2020.112199

Chihoub W, Dias MI, Barros L et al (2019) Valorisation of the green waste parts from turnip, radish and wild cardoon: nutritional value, phenolic profile and bioactivity evaluation. Food Res Int 126:108651. https://doi.org/10.1016/j.foodres.2019.108651

Choi IS, Cho EJ, Moon JH, Bae HJ (2015) Onion skin waste as a valorization resource for the by-products quercetin and biosugar. Food Chem 188:537–542. https://doi.org/10.1016/j.foodchem.2015.05.028

Christiaens S, Uwibambe D, Uyttebroek M et al (2015) Pectin characterisation in vegetable waste streams: a starting point for waste valorisation in the food industry. LWT – Food Sci Technol 61:275–282. https://doi.org/10.1016/j.lwt.2014.12.054

Ciric A, Krajnc B, Heath D, Ogrinc N (2020) Response surface methodology and artificial neural network approach for the optimization of ultrasound-assisted extraction of polyphenols from garlic. Food Chem Toxicol 135:110976. https://doi.org/10.1016/j.fct.2019.110976

Coe S, Ryan L (2016) White bread enriched with polyphenol extracts shows no effect on glycemic response or satiety, yet may increase postprandial insulin economy in healthy participants. Nutr Res 36:193–200. https://doi.org/10.1016/j.nutres.2015.10.007

Conidi C, Cassano A, Garcia-Castello E (2014) Valorization of artichoke wastewaters by integrated membrane process. Water Res 48:363–374. https://doi.org/10.1016/j.watres.2013.09.047

Connolly A, Cermeño M, Crowley D et al (2019) Characterisation of the in vitro bioactive properties of alkaline and enzyme extracted brewers' spent grain protein hydrolysates. Food Res Int 121:524–532. https://doi.org/10.1016/j.foodres.2018.12.008

Corell L, Armenta S, Esteve-Turrillas FA, de la Guardia M (2018) Flavonoid determination in onion, chili and leek by hard cap espresso extraction and liquid chromatography with diode array detection. Microchem J 140:74–79. https://doi.org/10.1016/j.microc.2018.04.014

Crea R (2002) Method of obtaining a hydroxytyrosol-rich composition from vegetation water. World Intellectual Property Organization. WO/2002/0218310. Crowely, Geneva

Cuco RP, Cardozo-Filho L, da Silva C (2019) Simultaneous extraction of seed oil and active compounds from peel of pumpkin (Cucurbita maxima) using pressurized carbon dioxide as solvent. J Supercrit Fluids 143:8–15. https://doi.org/10.1016/j.supflu.2018.08.002

Cumba HJ, Bellmer D (2005) Production of value-added products from meat processing cellulosic waste. In: 2005 ASAE Annual International Meeting. American Society of Agricultural and Biological Engineers. ASAE, St. Joseph

da Silva FV, dos Santos RL, Fujiki TL et al (2010) Desenvolvimento de sistema de controle para precipitação de bromelina a partir de resíduos de abacaxi. Cienc e Tecnol Aliment 30:1033–1040. https://doi.org/10.1590/S0101-20612010000400031

Da Silva LMR, De Figueiredo EAT, Ricardo NMPS et al (2014) Quantification of bioactive compounds in pulps and by-products of tropical fruits from Brazil. Food Chem 143:398–404. https://doi.org/10.1016/j.foodchem.2013.08.001

Dahmoune F, Madani K, Jauregi P et al (2013) Fractionation of a red grape marc extract by colloidal gas aphrons. In: Chemical engineering transactions. Italian Association of Chemical Engineering – AIDIC, Milano, pp 1903–1908

Dai Q, Yang Y, Chen K et al (2019) Optimization of supercritical CO_2 operative parameters to simultaneously increase the extraction yield of oil and pentacyclic triterpenes from artichoke leaves and stalks by response surface methodology and ridge analysis. Eur J Lipid Sci Technol 121:1800120. https://doi.org/10.1002/ejlt.201800120

Dang Y-Y, Zhang H, Xiu Z-L (2014) Microwave-assisted aqueous two-phase extraction of phenolics from grape (*Vitis vinifera*) seed. J Chem Technol Biotechnol 89:1576–1581. https://doi.org/10.1002/jctb.4241

Daraee A, Ghoreishi SM, Hedayati A (2019) Supercritical CO2 extraction of chlorogenic acid from sunflower (Helianthus annuus) seed kernels: modeling and optimization by response surface methodology. J Supercrit Fluids 144:19–27. https://doi.org/10.1016/j.supflu.2018.10.001

Davidov-Pardo G, McClements DJ (2015) Nutraceutical delivery systems: resveratrol encapsulation in grape seed oil nanoemulsions formed by spontaneous emulsification. Food Chem 167:205–212. https://doi.org/10.1016/j.foodchem.2014.06.082

de Moraes Crizel T, Jablonski A, de Oliveira Rios A et al (2013) Dietary fiber from orange byproducts as a potential fat replacer. LWT – Food Sci Technol 53:9–14. https://doi.org/10.1016/j.lwt.2013.02.002

Delgado-Moreno L, Sánchez-Moreno L, Peña A (2007) Assessment of olive cake as soil amendment for the controlled release of triazine herbicides. Sci Total Environ 378:119–123. https://doi.org/10.1016/j.scitotenv.2007.01.023

Derrien M, Aghabararnejad M, Gosselin A et al (2018) Optimization of supercritical carbon dioxide extraction of lutein and chlorophyll from spinach by-products using response surface methodology. LWT 93:79–87. https://doi.org/10.1016/j.lwt.2018.03.016

DeSilva K, Stockmann R, Smithers GW (2003) Isolation procedures for functional dairy components – Novel approaches to meeting the challenges. Austr J Dairy Technol 58:148–152

Dessie W, Zhang W, Xin F et al (2018) Succinic acid production from fruit and vegetable wastes hydrolyzed by on-site enzyme mixtures through solid state fermentation. Bioresour Technol 247:1177–1180. https://doi.org/10.1016/j.biortech.2017.08.171

Deydier E, Guilet R, Sarda S, Sharrock P (2005) Physical and chemical characterisation of crude meat and bone meal combustion residue: "waste or raw material?". J Hazard Mater 121:141–148. https://doi.org/10.1016/j.jhazmat.2005.02.003

Dhillon GS, Brar SK, Verma M, Tyagi RD (2011) Apple pomace ultrafiltration sludge – a novel substrate for fungal bioproduction of citric acid: optimisation studies. Food Chem 128:864–871. https://doi.org/10.1016/j.foodchem.2011.03.107

Dhillon GS, Brar SK, Verma M (2012) Biotechnological potential of industrial wastes for economical citric acid bioproduction by Aspergillus niger through submerged fermentation. Int J Food Sci Technol 47:542–548. https://doi.org/10.1111/j.1365-2621.2011.02875.x

Di Mauro A, Fallico B, Passerini A et al (1999) Recovery of hesperidin from orange peel by concentration of extracts on styrene-divinylbenzene resin. J Agric Food Chem 47:4391–4397. https://doi.org/10.1021/jf990038z

Díaz AB, De Ory I, Caro I, Blandino A (2012) Enhance hydrolytic enzymes production by Aspergillus awamori on supplemented grape pomace. Food Bioprod Process 90:72–78. https://doi.org/10.1016/j.fbp.2010.12.003

Díaz AI, Laca A, Laca A, Díaz M (2017) Treatment of supermarket vegetable wastes to be used as alternative substrates in bioprocesses. Waste Manag 67:59–66. https://doi.org/10.1016/j.wasman.2017.05.018

Dominguez-Perles R, Moreno DA, Carvajal M, Garcia-Viguera C (2011) Composition and antioxidant capacity of a novel beverage produced with green tea and minimally-processed byproducts of broccoli. Innov Food Sci Emerg Technol 12:361–368. https://doi.org/10.1016/j.ifset.2011.04.005

Drosou C, Kyriakopoulou K, Bimpilas A et al (2015) A comparative study on different extraction techniques to recover red grape pomace polyphenols from vinification byproducts. Ind Crops Prod 75:141–149. https://doi.org/10.1016/j.indcrop.2015.05.063

Drouet S, Leclerc EA, Garros L et al (2019) A green ultrasound-assisted extraction optimization of the natural antioxidant and anti-aging flavonolignans from milk thistle silybum marianum (L.) gaertn. fruits for cosmetic applications. Antioxidants 8. https://doi.org/10.3390/antiox8080304

Duba KS, Casazza AA, Ben MH et al (2015) Extraction of polyphenols from grape skins and defatted grape seeds using subcritical water: experiments and modeling. Food Bioprod Process 94:29–38. https://doi.org/10.1016/j.fbp.2015.01.001

Dumas Y, Dadomo M, Di Lucca G, Grolier P (2003) Effects of environmental factors and agricultural techniques on antioxidant content of tomatoes. J Sci Food Agric 83:369–382

Ebrahimzadeh SK, Navidshad B, Farhoomand P, Aghjehgheshlagh FM (2018) Effects of grape pomace and vitamin E on performance, antioxidant status, immune response, gut morphology and histopathological responses in broiler chickens. South Afr J Anim Sci 48:324–336. https://doi.org/10.4314/sajas.v48i2.13

Eyvazkhani R, Bahmanyar H, Mirdehghan Ashkezari SM, Najafipour I (2020) Extraction of essential constituents from effluent of hydro-distillation of fennel and investigation of hydrodynamic parameters using a rotary disc column (RDC). Chem Eng Commun. https://doi.org/10.1080/00986445.2020.1734577

FAO (2016) SAVE FOOD: Global Initiative on Food Loss and Waste Reduction. Food Agric. Organ. United Nations 01–02. http://www.fao.org/save-food/resources/infographic/en/ Accessed 11th Mar 2021

FAO (2020a) The State of Food Security and Nutrition in the World 2020. FAO, IFAD, UNICEF, WFP and WHO

FAO (2020b) FAOSTAT. http://www.fao.org/faostat/en/#home. Accessed 10 Mar 2021

Farhat A, Fabiano-Tixier AS, El Maataoui M et al (2011) Microwave steam diffusion for extraction of essential oil from orange peel: kinetic data, extract's global yield and mechanism. Food Chem 125:255–261. https://doi.org/10.1016/j.foodchem.2010.07.110

Farkas C, Rezessy-Szabó JM, Gupta VK et al (2019) Microbial saccharification of wheat bran for bioethanol fermentation. J Clean Prod 240. https://doi.org/10.1016/j.jclepro.2019.118269

Federici F, Fava F, Kalogerakis N, Mantzavinos D (2009) Valorisation of agro-industrial by-products, effluents and waste: concept, opportunities and the case of olive mill waste waters. J Chem Technol Biotechnol 84:895–900

Ferhat MA, Meklati BY, Smadja J, Chemat F (2006) An improved microwave Clevenger apparatus for distillation of essential oils from orange peel. J Chromatogr A 1112:121–126. https://doi.org/10.1016/j.chroma.2005.12.030

Ferioli F, Giambanelli E, D'Alessandro V, D'Antuono LF (2020) Comparison of two extraction methods (high pressure extraction vs. maceration) for the total and relative amount of hydrophilic and lipophilic organosulfur compounds in garlic cloves and stems. An application to the Italian ecotype "Aglio Rosso di Sulmona" (Sulmona Red Garlic). Food Chem 312. https://doi.org/10.1016/j.foodchem.2019.126086

Fernández-Bolaños J, Rodríguez G, Gómez E et al (2004) Total recovery of the waste of two-phase olive oil processing: isolation of added-value compounds. J Agric Food Chem 52:5849–5855. https://doi.org/10.1021/jf030821y

Fernández-López J, Fernández-Ginés JM, Aleson-Carbonell L et al (2004) Application of functional citrus by-products to meat products. Trends Food Sci Technol 15:176–185. https://doi.org/10.1016/j.tifs.2003.08.007

Ferreira AS, Nunes C, Castro A et al (2014) Influence of grape pomace extract incorporation on chitosan films properties. Carbohydr Polym 113:490–499. https://doi.org/10.1016/j.carbpol.2014.07.032

Ferreira DF, Barin JS, Binello A et al (2019) Highly efficient pumpkin-seed extraction with the simultaneous recovery of lipophilic and hydrophilic compounds. Food Bioprod Process 117:224–230. https://doi.org/10.1016/j.fbp.2019.07.014

Ferrentino G, Ndayishimiye J, Haman N, Scampicchio M (2019) Functional activity of oils from Brewer's spent grain extracted by supercritical carbon dioxide. Food Bioprocess Technol 12:789–798. https://doi.org/10.1007/s11947-019-02249-3

Ferrer J, Páez G, Mármol Z et al (2001) Agronomic use of biotechnologically processed grape wastes. Bioresour Technol 76:39–44. https://doi.org/10.1016/s0960-8524(00)00076-6

Ferri F, Bertin L, Scoma A et al (2011) Recovery of low molecular weight phenols through solid-phase extraction. Chem Eng J 166:994–1001. https://doi.org/10.1016/j.cej.2010.11.090

Fishman ML, Chau HK, Hoagland P, Ayyad K (1999) Characterization of pectin, flash-extracted from orange albedo by microwave heating, under pressure. Carbohydr Res 323:126–138. https://doi.org/10.1016/S0008-6215(99)00244-X

Frumento D, Paula A, Santo E et al (2013) Development of milk fermented with lactobacillus acidophilus fortified with vitis vinifera marc flour. Food Technol Biotechnol 51(3):370–375

Fuentes-Alventosa JM, Jaramillo-Carmona S, Rodríguez-Gutiérrez G et al (2013) Preparation of bioactive extracts from asparagus by-product. Food Bioprod Process 91:74–82. https://doi.org/10.1016/j.fbp.2012.12.004

Galanakis CM (2012) Recovery of high added-value components from food wastes: conventional, emerging technologies and commercialized applications. Trends Food Sci Technol 26:68–87

Galanakis CM (2017) Handbook of grape processing by-products: sustainable solutions. Academic, London

García Herrera P, Sánchez-Mata MC, Cámara M (2010) Nutritional characterization of tomato fiber as a useful ingredient for food industry. Innov Food Sci Emerg Technol 11:707–711. https://doi.org/10.1016/j.ifset.2010.07.005

Gargiulo V, Lanzetta R, Parrilli M, De Castro C (2009) Structural analysis of chondroitin sulfate from Scyliorhinus canicula: a useful source of this polysaccharide. Glycobiology 19:1485–1491. https://doi.org/10.1093/glycob/cwp123

Garrido MD, Auqui M, Martí N, Linares MB (2011) Effect of two different red grape pomace extracts obtained under different extraction systems on meat quality of pork burgers. LWT – Food Sci Technol 44:2238–2243. https://doi.org/10.1016/j.lwt.2011.07.003

Gbashi S, Njobeh P, Steenkamp P, Madala N (2017) Extracción con agua caliente presurizada y huella quimiométrica de flavonoides de Bidens pilosa mediante espectrometría de masas en tándem por UPLC. CYTA – J Food 15:171–180. https://doi.org/10.1080/19476337.2016.1230151

Gehring CK, Gigliotti JC, Moritz JS et al (2011) Functional and nutritional characteristics of proteins and lipids recovered by isoelectric processing of fish by-products and low-value fish: a review. Food Chem 124:422–431

Germec M, Ozcan A, Turhan I (2019) Bioconversion of wheat bran into high value-added products and modelling of fermentations. Ind Crops Prod 139. https://doi.org/10.1016/j.indcrop.2019.111565

Ghafoor K, Park J, Choi YH (2010) Optimization of supercritical fluid extraction of bioactive compounds from grape (Vitis labrusca B.) peel by using response surface methodology. Innov Food Sci Emerg Technol 11:485–490. https://doi.org/10.1016/j.ifset.2010.01.013

Ghanbari R, Anwar F, Alkharfy KM et al (2012) Valuable nutrients and functional bioactives in different parts of olive (Olea europaea L.)-A review. Int J Mol Sci 13:1291–1340

Gil M, Del Barrio-Galán R, Ubeda C, Peña-Neira Á (2018) Effectiveness of fibers from "Cabernet Sauvignon" (Vitis vinifera) pomace as fining agents for red wines. J Food Quality 2018:6408734. https://doi.org/10.1155/2018/6408734

Gomez-Guillen MC, Gimenez B, Lopez-Caballero ME, Montero MP (2011) Functional and bioactive properties of collagen and gelatin from alternative sources: a review. Food Hydrocoll 25:1813–1827

González-Centeno MR, Comas-Serra F, Femenia A et al (2015) Effect of power ultrasound application on aqueous extraction of phenolic compounds and antioxidant capacity from grape pomace (Vitis vinifera L.): experimental kinetics and modeling. Ultrason Sonochem 22:506–514. https://doi.org/10.1016/j.ultsonch.2014.05.027

Górecka D, Dziedzic K, Pachołek B, Górecka M (2010) Raspberry pomace as a potential fiber source for cookies enrichment. Acta Sci Pol Technol Aliment 9:451–462

Grabowska M, Wawrzyniak D, Rolle K et al (2019) Let food be your medicine: nutraceutical properties of lycopene. Food Funct 10:3090–3102

Guerra-Rivas C, Vieira C, Rubio B et al (2016) Effects of grape pomace in growing lamb diets compared with vitamin E and grape seed extract on meat shelf life. Meat Sci 116:221–229. https://doi.org/10.1016/j.meatsci.2016.02.022

Gustavsson J, Cederberg C, Sonesson U et al (2011) Global food losses and food waste: extent, causes and prevention. FAO, Düsseldorf

Gustafson S (2019) FAO SOFA report 2019: New insights into food loss and waste | IFPRI : International Food Policy Research Institute [WWW Document]. URL https://www.ifpri.org/blog/fao-sofa-report-2019-new-insights-food-loss-and-waste Accessed 11th Mar 2021

Hafid HS, Rahman NA, Md Shah UK, Baharudin AS (2015) Enhanced fermentable sugar production from kitchen waste using various pretreatments. J Environ Manage 156:290–298. https://doi.org/10.1016/j.jenvman.2015.03.045

Hall RM, Mayer DA, Mazzutti S, Ferreira SRS (2018) Simulating large scale SFE applied to recover bioactive compounds from papaya seeds. J Supercrit Fluids 140:302–309. https://doi.org/10.1016/j.supflu.2018.07.013

Han D, Row KH (2011) Determination of luteolin and apigenin in celery using ultrasonic-assisted extraction based on aqueous solution of ionic liquid coupled with HPLC quantification. J Sci Food Agric 91:2888–2892. https://doi.org/10.1002/jsfa.4553

Hao R, Li Q, Zhao J et al (2015) Effects of grape seed procyanidins on growth performance, immune function and antioxidant capacity in weaned piglets. Livest Sci 178:237–242. https://doi.org/10.1016/j.livsci.2015.06.004

Hataminia F, Farhadian N, Karimi M, Ebrahimi M (2018) A novel method for squalene extraction from pumpkin seed oil using magnetic nanoparticles and exploring the inhibition effect of extracted squalene on angiogenesis property. J Taiwan Inst Chem Eng 91:1–9. https://doi.org/10.1016/j.jtice.2018.05.017

He Q, Li Y, Zhang P et al (2016) Optimisation of microwave-assisted extraction of flavonoids and phenolics from celery (Apium graveolens L.) leaves by response surface methodology. Czech J Food Sci 34:341–349. https://doi.org/10.17221/266/2015-CJFS

Henkel M, Müller MM, Kügler JH et al (2012) Rhamnolipids as biosurfactants from renewable resources: concepts for next-generation rhamnolipid production. Process Biochem 47:1207–1219

Henríquez C, Speisky H, Chiffelle I et al (2010) Development of an ingredient containing apple peel, as a source of polyphenols and dietary fiber. J Food Sci. https://doi.org/10.1111/j.1750-3841.2010.01700.x

Herrera-Ponce AL, Alarcón-Rojo AD, Salmeron I, Rodríguez-Figueroa JC (2019) Physiological health effects of whey protein-derived bioactive peptides: a review. Rev Chil Nutr 46:205–214

Hollmann J, Elbegzaya N, Pawelzik E, Lindhauer MG (2009) Isolation and characterization of glucuronoarabinoxylans from wheat bran obtained by classical and ultrasound-assisted extraction methods. Qual Assur Saf Crop Foods 1:231–239. https://doi.org/10.1111/j.1757-837X.2009.00039.x

Huber GM, Rupasinghe HPV (2009) Phenolic profiles and antioxidant properties of apple skin extracts. J Food Sci. https://doi.org/10.1111/j.1750-3841.2009.01356.x

Hussein MHM, El-Hady MF, Shehata HAH et al (2013) Preparation of some eco-friendly corrosion inhibitors having antibacterial activity from sea food waste. J Surfactants Deterg 16:233–242. https://doi.org/10.1007/s11743-012-1395-3

Hỳsek Š, Neuberger P, Sikora A et al (2019) Waste utilization: insulation panel from recycled polyurethane particles and wheat husks. Materials (Basel) 12. https://doi.org/10.3390/ma12193075

Iravani S, Varma RS (2020) Greener synthesis of lignin nanoparticles and their applications. Green Chem 22:612–636

Ishida BK, Chapman MH (2009) Carotenoid extraction from plants using a novel, environmentally friendly solvent. J Agric Food Chem 57:1051–1059. https://doi.org/10.1021/jf8026292

Ishikawa S, Maruyama A, Yamamoto Y, Hara S (2014) Extraction and characterization of glucosinolates and isothiocyanates from rape seed meal. J Oleo Sci 63:303–308. https://doi.org/10.5650/jos.ess13170

Jafarian Asl P, Niazmand R, Yahyavi F (2020) Extraction of phytosterols and tocopherols from rapeseed oil waste by supercritical CO2 plus co-solvent: a comparison with conventional solvent extraction. Heliyon 6. https://doi.org/10.1016/j.heliyon.2020.e03592

Jaime L, Vázquez E, Fornari T et al (2015) Extraction of functional ingredients from spinach (Spinacia oleracea L.) using liquid solvent and supercritical CO2 extraction. J Sci Food Agric 95:722–729. https://doi.org/10.1002/jsfa.6788

Jara-Palacios MJ, Hernanz D, Escudero-Gilete ML, Heredia FJ (2016) The use of grape seed byproducts rich in flavonoids to improve the antioxidant potential of red wines. Molecules 21. https://doi.org/10.3390/molecules21111526

Javier H, Ángel SJ, Aida G et al (2019) Revalorization of grape marc waste from liqueur wine: biomethanization. J Chem Technol Biotechnol 94:1499–1508. https://doi.org/10.1002/jctb.5909

Jeantet R, Rodríguez J, Garem A (2000) Nanofiltration of sweet whey by spiral wound organic membranes: impact of hydrodynamics. Lait 80:155–163. https://doi.org/10.1051/lait:2000115

Kalli E, Lappa I, Bouchagier P et al (2018) Novel application and industrial exploitation of winery by-products. Bioresour Bioprocess 5:46. https://doi.org/10.1186/s40643-018-0232-6

Kalogeropoulos N, Chiou A, Pyriochou V et al (2012) Bioactive phytochemicals in industrial tomatoes and their processing byproducts. LWT – Food Sci Technol 49:213–216. https://doi.org/10.1016/j.lwt.2011.12.036

Kanatt SR, Chander R, Radhakrishna P, Sharma A (2005) Potato peel extract – a natural antioxidant for retarding lipid peroxidation in radiation processed lamb meat. J Agric Food Chem 53:1499–1504. https://doi.org/10.1021/jf048270e

Kang HY, Yang PY, Dominy WG, Lee CS (2010) Bioprocessing papaya processing waste for potential aquaculture feed supplement – economic and nutrient analysis with shrimp feeding trial. Bioresour Technol 101:7973–7979. https://doi.org/10.1016/j.biortech.2010.05.058

Kapusta I, Janda B, Szajwaj B et al (2007) Flavonoids in horse chestnut (Aesculus hippocastanum) seeds and powdered waste water byproducts. J Agric Food Chem 55:8485–8490. https://doi.org/10.1021/jf071709t

Karantonis HC, Tsantila N, Stamatakis G et al (2008) Bioactive polar lipids in olive oil, pomace and waste byproducts. J Food Biochem 32:443–459. https://doi.org/10.1111/j.1745-4514.2008.00160.x

Kardel M, Taube F, Schulz H et al (2013) Different approaches to evaluate tannin content and structure of selected plant extracts – review and new aspects. J Appl Bot Food Qual 86:154–166

Karp SG, Igashiyama AH, Siqueira PF et al (2011) Application of the biorefinery concept to produce l-lactic acid from the soybean vinasse at laboratory and pilot scale. Bioresour Technol 102:1765–1772. https://doi.org/10.1016/j.biortech.2010.08.102

Karuppiah H, Kirubakaran N, Sundaram J (2018) Genetic resources for arcelin, a stored product insect antimetabolic protein from various accessions of pulses of Leguminosae. Genet Resour Crop Evol 65:79–90. https://doi.org/10.1007/s10722-017-0510-8

Kazemi M, Khodaiyan F, Hosseini SS, Najari Z (2019) An integrated valorization of industrial waste of eggplant: simultaneous recovery of pectin, phenolics and sequential production of pullulan. Waste Manag 100:101–111. https://doi.org/10.1016/j.wasman.2019.09.013

Kehili M, Kammlott M, Choura S et al (2017) Supercritical CO2 extraction and antioxidant activity of lycopene and β-carotene-enriched oleoresin from tomato (Lycopersicum esculentum L.) peels by-product of a Tunisian industry. Food Bioprod Process 102:340–349. https://doi.org/10.1016/j.fbp.2017.02.002

Kelly PM, Kelly J, Mehra R et al (2000) Implementation of integrated membrane processes for pilot scale development of fractionated milk components. Lait 80:139–153. https://doi.org/10.1051/lait:2000101

Kerton FM, Liu Y, Omari KW, Hawboldt K (2013) Green chemistry and the ocean-based biorefinery. Green Chem 15:860–871

Khaidir REM, Fen YW, Zaid MHM et al (2019) Exploring Eu3+-doped ZnO-SiO2 glass derived by recycling renewable source of waste rice husk for white-LEDs application. Results Phys 15. https://doi.org/10.1016/j.rinp.2019.102596

Kim S-B, Ji C-I, Woo J-W et al (2012) Simplified purification of chondroitin sulphate from scapular cartilage of shortfin mako shark (Isurus oxyrinchus). Int J Food Sci Technol 47:91–99. https://doi.org/10.1111/j.1365-2621.2011.02811.x

Kiran EU, Trzcinski AP, Ng WJ, Liu Y (2014) Bioconversion of food waste to energy: A review. Fuel 134:389–399

Kitrytė V, Kavaliauskaitė A, Tamkutė L et al (2020) Zero waste biorefining of lingonberry (Vaccinium vitis-idaea L.) pomace into functional ingredients by consecutive high pressure and enzyme assisted extractions with green solvents. Food Chem 322. https://doi.org/10.1016/j.foodchem.2020.126767

Kock LB, Brummer Y, Exley T et al (2018) In vitro assessment of oat β-glucans nutritional properties: an inter-laboratory methodology evaluation. Carbohydr Polym 200:271–277. https://doi.org/10.1016/j.carbpol.2018.07.082

Kong J, Gao S, Liu Y et al (2019) Recycling of carbonized rice husk for producing high purity silicon by the combination of electric arc smelting and slag refining. J Hazard Mater 380. https://doi.org/10.1016/j.jhazmat.2019.120827

Korotych O, Mondal J, Gattás-Asfura KM et al (2019) Evaluation of commercially available styrene-co-maleic acid polymers for the extraction of membrane proteins from spinach chloroplast thylakoids. Eur Polym J 114:485–500. https://doi.org/10.1016/j.eurpolymj.2018.10.035

Kosseva MR (2011) Management and processing of food wastes. In: Comprehensive biotechnology, 2nd edn. Food and Agriculture Organization of the United Nations, Rome

Koushkbaghi M, Kazemi MJ, Mosavi H, Mohseni E (2019) Acid resistance and durability properties of steel fiber-reinforced concrete incorporating rice husk ash and recycled aggregate. Constr Build Mater 202:266–275. https://doi.org/10.1016/j.conbuildmat.2018.12.224

Koutinas AA, Papapostolou H, Dimitrellou D et al (2009) Whey valorisation: a complete and novel technology development for dairy industry starter culture production. Bioresour Technol 100:3734–3739. https://doi.org/10.1016/j.biortech.2009.01.058

Krasnoshtanova AA (2010) Obtaining enzymatic protein and lipid hydrolysates from byproducts of the meat processing industry. Catal Ind 2:173–179. https://doi.org/10.1134/S2070050410020133

Krisch J, Galgoczy L, Vágvölgyi C et al (2009) Antimicrobial and antioxidant potential of waste products remaining after juice pressing. J Eng Ann 42(2):483–490

Kroyer GT (1995) Impact of food processing on the environment-an overview. LWT – Food Sci Technol 28:547

Kryževičiute N, Kraujalis P, Venskutonis PR (2016) Optimization of high pressure extraction processes for the separation of raspberry pomace into lipophilic and hydrophilic fractions. J Supercrit Fluids 108:61–68. https://doi.org/10.1016/j.supflu.2015.10.025

Kulczyński B, Gramza-Michałowska A, Królczyk JB (2020) Optimization of extraction conditions for the antioxidant potential of different pumpkin varieties (Cucurbita Maxima). Sustain 12. https://doi.org/10.3390/su12041305

Kumar S, Kumar Walia Y (2017) Extraction of methylcellulose from wheat straw of Himachal Pradesh, India. Oriental J Chem 33:2625–2631. https://doi.org/10.13005/ojc/330560

Kumar S, Mor DR (2013) Utilization of rice husk and their ash: a review. Res J Chem Environ Sci 1(5):126–129

Kumar A, Kushwaha VS, Sharma P (2014) Pharmaceutical microemulsion: formulation, characterization and drug deliveries across skin. Int J Drug Dev Res 6(1):1–21

Kumar A, Sengupta B, Dasgupta D et al (2016) Recovery of value added products from rice husk ash to explore an economic way for recycle and reuse of agricultural waste. Rev Environ Sci Biotechnol 15:47–65

Kumar A, Kumar D, George N et al (2018) A process for complete biodegradation of shrimp waste by a novel marine isolate Paenibacillus sp. AD with simultaneous production of chitinase and chitin oligosaccharides. Int J Biol Macromol 109:263–272. https://doi.org/10.1016/j.ijbiomac.2017.12.024

Lafarga T, Rodríguez-Roque MJ, Bobo G et al (2019) Effect of ultrasound processing on the bioaccessibility of phenolic compounds and antioxidant capacity of selected vegetables. Food Sci Biotechnol 28:1713–1721. https://doi.org/10.1007/s10068-019-00618-4

Lante A, Nardi T, Zocca F et al (2011) Evaluation of red chicory extract as a natural antioxidant by pure lipid oxidation and yeast oxidative stress response as model systems. J Agric Food Chem 59:5318–5324. https://doi.org/10.1021/jf2003317

Lee PC, Lee WG, Kwon S et al (2000) Batch and continuous cultivation of Anaerobiospirillum succiniciproducens for the production of succinic acid from whey. Appl Microbiol Biotechnol 54:23–27. https://doi.org/10.1007/s002530000331

Lee SB, Jang HS, Hong IK (2017) Optimization of extraction process for antioxidant from persimmon leaf and thistle using response surface methodology. Appl Chem Eng 28:442–447. https://doi.org/10.14478/ace.2017.1037

Leite AC, Ferreira AM, Morais ES et al (2018) Cloud point extraction of chlorophylls from spinach leaves using aqueous solutions of nonionic surfactants. ACS Sustain Chem Eng 6:590–599. https://doi.org/10.1021/acssuschemeng.7b02931

Li L, Shao J, Zhu X et al (2013) Effect of plant polyphenols and ascorbic acid on lipid oxidation, residual nitrite and N-nitrosamines formation in dry-cured sausage. Int J Food Sci Technol 48:1157–1164. https://doi.org/10.1111/ijfs.12069

Li P, Zeng Y, Xie Y et al (2017) Effect of pretreatment on the enzymatic hydrolysis of kitchen waste for xanthan production. Bioresour Technol 223:84–90. https://doi.org/10.1016/j.biortech.2016.10.035

Lima PM, Rubio FTV, Silva MP et al (2019) Nutritional value and modelling of carotenoids extraction from pumpkin (Cucurbita Moschata) Peel Flour By-Product. Int J Food Eng 15. https://doi.org/10.1515/ijfe-2018-0381

Liu Y, Shi J, Langrish TAG (2006) Water-based extraction of pectin from flavedo and albedo of orange peels. Chem Eng J 120:203–209. https://doi.org/10.1016/j.cej.2006.02.015

Liu D, Vorobiev D, Savoire R, Lanoiselle J-L (2011) Extraction of polyphenols from grape seeds by unconventional methods and extract concentration through polymeric membrane. Food Process Eng a Chang World Proc 11th Int Congr Eng Food 2011:1939–1940

Liu W, Dong Z, Sun D et al (2019) Bioconversion of kitchen wastes into bioflocculant and its pilot-scale application in treating iron mineral processing wastewater. Bioresour Technol 288:121505. https://doi.org/10.1016/j.biortech.2019.121505

Löf D, Schillén K, Nilsson L (2011) Flavonoids: Precipitation kinetics and interaction with surfactant micelles. J Food Sci 76:35–39. https://doi.org/10.1111/j.1750-3841.2011.02103.x

López-Linares JC, García-Cubero MT, Lucas S et al (2019) Microwave assisted hydrothermal as greener pretreatment of brewer's spent grains for biobutanol production. Chem Eng J 368:1045–1055. https://doi.org/10.1016/j.cej.2019.03.032

Lorente A, Remón J, Budarin VL et al (2019) Analysis and optimisation of a novel "bio-brewery" approach: Production of bio-fuels and bio-chemicals by microwave-assisted, hydrothermal liquefaction of brewers' spent grains. Energy Convers Manag 185:410–430. https://doi.org/10.1016/j.enconman.2019.01.111

Lorenzo JM, Sineiro J, Amado IR, Franco D (2014) Influence of natural extracts on the shelf life of modified atmosphere-packaged pork patties. Meat Sci 96:526–534. https://doi.org/10.1016/j.meatsci.2013.08.007

Lou R, Ma R, Lin K-T et al (2020) Facile extraction of wheat straw by deep eutectic solvent (DES) to produce lignin nanoparticles. ACS Sustain Chem Eng 14:59. https://doi.org/10.1021/acssuschemeng.8b05816

Lozano-Sánchez J, Giambanelli E, Quirantes-Piné R et al (2011) Wastes generated during the storage of extra virgin olive oil as a natural source of phenolic compounds. J Agric Food Chem 59:11491–11500. https://doi.org/10.1021/jf202596q

Luft L, Confortin TC, Todero I et al (2019) Ultrasound technology applied to enhance enzymatic hydrolysis of Brewer's spent grain and its potential for production of fermentable sugars. Waste Biomass Valoriz 10:2157–2164. https://doi.org/10.1007/s12649-018-0233-x

Luguel C, Annevelink B, Laurmaa J et al (2011) European biorefinery joint strategic research roadmap. JRC Tech Rep 2011:68

Luo J, Zhu Y, Song A et al (2019) Efficient short-chain fatty acids recovery from anaerobic fermentation of wine vinasse and waste activated sludge and the underlying mechanisms. Biochem Eng J 145:18–26. https://doi.org/10.1016/j.bej.2019.02.010

MacHmudah S, Zakaria WS et al (2012) Lycopene extraction from tomato peel by-product containing tomato seed using supercritical carbon dioxide. J Food Eng 108:290–296. https://doi.org/10.1016/j.jfoodeng.2011.08.012

Mahboubi A, Ferreira JA, Taherzadeh MJ, Lennartsson PR (2017) Value-added products from dairy waste using edible fungi. Waste Manag 59:518–525. https://doi.org/10.1016/j.wasman.2016.11.017

Makris DP (2018) Green extraction processes for the efficient recovery of bioactive polyphenols from wine industry solid wastes – recent progress. Curr Opin Green Sustain Chem 13:50–55

Maneerat N, Tangsuphoom N, Nitithamyong A (2017) Effect of extraction condition on properties of pectin from banana peels and its function as fat replacer in salad cream. J Food Sci Technol 54:386–397. https://doi.org/10.1007/s13197-016-2475-6

Manso T, Nunes C, Raposo S, Lima-Costa ME (2010) Carob pulp as raw material for production of the biocontrol agent P. agglomerans PBC-1. J Ind Microbiol Biotechnol 37:1145–1155. https://doi.org/10.1007/s10295-010-0762-1

Maragkoudakis P, Nardi T, Bovo B et al (2010) Valorisation of a milk industry by-product as substrate for microbial growth. J Biotechnol 150:340–340. https://doi.org/10.1016/j.jbiotec.2010.09.361

Marchiani R, Bertolino M, Ghirardello D et al (2016) Physicochemical and nutritional qualities of grape pomace powder-fortified semi-hard cheeses. J Food Sci Technol 53:1585–1596. https://doi.org/10.1007/s13197-015-2105-8

Martin-Diana AB, Rico D, Frias J et al (2006) Whey permeate as a bio-preservative for shelf life maintenance of fresh-cut vegetables. Innov Food Sci Emerg Technol 7:112–123. https://doi.org/10.1016/j.ifset.2005.08.002

Martinez-Sena MT, de la Guardia M, Esteve-Turrillas FA, Armenta S (2017) Hard cap espresso extraction and liquid chromatography determination of bioactive compounds in vegetables and spices. Food Chem 237:75–82. https://doi.org/10.1016/j.foodchem.2017.05.101

Martins BC, Rescolino R, Coelho DF et al (2014) Characterization of bromelain from ananas comosus agroindustrial residues purified by ethanol factional precipitation. Chem Eng Trans 37:781–786. https://doi.org/10.3303/CET1437131

Massa TB, Stevanato N, Cardozo-Filho L, da Silva C (2019) Pumpkin (Cucurbita maxima) by-products: obtaining seed oil enriched with active compounds from the peel by ultrasonic-assisted extraction. J Food Process Eng 42. https://doi.org/10.1111/jfpe.13125

Matassa S, Verstraete W, Pikaar I, Boon N (2016) Autotrophic nitrogen assimilation and carbon capture for microbial protein production by a novel enrichment of hydrogen-oxidizing bacteria. Water Res 101:137–146. https://doi.org/10.1016/j.watres.2016.05.077

Mateos-Aparicio I, Redondo-Cuenca A, Villanueva-Suárez MJ et al (2010) Pea pod, broad bean pod and okara, potential sources of functional compounds. LWT – Food Sci Technol 43:1467–1470. https://doi.org/10.1016/j.lwt.2010.05.008

Matharu AS, de Melo EM, Houghton JA (2016) Opportunity for high value-added chemicals from food supply chain wastes. Bioresour Technol 215:123–130

Maumela P, van Rensburg E, Chimphango AFA, Görgens JF (2020) Sequential extraction of protein and inulin from the tubers of Jerusalem artichoke (Helianthus tuberosus L.). J Food Sci Technol 57:775–786. https://doi.org/10.1007/s13197-019-04110-z

Mauro RP, Agnello M, Rizzo V et al (2020) Recovery of eggplant field waste as a source of phytochemicals. Sci Hortic (Amsterdam) 261:109023. https://doi.org/10.1016/j.scienta.2019.109023

Mena-García A, Rodríguez-Sánchez S, Ruiz-Matute AI, Sanz ML (2020) Exploitation of artichoke byproducts to obtain bioactive extracts enriched in inositols and caffeoylquinic acids by Microwave Assisted Extraction. J Chromatogr A 1613. https://doi.org/10.1016/j.chroma.2019.460703

Mildner-Szkudlarz S, Bajerska J (2013) Protective effect of grape by-product-fortified breads against cholesterol/cholic acid diet-induced hypercholesterolaemia in rats. J Sci Food Agric 93:3271–3278. https://doi.org/10.1002/jsfa.6171

Minhalma M, Magueijo V, Queiroz DP, de Pinho MN (2007) Optimization of "Serpa" cheese whey nanofiltration for effluent minimization and by-products recovery. J Environ Manag 82:200–206. https://doi.org/10.1016/j.jenvman.2005.12.011

Mitić M, Janković S, Mašković P et al (2020) Kinetic models of the extraction of vanillic acid from pumpkin seeds. Open Chem 18:22–30. https://doi.org/10.1515/chem-2020-0001

Mizuno H, Usuki T (2018) Ionic liquid-assisted extraction and isolation of cynaropicrin and cnicin from artichoke and blessed thistle. ChemistrySelect 3:1781–1786. https://doi.org/10.1002/slct.201703063

Moate PJ, Williams SRO, Torok VA et al (2014) Grape marc reduces methane emissions when fed to dairy cows. J Dairy Sci 97:5073–5087. https://doi.org/10.3168/jds.2013-7588

Mohdaly AAA, Sarhan MA, Smetanska I, Mahmoud A (2010) Antioxidant properties of various solvent extracts of potato peel, sugar beet pulp and sesame cake. J Sci Food Agric 90:218–226. https://doi.org/10.1002/jsfa.3796

Mollea C, Chiampo F, Conti R (2008) Extraction and characterization of pectins from cocoa husks: a preliminary study. Food Chem 107:1353–1356. https://doi.org/10.1016/j.foodchem.2007.09.006

Monti L, Donati E, Zambrini AV, Contarini G (2018) Application of membrane technologies to bovine Ricotta cheese exhausted whey (scotta). Int Dairy J 85:121–128. https://doi.org/10.1016/j.idairyj.2018.05.007

Mouratoglou E, Malliou V, Makris DP (2016) Novel glycerol-based natural eutectic mixtures and their efficiency in the ultrasound-assisted extraction of antioxidant polyphenols from agri-food waste biomass. Waste Biomass Valoriz 7:1377–1387. https://doi.org/10.1007/s12649-016-9539-8

Murad HA, Abo-Elkhair AG, Azzaz HH (2019) Production of xanthan gum from nontraditional substrates with perspective of the unique properties and wide industrial applications. JSMC Microbiol 1:6

Murado MA, Fraguas J, Montemayor MI et al (2010) Preparation of highly purified chondroitin sulphate from skate (Raja clavata) cartilage by-products. Process optimization including a new procedure of alkaline hydroalcoholic hydrolysis. Biochem Eng J 49:126–132. https://doi.org/10.1016/j.bej.2009.12.006

Murthy PS, Madhava Naidu M (2012) Sustainable management of coffee industry by-products and value addition – a review. Resour Conserv Recycl 66:45–58

Naik A, Raghavendra SN, Raghavarao KSMS (2012) Production of coconut protein powder from coconutwet processing waste and its characterization. Appl Biochem Biotechnol 167:1290–1302

Naik M, Rawson A, Rangarajan JM (2020) Radio frequency-assisted extraction of pectin from jackfruit (Artocarpus heterophyllus) peel and its characterization. J Food Process Eng. https://doi.org/10.1111/jfpe.13389

Nath P, Varghese E, Kaur C (2015) Optimization of enzymatic maceration for extraction of carotenoids and total phenolics from sweet pepper using response surface methodology. Indian J Hortic 72:547–552. https://doi.org/10.5958/0974-0112.2015.00100.0

Nawaz H, Shad MA, Rehman N et al (2020) Effect of solvent polarity on extraction yield and antioxidant properties of phytochemicals from bean (Phaseolus vulgaris) seeds. Brazilian J Pharm Sci 56. https://doi.org/10.1590/s2175-97902019000417129

Nawirska A, Kwaśniewska M (2005) Dietary fibre fractions from fruit and vegetable processing waste. Food Chem 91:221–225. https://doi.org/10.1016/j.foodchem.2003.10.005

Nelson ML (2010) Utilization and application of wet potato processing coproducts for finishing cattle. J Anim Sci 88. https://doi.org/10.2527/jas.2009-2502

Nguyen M, Reynolds N, Vigneswaran S (2003) By-product recovery from cottage cheese production by nanofiltration. J Clean Prod 11:803–807. https://doi.org/10.1016/S0959-6526(02)00130-0

Nile SH, Nile AS, Keum YS, Sharma K (2017) Utilization of quercetin and quercetin glycosides from onion (Allium cepa L.) solid waste as an antioxidant, urease and xanthine oxidase inhibitors. Food Chem 235:119–126. https://doi.org/10.1016/j.foodchem.2017.05.043

Nishimura H, Tan L, Kira N et al (2017) Production of ethanol from a mixture of waste paper and kitchen waste via a process of successive liquefaction, presaccharification, and simultaneous saccharification and fermentation. Waste Manag 67:86–94. https://doi.org/10.1016/j.wasman.2017.04.030

Noda Y, Asada C, Sasaki C, Nakamura Y (2019) Effects of hydrothermal methods such as steam explosion and microwave irradiation on extraction of water soluble antioxidant materials from garlic husk. Waste Biomass Valoriz 10:3397–3402. https://doi.org/10.1007/s12649-018-0353-3

Oğuzkan SB (2019) Extraction of capsinoid and its analogs from pepper waste of different genotypes. Nat Prod Commun 14. https://doi.org/10.1177/1934578X19865673

Ohkouchi Y, Inoue Y (2006) Direct production of l(+)-lactic acid from starch and food wastes using Lactobacillus manihotivorans LMG18011. Bioresour Technol 97:1554–1562. https://doi.org/10.1016/j.biortech.2005.06.004

Okino-Delgado CH, do Prado DZ, Facanali R et al (2017) Bioremediation of cooking oil waste using lipases from wastes. PLoS One 12:e0186246. https://doi.org/10.1371/journal.pone.0186246

Okuno S, Yoshinaga M, Nakatani M et al (2002) Extraction of antioxidants in sweetpotato waste powder with supercritical carbon dioxide. Food Sci Technol Res 8:154–157. https://doi.org/10.3136/fstr.8.154

Ona JI, Halling PJ, Ballesteros M (2019) Enzyme hydrolysis of cassava peels: treatment by amylolytic and cellulolytic enzymes. Biocatal Biotransform 37:77–85. https://doi.org/10.1080/10242422.2018.1551376

Paes J, Dotta R, Barbero GF, Martínez J (2014) Extraction of phenolic compounds and anthocyanins from blueberry (Vaccinium myrtillus L.) residues using supercritical CO2 and pressurized liquids. J Supercrit Fluids 95:8–16. https://doi.org/10.1016/j.supflu.2014.07.025

Pagano E, Romano B, Izzo AA, Borrelli F (2018) The clinical efficacy of curcumin-containing nutraceuticals: an overview of systematic reviews. Pharmacol Res 134:79–91

Pap N, Pongrácz E, Myllykoski L, Keiski R (2004) Waste minimization and utilization in the food industry: processing of arctic berries, and extraction of valuable compounds from juice- processing by-products. Proc Waste Minimization Resour Use Optim Conf 2004:159–168

Papaioannou EH, Liakopoulou-Kyriakides M (2012) Agro-food wastes utilization by Blakeslea trispora for carotenoids production. Acta Biochim Pol 59:151–153

Papoutsis K, Grasso S, Menon A et al (2020) Recovery of ergosterol and vitamin D2 from mushroom waste – potential valorization by food and pharmaceutical industries. Trends Food Sci Technol 99:351–366

Parashar A, Jin Y, Mason B et al (2016) Incorporation of whey permeate, a dairy effluent, in ethanol fermentation to provide a zero waste solution for the dairy industry. J Dairy Sci 99:1859–1867. https://doi.org/10.3168/jds.2015-10059

Parate VR, Talib MI (2015) Characterization of tur dal (Cajanus cajan) husk carbon and its kinetics and isotherm study for removing Cu (II) ions. IOSR J Environ Sci 9:27–41. https://doi.org/10.9790/2402-09432741

Parfitt J, Barthel M, Macnaughton S (2010) Food waste within food supply chains: quantification and potential for change to 2050. Philos Trans R Soc B Biol Sci 365:3065–3081. https://doi.org/10.1098/rstb.2010.0126

Parthiba Karthikeyan O, Trably E, Mehariya S et al (2018) Pretreatment of food waste for methane and hydrogen recovery: a review. Bioresour Technol 249:1025–1039

Patel SR, Murthy ZVP (2010) Optimization of process parameters by Taguchi method in the recovery of lactose from whey using sonocrystallization. Cryst Res Technol 45:747–752. https://doi.org/10.1002/crat.201000139

Patsea M, Stefou I, Grigorakis S, Makris DP (2017) Screening of natural sodium acetate-based low-transition temperature mixtures (LTTMs) for enhanced extraction of antioxidants and pigments from red vinification solid wastes. Environ Process 4:123–135. https://doi.org/10.1007/s40710-016-0205-8

Pavlić B, Đurković AV, Vladić J et al (2016) Extraction of minor compounds (chlorophylls and carotenoids) from yarrow–rose hip mixtures by traditional versus green technique. J Food Process Eng 39:418–424. https://doi.org/10.1111/jfpe.12242

Pavlović MD, Buntić AV, Šiler-Marinković SS, Dimitrijević-Branković SI (2013) Ethanol influenced fast microwave-assisted extraction for natural antioxidants obtaining from spent filter coffee. Sep Purif Technol 118:503–510. https://doi.org/10.1016/j.seppur.2013.07.035

Pedras B, Salema-Oom M, Sá-Nogueira I et al (2017) Valorization of white wine grape pomace through application of subcritical water: analysis of extraction, hydrolysis, and biological activity of the extracts obtained. J Supercrit Fluids 128:138–144. https://doi.org/10.1016/j.supflu.2017.05.020

Pedroza MA, Carmona M, Salinas MR, Zalacain A (2011) Use of dehydrated waste grape skins as a natural additive for producing rosé Wines: study of extraction conditions and evolution. J Agric Food Chem 59:10976–10986. https://doi.org/10.1021/jf202626v

Pereira CD, Diaz O, Cobos A (2002) Valorization of by-products from ovine cheese manufacture: clarification by thermocalcic precipitation/microfiltration before ultrafiltration. Int Dairy J 12:773–783. https://doi.org/10.1016/S0958-6946(02)00070-5

Perotto G, Ceseracciu L, Simonutti R et al (2018) Bioplastics from vegetable waste: via an eco-friendly water-based process. Green Chem 20:894–902. https://doi.org/10.1039/c7gc03368k

Peschel W, Sánchez-Rabaneda F, Diekmann W et al (2006) An industrial approach in the search of natural antioxidants from vegetable and fruit wastes. Food Chem 97:137–150. https://doi.org/10.1016/j.foodchem.2005.03.033

Pfukwa TM, Fawole OA, Manley M et al (2019) Food preservative capabilities of grape (Vitis vinifera) and clementine mandarin (Citrus reticulata) by-products extracts in South Africa. Sustain 11. https://doi.org/10.3390/su11061746

Piccolella S, Bianco A, Crescente G et al (2019) Recovering cucurbita pepo cv. 'lungo fiorentino' wastes: UHPLC-HRMS/MS metabolic profile, the basis for establishing their nutra- and cosmeceutical valorisation. Molecules 24. https://doi.org/10.3390/molecules24081479

Písecký J (2005) Spray drying in the cheese industry. Int Dairy J 15:531–536

Piwowarek K, Lipińska E, Hać-Szymańczuk E (2016) Possibility of using apple pomaces in the process of propionic-acetic fermentation. Electron J Biotechnol 23:1–6. https://doi.org/10.1016/j.ejbt.2016.07.004

Plazzotta S, Manzocco L, Nicoli MC (2017) Fruit and vegetable waste management and the challenge of fresh-cut salad. Trends Food Sci Technol 63:51–59

Plazzotta S, Calligaris S, Manzocco L (2018a) Innovative bioaerogel-like materials from fresh-cut salad waste via supercritical-CO2-drying. Innov Food Sci Emerg Technol 47:485–492. https://doi.org/10.1016/j.ifset.2018.04.022

Plazzotta S, Calligaris S, Manzocco L (2018b) Application of different drying techniques to fresh-cut salad waste to obtain food ingredients rich in antioxidants and with high solvent loading capacity. LWT – Food Sci Technol 89:276–283. https://doi.org/10.1016/j.lwt.2017.10.056

Plazzotta S, Ibarz R, Manzocco L, Martín-Belloso O (2020) Optimizing the antioxidant biocompound recovery from peach waste extraction assisted by ultrasounds or microwaves. Ultrason Sonochem 63. https://doi.org/10.1016/j.ultsonch.2019.104954

Poltue T, Suddeepong A, Horpibulsuk S et al (2019) Strength development of recycled concrete aggregate stabilized with fly ash-rice husk ash based geopolymer as pavement base material. Road Mater Pavement Des. https://doi.org/10.1080/14680629.2019.1593884

Pontonio E, Dingeo C, Di Cagno R et al (2020) Brans from hull-less barley, emmer and pigmented wheat varieties: from by-products to bread nutritional improvers using selected lactic acid bacteria and xylanase. Int J Food Microbiol 313. https://doi.org/10.1016/j.ijfoodmicro.2019.108384

Prakash Maran J, Sivakumar V, Thirugnanasambandham K, Sridhar R (2014) Microwave assisted extraction of pectin from waste Citrullus lanatus fruit rinds. Carbohydr Polym 101:786–791. https://doi.org/10.1016/j.carbpol.2013.09.062

Prameela K, Venkatesh K, Immandi SB et al (2017) Next generation nutraceutical from shrimp waste: the convergence of applications with extraction methods. Food Chem 237:121–132

Prasad SK, Singh MK (2015) Horse gram- an underutilized nutraceutical pulse crop: a review. J Food Sci Technol 52:2489–2499

Rabiei Z, Naderi S, Rafieian-Kopaei M (2017) Study of antidepressant effects of grape seed oil in male mice using tail suspension and forced swim tests. Bangladesh J Pharmacol 12:397–402. https://doi.org/10.3329/bjp.v12i4.33520

Rahmanian N, Jafari SM, Galanakis CM (2014) Recovery and removal of phenolic compounds from olive mill wastewater. JAOCS, J Am Oil Chem Soc 91:1–18. https://doi.org/10.1007/s11746-013-2350-9

Ramos P, Santos SAO, Guerra ÂR et al (2013) Valorization of olive mill residues: antioxidant and breast cancer antiproliferative activities of hydroxytyrosol-rich extracts derived from olive oil by-products. Ind Crops Prod 46:359–368. https://doi.org/10.1016/j.indcrop.2013.02.020

Ravanfar R, Moein M, Niakousari M, Tamaddon A (2018) Extraction and fractionation of anthocyanins from red cabbage: ultrasonic-assisted extraction and conventional percolation method. J Food Meas Charact 12:2271–2277. https://doi.org/10.1007/s11694-018-9844-y

Ravindran R, Jaiswal AK (2016) Exploitation of food industry waste for high-value products. Trends Biotechnol 34:58–69. https://doi.org/10.1016/J.TIBTECH.2015.10.008

Rektor A, Vatai G (2004) Membrane filtration of mozzarella whey. Desalination 162:279–286. https://doi.org/10.1016/S0011-9164(04)00052-9

Riggi E, Avola G (2008) Fresh tomato packinghouses waste as high added-value biosource. Resour Conserv Recycl 53:96–106. https://doi.org/10.1016/j.resconrec.2008.09.005

Riva G (2013) Volume 1 – I sottoprodotti di interesse del DM 6.7.2012. Inquadramento, potenzialità e valutazioni 155:317

Rodríguez R, Jiménez A, Fernández-Bolaños J et al (2006) Dietary fibre from vegetable products as source of functional ingredients. Trends Food Sci Technol. https://doi.org/10.1016/j.tifs.2005.10.002

Rodríguez M, Nolasco S, Izquierdo N et al (2019) Microwave-assisted extraction of antioxidant compounds from sunflower hulls. Heat Mass Transf und Stoffuebertragung 55:3017–3027. https://doi.org/10.1007/s00231-019-02648-4

Rohm H, Brennan C, Turner C et al (2015) Adding value to fruit processing waste: innovative ways to incorporate fibers from berry pomace in baked and extruded cereal-based foods—a SUSFOOD Project. Foods 4:690–697. https://doi.org/10.3390/foods4040690

Rojas MC, Brewer MS (2007) Effect of natural antioxidants on oxidative stability of cooked, refrigerated beef and pork. J Food Sci 72. https://doi.org/10.1111/j.1750-3841.2007.00335.x

Romani A, Pinelli P, Galardi C et al (2003) Flavonoids in leaves of black cabbage (Brassica oleracea var. acephala DC. subvar. viridis cv. serotina) grown on different soils and at different elevations. Ital J Food Sci 15:197–205

Rosales Soto MU, Brown K, Ross CF (2012) Antioxidant activity and consumer acceptance of grape seed flour-containing food products. Int J Food Sci Technol 47:592–602. https://doi.org/10.1111/j.1365-2621.2011.02882.x

Sabeena Farvin KH, Grejsen HD, Jacobsen C (2012) Potato peel extract as a natural antioxidant in chilled storage of minced horse mackerel (Trachurus trachurus): effect on lipid and protein oxidation. Food Chem 131:843–851. https://doi.org/10.1016/j.foodchem.2011.09.056

Sagdic O, Ozturk I, Cankurt H, Tornuk F (2012) Interaction between some phenolic compounds and probiotic bacterium in functional ice cream production. Food Bioprocess Technol 5:2964–2971. https://doi.org/10.1007/s11947-011-0611-x

Saidi S, Saoudi M, Ben Amar R (2018) Valorisation of tuna processing waste biomass: isolation, purification and characterisation of four novel antioxidant peptides from tuna by-product hydrolysate. Environ Sci Pollut Res 25:17383–17392. https://doi.org/10.1007/s11356-018-1809-5

Saleh IA, Vinatoru M, Mason TJ et al (2017) Extraction of silymarin from milk thistle (Silybum marianum) seeds–a comparison of conventional and microwave-assisted extraction methods. J Microw Power Electromagn Energy 51:124–133. https://doi.org/10.1080/08327823.2017.1320265

Sánchez-Alonso I, Jiménez-Escrig A, Saura-Calixto F, Borderías AJ (2007) Effect of grape antioxidant dietary fibre on the prevention of lipid oxidation in minced fish: evaluation by different methodologies. Food Chem 101:372–378. https://doi.org/10.1016/j.foodchem.2005.12.058

Sánchez-García YI, Bhangu SK, Ashokkumar M, Gutiérrez-Méndez N (2018) Sonocrystallization of lactose from whey. In: Technological approaches for novel applications in dairy processing. InTech, London

Sánchez-Monedero MA, Cayuela ML, Mondini C et al (2008) Potential of olive mill wastes for soil C sequestration. Waste Manag 28:767–773. https://doi.org/10.1016/j.wasman.2007.09.029

Santamaría B, Salazar G, Beltrán S, Cabezas JL (2002) Membrane sequences for fractionation of polyphenolic extracts from defatted milled grape seeds. Desalination 148:103–109. https://doi.org/10.1016/S0011-9164(02)00661-6

Santos J, Oliveira MBPP, Ibáñez E, Herrero M (2014) Phenolic profile evolution of different ready-to-eat baby-leaf vegetables during storage. J Chromatogr A 1327:118–131. https://doi.org/10.1016/j.chroma.2013.12.085

Santos MIS, Fradinho P, Martins S et al (2019) A novel way for whey: cheese whey fermentation produces an effective and environmentally-safe alternative to chlorine. Appl Sci 9. https://doi.org/10.3390/app9142800

Sardari RRR, Sutiono S, Azeem HA et al (2019) Evaluation of sequential processing for the extraction of starch, lipids, and proteins from wheat bran. Front Bioeng Biotechnol 7. https://doi.org/10.3389/fbioe.2019.00413

Sareena C, Ramesan M, Purushothaman E (2012) Utilization of coconut shell powder as a novel filler in natural rubber. J Reinf Plast Compos 31:533–547. https://doi.org/10.1177/0731684412439116

Sarkar A, Kaul P (2014) Evaluation of tomato processing by-products: a comparative study in a pilot scale setup. J Food Process Eng 37:299–307. https://doi.org/10.1111/jfpe.12086

Sarkis JR, Michel I, Tessaro IC, Marczak LDF (2014) Optimization of phenolics extraction from sesame seed cake. Sep Purif Technol 122:506–514. https://doi.org/10.1016/j.seppur.2013.11.036

Satari B, Karimi K (2018) Citrus processing wastes: environmental impacts, recent advances, and future perspectives in total valorization. Resour Conserv Recycl 129:153–167. https://doi.org/10.1016/J.RESCONREC.2017.10.032

Schieber A (2017) Side streams of plant food processing as a source of valuable compounds: selected examples. Annu Rev Food Sci Technol 8:97–112. https://doi.org/10.1146/annurev-food-030216-030135

Scordino M, Di Mauro A, Passerini A, Maccarone E (2007) Highly purified sugar concentrate from a residue of citrus pigments recovery process. LWT – Food Sci Technol 40:713–721. https://doi.org/10.1016/j.lwt.2006.03.007

Sehm J, Treutter D, Lindermayer H et al (2011) The influence of apple- or red-grape pomace enriched piglet diet on blood parameters, bacterial colonisation, and marker gene expression in piglet white blood cells. Food Nutr Sci 02:366–376. https://doi.org/10.4236/fns.2011.24052

Seixas FL, Fukuda DL, Turbiani FRB et al (2014) Extraction of pectin from passion fruit peel (Passiflora edulis f.flavicarpa) by microwave-induced heating. Food Hydrocoll 38:186–192. https://doi.org/10.1016/j.foodhyd.2013.12.001

Sekifuji R, Tateda M (2019) Study of the feasibility of a rice husk recycling scheme in Japan to produce silica fertilizer for rice plants. Sustain Environ Res 29. https://doi.org/10.1186/s42834-019-0011-x

Selani MM, Contreras-Castillo CJ, Shirahigue LD et al (2011) Wine industry residues extracts as natural antioxidants in raw and cooked chicken meat during frozen storage. Meat Sci 88:397–403. https://doi.org/10.1016/j.meatsci.2011.01.017

Selmane D, Christophe V, Gholamreza D (2008) Extraction of proteins from slaughterhouse by-products: Influence of operating conditions on functional properties. Meat Sci 79:640–647. https://doi.org/10.1016/j.meatsci.2007.10.029

Shah NP, Ding WK, Fallourd MJ, Leyer G (2010) Improving the stability of probiotic bacteria in model fruit juices using vitamins and antioxidants. J Food Sci 75:M278–M282. https://doi.org/10.1111/j.1750-3841.2010.01628.x

Shalini R, Gupta DK (2010) Utilization of pomace from apple processing industries: a review. J Food Sci Technol 47:365

Sharma P, Gaur VK, Kim SH, Pandey A (2020) Microbial strategies for bio-transforming food waste into resources. Bioresour Technol 299. https://doi.org/10.1016/j.biortech.2019.122580

Sherazi STH, Mahesar SA, Sirajuddin (2016) Vegetable oil deodorizer distillate: a rich source of the natural bioactive components. J Oleo Sci 65:957–966

Shimamoto Y, Suzuki T (2019) Recycle of rice husk into agro-infrastructure for decreasing carbon dioxide. Paddy Water Environ 17:555–559. https://doi.org/10.1007/s10333-019-00752-z

Shoveller AK, Stoll B, Ball RO, Burrin DG (2005) Nutritional and functional importance of intestinal sulfur amino acid metabolism. J Nutr 135:1609–1612. https://doi.org/10.1093/jn/135.7.1609

Silvan JM, Mingo E, Martinez-Rodriguez AJ (2017) Grape seed extract (GSE) modulates *campylobacter* pro-inflammatory response in human intestinal epithelial cell lines. Food Agric Immunol 28:739–753. https://doi.org/10.1080/09540105.2017.1312292

Singh J, Jayaprakasha GK, Patil BS (2018) An optimized solvent extraction and characterization of unidentified flavonoid glucuronide derivatives from spinach by UHPLC-HR-QTOF-MS. Talanta 188:763–771. https://doi.org/10.1016/j.talanta.2018.06.025

Sivakumar G, Aliboni A, Antonini A, Bacchetta L (2007) Bioactive sulforaphane from in vitro propagated brassica seedlings. Eng Life Sci 7:275–277. https://doi.org/10.1002/elsc.200620192

Smilowitz JT, Dillard CJ, German JB (2005) Milk beyond essential nutrients: the metabolic food. Aust J Dairy Technol 60:77–83

Smith J, Hong-Shum L (2011) Food additives data book, 2nd edn. Wiley, New York

Smithers GW (2008) Whey and whey proteins-From "gutter-to-gold.". Int Dairy J 18:695–704

Solana M, Mirofci S, Bertucco A (2016) Production of phenolic and glucosinolate extracts from rocket salad by supercritical fluid extraction: process design and cost benefits analysis. J Food Eng 168:35–41. https://doi.org/10.1016/j.jfoodeng.2015.07.017

Sousa SC, Vázquez JA, Pérez-Martín RI et al (2017) Valorization of by-products from commercial fish species: extraction and chemical properties of skin gelatins. Molecules 22. https://doi.org/10.3390/molecules22091545

Spigno G, Amendola D, Dahmoune F, Jauregi P (2015) Colloidal gas aphrons based separation process for the purification and fractionation of natural phenolic extracts. Food Bioprod Process 94:434–442. https://doi.org/10.1016/j.fbp.2014.06.002

Stabnikova O, Wang JY, Bo Ding H, Tay J-H (2005) Biotransformation of vegetable and fruit processing wastes into yeast biomass enriched with selenium. Bioresour Technol 96:747–751. https://doi.org/10.1016/j.biortech.2004.06.022

Stojceska V, Ainsworth P, Plunkett A et al (2008) Cauliflower by-products as a new source of dietary fibre, antioxidants and proteins in cereal based ready-to-eat expanded snacks. J Food Eng 87:554–563. https://doi.org/10.1016/j.jfoodeng.2008.01.009

Stökle K, Kruse A (2019) Extraction of sugars from forced chicory roots. Biomass Convers Biorefinery 9:699–708. https://doi.org/10.1007/s13399-019-00374-9

Stökle K, Hülsemann B, Steinbach D et al (2019) A biorefinery concept using forced chicory roots for the production of biogas, hydrochar, and platform chemicals. Biomass Convers Biorefinery. https://doi.org/10.1007/s13399-019-00527-w

Strati IF, Oreopoulou V (2011) Effect of extraction parameters on the carotenoid recovery from tomato waste. Int J Food Sci Technol 46:23–29. https://doi.org/10.1111/j.1365-2621.2010.02496.x

Stumpf B, Künne M, Ma L et al (2020) Optimization of the extraction procedure for the determination of phenolic acids and flavonoids in the leaves of globe artichoke (Cynara cardunculus var. scolymus L.). J Pharm Biomed Anal 177. https://doi.org/10.1016/j.jpba.2019.112879

Suárez M, Romero MP, Ramo T et al (2009) Methods for preparing phenolic extracts from olive cake for potential application as food antioxidants. J Agric Food Chem 57:1463–1472. https://doi.org/10.1021/jf8032254

Sud D, Mahajan G, Kaur MP (2008) Agricultural waste material as potential adsorbent for sequestering heavy metal ions from aqueous solutions – a review. Bioresour Technol 99:6017–6027

Sui W, Xiao Y, Liu R et al (2019) Steam explosion modification on tea waste to enhance bioactive compounds' extractability and antioxidant capacity of extracts. J Food Eng 261:51–59. https://doi.org/10.1016/j.jfoodeng.2019.03.015

Sun Y, Xiu C, Liu W et al (2016) Grape seed proanthocyanidin extract protects the retina against early diabetic injury by activating the Nrf2 pathway. Exp Ther Med 11:1253–1258. https://doi.org/10.3892/etm.2016.3033

Tahergorabi R, Jaczynski J (2014) Isoelectric solubilization/precipitation as a means to recover protein and lipids from seafood by-products. In: Seafood processing by-products: trends and applications. Springer, New York, pp 101–123

Tahergorabi R, Sivanandan L, Jaczynski J (2012) Dynamic rheology and endothermic transitions of proteins recovered from chicken-meat processing by-products using isoelectric solubilization/precipitation and addition of TiO 2. LWT – Food Sci Technol 46:148–155. https://doi.org/10.1016/j.lwt.2011.10.013

Tanongkankit Y, Sablani SS, Chiewchan N, Devahastin S (2013) Microwave-assisted extraction of sulforaphane from white cabbages: effects of extraction condition, solvent and sample pretreatment. J Food Eng 117:151–157. https://doi.org/10.1016/j.jfoodeng.2013.02.011

Tesaki S, Tanabe S, Moriyama M et al (1999) Isolation and identification of an antibacterial compound from grape and its application to foods. Nippon Nogeikagaku Kaishi 73:125–128. https://doi.org/10.1271/nogeikagaku1924.73.125

Toldrá F, Aristoy MC, Mora L, Reig M (2012) Innovations in value-addition of edible meat by-products. Meat Sci 92:290–296

Tripathi A, Rawat Ranjan M (2015) Heavy Metal Removal from Wastewater Using Low Cost Adsorbents. J Bioremediation Biodegrad 06:1–5. https://doi.org/10.4172/2155-6199.1000315

Tseng A, Zhao Y (2013) Wine grape pomace as antioxidant dietary fibre for enhancing nutritional value and improving storability of yogurt and salad dressing. Food Chem 138:356–365. https://doi.org/10.1016/j.foodchem.2012.09.148

Ueno T, Ozawa Y, Ishikawa M et al (2003) Lactic acid production using two food processing wastes, canned pineapple syrup and grape invertase, as substrate and enzyme. Biotechnol Lett 25:573–577. https://doi.org/10.1023/A:1022888832278

Upadhyay A, Chompoo J, Araki N, Tawata S (2012) Antioxidant, antimicrobial, 15-LOX, and AGEs inhibitions by pineapple stem waste. J Food Sci 77. https://doi.org/10.1111/j.1750-3841.2011.02437.x

Urbonaviciene D, Viskelis P, Viskelis J et al (2012) Lycopene and β-carotene in non-blanched and blanched tomatoes. J Food Agric Environ 10:142–146

Uribe E, Lemus-Mondaca R, Vega-Gálvez A et al (2013) Quality characterization of waste olive cake during hot air drying: nutritional aspects and antioxidant activity. Food Bioprocess Technol 6:1207–1217. https://doi.org/10.1007/s11947-012-0802-0

Valcarcel J, Novoa-Carballal R, Pérez-Martín RI et al (2017) Glycosaminoglycans from marine sources as therapeutic agents. Biotechnol Adv 35:711–725

Valencia-Arredondo JA, Hernández-Bolio GI, Cerón-Montes GI et al (2020) Enhanced process integration for the extraction, concentration and purification of di-acylated cyanidin from red cabbage. Sep Purif Technol 238. https://doi.org/10.1016/j.seppur.2019.116492

Vasala A, Panula J, Neubauer P (2005) Efficient lactic acid production from high salt containing dairy by-products by Lactobacillus salivarius ssp. salicinius with pre-treatment by proteolytic microorganisms. J Biotechnol 117:421–431. https://doi.org/10.1016/j.jbiotec.2005.02.010

Vergara-Salinas JR, Vergara M, Altamirano C et al (2015) Characterization of pressurized hot water extracts of grape pomace: chemical and biological antioxidant activity. Food Chem 171:62–69. https://doi.org/10.1016/j.foodchem.2014.08.094

Verma N, Bansal MC, Kumar V (2011) Pea peel waste: a lignocellulosic waste and its utility in cellulase production BY Trichoderma reesei under solid state cultivation. BioResources 6:1505–1519

Villaescusa I, Fiol N, Martínez M et al (2004) Removal of copper and nickel ions from aqueous solutions by grape stalks wastes. Water Res 38:992–1002. https://doi.org/10.1016/j.watres.2003.10.040

Vincenzetti S, Cecchi T, Perinelli DR et al (2018) Effects of freeze-drying and spray-drying on donkey milk volatile compounds and whey proteins stability. LWT – Food Sci Technol 88:189–195. https://doi.org/10.1016/j.lwt.2017.10.019

Viuda-Martos M, Fernandez-Lopez J, Sayas-Barbera E et al (2011) Physicochemical characterization of the orange juice waste water of a citrus by-product. J Food Process Preserv 35:264–271. https://doi.org/10.1111/j.1745-4549.2009.00450.x

Vorobiev E, Lebovka N (2010) Enhanced extraction from solid foods and biosuspensions by pulsed electrical energy. Food Eng Rev 2:95–108. https://doi.org/10.1007/s12393-010-9021-5

Walia A, Guleria S, Mehta P et al (2017) Microbial xylanases and their industrial application in pulp and paper biobleaching: a review. 3 Biotech 7. https://doi.org/10.1007/s13205-016-0584-6

Wang Q, Wang X, Wang X, Ma H (2008) Glucoamylase production from food waste by Aspergillus niger under submerged fermentation. Process Biochem 43:280–286. https://doi.org/10.1016/j.procbio.2007.12.010

Wang K, Yin J, Shen D, Li N (2014) Anaerobic digestion of food waste for volatile fatty acids (VFAs) production with different types of inoculum: effect of pH. Bioresour Technol 161:395–401. https://doi.org/10.1016/j.biortech.2014.03.088

Wang Y, Li F, Zhuang H et al (2015) Effects of plant polyphenols and α-tocopherol on lipid oxidation, residual nitrites, biogenic amines, and N-nitrosamines formation during ripening and storage of dry-cured bacon. LWT – Food Sci Technol 60:199–206. https://doi.org/10.1016/j.lwt.2014.09.022

Watson RR, Preedy V, Zibadi S (2014) Wheat and Rice in Disease Prevention and Health, Wheat and Rice in Disease Prevention and Health. Elsevier Inc. https://doi.org/10.1016/C2012-0-00472-3

Weseler AR, Bast A (2017) Masquelier's grape seed extract: from basic flavonoid research to a well-characterized food supplement with health benefits. Nutr J 16:1–19

Westereng B, Michaelsen TE, Samuelsen AB, Knutsen SH (2008) Effects of extraction conditions on the chemical structure and biological activity of white cabbage pectin. Carbohydr Polym 72:32–42. https://doi.org/10.1016/j.carbpol.2007.07.017

Wolfe KL, Liu RH (2003) Apple peels as a value-added food ingredient. J Agric Food Chem. https://doi.org/10.1021/jf025916z

Wu Y, Ma H, Zheng M, Wang K (2015) Lactic acid production from acidogenic fermentation of fruit and vegetable wastes. Bioresour Technol 191:53–58. https://doi.org/10.1016/j.biortech.2015.04.100

Xu C, Yagiz Y, Marshall S et al (2015) Application of muscadine grape (Vitis rotundifolia Michx.) pomace extract to reduce carcinogenic acrylamide. Food Chem 182:200–208. https://doi.org/10.1016/j.foodchem.2015.02.133

Yaakob MA, Mohamed RMSR, Al-Gheethi A et al (2019) Optimising of Scenedesmus sp. biomass production in chicken slaughterhouse wastewater using response surface methodology and potential utilisation as fish feeds. Environ Sci Pollut Res 26:12089–12108. https://doi.org/10.1007/s11356-019-04633-0

Yadav JSS, Yan S, Pilli S et al (2015) Cheese whey: a potential resource to transform into bioprotein, functional/nutritional proteins and bioactive peptides. Biotechnol Adv 33:756–774

Yan N, Chen X (2015) Sustainability: don't waste seafood waste. Nature 524:155–157

Yang EJ, Kim SS, Oh TH et al (2009) Essential oil of citrus fruit waste attenuates LPS-induced nitric oxide production and inhibits the growth of skin pathogens. Int J Agric Biol 11:791–794

Yang QQ, Gan RY, Ge YY et al (2019) Ultrasonic treatment increases extraction rate of common bean (Phaseolus vulgaris l.) antioxidants. Antioxidants 8. https://doi.org/10.3390/antiox8040083

Yu J, Ahmedna M (2013) Functional components of grape pomace: their composition, biological properties and potential applications. Int J Food Sci Technol 48:221–237

Zacharof MP (2017) Grape winery waste as feedstock for bioconversions: applying the biorefinery concept. Waste Biomass Valoriz 8:1011–1025

Zeaiter Z, Regonesi ME, Cavini S et al (2019) Extraction and characterization of inulin-type fructans from artichoke wastes and their effect on the growth of intestinal bacteria associated with health. Biomed Res Int 2019:1083952. https://doi.org/10.1155/2019/1083952

Zhang Q, Zhou MM, Chen PL et al (2011) Optimization of ultrasonic-assisted enzymatic hydrolysis for the extraction of luteolin and apigenin from celery. J Food Sci 76:C680–C685. https://doi.org/10.1111/j.1750-3841.2011.02174.x

Zhang G, Wu F, Ma T et al (2019) Preparation and characterization of cellulose nanofibers isolated from lettuce peel. Cellul Chem Technol 53:677–684. https://doi.org/10.35812/CelluloseChemTechnol.2019.53.66

Zhao J, Liu Y, Wang D et al (2017) Potential impact of salinity on methane production from food waste anaerobic digestion. Waste Manag 67:308–314. https://doi.org/10.1016/j.wasman.2017.05.016

Zill-E-Huma A-VM, Elmaataoui M, Chemat F (2011) A novel idea in food extraction field: Study of vacuum microwave hydrodiffusion technique for by-products extraction. J Food Eng 105:351–360. https://doi.org/10.1016/j.jfoodeng.2011.02.045

Zuorro A, Fidaleo M, Lavecchia R (2011) Enzyme-assisted extraction of lycopene from tomato processing waste. Enzyme Microb Technol 49:567–573. https://doi.org/10.1016/j.enzmictec.2011.04.020

Zykwinska A, Boiffard MH, Kontkanen H et al (2008) Extraction of green labeled pectins and pectic oligosaccharides from plant byproducts. J Agric Food Chem 56:8926–8935. https://doi.org/10.1021/jf801705a

Chapter 3
Current State of Art of the Usual Food Waste Valorization

Carla De Carolis

3.1 Introduction

It is estimated that enormous population and high rate of economic growth in many countries are intensifying the food demand, therefore also land cultivation demand is increasing as well as the global concern for food security, by considering that about 60% of the world's population is still malnourished, (Pimentel et al. 2009).

At the same time, thematics concerning the environment, energy security, and employment generation are consolidating a more demand for biofuels (Contran et al. 2013; Soto et al. 2015) in contrast with traditional fossil fuels. Notably, growth of biofuels market is placing increasing demand for major agricultural factors of production (van Eijck et al. 2013), mostly for land and labor which are traditionally used for foods and feeds productions, with further competition of natural resources between biofuels with food and feed sectors.

This will further intensify the crucial debate among food versus biofuels growing in the coming years. Such demand for inputs and increasing growth of population necessarily result in higher demand for food and feed. As a result, agricultural commodity prices are starting to rise to the marketplace internationally, with further prices increases of a wide array of agricultural products (Naylor et al. 2010). Increasing international demand for biofuels affects many market factors principally on the production, price, and availability of staple commodities. According to (Dufey 2006), the development benefits of biofuels were not straightforward if produced by the same agricultural resources used in the food sector, until a few years ago.

During latest 10 years, EU Commission developed the RED I, a first Directive on the promotion of Renewable Energy including Biomass resources, (RED I, Directive 2009/28/EC). The RED I Directive fully implements the European policy for renewable energy. It includes the 20-20-20 mandatory goals for 2020: a 20% reduction in CO_2 emissions compared to 1990 levels, a 20% share of the energy market for renewables (at least 10% blending target for transport biofuels) and a 20% growth

T. Cecchi, C. De Carolis, *Biobased products from food sector waste*,
https://doi.org/10.1007/978-3-030-63436-0_3

referred to energy efficiency (Directive (EU) 2009/28/EC). In energy-driven biorefineries, biomass is used for the production of liquid (biodiesel or bioethanol) and/or gaseous (biomethane) biofuels. In a sustainable perspective, the European Commission also published its 'Clean Energy for all Europeans' initiatives during November 2016. Concrete efforts have been carried out by European Commission in order to improve the Renewable Energy Directive RED I toward a new version RED II. The new Directive RED II (RED II, Directive 2018/2001/EC) is entered in force as new package of regulations for renewable energy growth for the further period from 2021 to 2030, and aims to produce biofuels under sustainability criteria with particular emphasis to the II generation of Biofuels with promoting the valorization of food wastes as forerunners for more sustainable production of biofuels and biomaterials, (Eu Commission 2018; Flach et al. 2017, 2019). The EU regulations are thus driving the Eu member states toward a more sustainable resource's policy, as well as other countries. In other terms, biofuels production must be driven by sustainable criteria, by using a replicable approach internationally based on the valorization of biomass residues and food waste to new resource, otherwise will create demanding challenges for food production, both globally than domestically.

3.2 Waste Hierarchy Applied to Biomass Residual and Food Wastes

Firstly, the difference between co-products and wastes: "Co-products are all materials that can be re-used while waste is defined as material that reached the end of its production cycle and cannot be recycled", (de Carolis et al. 2012).

The valorizing of food wastes and rural residues are mostly generated during the production and manufacturing processes by farms, food industries; the typical energy content is comprised of 10.5–12.00 MJ/kg. Food-waste management practices often differ among urban and rural areas, as well as for several residential and industrial producers, at last, for developed and developing nations.

The waste management situation of a developing country during the starting period differs from that of industrialized countries. The transfer of proven technology from one country to another can be quite inappropriate even if technically viable or affordable. It's thus very important to understand the local factors such as:

- Waste characteristics and seasonal variations in climate
- Social aspects, cultural attitudes towards solid waste, and political institutions
- Awareness of the more obvious resource limitations that often exist.

As the first concept, the role of sustainable food waste management is to reduce the amount of waste that is discharged into the environment by reducing the amount of waste generated. However, the environmental impact can be reduced by making more sustainable use of the food waste by investigating their possible uses (as biobased products and biofuels).

This is known as the "*Waste Hierarchy*". The waste hierarchy implies reducing, reuse, recycle waste and classify the waste management strategies in terms of waste minimization. The aim of the waste hierarchy is to produce the minimum amount of waste and maximizing as many practical benefits from new product generation, (Demirbas 2011).

It seems simple to apply the "*Waste Hierarchy*" concept to all wastes or residuals included those to the biomass sector. Thus part of food waste is classified as biomass deriving from agroindustrial, agricultural, and urban activities.

The rural and agroindustrial residues, as well as food wastes, represent potential biomass resources like straw, vegetable fruit peels, and organic fraction of municipal wastes.

Energy, new materials and added value products can be extracted o converted from wastes to new BBPs – products and Biofuels due to the several billion tonnes of biomass generated globally, (Lizardi-Jiménez and Hernández-Martínez 2017).

There are various options available to convert residues or food waste to biofuels, energy, or animal feed and fertilizers. The technologies traditionally used by advanced and developing countries are principally based on sanitary landfills, incineration, pyrolysis, and gasification (implementing plants) for specific waste, and anaerobic digestion or fermentation for organic wastes, Table 3.1. Further sections will take into consideration the important technologies and processes principally focused on food wastes and agroindustrial residues with particular attention to the valorization processes for producing Animal feeds, biofertilizers, and several generations of biofuels.

The choice of technology depends on the waste typology, its quality, and the local conditions; but their classification and assignment of different wastes are not easy. In UE countries, the wastes are classified with the "EWC Code", (EPA 2002).

The best compromise would be to choose a technology that has the lowest life cycle cost, causes lesser pollutions (air, water and soil), produces more power with less waste, and causes maximum volume reduction, (Demirbas 2011).

To obtain energy in a clean and cost-effective manner is a major target yet to be reached. Actually, one of the biggest problems is to find how to convert quickly and economically the lignocellulosic components of these wastes into simpler sugars, to

Table 3.1 Waste processes

Type of Waste		**Waste disposal and valorization pathways**	
Partially combustible waste streams	**Organic wastes/residues** (food wastes – agroindustrial byproducts)	**Food wastes for animal feeds**	
		Composting for biofertilizers	
		Biofuels *(I° II° III° Generations)*	Anaerobic digestion for gaseous biofuels
			Fermentation for liquid biofuels
			Transesterification for liquid biofuels (fats and oils)

enable their subsequent biochemical conversion to clean biofuels or BBPs – Biobased Products, (IRENA 2016).

Recently, producing energy and biofuels by wastes and residues (both conventional that advanced biofuel) is obtaining considerable importance for positive environmental and economic effects. Using organic urban wastes for energy purposes would avoid the enlargement of urban landfills with a consequent reduction of GHG emissions and more independence from fossil fuel utilization. Part of organic food waste is becoming principal biomass in Biogas plants for biomethane production.

At the latest consideration, it is also important to recognize that wastes often contain both energy and nutrient components. A basic rule for ecological sustainability is that energy may be extracted by production/consumption systems but nutrients must be recycled, (de Carolis et al. 2012). This is the base for a new and more efficient concept of "*Biorefinery*".

3.3 Food Wastes for Animal Feed

Boosting agricultural practices has caused a significant improvement in crop cultivations and ensured the food available to the increasing populations at the international level. But the increasing food demand and arising sophisticating agroindustrial processes create huge volumes of wastes called *'food chain supply wastes'* (Matharu et al. 2016), which are generated during different stages of the supply chain: from production in the farm, through the food industry, and retailing, to households and restaurants, with a generation of a huge variability of organic wastes, (Ong et al. 2018).

Furthermore, the great food waste generation is also resulting from the decision to discard food often no attractive with still has nutritive value, (Aschemann-Witzel et al. 2015).

This amount of food wastes can be diverted to several pathways, from energy production to chemical and material potentials due to the functionalized molecules stored in them, (Ki Lin et al. 2013).

The selection of process stream for food waste valorization is often depending on priorities of the countries and cost-effective, the readiness of technology used to valorize the food waste or by-product. In the countries where growth population is increasing, the food demand, as well as the energy consumption, have greater priorities if compared with other innovative and business pathways, as the BBPs sector in part, despite the idea of harvesting high value-added products from food waste streams align with the current concept of sustainable development, Implications for food and energy security have raised the concern of researchers, governments, and grass-root movements. Therefore to systematize process schemes for several food waste streams to high value-added products is more crucial than ever, including the production processes market-oriented to animal feed. In addition, legislation aspects implicated for food waste disposal and management should be also considered along with valorization processes (Ong et al. 2018).

In addition to valorize of food waste or byproducts for feed formulation, it would also maximize resource efficiency which further reduction of dependence by current raw materials (Myer et al. 1999; Westendorf 2000).

Despite Food wastes deriving from different agroindustrial and urban sector (from fruits and vegetable waste or bakery products, fish waste, meat waste, extra cooked food, milk, and dairy waste) are of high nutrition value which could be destined to the animal feedingstuffs formulation and thus deserving of great attention, it's necessary to select theirs on the basis of different needs of animal breedings, and seasonality of food wastes and their shelf-life (Martin et al. 2016).

Their inapt use and bad keeping of raw matter can create major health problems if delivered as animal feed to the animals. For this reason, most countries have developed specific regulations for using food waste as animal feeds (Dahiya et al. 2018).

Indeed, the use of food waste as animal feed has some drawbacks that can limit its feasibility. Key factor is its high water content, which often exceeds 80%, which can accelerate the growth of microbiological contamination, and make handling more difficult and (García 2005). Therefore a drying process represents an essential prerequisite. Furthermore, the analytical composition of such waste can vary markedly throughout the year (Westendorf 2000), and animal feed manufacturers are forced to change their feed formulations periodically depending on the composition in order to maintain constantly high the quality standard.

Several investigations have been carried out on different kinds of food wastes and byproducts. Garcia et al. (García 2005) has analyzed different organic wastes,[1] such as meat, fish, fruit, and vegetables, restaurant and household wastes, with the aim to determinate the chemical composition, undesiderable substances like nitrites, mycotoxins (aflatoxin B1, B2, G1 y G2, zearalenone, deoxynivalenol, ochratoxin A, fumonisin B1 and B2, T-2 toxin, HT-2 toxin and patuline), pesticides (aldrin, chlordane, DDT, endosulfan, endrin, HCH, hexachlorocyclohexane, heptachlor and hexachlorobenzene), heavy metals (lead, cadmium, arsenic and mercury), microbiological characteristics (Salmonella, Listeria, Escherichia coli, Staphylococcus aureus, Clostridium perfringens, sulfite-reducing clostridia, aerobic mesophiles, enterobacteriaceae, total coliforms), molds-yeast, dioxin, furan, PCB and sand mineral contents for every type of waste fraction.

Its results confirmed that vegetable and fruit byproducts deriving from agroindustrial and rural sectors seem to be suitable wastes for animal feed formulations. Notably, analysis of the undesirable substances found in these byproducts showed that nitrite levels were above the maximum legal limit[2] of 15 mg/kg for raw materials for animal feed formulations where coffee husk (33.3 mg/kg) and grape press-

[1] **Food Wastes and agroindustrial byproducts**: fruit/vegetables, potatoes, apple pressing, grape pressing/grapes, agriculture, bread and coffee husks. Byproducts generated in the retail: potato, cider, wine, horticulture, baking and coffee grinding sectors.

[2] **Directive 2002/32/EC of the European Parliament and of the Council of 7 May 2002** on undesirable substances in animal feed and its transposition into the legislation of the EU countries, and feed-market requirements that raw material for animal feed must fulfill.

ing/grapes (28 mg/kg)were used. For the remaining sectors, all nitrite levels were below the maximum legal limit. For what concerning the analyses results for mycotoxins, pesticides, heavy metals, dioxins, and microbiological tests, all values obtained were well below the maximum permitted limits (Martin et al. 2016).

Along the same lines, further investigations carried out by Esteban and colleagues (Esteban et al. 2007) confirmed that similar byproducts can be a sustainable alternative to traditional raw materials used for animal feed, and furthermore demonstrated to have high digestibility.

Despite the eligibility of vegetable wastes to being used for animal feed, these investigations confirmed that fruit/vegetable wastes have a too high water content (approximately 86%) which compromises the feed quality life and preservation and thus should be subjected to drying processes like Microwave (Motevali et al. 2011); Pulse Combustion Drying (PCD), (Zbicinski 2002); Static and Rotary Oven (Vega-Gálvez et al. 2009) and others (Besombes et al. 2010; Maache-Rezzoug et al. 2009).

The study of Esteban also highlights that nutritional composition on a dry matter basis showed 65% carbohydrates, 13% crude fiber, 12% crude protein, 8% ash, and 2% ether extract. In conclusion, reusing a mixture of vegetable and fruit byproducts as a new raw material for animal feed is feasible. There are no important differences in the composition of the different vegetable byproducts, thus they can be managed as a single by-product, although their high content of water can compromise the perishability and their content of proteins, as regards the recommended percentage of the most important nutritional parameters, limits their inclusion in feed formulations into the range of 3–6%, (Esteban et al. 2007; García 2005; Martin et al. 2016).

3.4 Composting for Biofertilizers

Composting is a biological practice where microorganisms thrive on waste material in a controlled manner, breaking down the organic fraction; the end product, called compost, is rich in soil nutrients and represents a fertilizer. Food Waste can be composted aerobically or anaerobically and result in aerobic and anaerobic compost, respectively. Vermicomposting is one of the strategy for the Food Waste Management that makes use of earthworms that would convert kitchen waste to high-quality compost, (Mousavi et al. 2017). Adi and colleagues described the composting for 49 days, after 3 weeks of composting, of coffee grounds and kitchen waste using Lumbricus rubellus. Different combinational treatments were carried out and the study revealed that coffee grounds stabilize kitchen waste and produce high-quality vermicompost, rich in nutrients (Adi and Noor 2009).

The production and quality of the compost depend on the environmental conditions, moisture content, composition of organic matter, pH, temperature, aeration and carbon to nitrogen ratio. Anyhow, bio-fertilizers from food waste can sustainably be produced also by using digestates of Anaerobic Digestion which are inexpensive and renewable sources of plant nutrients. Microorganisms like Pseudomonas, Klebsiella, Salmonella, Penicillium, Shigella, Bacteriodes, Aspergillus, and Bacillus which grow in acidogenic effluents are particularly useful in this respect (Alfa et al.

2014). Klebsiella and Clostridium are nitrogen-fixing vectors, (Bagyaraj and Rangaswami 2009; Panesar and Marwaha 2017) while Bacillus and Pseudomonas are involving as phosphate-solubilizing microrganisms (Sharma et al. 2013). These micro-organisms increase the availability of nutrients to the plants. Their presence avoids the amendment of soils with nitrogen and phosphorous fertilizers, respectively. Liquid effluents comprise are rich in degraded organic matter, available nutrients, and microbial biomass, in comparison to synthetic fertilizers, which increase soil organic content and hence its fertility (de Albuquerque et al. 2015).

Despite the composting of food wastes represent a viable pathway to avoid the landfills end-uses, compost processing as an option of food waste valorization highlights the largest environmental impact, with decomposition emissions contributing the most to global warming potential, acidification, and eutrophication impact categories under the average and maximum emissions scenarios (LCA).

However, composting processes would avoid environmental impacts due to peat mining and transport. The avoided impacts from peat replacement were demonstrated higher than the impacts from composting for all categories, illustrating that using compost instead of peat results in net environmental gains. For some aspects, composting process can be considered as eco-friendly pathway if compared with landfill treatment, despite more sustainable routes (i.e. bioenergy, animal feed, and BBPs) could be considered and some impacts associated with CH_4, N_2O and NH_3 emissions during the decomposition process of food wastes need for more consensus in the literature on emission values from composting processing (Saer et al. 2013).

3.5 Biofuels: First, Second and Third Generation

The biofuels are considered liquid and gaseous biofuels derived from biomass, used mainly in the transport sector. The most common biofuels are Bioethanol, Biodiesel (as substitutes of petrol and diesel, respectively) and at last Biomethane from Biogas (as substitute for Methane Natural Gas). Their identification of "*First, Second and Third Generation Biofuels*" depends on their feedstock used and generally the technology processes applied.

Sustainable conditions for biofuels production are the key to the progress and represent one of the most important challenges to tackle in the coming years. The development of European Directives lets to implement better sustainability criteria shifting, thus, from I° generation of biofuels production (Conventional Biofuels) to II° and III° generation of Biofuels (non-Food Feedstock Biofuels) with priority to rural biomass residues and food wastes rather than food commodities.

Conventional Biofuels or I° generation of Biofuels are considered: all biofuels derived from crops, which can also be used for food and feed (such as sugar, starch and vegetable oils (Bourguignon and Vandenbussche 2017).

A 'first generation' biofuels like biodiesel (bio-esters), bio-ethanol, and biogas are characterized for their Options manifold uses:

by its ability to be blended with petroleum-based fuels (max 10% blending) on the basis of FQD Directive, distributed through existing infrastructure or combusted in existing internal combustion engines, by the use of existing alternative vehicle technology like FFVs ("Flexible Fuel Vehicle") or natural gas vehicles.

"First generation" biofuels can offer some emissions benefits and can help to improve Transport and domestic energy security. But concerns exist about the sourcing of feedstocks, including the impact it may have on biodiversity and land use (ILUC) and competition with food crops, (Naik et al. 2010; Bourguignon and Vandenbussche 2017).

The main disadvantage of first-generation biofuels is the competition food-versus-fuel debate. One of the reasons for rising food prices is due to the increase in the production of these fuels, (Naik et al. 2010).

The second and third-generation biofuels, are also named advanced biofuels because do not directly compete with food and feed crops.

Notably, **II° generation biofuels** derive from waste and rural residues (i.e. municipal food waste and cereal residues, vegetable residues) or non-food crops (i.e.short rotation crops like miscanthus).

While **III° generation biofuels** are generally referred to biofuels produced by innovative pathways not always available on the commercial large scale because still in progress with technology still fine-tuning, (i.e. algae, hydrogen from gasification of biomass), (Bourguignon and and Vandenbussche 2017).

EU legislation contains a set of mandatory targets specific to the transport sector. The blending rates for advanced biofuels will be stepwise increased between 2020 and 2030 as reported in the RED II (Directive 2018/2001/EC). The aim is to boost the market for sustainable biofuels in the EU Market. As a part of the EU sustainability framework for biofuels and bioliquids the EU Renewable Energy Directive (Directive 2018/2001/EC) and the Fuel Quality Directive (FQD, Directive 2009/30/EC), contain also harmonized minimum greenhouse gas emission requirements which are mandatory for biofuels accounted towards their targets and/ or eligible for public support.

A key requirement of the FQD is that all fuel suppliers must meet at least of 60% reduction in GHG emissions by 2020 across all fuel categories supplied to the market. This is designed to be consistent with the 10% use of biofuels and shift demand towards biofuels with higher GHG savings, meaning advanced biofuels by residues and wastes. In addition, the FQD places limits on the content of palm oil and soy content of biodiesel. At last, the directive limits ethanol blends to 10% or less when ethanol is used as an oxygenate.

These Directives do not only include the quantifiable targets but also include specific environmental requirements to be considered for several biofuel production pathways.

The national scheme and regulation on biofuels sustainable criteria is based on the existing inputs according to the mentioned EU Directives with particular attention to the set of the sustainability criteria based on the latest scientific evidence. To this end, the several national Governments should monitor the supply for biofuels and respective markets, and they should, when appropriate, propose relevant measures to achieve a balanced approach between domestic production and imports, taking into account, inter alia, the development of the energy supply especially for transport sector, (Directive 2018/2001/EC) of bilateral trade negotiations, environmental, social and economic considerations and the security.

3.5.1 Biofuels Production and Technology Overview

The wide range of available biomass resources associated with human activities (rural residues, food wastes) is generally depicted by solid unprocessed organic material, with moisture content generally around 40–50%.

The biofuels can be produced from a number of feedstocks, giving priority to the agricultural residues (such as vegetable residues, straw, and sawdust), residues from agro-food industrial processes (such as vegetable food residues, vegetable oil mill effluent, organic wastes from vineries, cattle breedings, dairies, bakeries), food wastes (like organic fraction of urban wastes and used cooking oils), non-food energy crops (such as rapidly growing grasses like switchgrass and miscanthus, or short-rotation forestry like poplar and eucalyptus), and at last algae.

These feedstocks can be converted through different technologic processes (hydrolysis, pyrolysis, anaerobic digestion, esterification, gasification and hydrothermal upgrading) for several bioenergy pathways included Biofuels production such as biodiesel, bioethanol, methanol or butanol, Fig. 3.1.

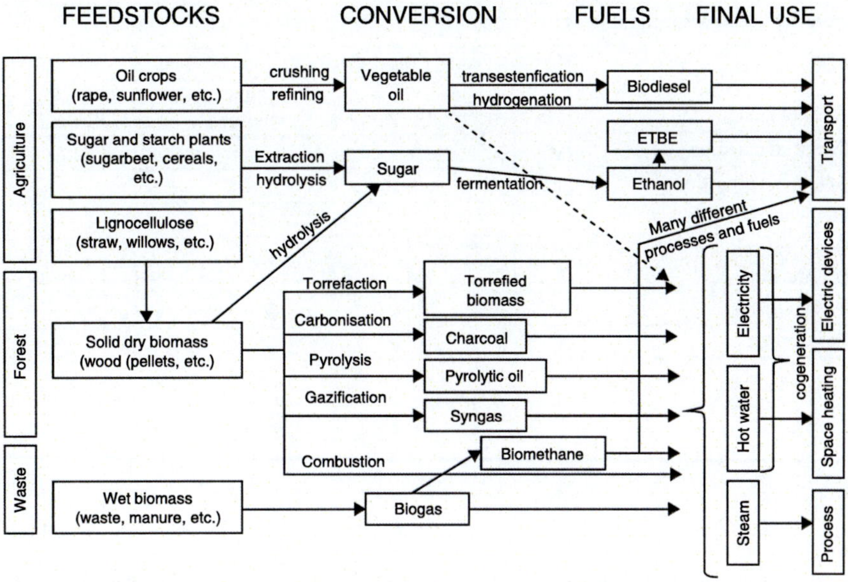

Fig. 3.1 Bioenergy routes, (with permission from. (AEBIOM 2006). (Reprinted from Boosting Bioenergy in Europe Report, AEBIOM, 2006, with permission from AEBIOM)

Several technology processes can also be followed by an array of secondary treatments (i.e.: stabilization, dewatering, upgrading, refining) depending on specific final products. The versatility of biomass processing technologies can produce material and energy like electricity, heat or CHP system, Biofuels both in solid, liquid and gas forms, Table 3.2.

The energetic density and physical properties of the biomass are critical factors and need to be analyzed in order to rightly match a specific feedstock to processing technology. The selection of processing technologies depends on the nature and structure of the biomass feedstocks and the desired Biofuels output. Anaerobic digestion, fermentation, and oil extraction are suitable processes for biomass that have easily extractable oils, sugars, or with high water contents (Table 3.3).

The lignin and cellulose are the most plentiful constituents of biomass with higher volume than other components; therefore, thermal processing and cellulose fermentation, when tested, have represented one of the most important sources for bioenergy until so far. Other specific technological processes, anaerobic digestion, oil extraction, and fermentation, are generally used as principal treatments for biofuel production by biomass sources with significant extractable non-lignocellulosic values, that are lipids, and starch and other carbohydrates, (de Carolis et al. 2012).

Energy directly converted (intended as heat, electricity or biofuels) doesn't only depend on available biomass, but also by energy conversion technology. More efficient technologies are able to convert more energy from the same amount of available biomass. Table 3.4 shows the different links among several parameters between biomass feedstock, technology uses and Technology Readiness Level.

Table 3.2 Biomass technologies overview

	Outputs: energy (heat/electricity) and biofuels				
Technologies	Heat	Electricity	Gas biofuels	Liquid biofuels	Solids biofuels
Direct combustion	✓	✓			
Anaerobic digestion	✓	✓	✓		
Fermentation				✓	
Oil extraction and esterification				✓	
Pyrolysis	✓	✓	✓	✓	✓
Gasification	✓	✓	✓	✓	

Table 3.3 Chemical characterization and technology processes

		Fats, Oils	Protein	Sugars/ Starch	Ligno-cellulose
Biomass sources					
Residues & wastes	Green waste				✓
	Animal waste	✓	✓		
	Organic fraction of urban waste	✓	✓	✓	✓
Processing technology					
Direct combustion		✓			✓
Anaerobic digestion		✓	✓	✓	Cellulose only
Fermentation			✓	✓	Cellulose only
Oil extraction and esterification		✓			
Pyrolysis		✓	✓	✓	✓
Gasification		✓	✓	✓	✓

Table 3.4 Technologies overview for bioenergy route and biofuels production

Biomass processing technology	Possible scale[a]	Feedstock flexibility	Conversion efficiency[b]	Output flexibility	Market value of product	TLR development status
Direct combustion	Large	High	Low	Low	Low	Established
Anaerobic digestion	Small	Medium	Medium	Low	Medium	Established
Fermentation	Medium[c]	Medium[d]	Medium	Low	High	Established
Oil extraction/ esterification	Small	Low	High	Low	High	Established
Pyrolisis	Large	High	Medium	High	Medium	Early commercial
Gasification	Large	Medium	Medium	Medium[e]	Medium	Early commercial

[a]Scale of possible industry is dependent on the scale of the available biomass resource. Those technologies able to use lignocellulosic biomass are at an advantage
[b]Energy efficiency measures the amount of energy in the feedstock retained in the products
[c]De-polymerisation of cellulose to sugars will allow access to a larger biomass pool; however, this technology is not commercially established
[d]This may be higher if technologies that generate sugar feedstocks from cellulose become mature
[e]The direct products of gasification are low, but this is the basis to a vast array of fuel and chemical products via synthesis reactions

Further investigations on the different technologies for biofuels production point out the greater inclination to use anaerobic digestion, fermentation, and esterification rather than pyrolysis and gasification due to development status of technology, even if that mentioned latest technologies provide higher flexibility of biofuels produced with high flexibility of biomass feedstock, (de Carolis et al. 2012; IRENA 2013), Fig. 3.2.

Conventional biofuels dominate most of total production, as well as the overall contribution of renewables to road transport.

The conventional biofuels derived from food crops have net benefits in terms of emissions reductions and energy balance, but they are affect the food price competition, and there is little opportunity for cost reduction, because the technology is relatively mature and only incremental improvements in process economics can be expected. The most available technologies commercially used for Biofuels production in the transport sector (First and Second generation) are still actually based on transesterification for biodiesel, fermentation for bioethanol, and anaerobic digestion for biogas and biomethane.

Looking further into the future, electric vehicles are gaining a more greater part of the transport sector, and the reliance on food crops will limit the potential contribution of conventional biofuels to total transport demand. Secondly, new renewable solutions are emerging and advanced biofuel plants are being started to be built at

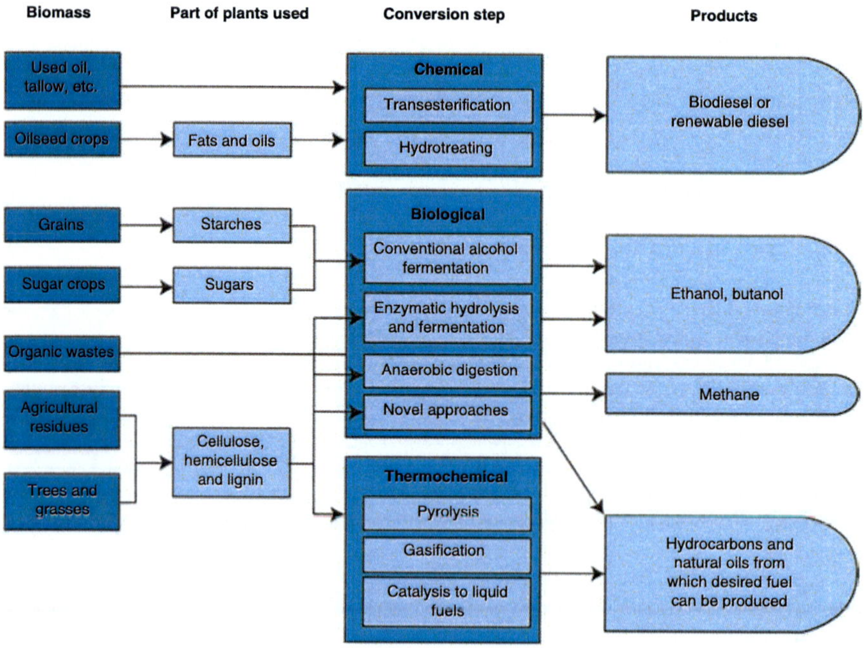

Fig. 3.2 Technology Pathways Overview from biomass feedstock to conventional and advanced Biofuels, (IRENA 2013). (Reprinted from Road Transport: the cost of renewable solutions. Report, IRENA 2013, with permission from IRENA)

commercial scales, during the recent decade. Further questions mark remain about which most promising patwhays will offer the least cost of advanced biofuels.

The research impact on the technology innovation on advanced biofuels is fastly proceeding with a reduction of the cost-technologies close to traditional pathways. But the advanced biofuels are still expensive, and only some types being commercialized today. It's the case of second-generation biofuels like Biodiesel by UCO (Used Cooking Oils), Bioethanol by cellulosic residues (Agro-Food Residues), Biogas/Biomethane from Organic Fraction wastes from Anaerobic Digestion.

3.5.2 Biodiesel (FAME/FAEE) by UCOs – Used Cooking Oil

Used Cooking Oils (UCOs) are oils and fats that are being used for cooking or frying in the food processing industry, restaurants, fast foods, and the household. UCO can be originated from both vegetable and animal fats and oils. It is estimated that currently around 90% of cooking oils and fat used in the EU are produced from vegetable oils, (Peters et al. 2013).

According to moderate estimations in the EU, the potential UCO to be collected is around 8 L UCO/capita/year. Extrapolated to the total EU population of around 500 million, 4 Mton of UCO is the annual capacity of production. This potential increases around 2% per year, following the annual increase in cooking oil usage. In order to achieve this level of collection, the collection and pre-treatment infrastructures should be improved by country to country, (Anderssen et al. 2007).

The principal disadvantage in using UCO as a biodiesel resource is its content of impurities represented by Free Fatty Acid (FFA) and water. The quality (and their price as raw material) are determined by their acid and saponification values. If the FFA content of the UCO is above 3.0% then significant problems may occur in the transesterification process.

Large amount of FFA in combination with moisture can lead to a large soap formation and hydrolysis respectively. This is attributed to the fact that the properties of UCOs are different from those of refined and crude oils. As a result of the high temperature during cooking and the water from the food, triglycerides in the oils are hydrolysed and this increases their free fatty acid (FFA) content. During the transesterification process, FFAs reacting with the alkaline catalyst to produce soap, which inhibits the reaction and the results in the reduction of biodiesel yield. The FFAs of UCO are sensitive to alkali catalyst, therefore they must be removed before transesterification. Both impurities are resulting in low biodiesel yield and in high catalyst consumption. Therefore, a pre-treatment before the transesterification process is required due to their significant adverse effects on the process, with a consequent arising of their cost, (Leung and Guo 2006; Banerjee and Chakraborty 2009).

Transesterification is the most common method to produce biodiesel, methanol is the normally used alcohol in the process due to its low cost and physical and chemical advantages, and the reaction can be carried out either discontinuously (batch) or continuously.

Transesterification is a relatively simple process in order to produce biodiesel, or more commonly Fatty Acid Methyl Esters – FAME, and Fatty Acid Ethyl Esters – FAEE.

The most common processes are:

- Classic transesterification process, Homogeneous - catalyzed transesterification, (Tsoutsos et al. 2016; Naik et al. 2010)
- Heterogeneous – catalyzed transesterification, (Navajas et al. 2013; Naik et al. 2010).
- Enzymatic, Enzyme – catalyzed transesterification (Yan et al. 2011; Oliveira and Rosa 2006),
- Supercritical methanol, Non – catalyzed transesterification, (Yin et al. 2008).

After the transesterification reaction, glycerol is produced as a by-product and separated by settling or centrifuging.

Biodiesel obtained from UCO consists of long-chain fatty acid methyl esters (FAMEs/FAEEs), (Atadashi et al. 2011). The UCO-Biodiesel is highly biodegradable and has minimal toxicity; it can replace petrodiesel fuel in many different applications including internal combustion engines without major modifications. The biodiesel is then purified before being used as diesel fuel in compliance with the standards EN 14214 for Europe and ASTM D 6751–12 for the USA, and other national quality standards and technical norms.

Several investigations indicate that trans-esterification reaction is the "heart" of the full chain, therefore optimization is critical for the full chain feasibility. Furthermore, good processes and pre-treatments of UCOs demonstrated a good yield of Biodiesel production, especially in EU Countries (Tsoutsos et al. 2016). UCO's full chains can also provide a dynamic business activity for each country with essential economic and environmental benefits, (Tournaki and Tsoustsos 2013). Their collection and conversion in Biodiesel can strongly reduce the negative impacts caused by FOG (Urban Fats, Oils and Greases) disposal in the environment, especially in wastewater treatment, preserving river and aquatic ecosystems as well as the underground urban sewage systems, (Arthur and Blanc 2013).

3.5.3 Bioethanol by Fermentation of Carbohydrates

A plenty variety of rural biomass and food wastes contain carbohydrates that represent the raw materials for producing Bioethanol by fermentation, (Shady et al. 2019). Raw materials are classified under three major categories:

(a) Sugar-based Biomass and Wastes: Sugar cane, Cereals, beetroot, fruits, palm juice, etc.
(b) Starch-based Biomass and Wastes: Cereals and derivates, including rice and corn, sweet sorghum, root plants and waste potato-based, cassava, etc.
(c) Cellulosic-based Biomass and Wastes: Wood and wood waste, cedar, pine, wood, etc. (agricultural, garden residues and fibers).

In the past decade, most Bioethanol plants were fuelled by food crops like corn, wheat, barley, sweet sorghum contributing to an increase in the growth of the First Generation of Biofuels.

Nowadays, the new trend of sustainable biofuels in accordance with new Directive (i.e. RED II Directive of EU Commission) follows Bioethanol produced from lingo-cellulosic biomass such as agroindustrial residue and food waste (i.e. rice straw, wheat straw), grasses (switchgrass, potato and corn wastes, starch-based food wastes). Both these alcohols are produced through biochemical process, (Shelley 2006).

Starch-based food wastes and biomass residues seem to be more promising due to high content of starch. The starch being considered a forerunner of excellent chemical building block, it consists of a long-chain polymer of glucose. By using conventional fermentation technology, macromolecular starch cannot be directly fermented to ethanol. The starch structure has to be first broken down into simpler and smaller glucose. In this process, starch feedstocks are grounded and mixed with water to produce a mash typically contained 15–20% starch.

The mash is then cooked, and subsequently treated with two enzyme preparation. The starch molecules are hydrolyzed by first enzyme, to short chains to glucose.

The first enzyme is amylase, that liberates "maltodextrin" oligosaccharides by liquefaction process. The dextrin and oligosaccharides are further hydrolyzed by pullulanase and glucoamylase in the saccharification process. During saccharification all dextrans are converted to glucose, maltose, and isomaltose. The mash is then cooled to 30 °C and yeast is added for fermentation, (Lee et al. 2007).

Ethanol production is usually obtained via enzymatic hydrolysis of starch-containing biomass. Actually, several Bioethanol plants is fed both by sugar beet and sugarcane residues especially in USA, thus part of food market competition with arising the price of commodities should be avoided.

Bio-ethanol production facilities can be classified into two groups: i.e. wet & dry mill processes (Shelley 2006).

Dry mills are built primarily to manufacture ethanol only.

According to Naik et al. (2010), modern wet milling plants are able to produce a plenty variety of valuable added value coproducts for nutraceutical and pharmaceutical sectors, as well as organic acids, solvents, and Bioethanol of second generation.

Two basic types of fermentation are used in the bioethanol plants: (a) aerobic and (b) anaerobic, therefore depends upon oxygen needed in the process or not and kind of fermentation carried out by several enzymes secreted by micro-organisms. Thousands of micro-organisms in nature are capable of providing fermentative changes. Most of interesting ones are capable of producing ethanol from sugar and starch. The microorganisms employed for ethanol production are classified into three categories: yeast (saccharomyces species), bacteria (zymomonas species), and mold (mycelium). These micro-organisms are often found in nature, and are very selective in their fermentations. Some Species are specific to hexoses, pentose, or mixtures of both. Efforts are being made by various researchers to produce ideal micro-organisms, which will be able to produce ethanol from any carbohydrate. The Bioethanol production by different biomass feedstock, (including ligneo-cellulosic wastes), represents the challenge for the coming years.

Further forerunner product currently used for bioethanol production is represented by starch. Starch is a polymer of glucose that cannot be directly fermented to ethanol by conventional fermentation technology. In the process of fermentation, starch is decomposed in glucose molecules after treated with yeast, (i.e. saccharomyces cerevial, saccharomyces uvarum, schizosaccaharomyces pombe, and kluyveromyces species). Under anaerobic conditions, yeasts metabolize glucose to ethanol. During the fermentation process, part of the glucose is used by the yeast for the production of cell mass and for metabolic products other than ethanol, thus reducing the conversion efficiency from its theoretical maximum (51%). In practice, the bioethanol conversion is estimated to be between 40 and 48% of glucose with 46% fermentation efficiency, (i.e. 1000 kg of fermentable sugar would produce about 583 lit of pure ethanol, (Lee et al. 2007; Naik et al. 2010).

3.5.4 *Anaerobic Digestion of Biomass Residues and Food Wastes*

The organic residuals and food wastes are characterized for having a humidity level over 50% that lets using the anaerobic digestion as principal process for conversion of wastes into the Biogas and further converted into biofuels (biomethane or methanol), and production of digestate, forerunner product for potential biofertilizers, (de Carolis et al. 2012).

During anaerobic digestion process, anaerobic bacteria in the absence of oxygen are used to break down the organic matter of biomass, and during the conversion, a mixture of methane and carbon dioxide gases is produced. The typical ratio of the gas mixture is 60 CH_4 and 30% CO_2.

The ratio between CH_4 and CO_2 gases is strictly depending on the raw biomass used in the biogas plant. In rural areas, the typical biomass is often represented by slurry and manure of several animal breeding. Their recovery is variable from time to time and maintaining a high and constant level is essential for a biogas plant productivity with sustainable conditions. The relation among manure, biogas, and biomethane is considered (Table 3.5).

Deepanraj et al. (2017) observed that substrate pretreatment has significant effects on biogas production from food wastes. Different pretreatments like autoclave, microwave and ultra-sonication of food waste were carried out, and anaerobic digestion was carried out with poultry manure. Maximum biogas production (10.12%) and yield (9926 ml) were observed with ultrasonication pre-treated samples. 41.96–46.52 g/l of volatile solids were also removed during the process. Wu et al. (2016) developed an improved biogas production from food waste by co-digestion with de-oiled grease trap waste. The study was carried out in different digesters like mesophilic digester (MD), temperature-phased anaerobic digester (TPAD), and temperature-phased anaerobic digester with recycling (TPAD-R). Mono-digestion of food waste as well as co-digestion with de-oiled grease trap

Table 3.5 Typical zootecnical wastes for biogas plants

Animal category	Live animal mass (kg)	Total fresh manure ($kg_{m.}$)[a]	Moisture (%)	TS Total Solids (% on $Kg_{m.}$)	VS Volatile Solids (% on TS)	Biogas Production (m^3/t_{sv})	CH_4 in Biogas (%)	References
Bovine	640	50–55, 51	83–88 86	11–15, 12	80–85	300–450	60–65	ASAE D384.1; ASAE D 384.2; Ohio State University (2006); Bordoni (2009)
Swine	60	5–6, 5.2	90	6–9, 8	75–90	450–550	60–65	
Horse	500	20–24.5, 23.6	85	14–15, 15	75	250–500	60–65	
Broiler	1.6–3.5	0.52–0.72	75	19–25, 23	75	300–500	60–65	ASAE D384.1; ASAE D 384.2 Ohio State University (2006): Bordoni (2009)
Turkey	6–15	0.48–1.2	74	19	95–98	300–500	60–65	
Duck	6.5–8	0.52–0.64	74	49	33	300–500	60–65	
Ovine	70–80	5.6–6.4	–	22–40	70–75	300–500	60–65	

[a]Fresh manure is referred for live weight animal indicated

waste was carried out. The results indicate that co-digestion increased the biogas yield by 19% in MD and TPAD-R with a biogas yield of 0.60 L/g of volatile solids.

Anaerobic Digestion (AD) is probably the most demanding process for the valorisation of Food Waste towards methane (CH_4), hydrogen (H_2) and volatile fatty acids (VFA), acetic acid (HAc), propionic acid (HPr), butyric acid (HBu), isobutyric acid (HIBu), valeric acid (HVa) and iso-valeric acid (HIVa) production.

Anaerobic fermentation of several food wastes is obtaining great interest not only for biofuels bio-platforms but also for bio- commodity chemical production (Dahiya and Joseph 2015; Sarkar et al. 2016). Anaerobic fermentation includes connected processes starting with hydrolysis, and followed by acidogenesis, acetogenesis, dehydrogenation and at last methanogenesis thanks by a syntrophic association of various bacteria activities (i.e. hydrolytic, acetogenic, hydrogen-producing, acetate-forming microbes, homoacetogens, acetoclastic methanogens, and hydrogenoclastic methanogens), (Venkata Mohan 2009). Each stage of Anaerobic fermentation produces important bio-products not valorized yet as biobased products but directly converted into the biogas, by using classic anaerobic digestion plants. Hydrolysis yields produce sugars, Acidogenesis yields VFA's and H_2, acetogenesis yields HAc, and H_2 while the terminating process methanogenesis yields CH_4. For Biobased products platform, it is important to enhance the production of Volatile Fatty Acids – VFAs and H_2, at the expense of the methanogenesis which can be suppressed by implementing pretreatments (Venkata Mohan et al. 2008) VFAs represent the important chemical building blocks that can be biologically and/or chemically converted to biobased products (i.e alcohols, polymers, pharmaceuticals products, etc). Anaerobic Digestion lets to use of several Food wastes as feedstock (i.e. carbohydrate, lipids, and proteins), so makes the process relatively inexpensive in comparison to other technologies, (Sarkar et al. 2016).

The calorific values of Food Waste under STP conditions is 5.35 MJ/kg, which signifies its potentiality as feedstock to harvest bio-energy. Similarly, 1 kg of COD of FW can produce 20.83 mol H_2 (466 L H_2) through dark fermentation pathway with an energy output of 5906.97 kJ and which equals 1.6 kWh. But H_2 produced through butyric acid pathway is 10.417 mol H_2 (233.33 L H_2; 2953.49 kJ). Possible problems can be caused by inhibition caused by high concentrations of VFAs and NH-N production (Akroum-Amrouche et al. 2013). One kg of FW (5.208 mol of glucose) can produce 624.96 g of acetic acid (0.875 MJ/mole), 916 g of butyric acid (2.192 MJ/mole), and 385 g of propionic acid (1.534 MJ/mole) (Mohanakrishna and Venkata Mohan 2013). The individual calorific values of the harnessed biobased products from food wastes directly contribute to the commercial value at the industrial scale. The specific yields of biogas (CH_4, H_2) depends on the actual conversion efficiencies that can be calculated to evaluate the bioprocess performance. The utilization of energy-intensive and commercially viable biobased products from FW through integration of bioprocesses can lead towards environmental sustainability and significantly contribute to the development of the bioeconomy. This also reduces the negative impact of FW on the environment to a certain extent by integrating various bio-processes towards maximum resource recovery with simultaneous revenue generation.

Due to the increased cost of energy, likewise, Anaerobic Digestion of organic biomass, and food wastes seem to be an attractive alternative for Biofuels production and biofertilizer for available for organic cultivations. Thus anaerobic digestions in landfills are becoming potential ponds of methane production from organic urban wastes. Their efficiency conversion is around 50% of CH_4 and CO_2 production. These components together with atmospheric nitrogen, oxygen and traces of organic compounds are known as landfill gas (LFG).

According to the Environmental Protection Agency (EPA), each pound of biodegradable organic food waste can produce 10–12 standard cubic feet of gas. This gas can be compared to low-quality natural gas. If part of volatile compounds and contaminants, especially vinyl chloride and hydrogen sulfide, and CO_2 will be removed (Lee et al. 2007), the quality of biogas will be commercially valorized and potentially injectable into the Gas grid without energy losses.

The removal of contaminants from LFG requires efficient separation technology to use it as a substitute for natural gas with potential and different uses: from biofuel used in transport directly used into the internal combustion engine, in the turbine of gas power plant, microturbine, or directly used into the boiler, greenhouse and in common CHP systems.

During recent decades, some options for energy conversion from biomass and organic waste to biofuels are growing through anaerobic digestion. The current research of landfill gas by anaerobic digestion focuses on the generation of liquid fuels instead of Biogas and biomethane due to the high cost associated with meth-

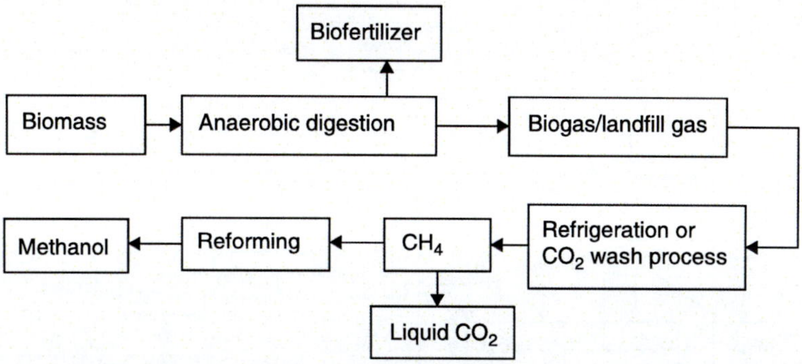

Fig. 3.3 Biorefinery concept for biomass residues and food wastes by using Anaerobic Digestion, (Naik et al. 2010). (Reprinted from, Renewable and sustainable Energy Reviews, 14 (2010), Naik S.N., Vaibhav V.G. Presant K. Rout Ajay K. Dalai, "Production of first and second generation biofuels: A comprehensive review", copyright 2010, with permission from Elsevier)

ane purification and collection (Lee et al. 2007). The promising option is represented by methanol production as liquid biofuel because has several advantages. First, it contains low sulfur and low ash and it can be made for commercial use. Second, liquid fuel is much easier to handle, store, and transport than gaseous products if the pipeline infrastructure is not present in certain territories.

The biorefinery concept of anaerobic digestion is shown in Fig. 3.3.

3.6 Technology Progress for Advanced Biofuels

Recently, other interesting technologies are delivering promising results for a more competitive biofuels production by using lignocellulosic based products from dedicated crops (miscanthus, poplar, switchgrass, and reed canary grass), rural residues (i.e. straw, woody prunings), forestry (i.e. poplar, eucalyptus, willow), agroindustrial wastes (wood shells, nut kernels), and urban wastes (i.e. garden prunings).

Lignocellulosic materials are a collection of feedstock for "advanced biofuels", and can be either through several conversion processes: (physical, thermal, biological and chemical). The traditional processes are hydrolysis and fermentation (i.e. Bioethanol), and then to a more advanced process like pyrolysis and gasification (i.e. Fisher-Tropsch, bio-diesel, bio-DME, and bio-SNG) (Fig. 3.4), for which commercial utilization is still under development.

The hydrolysis, sugars are extracted from the lignocellulosic feedstock, after which the sugars are fermented to ethanol (Fig. 3.5). BTL (Biomass-to-Liquids) or Fischer–Tropsch diesel (FT-diesel) can be considered a full substitute of diesel. The methane can be also turned by the gasification of lignocellulosic biomass. By using the same process, the lignocellulosic matter can also produce Bio-DME (Dimethyl Ether), which can be used in diesel vehicles after slight adaptations, (ETIP Bioenergy 2020).

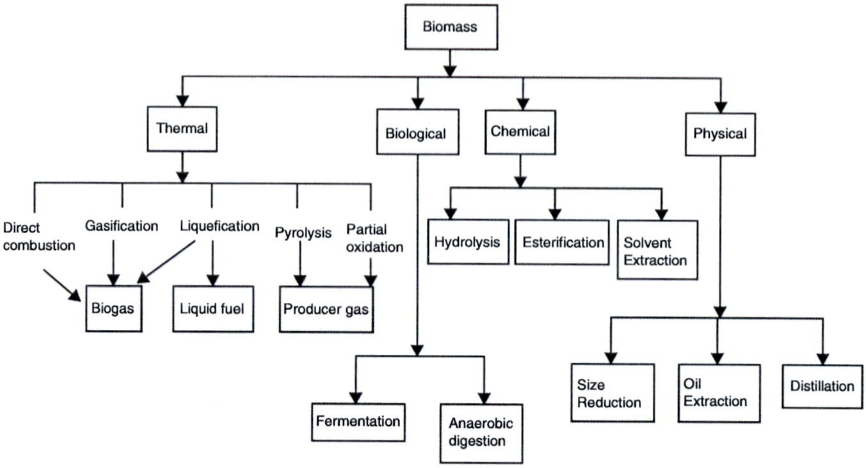

Fig. 3.4 Thermo-chemical and biological conversion processes for several types of biomass. (Naik et al. 2010)

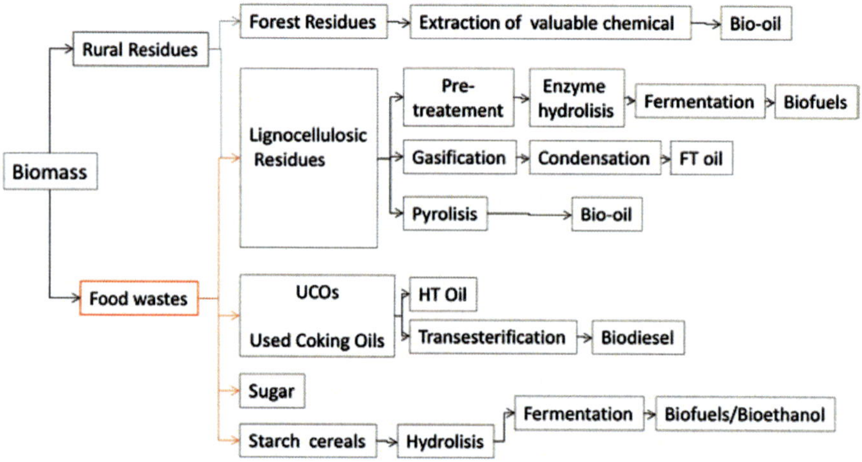

Fig. 3.5 II° generation biofuel production from biomass

All biofuels pathways are promising to strengthen the whole biofuel production but many advanced biofuel production routes are still at the prototype or demonstration stage. Only two – ethanol from lignocellulosic material and methanol produced through gasification – being considered ready for commercialization, (Fig. 3.6), (IRENA 2016).

Conventional biofuels dominate total biofuels production, as well as the overall contribution of renewables to road transport. However, new renewable solutions are emerging as advanced biofuel plants have started to be built at commercial scales. II generation biofuels offer the potential for significantly lower and more stable feedstock costs and could meet a much larger proportion of transport demand in

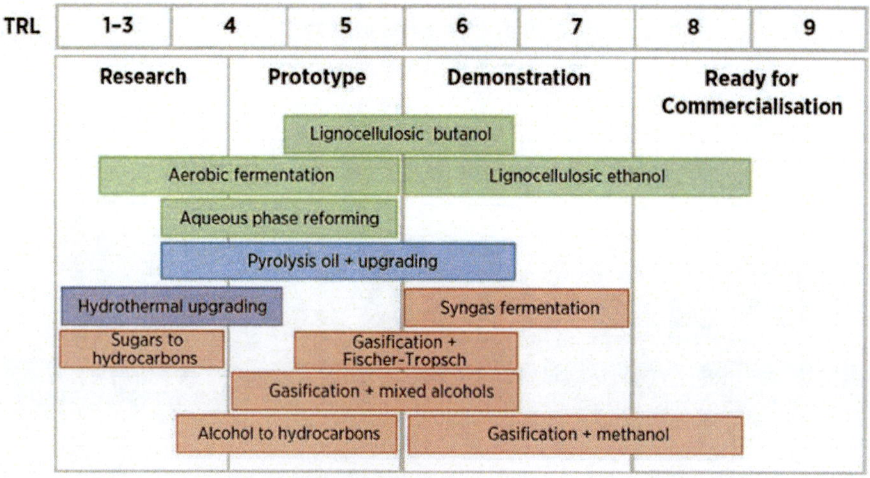

Fig. 3.6 Advanced biofuels technology readiness level close to commercialization, (Colours represent the principal conversion process, Hydrolysis (green), Pyrolysis (blue) and Hytdrotermal upgrading (purple) and Gasification (red)) (IRENA 2016). (Reprinted from Innovation Outlook, Advanced Liquid Biofuels Report, IRENA 2016, with permission from IRENA)

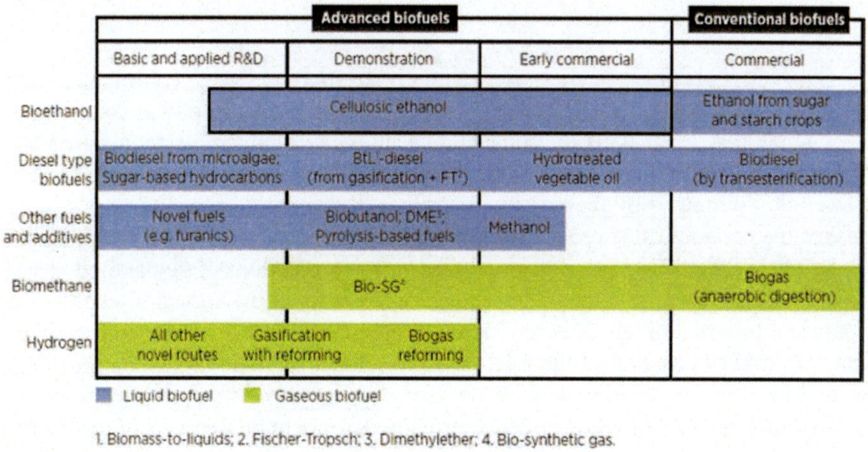

Fig. 3.7 Readiness of different Biofuels pathways, (IRENA 2013). (Reprinted from Road Transport: The cost of Renewable Solutions Report, IRENA 2013, with permission from IRENA)

short and medium time, given that feedstocks can be sourced from a wide range of biomass sources. This is due to advanced biofuels use biomass feedstocks from wastes/residues (with a focus on lignocellulosic material) that do not have to be grown on pasture or arable land. The challenge is still to prove the efficiency, reliability, and commercial attractiveness of the different pathways for advanced biofuels (Fig. 3.7).

Question marks remain about which advanced biofuels pathways will offer the least cost fuels (Fig. 3.8)

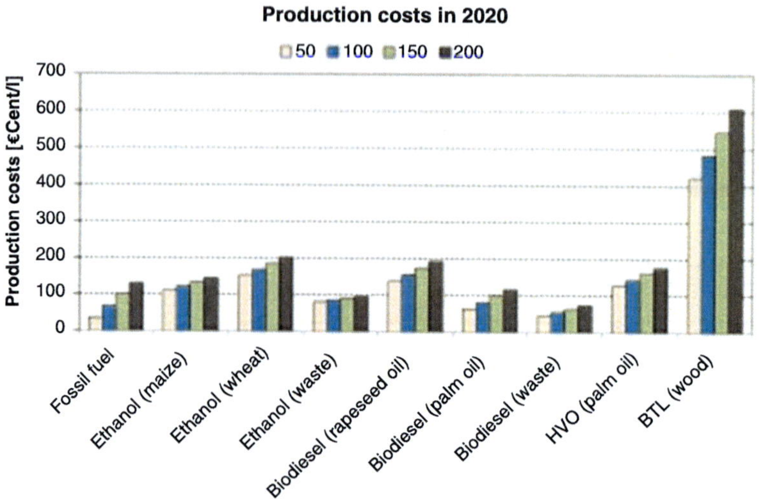

Fig. 3.8 Cost of Biofuels production by different feedstocks by 2020, (Festelab et al. 2014). (Reprinted from ZEW – Zentrum Fur Europaische Wirtschaftsfornschung GmbH, "Modelling production cost scenarios for biofuels and fossil fuels in Europe" Report, 2013. With permission from principal author Festel G.)

Conventional biofuels, derived essentially from food-based feedstocks, have seen their production costs increase in recent years due to high food prices.

The outlook after 2020 for conventional biofuels is mixed, as food prices are projected to remain high. Total production costs for conventional ethanol and biodiesel are dominated by feedstock costs rather than plant management costs. This makes the economics of production heavily dependent on movements in the local and global markets for the feedstock used. Current projections of global food prices to 2020 – and hence also the main production costs for conventional biofuels – are for prices to remain high and even increase for some food crops. This forecast will not help to decrease part of the price of conventional biofuels in the coming years (Fig. 3.9).

But the price of biofuels is not only strongly dependent by the price of respective feedstocks but also on the efficiency of technology conversion (Fig. 3.10). The research impact on the technology innovation on advanced biofuels is going on fastly with a reduction of the cost-technologies close to traditional pathways.

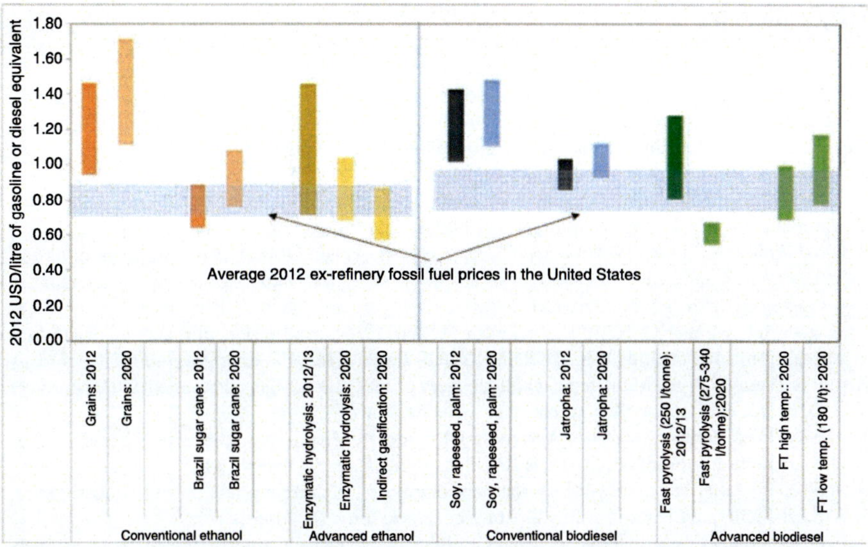

Fig. 3.9 Cost production of conventional and advanced biofuels, (IRENA 2013). (Reprinted from Road Transport: The cost of Renewable Solutions Report, IRENA 2013, with permission from IRENA)

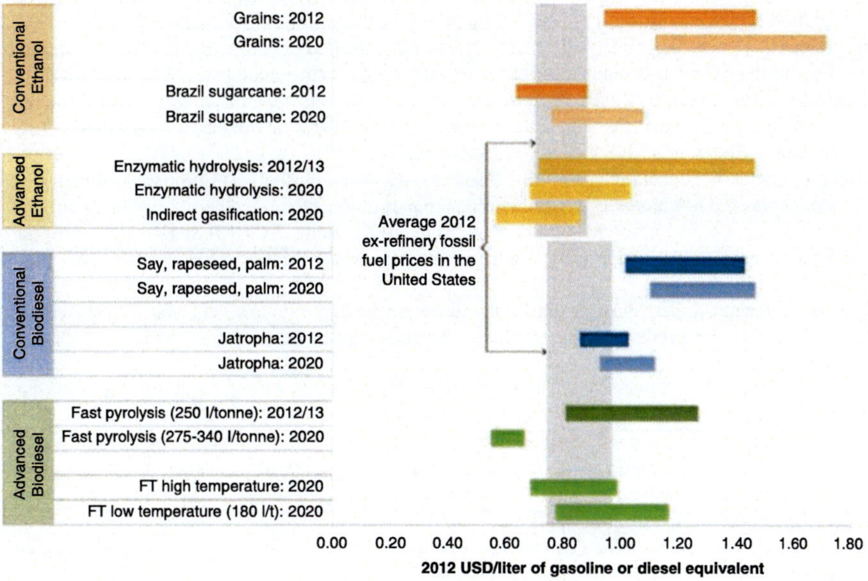

Fig. 3.10 Technologies processes and cost of biofuels. (IRENA 2013)

Bibliography

Adi AJ, Noor ZM (2009) Waste recycling: utilization of coffee grounds and kitchen waste in vermicomposting. Bioresour Technol 100:1027–1030

AEBIOM (2006) Boosting bioenergy in Europe. www.aebiom.org.

Akroum-Amrouche D, Abdi N, Lounici H, Mameri N (2013) Biohydrogen production by dark and photo-fermentation processes. Renewable and Sustainable Energy Conference (IRSEC), International, IEEE. https://doi.org/10.1109/irsec.2013.6529679

Alfa MI, Adie DB, Igboro SB, Oranusi US, Dahunsi SO, Akalia DM (2014) Assessment of biofertilizer quality and health implications of anaerobic digestion effluent of cow dung and chicken droppings. Renew Energy J 63:681–686

Anderssen IH, Webber C, Kelly R, Andersen O (2007) Localised production and supply of biodiesel from used cooking oils. BIODIENET UE Project: IEE Funding Programme

Arthur S, Blanc J (2013) Management and recovery of FOG (fats, oils and greases), CREW project CD2013/6. Available online at: crew.ac.uk/publications

ASAE (2003) American Society of Agricultural Engineers, Manure production and characteristics. ASAE Technical Standard D.384.1. Am. Soc. Agric. Eng. St. Joseph, MI.

ASAE (2005) American Society of Agricultural Engineers, Manure production and characteristics. ASAE Technical Standard D.384.2. Am. Soc. Agric. Eng. St. Joseph, MI.

Aschemann-Witzel J, De Hooge I, Amani P, Bech-Larsen T, Oostindjer M (2015) Consumer-related food waste: causes and potential for action. Sustain J 7(6):6457–6477

Atadashi IM, Aroua MK, Abdul AA (2011) Biodiesel separation and purification: a review. Renew Energy J 36(2):6457–6477

Bagyaraj DJ, Rangaswami G (2009) Agricultural microbiology (EDN2). In: PHI learning. ISBN 10:8120306686

Banerjee A, Chakraborty R (2009) Parametric sensitivity in transesterification of waste cooking oil for biodiesel production – a review. Resour Conserv Recycling J 53(9):490–497

Besombes C, Berka-Zougali B, Allaf K (2010) Instant controlled pressure drop extraction of lavandin essential oils: fundamentals and experimental studies. J Chromatogr A 1217(44):6807–6815

Bordoni A, Romagnoli E, Foppa Pedretti E, Toscano G, Rossini G, Cozzolino E, Riva G (2009) La filiera del biogas. Aspetti salienti dello stato dell'arte e prospettive. ASSAM ISBN 9788882490904. http://hdl.handle.net/11566/54451.

Bourguignon D, Vandenbussche T (2017) Briefing – advanced biofuels, © European Union. http://www.eprs.ep.parl.union.eu: http://www.eprs.ep.parl.union.eu

Contran N, Chessa L, Lubino M, Bellavite M, Roggero PP, Enne G (2013) State-of-the-art of the *Jatropha curcas* productive chain: from sowing to biodiesel and by-products. Ind Crop Prod 42:202–215

Dahiya S, Joseph J (2015) High rate biomethanation technology for solid waste management and rapid biogas production: an emphasis on reactor design parameters. Bioresour Technol J 188:73–78

Dahiya S, Kumar AN, Sravan JS, Chatterje S, Sarkar O, Venkata S (2018) Food waste biorefinery: sustainable strategy for circular bioeconomy. Bioresour Technol J 248(A):2–12

de Albuquerque Nunes VAG, Scherrer Menezes JF, de Melo Benites V, de Lima Junior SA, dos Santos Oliveira A (2015) Use of organic compost produced from slaughterhouse waste as fertilizer in soybean and corn crops. Soils Plant Nutri 72(4). https://doi.org/10.1590/0103-9016-2014-0094

de Carolis C, Foppa PE, Riva GL (2012) Handbook on renewable energy – biomass section. http://ener-supply.eu/en/news_27.php: ENER SUPPLY Project, Southeast Europe Funding Programme

Deepanraj B, Sivasubramanian V, Jayaraja S (2017) Effect of substrate pretreatment on biogas production through anaerobic digestion of food waste. Int J Hydrog Energy 42(42). https://doi.org/10.1016/j.ijhydene.2017.06.178

Demirbas A (2011) Waste management, waste resource facilities and waste conversion processes. Energy Conversion Manag J 52(2):1280–1287

Directive (EU) 2009/28/EC of the European Parliament and of the Council of 23 April 2009 on the promotion of the use of energy from renewable sources and amending and subsequently repealing Directives 2001/77/EC and 2003/30/EC (2009), RED I. CELEX n. 32009L0028.

Directive 2009/30/EC of the European Parliament and of the Council of 23 April 2009 amending Directive 98/70/EC as regards the specification of petrol, diesel and gas-oil and introducing a mechanism to monitor and reduce greenhouse gas emissions and amending Council Directive 1999/32/EC as regards the specification of fuel used by inland waterway vessels and repealing Directive 93/12/EEC (2009), FQD. CELEX n. 32009L0030.

Directive (EU) 2018/2001 of the European Parliament and of the Council of 11 December 2018 on the promotion of the use of energy from renewable sources (2018), RED II. CELEX n. 32018L2001.

Dufey A (2006) Biofuels production, trade and sustainable development: emerging issues. Sustainable markets discussion. IIED – International Institute for Environment and Development, Earthprint Limited, Orders Department, London. ISBN: 978-1-84369-643-8

EPA (2002) European waste catalogue and hazardous waste list. EPA, Environmental Protection Agency, Ireland

Esteban MB, García AJ, Ramos P, Márquez MC (2007) Evaluation of fruit–vegetable and fish wastes as alternative feedstuffs in pig diets. Waste Manag J 27(2). https://doi.org/10.1016/j.wasman.2006.01.004,2007

ETIP Bioenergy (2020) FT_liquid and Biomass to Liquid (BtL). European Technology and Innovation Platform. www.etipbioenergy.eu

EU Commission (2017) Research and innovation perspective of the mid-and long-term potential for advanced biofuels in Europe. Luxembourg, POEU 2017. ISBN 978-92-79-70565-6. https://doi.org/10.2777/05471

EU Commission (2018) Energy modelling – interactive graphs in transport sector. https://ec.europa.eu/energy/en/content/energy-modelling-interactive-graphs?type=radar&themes=s_27_biofuels-in-total-fuels-exclhydrogen-and-electricity&second_scenario=undefined&index_year=#container-charts-controls

Eu Directive 2009/28/EC of the European Parliament and of the Council of 23 April 2009 on the promotion of the use of energy from renewable sources and amending and subsequently repealing Directives 2001/77/EC and 2003/30/EC (2009)

Festelab G, Würmseher M, Rammer C, Boles E, Bellof M (2014) Modelling production cost scenarios for biofuels and fossil fuels in Europe. J Clean Prod 66:242–253

Flach B, Lieberz S, Rossetti A (2017) EU biofuels annual. Global Agricultural Information Network – USDA Foreign Agricultural Service, The Hague

Flach B, Lieberz S, Bolla S (2019) EU biofuels annual 2019. GAIN report number: NL8027. The Hague. http://gain.fas.usda.gov/Pages/Default.aspx

García AJ (2005) Biodegradable municipal solid waste: characterization and potential use as animal feedstuffs. Waste Manag 25:780–787

IRENA (2013) Road transport: the cost of renewable solutions. IRENA – International Renewable Energy Agency, Abu Dhabi

IRENA (2016) Innovation outlook, Advanced liquid biofuels. ISBN 978-92-95111-52-3

Ki Lin CS, Pfaltzgraff LA, Herrero-Davila L, Mubofu EB, Abderrahim S, Clark JH, Koutinas AA, Kopsahelis N, Stamatelatou K, Dickson F, Thankappan S, Mohamed Z, Brocklesby R, Luque R (2013) Food waste as a valuable resource for the production of chemicals, materials and fuels. Current situation and global perspective. J Energy Environ Sci 2:426–464

Lee S, Speight GJ, Loyalka SK (2007). Handbook of alternative fuel technologies. CRC Taylor and Francis Group. ISBN-13: 978-0-8247-4069-6

Leung DYC, Guo Y (2006) Transesterification of neat and used frying oil: optimization for biodiesel production. Fuel Process Technol J 87(10):883–890

Li Y, Khanal SK (2016) Bioenergy: principles and applications. Wiley-Blackwell. ISBN: 978-1-118-56831-6

Lizardi-Jiménez MA, Hernández-Martínez R (2017) Solid state fermentation (SSF): diversity of applications to valorize waste and biomass. 3Biotech 7(44)

Maache-Rezzoug Z, Maugard T, Zarguil I, Bezzine E, El Marzoukid M-N, Loisel C (2009) Effect of instantaneous controlled pressure drop (DIC) on physicochemical properties of wheat, waxy and standard maize starches. J Cereal Sci 49(3):346–353

Martin SD, Ramos S, Zuffa J (2016) Valorisation of food waste to produce new raw materials for animal feed. Food Chem 198:68–74

Matharu AS, de Melo EM, Houghton JA (2016) Opportunity for high value-added chemicals from food supply chain wastes. Bioresour Technol J 215:123–130

Mohanakrishna G, Venkata Mohan S (2013) Multiple process integrations for broad perspective analysis of fermentative H2 production from wastewater treatment:technical and environmental considerations. Appl Energy J 107:244–254

Motevali A, Minaei S, Khoshtagaza MH (2011) Evaluation of energy consumption in different drying methods. Energy Convers Manag 52(2):1192–1199

Mousavi SA, Faraji M, Janjani H (2017) Recycling of three different types of rural wastes employing vermicomposting technology by Eisenia fetida at low temperature. Global NEST J 19(4). https://doi.org/10.30955/gnj.002259

Myer RO, Brendemuhl JH, Johnson DD (1999) Evaluation of dehydrated restaurant food waste products as feedstuffs for finishing pigs. J Anim Sci 77(3):685–692

Naik SN, Goud VV, Rout PK, Dalai AK (2010) Production of first and second generation biofuels: a comprehensive review. Renew Sust Energ Rev 14(2):578–597

Navajas A, Issariyakul T, Arzamendi G, Gandìa LM, Dalai AK (2013) Development of eggshell derived catalyst for transesterification of used cooking oil for biodiesel production. Asia-Pacific J of Chemical Eng. 8(5): 742-748. https://doi.org/10.1002/apj.1715.

Naylor RL, Liska AJ, Burke MB, Falcon WP, Gaskell JC, Rozelle SD, Cassman KG (2010) The ripple effect: biofuels, food and security and the environment. Environ Sci Policy Sustain Dev 49(9):30–43

Ohio State University (2006) Ohio livestock manure management guide. Copyright © 2006, Ohio State University Extension, TDD No. 800-589-8292.

Oliveira AC, Rosa MF (2006) Enzymatic transesterification of sunflower oil in an aqueous-oil biphasic system. J Am Oil Chem Soc 83(1):21–25

Ong KL, Kaura G, Pensupa N, Uisan K, Ki Lin CS (2018) Trends in food waste valorization for the production of chemicals, materials and fuels: case study South and Southeast Asia. Bioresour Technol J 248(A):100–112

Panesar PS, Marwaha SS (2017) Biotechnology in agriculture and food processing: opportunities and challenges. CRC Press. ISBN 9781138073265

Peters D, Toop G, van den Bos A, Spöttle M (2013) Assessing the EC ILUC proposal Dutch national ILUC Impact Assessment. ECOFYS Netherlands B.V., Utrech. Project number: B1ENL13265

Pimentel D, Marklein A, Toth MA, Karpoff MN, Paul GS, Mccormack R, Kyriazis J, Krueger T (2009) Food versus biofuels: environmental and economic costs. Hum Ecol J 37(1)

Saer A, Lansing S, Davitt NH, Graves RE, Graves RE (2013) Life cycle assessment of a food waste composting system: environmental impact hotspots. J Clean Prod 52:234–244

Sarkar O, Amradi NK, Dahiya S, Vamshi KV, Yeruva DK, Mohan SV (2016) Regulation of acidogenic metabolism towards enhanced short chain fatty acid biosynthesis from waste: metagenomic profiling. RSC Adv J 6(22):18641–18653

Shady HS, Gwilym A, Amit K (2019) Moving towards the second generation of lignocellulosic biorefineries in the EU: drivers, challenges, and opportunities. Renew Sust Energ Rev 101(C):590–599

Sharma SB, Sayyed RZ, Trivedi MH, Gobi TA (2013) Phosphate solubilizing microbes: sustainable approach for managing phosphorus deficiency in agricultural soils. Springer Plus 2:587

Shelley M (2006) Alcoholic fuels. CRC Taylor and Francis Group. ISBN-0-8493-3944-8

Soto I, Achten WMJ, Muys B, Mathijs E (2015) Who benefit from energy policy incentives? The case of Jatropha adoption by smallholders in Mexico. Energy Policy J 79:37–47

Tournaki S, Tsoustsos T (2013) Promotion of used cooking oil recycling for sustainable biodiesel production. www.recoilproject.eu: EU Project RECOIL, IEE/11/091/SI2.616369

Tsoutsos TD, Tournaki S, Paraíba O, Kaminaris SD (2016) The used cooking oil-to-biodiesel chain in Europe assessment of best practices and environmental performance. Renew Sustain Energy Rev 54:74–83

Van Eijck J, Romijn HA, Smeets E, Bailis R, Rooijakkers M, Hooijkaas N, Verweij PA, Faaij APC (2013) Comparative analysis of key socio-economic and environmental impacts and smallholder and plantation based Jathropa biofuel production systems in Tanzania

Vega-Gálvez A, Di Scala K, Rodríguez K, Lemus-Mondaca R, Mirandaa M, López J, Perez-Won M (2009) Effect of air-drying temperature on physico-chemical properties, antioxidant capacity, colour and total phenolic content of red pepper (Capsicum annuum, L. var. Hungarian). Food Chem J 117(4):647–653

Venkata Mohan S (2009) Harnessing of biohydrogen from wastewater treatment using mixed fermentative consortia: process evaluation towards optimization. Int J Hydrog Energy 34(17):7460–7474

Venkata Mohan S, Lalit BV, Sarma PN (2008) Effect of various pretreatment methods on anaerobic mixed microflora to enhance biohydrogen production utilizing dairy wastewater as substrate. Bioresour Technol J 99(1):59–67

Westendorf ML (2000) Food waste as animal feed: an introduction. Wiley-Blackwell. ISBN: 978-0-813-82540-3

Wu L-J, Kobayashi T, Kuramochi H, Li Y-Y, Xu K-Q (2016) Improved biogas production from food waste by co-digestion with de-oiled grease trap waste. Bioresour Technol J 44. https://doi.org/10.1016/j.biortech.2015.11.061

Yan J, Yan Y, Liu S, Hu J, Wang G (2011) Preparation of cross-linked lipase-coated micro-crystals for biodiesel production from waste cooking oil. Bioresour Technol J 102(7):4755–4758

Yin J-Z, Xiao M, Wang A-Q, Xiuc Z-L (2008) Synthesis of biodiesel from soybean oil by coupling catalysis with subcritical methanol. Energy Convers Manag J 49(12):3512–3516

Zbicinski I (2002) Equipment, technology, perspectives and modeling of pulse combustion drying. Chem Eng J 86:33–46

Part II
Innovative Food Waste Upcycling

Chapter 4
Biocascading: General Strategy for the Recovery of Valuable Substances from Food Waste

Teresa Cecchi

4.1 Introduction

Food waste (FW) is a precious source of organic carbon which can be utilized for the extraction and synthesis of several high value chemicals and represents a huge opportunity for eco-friendly green processes. The conversion of FW to value added products is a hot research field. Technological hindrance is obviously due to the heterogeneous nature of the waste and its high moisture content. Anyhow, fine-tuning of the available strategies can pave the way to proper and economically viable management of FW.

Green extraction of target compounds should be characterized by reduced energy demands, the use of reusable and nontoxic solvents, and environmentally benign physico-chemical technologies. These issues represent crucial prerequisites toward sustainable industrial production.

According to the waste hierarchy concept detailed in Chap. 1 (2008/98/EC 2008), waste prevention is mandatory, and this is particularly true for FW, because its management may lead to several health hazards as well as environmental issues. Generally, FW contributes almost half of the total municipal wastes. Organic components of food wastes include fruits, vegetables, cooked food wastes, meat etc.

According to the US Environmental Protection Agency there is a hierarchy for FW recovery to prevent environmental impact (https://www.flickr.com/photos/usepagov/3678814719).

Different types of waste management can be adopted according to the properties of the specific FW. The current legislation on FW treatment prioritizes the prevention of waste generation and least emphasizes disposal. Landfill and or incineration are obviously the worst options because their environmental impact is the highest. Regardless of the existence of different traditional methods, detailed in Chap. 3, used to harness FW energy, effective conversion of FW to valuable resources is

T. Cecchi, C. De Carolis, *Biobased products from food sector waste*,
https://doi.org/10.1007/978-3-030-63436-0_4

challenged by its heterogeneous nature and high moisture content. Production of animal feed, anaerobic digestion to get biogas, and composting to get bio-fertilizers are well-established industrial practices and represent the current state of the art of FW valorization within the waste to wealth strategy, currently practiced in most industrialized countries. North America and Europe have efficient operational FW-to-energy plants where waste management results in the energy supply of the country. Though composting is an effective method for bioconversion of FW, impurities in the compost and gaseous emissions of nitrous oxide and methane, which have a higher impact on global warming compared to carbon dioxide, are challenging (Nasini et al. 2016); anyhow, similar waste management strategies are lacking in most of developing countries (Sharholy et al. 2008).

The real challenge is actually the recovery of important chemicals from FW, before being treated according to its current valorization strategies. FW is mainly composed of the organic fraction that includes carbohydrates, proteins, fats, lipids as well as inorganic components. Macromolecules can be further digested into simpler organic compounds viz., glucose, amino acids, fatty acids (Dahiya and Joseph 2015), etc. A common strategy cannot be adopted for all kinds of FW. The biorefinery concept, described in Chap. 1, is one of the major stories of the last decade with an ever-increasing number of scientific publications, with most studies focusing on the development of cost-effective processing methods for biorefinery operations (Bauer et al. 2017). The global biorefinery products market is predicted to reach US$1128 billion by 2022 with the United States, Brazil, EU, and Canada as key players in this sector (Hassan et al. 2019).

Biorefining is the biobased analog to petroleum refineries. Typically, biorefineries outputs have been energy and biofuels, but integrated biorefining that includes biobased products output is the preferred valorization approach within circular economy models (de Jong et al. 2012).

In Chap. 1, we classified biorefineries according to the raw material used as feedstock in first, second, third, and fourth generation biorefineries. FW can be used in second generation biorefineries which are instrumental for advocating circular bioeconomy (Venkata Mohan and Sarkar 2017) and can deploy strategic waste conversion to economically beneficial product recovery. The diversion of FW as a biorefinery feedstock has the potential to produce a gamut of biobased products, simultaneously reducing the carbon footprints according to the cascading concept (Annevelink et al. 2017) (European Commission 2015) applied to biomass (biocascading). Cascading is a relevant strategy to generate valuable substances from products that supposedly reached the end of their life-cycle, thereby prolonging value chains. Cascading is exemplified by a cascading pyramid illustrated in Fig. 4.1. It is actually the core concept of biorefining: biomass, including that coming from FW, can be conceived as the raw material of many consecutive processing operations, where quality products derive after all and fit seamlessly into the carbon cycle. Cascading is a design tool; it prescribes that biomass is used sequentially as often as possible and finally for energy, including that necessary to fuel the process. In a cascading approach, the most valuable components are extracted from biomass first, in order to be utilized in complex chemical processes, such as the production of

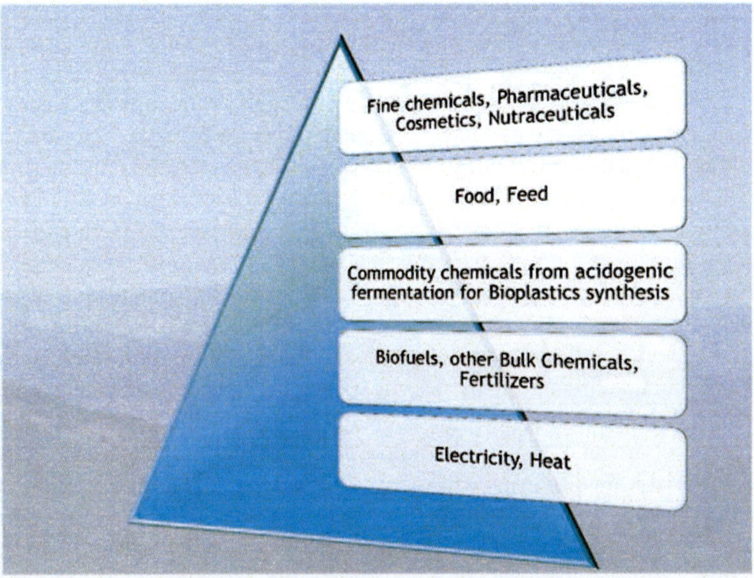

Fig. 4.1 Cascading Pyramid (Annevelink et al. 2017) (European Commission 2015), (Girotto et al. 2017)

medicines or chemicals. Although the dominant biorefining outputs remain energy and fuels, product-driven biorefining is an expanding business. The use of valuable biomass components for electricity or heat production would be a waste. Production of bulk chemicals from biomass waste was demonstrated to be 3.5 times more profitable than converting it into biofuel (Tuck et al. 2012). Focusing on the production of a single product from food and kitchen waste does not compare, economically, to the conversion of heterogeneous FW to multiple value added products (Dahiya et al. 2018). Figure 4.1 underscores a hierarchical, smart, and efficient use of biomass, to value-adding applications whose revenue can increase three orders of magnitude from the base to the top of the pyramid (European Commission 2015).

In the following, we will discuss the technologies needed by second generation biorefineries that make use of FW as feedstock. FW is interesting as potential feedstock because it would not trigger concerns in terms of land use for non-food purposes.

Physical, chemical, biochemical, and thermochemical conversion methods are instrumental technologies that can also be integrated to make the most of the biomass cascade, shown in Fig. 4.1, according to the product value hierarchy.

The operational cost and the value of the output target products are the two main factors that determine whether FW conversion process is sustainably feasible. In the following, we will discuss the current trends and recent developments of technology in the conversion of FW into outputs different from biofuels, biofertilizers, and animal feed.

4.2 Chemical Methods

Within the framework of a general recovery strategy of valuable substances from FW, chemical methods are particularly eligible for the recovery of specific chemical building blocks, nutraceuticals, bioactive substances, cosmetics and, pharmaceuticals. It is clear that FW is a precious raw material for the production of a wide range of high-added value chemicals and materials (Esteban and Ladero 2018).

In the following, we will discuss classical and smarter current conversion strategies that capitalize on the principles of analytical chemistry. The aims of the whole procedure are (i) target compound extraction, purification from impurities and toxic compounds, and isolation (ii) yield maximization, (iii) prevention of deterioration and loss of functionality during processing, (iv) use of a healthy process to maintain the food-grade nature of the final product without affecting the functional properties of the target compound, (v) concentration of the target compound, and, possibly, development of a final food-grade product that meets the consumers' high-quality standards for safety and organoleptic characteristics.

Processing often progresses from the macroscopic pretreatment to the separation of macro and micromolecules and afterward to the extraction of the target compound, followed by the purification and, eventually, encapsulation of the high-added value compound.

Depending on the characteristics of the FW matrix and on the kind of the target compound to be extracted, the techniques involved are known to vary.

Some of the initial FW treatment steps such as pretreatment, separation of small molecules from macromolecules, and even extraction can be skipped in case a single step technology is used for the isolation of the target compound. Moreover, it has to be emphasized that a specific technology can be used for different purposes during the whole FW treatment (for example, drying can be necessary as a pretreatment step but also for the final product formation).

4.2.1 *Pretreatment*

Pretreatment is a major initial step; it aims at adjusting the FW matrix according to the water content, enzymatic activity, and permeability of the bioresource tissues to the extracting solvent.

This initial step is also related to the protection from microbial growth and control of the enzymatic activity, which in turn avoids loss of functionality down the processing chain and increases the stability of the final product.

The pretreatment strategy depends on the nature and structure of the substrate (i.e., solid, sludge, or wastewater).

4.2.1.1 Solid Samples

If the substrate is solid (e.g., fruit or vegetable by-product) a size reduction step is often necessary. Hammer mills are impact mills used to fracture hard materials. Knives mills are preferred for soft materials, such as fruit and vegetable waste. To avoid friction heating, in cryogenic milling, solid carbon dioxide or liquid nitrogen is mixed with the feed. Since the majority of food wastes have a high moisture level, wet milling step is particularly useful: it mechanically disrupts the waste matrix and improves the yield of the subsequent separation and extraction stages, increasing the diffusion of extractants inside the food matrix (Oreopoulou and Russ 2007). This is often accomplished via swelling and tissue softening. For solid lignocellulosic biomasses, a disruption of the intricate supramolecular structure is a crucial first step in making carbohydrates more available for the conversion of the different components into multiple platform molecules, as detailed in Chap. 5 (Seidl and Goulart 2016). An assortment of approaches has been investigated to disarray the layers and disintegrate lignins, cellulose, and hemicellulose so that the polymers are converted into smaller fragments prior to enzymatic hydrolysis or other treatments (Fig. 4.2). Each pretreatment method has its own pros and cons. Mechanical breakdown of the lignocellulosic biomass, to reduce the particle size can be obtained not only via milling but also via grinding, chipping, and ultrasound treatment. Explosion physico-chemical methods make use of steam and hot water, ammonia, CO_2, and SO_2; in some cases, the efficiency of thermochemical conversion of lignin may be compromised. Chemical methods rely on acid hydrolysis, aqueous-organic solvent pretreatment, and oxidative delignification with ozone and hydrogen peroxide; the process is fast, but toxic substances might be involved in the process. Biological pretreatment makes use of bacterial and fungal monocoltures, mixed co-coltures, and genetically engineered bacteria and fungi. The microbial consortia secrete lig-

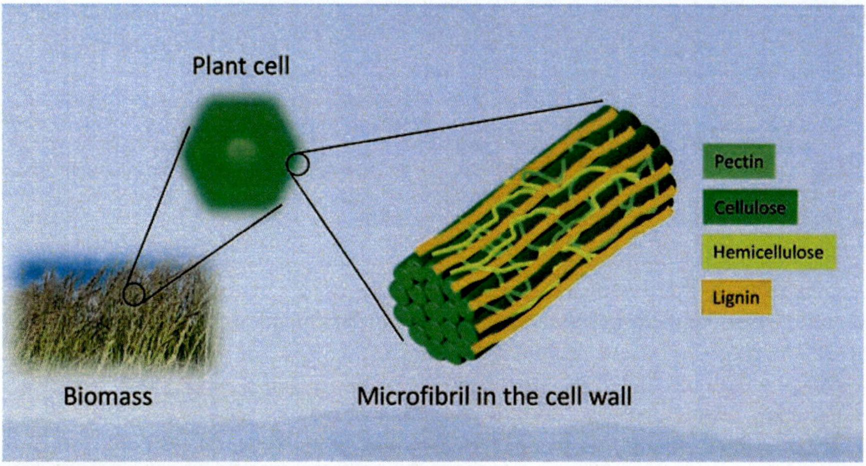

Fig. 4.2 Structure of the plant cell wall microfibril

ninolytic enzymes, which depolymerize lignin and open up the cell wall, enabling cellulolytic and hemicellulolytic microorganisms to hydrolyze, respectively, cellulose and hemicellulose into monomeric sugars. The biological pretreatment is slower than the chemical one, but it is a cost effective and environmentally friendly technique. Combinations of methods (e.g., fungal and physicochemical) were also explored (Sharma et al. 2019, Hassan et al. 2019).

4.2.1.2 Watery Food Waste or Food Wastewaters

In case of watery food waste or food wastewaters, liquid/solid, or liquid/liquid separations are needed.

Dewatering and concentration of valuable components in the liquid phase is mandatory for reduction of transport, storage and disposal expenditures as well as for energy savings in subsequent treatments. Moreover, reduced water activity results in improved microbiological stability of the processed products. The amount of water in FW varies and can be bound in the biomass with increasing binding forces from capillary sorption and adhesion to chemisorption.

Removal of water is commonly accomplished by thermal evaporation. On the other hand, part of the water from wet wastes can be removed by less expensive mechanical dewatering processes, such as centrifugation, filtration, expression, osmotic dehydration, and reverse osmosis, where liquid water is squeezed out of the sample, or by other emerging technologies.

Mechanical separative methods are widely used, but it has to be taken into account that bound water is not even removable by the maximum mechanical strain (Galanakis 2012). Subtraction of solids facilitates the substrate flow, mixing, and homogenization and reduces the presence of fats susceptible to autoxidation. Mechanical dewatering procedures avoid thermal-induced loss of functionality.

Centrifugation is useful to remove solids from liquids since the centrifugal force separates particles from a solution according to their size and density, rotor speed (usually expressed as revolutions per minute and related to the centrifugal acceleration), and viscosity of the medium. Centrifugation is also advantageous to separate immiscible phases with different densities. It is used, for example, to remove fat from olive mill wastewaters (Galanakis et al. 2010a), to recover protein, chitin, carotenoids, and glycosaminoglycans from shrimp processing waste (Cahú et al. 2012).

Pressure-driven operations include expression and filtration.

Mechanical expression or pressing is widely used for juice recovery and oil expelling. The applied pressure results in the release of liquid from disrupted cells of the matrix. It was used for the separation of onion waste into juice and paste, (Roldán et al. 2008), the expression of oil from grape seeds (Fernández et al. 2010), or pumpkin seeds (Jarienė et al. 2008).

Filtration, a pressure-driven membrane operation, includes microfiltration (MF), ultrafiltration (UF), nanofiltration (NF), and reverse osmosis (RO). Figure 4.3 illustrates, for each pressure-driven process, which permeates passes through the membrane.

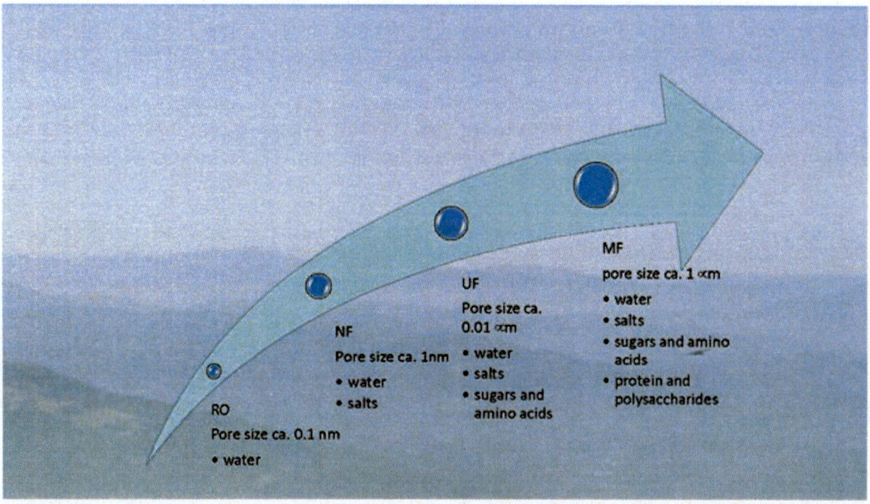

Fig. 4.3 Permeates in membrane filtration technology according to the pore size

Filtration operations are gaining momentum in FW processing. These processes are based on the application of a specific hydrostatic pressure on a fluid transported through the membrane. The feed solution is then transformed into a permeate containing all compounds that have permeated the membrane and a retentate comprising all components retained by the membrane. Pressure-driven membrane operations are used mostly in a sequential form. MF is particularly used in the pretreatment step, and it makes use of membranes characterized by pore sizes around 1 μm which are able to retain bacteria, micelles, and suspended compounds. All filtration operations are particularly useful for whey processing, as illustrated in Chap. 2, including the separation of the precipitate resulting from the removal of lipids by calcium-binding by heat treatment and/or pH adjustment (thermocalcic precipitation) (Quemerais and Daufin 1995). These technologies do not affect the functional properties of proteins due to the low applied temperatures (Lipnizki 2010). Oil droplets and fats in spent wastewaters derived from (i) margarine production (Chmiel et al. 2003), (ii) a fish meal factory (Afonso and Bórquez 2002), or (iii) olive mill (Rahmanian et al. 2014) can be similarly recovered by MF.

Unfortunately, colloidal fluids are very difficult to dewater efficiently by these mechanical dewatering methods (Yoshida 1993). Electro-osmotic drying, also called electro-dewatering or electrically-assisted mechanical dewatering, combines mechanical pressure with electrostatic effects induced by electrochemical double layers, which are formed at the particle-water interface of colloidal aqueous suspensions. It is advisable to lower the electrical contact resistance between the electrode and the colloidal fluid bed. The principal benefit of this technique is that it does not overdry the surface of the substrate. However, its selectivity towards different colloidal components might not be satisfactory. The method was tested for the dewatering of a range of food byproducts (brewer's spent grain, cauliflower trimmings, mango peel, orange peel, and melon peel) (Ng et al. 2011) and for highly viscous wastes having various kinds of colloidal components like tomato pomace (Jumah et al. 2005).

Thermal processing is, however, the most common method to induce the evaporation of water for its subsequent removal (Ma et al. 2018). It provides heat transfer for supplying the latent heat of evaporation, and for this reason, it is more energy-intensive than mechanical drying. For FW concentration, vaporization may be obtained using single-stage or multiple effect evaporators. If the liquid stream contains heat-sensitive compounds, it may be necessary to reduce the boiling temperature by operating under vacuum conditions. The vacuum operations increase the cost of the process. Thermal processing provides microbial pasteurization but also causes activation or de-activation of key enzymes, which subsequently affect the yield and the quality of target compounds (Galanakis et al. 2010b), (Galanakis 2012). Thermal air-drying at high temperatures has been applied in household FW for important mass reduction (87% w/w) and hygienic moisture removal (Sotiropoulos et al. 2016). Thermal drying was also useful in the treatment of exhausted sugar beet pulp, from which ferulic acid can be recovered for subsequent transformation into vanillin (Schieber et al. 2001), and for eliminating pathogens while removing water from animal food residues (Lin et al. 2013).

Thermal drying suffers from nonuniform heating patterns due to limited heat and mass transfer in relation to the size of the product to be dried: the slowest heating or drying regions are found in the thermal center of the treated product. This fact leads to overheating of surface regions. This drawback could be overcome using dielectric drying; this process makes use of the electromagnetic energy associated whit radio-frequency or nonionizing microwave radiations, to obtain rapid and volumetric heating, deep penetration, moisture self-balance effects, and the evaporation of liquids, removed by a carrier gas. These radiations cause migration of ions and rotation of dipoles. The dielectric properties of the materials have a significant impact on the transformation of electromagnetic into thermal energy, which in turn results in the evaporation of the liquid. Dielectric drying is not common in the field of FW, it was explored for a starch-water model system (Jiang et al. 2016), but it could bridge the gap between academia and industry also in this field.

Freeze drying, also known as lyophilization, aims at water removal to preserve perishable materials and/or to reduce transport costs. The process involves freezing the material below the triple point of water, lowering pressure, then removing the ice adding heat to allow the frozen water in the material to sublimate. This technique avoids enzyme induced alteration, and it has a minimum impact on sensitive compounds, colors, and textures of food-related materials, which can be dried to 1–5% residual moisture and subsequently rehydrated; unfortunately, it is energy demanding due to the presence of vacuum conditions (Dennis et al. 2009). It follows that pretreatment by freeze drying can be an alternative for thermal drying when the high operational cost is justified by the importance of the target compounds. This is the case of tomato peel processing for lycopene recovery (Sarkar and Kaul 2014), red wine grape skins drying with retention of volatile compounds (terpenes, sesquiterpenes, norisoprenoids, C6 alcohols) without increasing the number of browning derivatives (de Torres et al. 2010), and pretreatment of lemon byproducts (Masmoudi et al. 2008) and goldenberry pomace (Ramadan et al. 2008).

An innovative macroscopic pretreatment capitalizes on an old technology (developed in the early 70s), that is foam-mat drying. It has been lately considered attractive because it enables the conversion of a liquid or semiliquid FW to a stable foam using inert gases, foaming agents as well as a rapid supply of hot air. The foamed material usually has a density of almost 0 to 800 kg/m^3 depending on the degree of gas bubbles incorporated into the matrix (Ratti and Kudra 2006) and it should be mechanically and thermally stable so as not to collapse throughout treatment for at least 1 hour. Egg white is an excellent choice of food-grade foaming agent. This process offers an effective protection against deteriorative microbial, chemical and biochemical reactions. Xanthan gum or methyl cellulose are common foam stabilizers. Moreover, lower temperatures and shorter drying time are particularly precious for heat-sensitive and viscous substrates such as labile antioxidants in high-sugar fruit byproducts (Rajkumar et al. 2007). Low drying throughput and some losses in aromatic components and thermolabile compounds are likely to occur (Marques et al. 2006). Foam-mat-dried products would show good reconstitution properties due to their spongy nature.

A chemical pretreatment is sometimes necessary to free the target components from specific chemical bonds. The chemistry of this pretreatment step is obviously dependent on the specific target molecule and matrix. For example, the recapture of phytosterols in raw materials necessitates a hydrolysis step under high pressure and temperature or saponification with an alkaline solution (Fernandes and Cabral 2007). Using enzymes like cellulase, α-amylase, and pectinase (possibly obtained from FW, as explained in Chap. 5) prior to extraction in order to break the cell walls and hydrolyze the structural polysaccharides makes the FW matrix more accessible to the solvents, thereby enhancing the target compound yield and decreasing the extraction time (Nayak and Bhushan 2019).

A microbial inactivation pretreatment via more sophisticated technologies is needed if a non-thermal microbial or enzymatic inactivation is mandatory (Wan et al. 2009). Low-temperature ("cold") plasma is produced by electrical discharges under various vacuum levels; their antimicrobial efficacy stems from the myriad of the generated reactive species, such as reactive oxygen species, UV radiation, energetic ions, and charged particles. Again, this promising technology still has to be applied to FW treatment, but its usefulness may be impaired if the substrate is oxidizable.

4.2.2 Separation of Small Molecules from Macromolecules

Separation of small molecules (i.e., antioxidants, acids, or ions) from macromolecules is often necessary as a preextraction step. Macromolecules include proteins, hydrocolloids, and dietary fiber resistant to digestion by human alimentary enzymes (soluble such as pectin and β-glucans or insoluble such as lignin, cellulose, and hemicellulose). The armory for tackling this kind of separation relies on both classical and emerging technologies.

4.2.2.1 Alcohol Precipitation

Alcohol precipitation is a very popular separative method because it is easy, cheap, and nontoxic; macromolecules are collected in the alcohol insoluble residue possibly free from low molecular weight compounds (Galanakis 2012) (Koubala et al. 2008). Unfortunately, it is not always selective since complexes between the smaller and the larger molecules can impair the quantitative recovery of the former in the presence of the latter.

Pectin can be recovered, via the addition of ethanol as a precipitant, from raw papaya peel (Boonrod et al. 2006), from citrus waste (Bafrani 2010), from olive mill wastewater (Galanakis et al. 2010c) and during the extraction of polyphenolics from Mango peel (Berardini et al. 2005).

Alcohol precipitation (eventually in combination with further purification steps) was employed to recover a pharmaceutical compound (i.e., chondroitin sulfate) from scapular cartilage of Shortfin mako shark (Kim et al. 2012), waste of *Scyliorhinus canicula*, (Gargiulo et al. 2009), and skate cartilage (Murado et al. 2010).

Recovery of bromelain, which is a group of proteolytic enzymes from pineapple (*Ananas comosus*) waste, was obtained using ethanol precipitation at low temperature (da Silva et al. 2010). These enzymes have important anti-inflammatory, antithrombotic, and fibrinolytic properties and the low-temperature precipitation resulted in the recovery of 98% of the total enzymatic activity (Martins et al. 2014).

4.2.2.2 Isoelectric Precipitation

Isoelectric precipitation of proteins is due to protein-protein hydrophobic attractions stronger than protein-water electrostatic affinity when the medium pH is close to the isoelectric point (Bourtoom et al. 2009). Conversely, when the pH value is far from the isoelectric point, protein solubilization occurs. This procedure enables the separation of protein from muscular lipids or food byproducts such as bones, and skins and it is particularly exploited for fish byproducts. Typically, the homogenization of byproducts in water is followed by protein solubilization under acidic or alkaline conditions. A centrifugation step enables the recapture of crude fish oil (lightest fraction) that is rich in ω-3 fatty acids, and that can be further processed for nutraceutical applications. The heaviest fraction (bones, scale, skin) rich in minerals is useful as animal feed. The pH of the medium fraction is adjusted to the medium isoelectric point of the protein of interest, and the method was used for fish (Gehring et al. 2011), (Tahergorabi et al. 2011a), (Tahergorabi and Jaczynski 2014), (Chen and Jaczynski 2007), (Chen et al. 2007) and chicken byproducts (Tahergorabi et al. 2011b).

4.2.2.3 Ultrafiltration

Among pressure-driven membrane operation, ultrafiltration is the technology of choice to recapture protein as retentate, on the basis of the pore size detailed in Fig. 4.3. Water, salts, and small molecules, such as sugars and amino acids, permeate the membrane. Watery FWs are good candidates for ultrafiltration. Proteins from whey (Barba et al. 2001) (see Chap. 2), fishery effluents (Afonso and Bórquez 2002), soybean processing wastewaters (Jiang and Wang 2013), and poultry processing wastewater (Lo et al. 2005) were recaptured by ultrafiltration. This technology was also used, often with a sequential nanofiltration, to recapture polyphenols from wine industry waste (Galanakis et al. 2013), (Santamaría et al. 2002), artichoke wastewaters (Conidi et al. 2014), or olive mill wastewater (Cassano et al. 2013), (Rahmanian et al. 2014) (Galanakis et al. 2010d), as well as to recover pectin from olive mill wastewater (Galanakis et al. 2010b). The application of sequential membranes could eventually increase target compound selectivity. Membrane procedures are efficient and easy-going, but are fraught with potential fouling problems that restrict their applications.

4.2.2.4 Gas Aphrons

Among emerging technologies colloidal gas aphrons have raised scientific interest. They are surfactant stabilized microbubbles (10–100 μm) produced by stirring of a surfactant solution at high speeds (i.e., >5500 rpm). These microfoams have an increased interfacial area and higher stability compared to conventional foams. According to the nature of surfactant (i.e., cationic, anionic, or non-ionic) the external surface of the micro-bubble may be charged and able to adsorb reversely charged target compounds, resulting in their selective separation from the bulk liquid without mechanical aid.

Interestingly, after contacting and adsorbing the target compound, the aphrons separate in a few seconds from the bulk liquid as a top phase, and then they are allowed to collapse. They provide an improvement to conventional flotation techniques. However, the main drawback of this separation method is the presence of surfactant in the product stream; biodegradable and nontoxic surfactants are mandatory if the final product is meant for human consumption; in this way, the removal of the surfactant can be avoided.

The separation of proteins from whey using colloidal gas aphrons generated with an anionic surfactant (dioctyl sulfosuccinate, sodium salt, or Aerosol OT) has been investigated (Fuda and Jauregi 2006). At low pH, ionic interactions between the negatively charged surfactant and the charged proteins resulted in the selective separation of lactoferrin and lactoperoxidase. In fact, ionic surfactants act as ion exchangers, and hence selectivity can be manipulated by the type of surfactant (cationic or anionic), pH, and ionic strength of the solution. Colloidal gas aphrons were also used to separate phenolic compounds from nonphenolic compounds (such as sugars and minerals), but also to fractionate different classes of phenolic compounds that can be obtained from wine-making waste (Löf et al. 2011), (Spigno et al. 2015), (Dahmoune et al. 2013). Gas aphrons have also been applied to clarify a palm oil mill effluent (Subramaniam et al. 2007) and to recover proteins such as lysozyme and β-casein from food processing plants wastewaters (Jarudilokkul et al. 2004).

4.2.2.5 Ultrasound-Assisted Crystallization

Ultrasound-assisted crystallization is a technique that influences both nucleation and crystal growth (Luque de Castro and Priego-Capote 2007). Ultrasound is known to generate cavitation due to sudden changes of pressure in the bulk liquid; where the pressure is low, the development of small vapor-filled cavities is enabled. When these small bubbles dissolved in the liquid being sonicated, collapse, they locally produce very high pressure (100 MPa) and temperature (5000 K) (Deora et al. 2013); this condition, in turn, results in supersaturation, leading to spontaneous nucleation. Shockwaves facilitate the solute transfer process to the crystal face, thereby enhancing the growth rate (Bund and Pandit 2007). Sonocrystallization provides uniform primary nucleation that results in the production of smaller, purer, and uniform crystals; hence it can be useful for the separation of small molecules from larger ones. Ultrasound-assisted crystallization has been used to separate and recover lactose from whey and for proteins segregation (Galanakis 2012) without seeding (addition of preformed lactose fibrils) and with impressive lactose recovery yield (ca. 90%) (Patel and Murthy 2010) (Sánchez-García et al. 2018).

4.2.2.6 Extrusion

Extrusion is a technique commonly used to convert raw food material into a series of dies; its ability to modify the physicochemical properties of macromolecules and micromolecules may be useful for their separation in FW. The shear energy exerted by a rotating screw on the feed materials, heated to their plasticizing or melting point by the barrel, can be exploited for the macromolecule recovery; for example, this approach increased the solubility of the cell wall pectic polymers and hemicelluloses in outer scale leaves of onion waste which is a potential source of dietary fibers, especially rich in pectic polymers, with little starch, for fiber-enriched foodstuffs (Ng et al. 1999).

4.2.3 *Extraction*

Extraction is the crucial stage of downstream processing. The physicochemical characteristics of the extractants (i.e., polarity, solubility, or volatility) influence the choice of the extraction technology and the procedure parameters. Classical extraction procedures are still largely employed. Emerging extraction technologies, such as microwave-assisted extraction (MAE), ultrasound-assisted extraction (UAE), and pulsed electric field (PEF) extraction, among others, are the focus of interest of academia and industry. Their advantages include shorter extraction time, improved selectivity, energy efficiency, reduced thermal decomposition of thermo-labile components, and their inherent "green" nature as regards the use of solvents (and their removal from the extracts) compliant with the green chemistry standards set by Environmental Protection Agency, USA, detailed in Chap. 10.

In the following, we will detail the available technologies from the oldest to the newest approaches.

4.2.3.1 Solvent Extraction

Solvent extraction is the oldest approach to the extraction step. The extractants are selectively transferred from the aqueous FW matrix to the solvent on the basis of their intrinsic polarity, their solubility, and chemical affinity to the solvent.

Judicious choice of the solvent and optimization of extraction parameters like temperature, time of contact, pH, solid to liquid ratio, particle size, stirring rate, etc. maximizes the yield of the target compound extraction. Classical solvent extraction is a time-consuming technique, and expensive solvents are needed; if toxic solvents are used, their further elimination from the extract is needed.

Phenols are easily solubilized in polar protic mediums like hydroalcoholic mixtures, while corresponding fractions can be obtained on the basis of polarity by varying alcohol concentrations (Tsakona et al. 2012). Ethanol is the most used solvent extractor because it is rather cheap, and it is recognized as safe according to American Food and Drug Administration. Apolar target compounds such as carotenoids and tomato lycopene, in particular, are more liposoluble and non-polar solvents are necessary for their extraction; classical apolar solvents, such as dichloromethane, hexane, chloroform, etc. must be removed completely from the extract for its food-grade use; ethyl lactate was successfully used and had the double advantage of being environmentally friendly and giving the highest carotenoid yield (243.00 mg/kg dry tomato waste) at 70 °C (Strati and Oreopoulou 2011). Sequential extraction steps increase the extract yield, but also the consumed time and cost of the process (Oreopoulou and Tzia 2007). Traditional Soxhlet methodology, extraction in a packed bed reactor operating under reflux of the solvent or in an agitated vessel, followed by filtration of the extract are classical solutions. Extraction under non-classical conditions is an enthusiastically developing area in applied research, and industrial scale-up is a hot topic.

4.2.3.2 Pressurized Liquid Extraction

Pressurized-liquid extraction (PLE) uses a pressurized liquid to achieve good extraction yields of bioactive compounds in a shorter time. For example, it was used for the extraction of antioxidants, anthocyanins, and proanthocyanidins from lingonberry (*Vaccinium vitis-idaea L.*) pomace (Kitrytė et al. 2020). The "pressurized liquid" can also be "subcritical water" (that is, pressurized hot water) at temperatures between 100 and 374 °C, kept at pressures higher than its vapor pressure, so that water is maintained in its liquid state, with reduced viscosity and dielectric constant. This way, water mimics a good solvent for medium and low-polarity substances, such as polyphenols, allowing for the extraction of compounds that are not typically water-soluble. The crucial advantage is the safety of water compared to organic solvents and the fast process. Moreover, although the temperature is relatively high for water-based operations, the extraction method is so fast that volatiles and thermally sensitive components are retained, at variance with steam distillation procedures. The temperature is the key parameter to optimize to get the highest recovery of polyphenols from wine grape pomace (Vergara-Salinas et al. 2015), (Pedras et al. 2017), (Duba et al. 2015). Subcritical water extraction is generally less expensive than supercritical carbon dioxide extraction because increasing the temperature of a system is cheaper than increasing its pressure.

4.2.3.3 Steam Distillation and Hydrodistillation

Steam distillation and hydrodistillation are used to enhance the extractability of volatile compounds (Farhat et al. 2011). In direct steam distillation, the waste matrix is supported on a perforated grid and is not in contact with the liquid; the heat transfer via the vapor phase induces milder treatment of the waste matrix to prevent overheating of the essential oils; in hydrodistillation the material (e.g., FW rich in essential oils) is completely submersed in boiling water, and the boiling point is always below 100 °C, because the resulting vapor pressure of the two immiscible liquids (i.e., water and essential oil) is the sum of the individual vapor pressures. These safe and easy techniques generate organic solvent-free products, but they are time-consuming and energy-intensive, hence, coupling hydrodistillation with microwave heating could reduce energy cost. Steam distillation and hydrodistillation are first choice methods to obtain essential oil (e.g., from citrus waste), which are used as natural flavorings and/or preservatives, in food, cosmetics, and pharmaceutical sectors, due to their antimicrobial, antioxidant, and anti-inflammatory properties (Farhat et al. 2011), (Ferhat et al. 2006), (Yang et al. 2009).

4.2.3.4 Supercritical Fluid Extraction

Supercritical fluid extraction (SFE) is a modern technique that involves the use of a fluid (i.e. CO_2) under pressure and temperature greater than the critical point; above the critical point, well defined and recognizable liquid and gas phases do not exist, and the fluid exhibits physicochemical properties intermediate between liquid and gas whilst being a single phase with a wide range of solvent properties, depending on its temperature and pressure. The choice of supercritical CO_2 is expedient due to its moderate critical conditions of temperature (31 °C) and pressure (73.8 MPa), nontoxic nature, and high chemical stability. Further, its penetration into the food matrix is faster due to its greater diffusion coefficient and lesser viscosity and surface tension, as compared to the use of typical organic solvent. It holds the promise to be environmentally benign. CO_2 is low-cost, uninflammable, inert, and easy to recycle. The density of the supercritical fluid is similar to that of liquids. Solubility increases with density, which in turn depends on pressure and temperature; hence supercritical fluids have a large absorption capacity. On the other side, the diffuse nature of the supercritical fluid, similar to that of gases allows for efficient extraction. A pressure release after the complete extraction enables easy recovery of target compounds: as the pressure decrease in the collection vessel, the supercritical CO_2 undergoes a phase change and becomes a gas again, negating the need for solvent removal through more conventional methods (McHugh and Krukonis 2013) and making SFE a green technology.

Food grade clean extracts, in yields comparable to those that can be achieved with conventional solvents and extraction techniques, are guaranteed by the release of CO_2 once the extraction procedure is over and the absence of organic solvent residues in the extract. Rapid extraction, easy concentration, and high selectivity are additional advantages. It is employed for difficult separation processes based on low quantity of valuable products such as flavorings and colorants from plant materials (Sowbhagya and Chitra 2010). SFE extractions from red grape pomace (Martinez et al. 2016) and white grape seeds (Da Porto and Natolino 2017) gave extracts with exceptionally high polyphenols content as gallic acid equivalents per 100 g of dry matter.

Table 4.1 details many recent applications of this green extraction procedure for the extraction of bioactives and other important compounds from a wide range of FW matrixes.

However, the high energy demand of SFE and higher capital costs still impair its wider applicability on the industrial level (Todd and Baroutian 2017).

4.2.3.5 Microwave-Assisted Fluid Extraction

Microwave-assisted extraction (MAE) is an environmentally benign extraction technique that is gaining momentum over the last years. Electromagnetic radiation in the frequency range from 300 MHz to 300 GHz is known as microwave. Microwave energy efficiently penetrates into the solid matrix with direct generation of heat via (i) resistance of the medium to the ionic migration and (ii) thermal agitation due to molecular dipole rotation and reorientation in the direction of the electric field inside the solid. Polar solvents and the sample moisture are easily heated due to the high dielectric constant; the increasing vapor pressure finally breaks the cell walls, thereby promoting the leaching of target compounds, the transfer of analytes from the sample matrix into the solvent, with short extraction time. For the extraction of wine grape pomace polyphenols, MAE with water as the solvent gave significantly higher yields, compared with the traditional Soxhlet methodology (Drosou et al. 2015). This easy-to-handle technique has recently been applied for

Table 4.1 Target compounds extracted by SFE from specific food waste

Food waste	Target compound	Reference
Grape peels	Fragrance compounds, anthocyanins	Ghafoor et al. (2010)
Blackberry (*Rubus sp.*) bagasse	Antioxidant compounds	(Pasquel Reátegui et al. 2014)
Blueberry (*Vaccinium myrtillus L.*) residues	Phenolic compounds and anthocyanins	Paes et al. (2014)
Raspberry pomace	Antioxidants	Kryževičiute et al. (2016)
Lingonberry (*Vaccinium vitis-idaea L.*) pomace	Linolenic and linoleic fatty acids	Kitrytė et al. (2020)
Waste citrus seeds	Edible fixed oils	Rosa et al. (2019)
Spent coffee grounds	Lipids	Ahangari and Sargolzaei (2013)
Brewery waste stream	Antioxidants	Barbosa-Pereira et al. (2013)
Olive mill wastewater	Hydroxytyrosol	Crea (2002)
Olive waste	Phenolic compounds	Caballero et al. (2020)
Feijoa (*Acca sellowiana (O. Berg) Burret*) peel	Phenolic compounds	Santos et al. (2019)
Brown onion skin by-product	Phenolic compounds	(Campone et al. 2018)
Mushroom waste	Ergosterol and vitamin d2	Papoutsis et al. (2020)
Rapeseed oil waste	Phytosterols and tocopherols	Jafarian Asl et al. (2020)
Tomato (*Lycopersicum esculentum L.*) peels by-product	Lycopene and β-carotene-	Kehili et al. (2017)
Filter-tea by-product (yarrow–rose hip mixtures)	Chlorophylls and carotenoids	Pavlić et al. (2016)
Papaya seeds	Bioactive compounds such as benzyl isothiocyanate	Hall et al. (2018)
Sea products waste	Lipids and pigments from algae, and oils from fish-processing plant waste streams	Kerton et al. (2013)

the extraction of pectin and phenols from apple pomace and potato byproducts (Oreopoulou and Tzia 2007), (Dugmore et al. 2017). Better results as regards the recovery of total phenols from berries pomace were obtained (Périno-Issartier et al. 2011) also because MAE is less harsh than conventional extraction methods. It was also used to recover triterpene saponins from yellow horn kernel (*Xanthoceras sorbifolia Bunge*) (Li et al. 2010), polyphenols from *Vitis vinifera* seeds (Casazza et al. 2010), (Dang et al. 2014), spent filter coffee (Pavlović et al. 2013), vegetable solid wastes (Baiano et al. 2014), total quercitin from yellow onion (*Allium cepa L.*) wastes (Zill-E-Huma et al. 2011), pectin from watermelon (*Citrullus lanatus*) fruit rinds (Prakash Maran et al. 2014) and passion fruit peels (Seixas et al. 2014).

Pressurized microwave-assisted extraction capitalizes on synergies between mechanical pressure and microwave-related heating in a sealed vessel containing the sample. The controlled pressure (below the working pressure of the vessel) and temperature (above the normal boiling point of the extraction solvent) allow fast and efficient extraction. In the solvent-free version of this technique, the vaporization of the in situ water results, again, in a pressure increase inside the cell and in the disruption of its wall, that favors the release of the target molecules (Li et al. 2013). Noteworthy, mass and heat transfers follow the same direction from inside the sample matrix to the outside, at variance with conventional separations.

This strategy has been proposed for the recapture of pectin from orange albedo and peels with a better recovery yield of pectin compared to conventional extractions (Fishman et al. 1999), or higher extraction rate (Liu et al. 2006). Pressurized microwave-extraction with superheated ethanol efficiently extracted phenols from olive mill wastewater (Galanakis 2012). The main limitations of pressurized processes are related to the possible degradation of thermolabile ingredients as well as to the demanding variables control associated with the operation.

4.2.3.6 Ultrasound-Assisted Extraction

The energy of ultrasound may be used either as a pretreatment step or in an effective solid–liquid extraction procedure that is called the ultrasound-assisted extraction (UAE). Ultrasound energy leads to cavitation phenomena which, in turn, result in effective micro-mixing. In ultrasonication, sound waves having frequencies higher than 20 kHz provide greater penetration of solvent into the waste matrix, improved mass transfer, and disruption of the cell walls. UAE is regarded as a fast and environmentally-friendly extraction technology, offering several advantages over other conventional and non-conventional procedures, such as lower cost, versatility, and ease of deployment (Tiwari 2015). UAE proved to be an effective and fast way to recover red grape pomace polyphenols (González-Centeno et al. 2015), (Bonfigli et al. 2017), (Caldas et al. 2018). The milder operational conditions of UAE compared to those of PLE may result in a less efficient extraction, but they are precious for sensitive polyphenols recovery (Paini et al. 2016). Ultrasound waves accelerate disruption of the plant cell walls and facilitate the release of extractable compounds; hence this approach is crucial for recovering target compounds from cellulosic byproducts, such as polyphenols from citrus peel, coconut shells, wheat bran (Chemat et al. 2011), and hemicelluloses from wheat straw (Sun and Tomkinson 2002).

4.2.3.7 Pulsed Electric Field Extraction

Pulsed electric field (PEF) extraction is a non-thermal approach of high potentiality since the electric field can induce pores development in the cell membrane (electroporation), its breakdown, and increased permeability and mass transfer. The application of PEF involves the discharge of electric pulses having high voltage for a few microseconds into the solid food matrix placed in between the two electrodes. Electric field exponential decay pulses and square wave pulses of ca.15–80 kV/cm result in a temperature elevation less than 3–5 °C, with a clear benefit for target compound preservation. Extraction of phenols from grape seeds, betalains pigments from red beetroot, and pectin from apple pomace (Liu et al. 2011) (Vorobiev and Lebovka 2010) relied on the application of the pulsed electric field. Besides, this technology was also used to deactivate enzymes (Wan et al. 2009). Other electrically induced extraction technologies include (i) pulsed ohmic heating, in which heat transfer accelerates by placing the sample between two electrodes and exposing it to a direct or alternative current, and (ii) high-voltage electrical discharges characterized by the use of short high voltage pulses. For solid FW, a dielectric liquid (i.e., tap water) has to be added into the chamber (Vorobiev and Lebovka 2010). This method has been assayed for the extraction of phenols from grape pomace and seeds (Boussetta et al. 2009).

4.2.3.8 Other Emerging Extraction Technologies

Eutectic (low-melting point) mixtures have been currently attracted attention as green extraction media (Płotka-Wasylka et al. 2017) for deep eutectic solvent (DES) extraction.

These low vapor pressure liquid mixtures can be formed by food-grade components, including organic acids (e.g., citric acid), polyols (e.g., glucose), and salts (e.g., choline chloride, sodium citrate). Polyphenol-enriched extracts produced with DES were obtained from red grape pomace, and lees (Mouratoglou et al. 2016) (Patsea et al. 2017) (Bosiljkov et al. 2017), and these extract were demonstrated to display superior bioactivities (Radošević et al. 2016).

Laser ablation is a solvent-free technique that intensifies heat and mass-exchange processes, accelerates chemical reactivity, induces photochemical changes, modifies macromolecules, and has an antimicrobial effect. Laser ablation is the result of focusing a pulsed laser beam onto a sample to subtract a small amount of its mass. The target material releases particles according to its thermo-optical properties and laser power density. If the laser flux is low, the material is heated by the adsorbed energy and evaporates or sublimates; on the converse, at high laser flux, it is converted to plasma. The photodynamic effect induced by laser irradiation has been utilized in order to extract active substances (aromas, anthocyanins, polysaccharides, and proteins) from biological matrixes. Laser ablation of citrus fruit peel led to an increase in the yield of pectin in the subsequent classical extraction with a solution of hydrochloric acid. Laser ablation of fruits and vegetables enables the production of edible films from the evaporated matter since peels are rich in waxes. It may find application in extracting aroma substances from the essential oils evaporated from the superficial oil glands (Panchev et al. 2011). Biocompatible coatings with strong adhesion from fish processing wastes were also produced by laser technology (Lusquiños et al. 2006).

For the smart extraction of volatiles, pervaporation (PV) is a promising semi-selective membrane-based separation process. In PV, a liquid stream undergoes selective evaporation through ("per") the membrane. The volatiles enriched vapor phase (permeate) is removed from the opposite side of the membrane. The remaining liquid is the retentate. If the feed phase is vapor, the technique is called vapor permeation (Néel 1995). The gradient in the chemical potential of each solute is maintained owing to the vacuum. PV is particularly useful for the safe extraction of aromas. It is a solventless and low-energy technique (Aroujalian and Raisi 2007), but if the feed contains suspended solids or dissolved salts, membrane fouling is the main drawback. PV was successfully used as a deodorization process of food industry effluents (i.e., cauliflower blanching water) with a concurrent valorization of its flavor compounds that have a particularly low detection threshold (about a few micrograms per cubic meter) (Souchon et al. 2002). Pervaporations finds application in the recovery of volatile fatty acids from anaerobic digestion of FW (Aghapour Aktij et al. 2020).

The use of membrane contactors for a membrane-mediated extraction is a promising alternative to established extraction methods. Membrane contactors are microporous hydrophobic membranes whose only role is to stabilize an interface between two or more fluids (usually including an aqueous phases), inhibit their mixing, and control the mass transfer between both phases. The separation is based on a gadient of the target compound concentration between the feed phase and the receptor (stripping) phase. The extraction of aromas from fruit, (San Román et al. 2010) and the recovery of volatile fatty acids, from anaerobic fermentation of FW (Aghapour Aktij et al. 2020), via liquid membrane technology, have been proposed. The low stability of liquid membranes and their high cost has so far limited the industrial exploitation even if their physical stability for practical use can be increased by their immobilization in porous structures.

Radio-frequency assisted extraction is another novel time-saving, and eco-friendly extraction technology; it was used, for example, to extract high methoxyl pectin from Jackfruit waste (peel). Compared to the conventional method of extraction, the yield of pectin was more than twofold (Naik et al. 2020).

4.2.4 *Purification and Isolation of Target Compounds*

This recovery stage aims at the purification and, if possible, isolation of the valuable component from co-extracted impurities. A number of assorted strategies are available.

4.2.4.1 Chromatographic Methods

Chromatographic methods are separative in their own right and can ensure the recapture of target molecules in pure forms destined for pharmaceutical and functional applications; chromatography can separate different compounds because of the different affinities of individual chemical species for two immiscible phases:

the stationary and the mobile phases. Molecules are set in motion by the mobile phase (eluent) and flow through the stationary phase. The higher the affinity for the stationary phase, the longer the time spent in it, the higher the time needed by a chemical species to exit the chromatographic system. In this way, each target compound is characterized by its own retention time and can be separated by other substances, such as impurities. Chromatographic methods are crucial during the purification and isolation of target compound from FW, but they are laboratory-intensive, solvent and time consuming and the maintenance cost are high. Target compounds can be isolated and characterized according to the following chromatographic mechanisms:

1. size-exclusion chromatography distinguishes the components based on their different sizes
2. ion exchange chromatography (anion- or cation-exchange chromatography) provides the separation through an exchange of ionic analytes between the stationary phase (a resin bearing groups oppositely charged to the target compounds) and the mobile phase (containing competing co-ions similarly charged as the target solutes); electrostatic interactions provide selectivity
3. affinity chromatography relies on biorecognition or interaction of solutes with a bio-specific ligand coupled to a chromatographic stationary phase
4. adsorption chromatography capitalizes on intermolecular forces established by the analyte with both the mobile and the stationary phase that is represented by an adsorbent. Apolar adsorbents characterize reversed phase chromatography, whereas polar stationary phases are used in normal phase chromatography; the type of the selected adsorbent and mobile phase, the temperature, and the flow rate, influence the selectivity and the efficiency of the method.

Mixed mechanisms are often possible.

Ion exchange and size exclusion chromatography have been applied for the separation of polysaccharides or their derivatives from Chinese sturgeon waste cartilage (Zhao et al. 2013), bean hulls (Zhong et al. 2012), fermented soybean curd residue (Shi et al. 2013), and underutilized tea plant waste (Quan et al. 2011). The recovery of phenolic compounds from a gamut of agricultural residues, wastewaters, and FW relied on all the different chromatographic mechanisms (Wang et al. 2014), (Prapakornwiriya and Diosady 2014), (Cardinali et al. 2012), (Lama-Muñoz et al. 2013), (Perera et al. 2012), (Caderby et al. 2013), (Weisz et al. 2013). Ion exchange or affinity chromatography was used for the fractionation of charged whey proteins, as illustrated in Chap. 2 (El-Sayed and Chase 2011). Whey proteins were also isolated by the mechanism of ion-exchange chromatography with ion exchange resin particles embedded within a membrane. Bhattacharjee et al. purified whey proteins (up to 90%) obtained via ultrafiltration via ion-exchange chromatography. The separation is based on the reversible electrostatic interaction between a charged protein molecule and the oppositely charged ion exchanger. Elution is modulated by fluid ionic strength and the presence of additives. Membrane ion-exchange chromatography is highly selective but slower than conventional membrane separations (Bhattacharjee et al. 2006).

Adsorption is the transfer of one or more components contained in a fluid phase to a porous solid sorbent, usually packed in a column; the liquid flows through synthetic polymeric resins, activated carbons, zeolites, clays, siliceous materials, lignin, or polysaccharide-based materials (*e.g.*, chitin, chitosan, cyclodextrin, and starch among others) that act as sorbents and form a stationary phase; the selective adsorption/desorption process, which can be tailor-made via the optimization of the experimental parameters, enables purification and isolation of target compounds. Soya waste and other FW, such as chicken feathers, rice husk, pomegranate peel, sugarcane bagasse, banana peel, and FW derived activated carbon, were also proposed as sorbents (Soto et al. 2011), (Gupta et al. 2009). Adsorption requires a relatively simple design, operation and scale up, and adsorbents can be easily regenerated and re-utilized. Nevertheless, this process is time consuming and demands further research with regard to the sorption behavior of individual components in complex mixtures since the thermodynamics of the adsorption equilibrium of each target compound is influenced by potential adsorption competitions. Adsorption has typically been applied for the isolation of phenolic compounds from a wide range of FW (Soto et al. 2011) e. g., hesperidin from orange peel (Di Mauro et al. 1999), and phenolic acids (e. g., ferulic, cinnamic, gallic) from olive mill wastewater (Ferri et al. 2011).

4.2.4.2 Nanofiltration and Reverse Osmosis

Generally speaking, membrane separation strategies offer high separation effectiveness, low energy supplies, mild operating conditions, no addition of chemicals, and easy scale-up for industrial purposes due to simple equipment design. Among pressure-driven membrane operations, nanofiltration (NF) is the technology of choice to recover small target molecules on the basis of the pore size detailed in Fig. 4.3. Reverse osmosis allows only water molecules to pass through the membrane pores, while nanofiltration is necessary when monovalent salt permeation is desirable (Galanakis et al. 2012).

Sieving is not the only mechanism to influence solutes transport or rejection through the membrane: the charge status of the membrane and that of the target compound determine electrostatic Donnan exclusion effects and dielectric effects; hence the pH and ionic strength of the medium are particularly important.

The application of sequential membranes is crucial in the whole recovery strategy from the pretreatment to the isolation of the target compound via increasingly selective separations. The final nanofiltration enabled the recovery and fractionation of different phenolic classes of compounds from wine industry waste (Galanakis et al. 2013), (Santamaría et al. 2002), artichoke wastewaters (Conidi et al. 2014), or olive mill wastewater (Cassano et al. 2013), (Rahmanian et al. 2014) (Galanakis et al. 2010d). Nanofiltration has also been suggested for the clarification of a phenol-containing beverage derived from olive mill wastewater by removing polymerized phenolic fractions (Galanakis 2012). A factorial design led to the optimization of the nanofiltration performance for the recapture of phenols from olive-oil-washing wastewater (Ochando-Pulido et al. 2020), which is also beneficial from the environmental point of view.

The high transmission of monovalent salts through NF membranes is broadly used to partially demineralize a number of final products, including whey, as detailed in Fig. 4.4. Nanofiltration of the ultrafiltration permeate enables the separation of lactose as retentate. Figure 4.4. depicts a typical example of target compounds complete valorization, and whey was chosen since it represents one of the most investigated food byproducts as regards FW valorization. The reader is referred to Chap. 2 for the sequential use of membrane technology in whey valorization.

Lactose obtained as NF retentate is often fermented to obtain lactic acid, ethanol, vitamin B12, and penicillin; other different outputs are possible; moreover, it can also be hydrolyzed to obtain galactose and glucose as illustrated in Fig. 4.4.

In lactic acid production, NF was combined with conventional fermenters after a MF step, necessary to remove cells and reduce membrane fouling (Pal and Dey 2013).

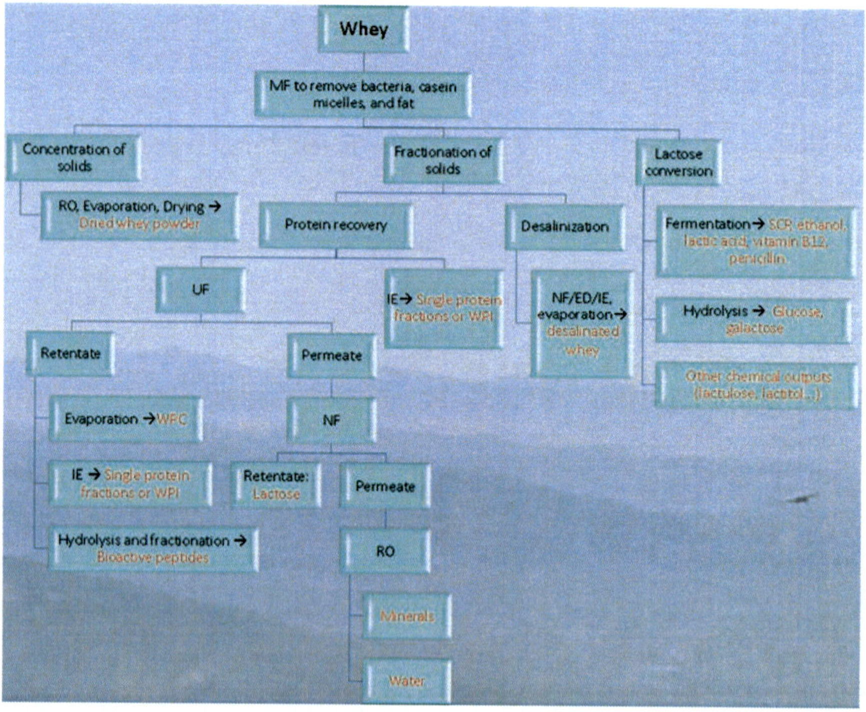

Fig. 4.4 Recovery and valorization of high-added value components from cheese whey. Outputs are in red. *MF* microfiltration, *UF* ultrafiltration, *NF* nanofiltration, *RO* reverse osmosis, *ED* electrodialysis, *IE* ion exchange, *SCP* single-cell protein, *WPC* whey protein concentrate, *WPI* whey protein isolate

Ultrafiltration and nanofiltration membrane cascades are often applied sequentially, for the recovery of peptides and bioactive peptides not only from cheese whey (Monti et al. 2018) but also from fish wastes (Abejón et al. 2016) hydrolysates. They are complex mixtures of nonhydrolyzed proteins, peptides, amino acids, enzymes, carbohydrates, and salts. For example, the membrane cascade enabled the isolation and purification of four antioxidant peptides from tuna by-product hydrolysate (Saidi et al. 2018) and resulted in the valorization of tuna processing waste biomass. Diluted solutions are needed to prevent fouling and electrostatic effects, which might impair the performance of this efficient and easy-going isolation technology.

4.2.4.3 Electrodialysis

Electrodialysis is an eco-friendly membrane technology widely used in the food sector (Serre et al. 2016). Electrodialysis separates charged particles according to their migration through ion-exchange membranes under an electric field. Anions and cations migrate towards the anodes and cathode, respectively, and the ion-exchange membranes prevent the passage of co-ions via electrostatic (Donnan) repulsion. Bipolar membranes containing both anion and cation exchange layers are useful but more expensive. Owning to its prominent properties, it has been utilized to demineralize solutions and concentrate the contained macromolecules, as in the case of oligosaccharides extract from soybean sheet slurry (Wang et al. 2009a) and separate peptides from a snow crab by-product hydrolyzate (Doyen et al. 2011). Electrodialysis offers an alternative to reverse osmosis as regards whey demineralization (Villeneuve et al. 2019). It enabled the recovery of lactoferrin, a nutraceutical iron-binding glycoprotein, with anti-inflammatory, antimicrobial, and immunostimulatory activities. One deeply studied application of electrodialysis is the separation and purification of volatile fatty acids produced from fermentation/anaerobic digestion (Dai et al. 2019), (Pan et al. 2018). Negatively charged carboxylates move through an anion exchange membrane towards the anode in the electrodialyzer thanks to the electric field (Aghapour Aktij et al. 2020), and this results in high concentration efficiency compared to conventional processes (distillation, precipitation, adsorption, and extraction) used for their recovery. The drawbacks of this technology are mainly related to stability and lifetime of the materials.

4.2.4.4 Magnetic Fishing

The magnetic field promoted separation technique, known as magnetic fishing, is another remarkable technology that implicates the use of magnetic particles bearing either ion-exchange groups or hydrophobic or affinity ligands. The particles are exposed to the crude sample. During an incubation period, target compounds bind to these particles according to a gamut of different adsorption mechanisms; the magnetic complex is then removed from the sample medium using a magnetic separator; target molecules can be eluted, after washing out any impurities, via pH or ionic strength modification (El-Sayed and Chase 2011). It is receiving great attention because it provides highly purified target compounds, also from very diluted feedstock, with impressive selectivity and no need for pretreatment.

For example, supermagnetic ion-exchangers enabled the fractionation of bovine whey proteins. After batch binding, magnetic particles could be easily recovered from the solution and rapidly eluted; the procedure is very gentle to the target compounds (Heebøll-Nielsen et al. 2004), (El-Sayed and Chase 2011).

Magnetic separation technique is straightforward, and it is becoming popular, especially for the separation of large biomolecules from FW, because it avoids many recovery steps, and it could be used in continuous separation processes with magnetically stabilized fluidized beds (Nath et al. 2015).

4.2.4.5 Aqueous Two-Phase System

Aqueous two-phase system (ATPS) is an interesting technique to recover biomolecules from FW extracts. It is based on the incompatibility of the components of two different aqueous phases (polymers, surfactant, salt, alcohol, and ionic liquid). Its separation efficacy depends on the hydrophobic and electrostatic interactions between the target compound and the phase components. The two-phase separation was used for the recovery of biologically relevant protein from whey (El-Sayed and Chase 2011). It has been employed for the partitioning of whey β-lactoglobulin and α-lactalbumin, simply mixing whey protein concentrate and aqueous hydroxypropylmethylcellulose at pH 6.5: the thermodynamic incompatibility which arises from mixing could be used as a method for fractionation of whey proteins (Jara and Pilosof 2011). It was also used for the isolation of citrus ascorbic acid with an alcohol and potassium phosphate salts-based ATPS (Reis et al. 2011). In some cases, it can represent a single step recovery procedure. ATPS based on ethanol/ammonium sulfate was proposed to extract and recover proanthocyanidins, well-known natural antioxidants, from grape seeds; ionic liquids served as adjuvants in the process, and extraction yields were higher (up to 99%), compared with conventional organic solvents isolation (Ran et al. 2019). Valuable proteins were recovered from shrimp (*Litopenaeus vannamei*) waste in a single-step ATPS extraction and purification, using polyethylene glycol and trisodium citrate system (Saravana Pandian et al. 2020). The same ATPS was used for the single-step extraction of xylanase after solid-state fermentation (SSF) of castor press cake (*Ricinus communis L.*) by *Aspergillus japonicus* URM5620 (Herculano et al. 2016). An ATPS composed of polyethylene glycol and potassium phosphate buffer was used for the purification of polyphenol oxidase from waste potato peel (Niphadkar et al. 2015). Extraction of rutin from acerola waste relied on an alcohol-salt-based ATPS (Reis et al. 2014).

Aqueous two-phase systems provide mild conditions that preserve labile compounds. However, the main drawbacks include long separation time and numerous required processing steps even if a continuous process can be envisioned; recycling the phase-forming components would reduce chemicals consumption and downstream pollution.

4.2.5 *Product Formation*

Many uses of food waste components are possible since they can serve as platform chemicals (Chap. 5), monomers for bioplastic production (Chap. 6), fillers of biocomposites (Chap. 7), or as new ingredients in diverse food products. Food-grade product manufacturing from FW sources should not affect the bioactivity of target compounds that might be sensitive to oxygen, temperature, pH, and light. If target compounds recovered from FW need to be processed to a food-grade state, they have to be free from toxic compounds and contaminants and in a specific physical state such as liquid (solution or emulsion), semisolid, or solid. The final step of downstream processing is product formation to deliver it in the market, ensuring that the valuable compounds will preserve their properties within a reasonable period.

A wide range of technologies are nowadays applied to recover macronutrients (polysaccharides, lipids, and proteins) as well as micronutrients (e.g., vitamins, bioactive compounds, mineral salts, pigments, volatile compounds). Freeze-drying (or lyophilization) was described among pretreatment technologies but can also be used for the final product formation. It is preferable as regards labile compounds preservation, but low temperatures and vacuum conditions make it an expensive, long, and energy-consuming process. Melt extrusion is used for polysaccharides, flavors, and nutrients. They are dry-fed and melted into a sheath of coating material solution, even if this process is prone to be solute degrading. In the following, common techniques used for product formation will be described.

4.2.5.1 Emulsification

Emulsification is the physical process of mixing two or more immiscible liquids to form a semistable blend, thanks to the action of a specific agent, the emulsifier, which has affinity for both phases. Its use is critical for the dispersion of ingredients with no reciprocal phase affinity. An emulsion is a mixture of immiscible liquids where one liquid (the oily phase) is dispersed as small spherical droplets (between 100 nm and 100μm) in the other. Emulsions are thermodynamically unstable systems, and, over time, small dispersed droplets can flocculate and coalesce to form large aggregates until the separation of the organic and aqueous phases is clearly visible. Emulsifiers create an emulsion stable within the time of observation, slowing down the destabilizing phenomena and preventing the droplets from coming into close proximity and from coalescing.

FW are rich in both emulsifiers and lipids. FW emulsifiers include soybean lecithin, phospholipids from oil refinement, vegetal proteins, gelatin from the denaturation of animal skins, bone or skeletal muscles collagen, whey proteins, biosurfactants, polysaccharides, and pectin. Lipids extracted from FW can be used to produce emulsified food products. Emulsification with the aforementioned natural emulsifier can be a smart way to deliver high value compounds from FW into a new food chain (Shaw et al. 2007), (McClements et al. 2017). Other non-drying methods, like liposome and emulsion entrapment, are typically used in more specific applications, i.e. to entrap hydrophilic phenols from potato peel in fish-rapeseed oil mixtures to prevent its peroxidation (Koduvayur Habeebullah et al. 2010).

4.2.5.2 Microencapsulation and Nanoencapsulation

Microencapsulation is a technique that entraps liquid droplets or solid particles of high-added value compounds into thin films of a food-grade microencapsulating agent; the film is a physical barrier that preserves components stability and eventually masks undesirable organoleptic features.

Microencapsulation facilitates the handling of the target compounds, limits their reactivity with environmental factors, and controls their release, eventually masking their taste.

When the compounds of interest are macromolecules such as polysaccharides, dietary fibers, or proteins, encapsulation is replaced by a direct drying process.

Spraying methods are the most common encapsulation methods even if air suspension coating, extrusion, centrifugal extrusion, freeze drying, coacervation, rotational suspension separation, cocrystallization, liposome entrapment, interfacial polymerization, and molecular inclusion have also been reported (Fang and Bhandari 2010).

Spray drying is a unit operation by which a liquid feed is atomized in a hot gas flow to instantly get a dry powder with particle size from 10μm to 3 mm. Air is the common heated drying agent but, more rarely, for oxidable compounds inert nitrogen is used as drying gas. The liquid feeding the sprayer can be a solution, an emulsion, or a suspension. The microbiological stability of products, reduced storage and transport costs, instantaneous solubility are the main advantages of this low cost, adaptable and continuous process. The cost of spray drying is 30–50 times cheaper than freeze-drying (Desobry et al. 1997). The mixture to be atomized is prepared by dispersing the target compound, which is usually of hydrophobic nature, into a solution of the coating agent and eventually a food-grade emulsifier, whose surface-activity facilitates droplet disruption, and prevents their aggregation; the evaporation of the solvent after the liquid passage in the atomization nozzle is due to the heat exchange with the drying gas, flowing either co-current or counter-current to the small drops, and leads to the formation of microcapsules. For thermolabile substances, the mixture can be sprayed into a partial vacuum to favor drying.

The coating material is often made of polysaccharides (starch, cellulose, cyclodextrin, inulin, pectin, gums, carrageenans, alginate) or proteins, and it is a useful tool to improve storage shelf life and target compound delivery. Whey protein gels and dietary fibers from soybean solid residue, palm trunk, and oil palm frond have been valorized as coating materials; for example, bilberry extract was encapsulated in whey protein gels at pH 1.5 (Betz and Kulozik 2011). Whey proteins acted as a wall system to encapsulate anhydrous milk fat by spray drying, with an encapsulation yield greater than 90% (Young et al. 1993). Microcapsules of gliadin containing polyunsaturated lipids were resistant to oxidative deterioration (Iwami et al. 1988). Soybean soluble polysaccharide retained microencapsulated ethyl butyrate during spray drying (Yoshii et al. 2001). Spray drying was employed for the preservation of tomato waste and carrot waste carotenoids (lycopene, lutein, α-carotene, and β-carotene) (Eun et al. 2020) and for the encapsulation of phenolics, such as myricetin, quercetin, quercetin-3-β-glucosi9e, caffeic acid, and p-coumaric acid, from wine lees (Pérez-Serradilla and Luque de Castro 2011). Unfortunately, spray drying is deleterious for thermolabile antioxidants, and volatile low molecular weight compounds. Rod-like chitin nanocrystals adsorption at the oil-water emulsion interface were effective against coalescence; interestingly, these nanocrystals were produced during acid hydrolysis of crude crab shells (Tzoumaki et al. 2011).

Innovative encapsulation makes use of nanoemulsions (with a droplet size of ca. 100–300 nm) more stable to droplet aggregation and gravitational separation; the risks associated with their oral ingestion is still an open question since the nanometric dimensions of the droplets influence their fate in the gastrointestinal tract. Their fabrication usually relies on the use of disruptive mechanical energy (i.e., high shear stirring, high-pressure homogenization, or ultrasound). They are able to protect and deliver lipophilic target compound, such as polyunsaturated fatty acids and flavors, vitamins, preservatives, and nutraceuticals (McClements and Rao 2011a). Nanocapsules have also been applied to mask the taste and odor of tuna fish oil (rich in ω-3 fatty acids) integrated into bread (Neethirajan and Jayas 2011) and the bitter taste of polyphenols (Fang and Bhandari 2010). For lipophilic polyphenols that are scarcely water soluble, nanoliposomes were obtained by a pH-driven method since water solubility increases with increasing pH (Peng et al. 2019). Encapsulation of betacyanins and polyphenols, extracted from leaves and stems of beetroot waste, made use of Ca(II)-alginate beads without altering the antioxidant activity (Aguirre Calvo et al. 2018). Nanoemulsion was exploited to enable moisture- and pH-triggered controlled release (McClements and Rao 2011b) of β-carotene, used as natural coloring agent, thereby enhancing its intestinal bioavailability, its dispersion ability in water and coloring strength (Silva et al. 2011).

4.3 Bioconversion

Bioconversion of FW into high-added value compounds is attracting increasing attention of the research fraternity around the globe; the valorization of this well established method was described in Chap. 3.

4.3.1 Anaerobic Fermentation/Digestion

Anaerobic fermentation (AF) is the process by which biomass is broken down by microorganisms in the absence of oxygen to form biogas (Coma et al. 2017) that is methane (CH_4), hydrogen (H_2) and volatile fatty acids (VFA). AF is an established technology for FW valorization (Dahiya and Joseph 2015), (Sarkar et al. 2016). FW is regarded as a favorable substrate due to its biodegradable nature, rich nutrient

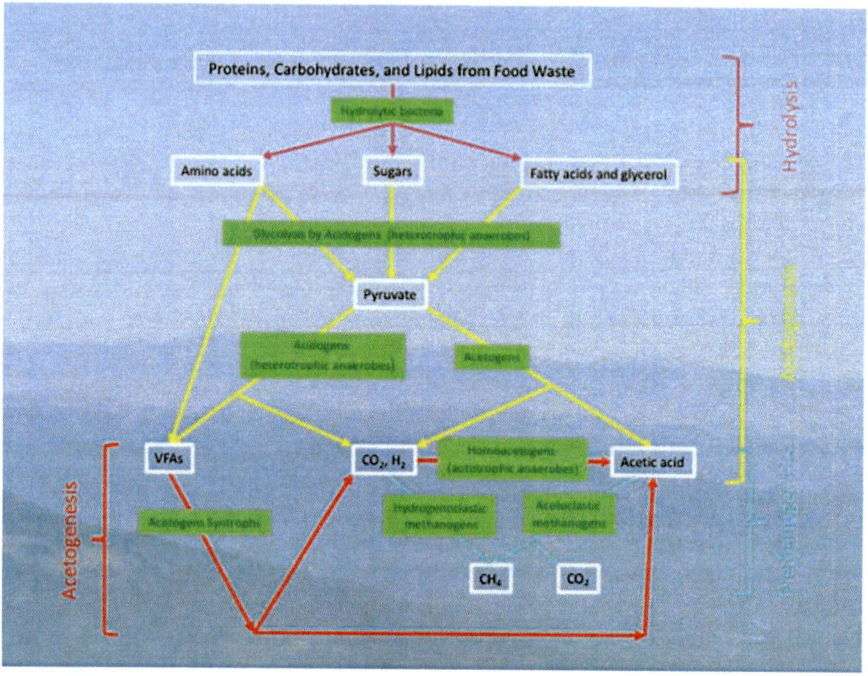

Fig. 4.5 Schematic pathway of the four steps of anaerobic fermentation for biobased products generation. VFA degrading bacteria and methanogens (hydrogenotrophs and aceticlasts) are interrelated in a "syntrophic interaction"

availability, and high moisture content. AF accomplishes waste disposal, energy and biobased products recovery (Zhou et al. 2018).

FW can be pretreated via physical (e.g., shredding, ultrasonic energy), thermal, chemical (e.g., alkali treatment, lipid extraction), and biochemical (use of enzymes) methods before AF. There are four sequential steps in AF, clearly detailed in Fig. 4.5; each stage yields important bio-products.

During the first hydrolytic step, sugars, amino acids, and long-chain fatty acids are respectively obtained from carbohydrates, proteins, and lipids thanks to hydrolytic bacteria. From them pyruvate, a pivotal molecule, is produced. For example, if hexoses are considered, all organisms have glycolysis, a sequence of nine reactions, each catalyzed by a specific enzyme, occurring in their cytoplasm, and the final product is pyruvate.

In the second step, namely, acidogenesis, acidogenic bacteria (heterotrophic anaerobes) yield VFAs, acetic acid, CO_2, and H_2; acetogens produce acetic acid. The acidogenic biochemistry relies on the acetate-ethanol type, propionate-type, butyrate-type, mixed-acid type, lactate-type, and homoacetogenic metabolic pathways (Zhou et al. 2018).

In the subsequent acetogenic step, VFA are degraded and this results in the production of acetic acid, CO_2, and H_2. Anyhow acetic acid can also be produced autotrophically by homoacetogens from CO_2, and H_2.

During the terminating methanogenesis process, hydrogenoclastic methanogens yield CH_4 from CO_2 and H_2 while acetoclastic methanogens produce CH_4 and CO_2 from acetic acid. If the last step, namely methanogenesis, is suppressed by specific inoculum pretreatment strategies (Sarkar et al. 2016), there is an enhanced production of H_2 and VFA that can be biologically and/or chemically converted to other biobased products.

These steps are achieved by the action of mixed syntrophic associations and bacteria consortia, including hydrolytic, acidogenic, hydrogen-producing, acetate-forming microbes, homoacetogens, acetoclastic methanogens, and hydrogenoclastic methanogens (Venkata Mohan 2009). Beneficial interactions between various trophic levels and syntrophic interactions are crucial in the biotransformation of complex and impure substrates like FW.

In terms of optimizing the bioconversion biomass, the priority mainly lies in the development of effective pretreatment tools, along with cost-effective hydrolytic enzymes and improved strains of microorganisms (DOE/EERE 2011).

In single-stage anaerobic digestion all of 4 steps take place simultaneously in a single reactor with less investment costs. In two-stage anaerobic digestion hydrogen and methane are produce sequentially in two separate reactors: in the first step, acidogens and hydrogen producing microorganisms produce VFA and H2, while in the second step acetogens and methanogens degrade VFAs to methane and carbon dioxide (Uçkun Kiran et al. 2014). Figure 4.5 depicts the schematic pathway of anaerobic fermentation for biobased products generation with the microorganisms involved in the process (Dahiya et al. 2018), (Amani et al. 2011), (Li et al. 2012a) (Liu and Whitman 2008), (Dahiya et al. 2015).

In the following, we will detail the biobased products which can be recovered from the AF of FW.

4.3.1.1 Sugars

Production of sugars or sugar-rich effluents from the FW can be obtained from the hydrolytic pretreatment of the FW (Hafid et al. 2015), as illustrated in Fig. 4.5. Sugars have industrial importance, and the concentration of produced sugars in the hydrolysate is dependent on the adopted pretreatment strategy. Sugars like glucose, fructose, galactose, and ribose are crucial compounds that can be obtained from FW (Li et al. 2012b). They come as a mixture whose composition, production rate, and degraded compounds formation or losses may vary with respect to FW kind and origin as well as the adopted pretreatment strategy.

Pretreatment can be physical, chemical, physico-chemical, biological, and enzymatic, and a combination of strategies is also possible (Hafid et al. 2015). Dilute hydrochloric acid and sulphuric acid pretreatment is the best strategy for maximizing sugar yield and avoiding the formation of sugar degraded compounds (furfurals/

hydroxyl methyl furfurals). A mixture of α-amylase, β-amylase, and glucoamylase enzymes is also used for the transformation of starch into small monomeric sugars, and an ultrasonic pretreatment increased the amount of reducing sugars (Jiang et al. 2014).

4.3.1.2 Biohydrogen

Biohydrogen (H_2) is regarded as a sustainable green fuel, and its production from FW has seen a spurt of research activities that are centered on optimizing this process.

H_2 is formed by the action of obligatory H_2 producing acetogenic bacteria (AB) on a gamut of FW and composite FW (Venkata Mohan et al. 2009) (Sarkar et al. 2016) during the acidogenic step of the AF, as illustrated in Fig. 4.5. Semi-pilot and pilot-scale studies modeled its production (Pasupuleti et al. 2014) (Venkata Mohan and Sarkar 2017) and up-scaling feasibility was also studied (Pasupuleti et al. 2014). FW is a highly eligible substrate for H_2 production as it is a carbon-rich substrate with high levels of easily degradable organic material, which favors a positive economic balance. The H_2 production yield in AF depends on the pretreatment procedure, the selected mixed microbiota, feed composition, medium pH and temperature, co-substrate addition, and reactor configuration (Venkata Mohan 2009). Biologically, H_2 can also be produced using photo-fermentation or bio-photolysis (Venkata Mohan 2009). Combining dark AF with photo-fermentation both in two-phase operation or single-stage operation (Chandra et al. 2015b) is a strategy to maximize the theoretical H_2 yield; the metabolites produced during AF can also be further utilized. The advancements in the biological H_2 production from FW may lead to worldwide usage of H_2 as a major transportation fuel.

4.3.1.3 Biomethane

Biogas is obtained as the final step of the AF, as illustrated in Fig. 4.5. It is a mixture of methane (CH_4), carbon dioxide (CO_2), nitrogen (N_2), moisture, and traces of hydrogen sulfide (H_2S). CH_4 production is perhaps the most studied and reviewed technology for the generation of bio-energy utilizing FW as a substrate with AF. Maximizing CH_4 production from FW involves the use of a proper reactor design, correct pretreatment (Krishna and Kalamdhad 2014), a judicious choice of the inoculum type (Deepanraj et al. 2017), and optimization of temperature, pH, carbon/nitrogen ratio (Zeshan et al. 2012), volatile fatty acid coproduction (Xu et al. 2014), organic load and volatile solids (VS), and codigestion (Agyeman and Tao 2014). The major limitation in CH_4 production is the lowering of pH during anaerobic digestion due to the production of volatile fatty acids, which inhibit the growth of methanogenic microbes. The advancements in process efficiency (Dahiya and Joseph 2015) have shifted the production of biomethane from a lab-scale to a commercial level as several industries have established full-scale technologies. The incorporation of some conductive materials into anaerobic digestion reactors treating complex organic matter was studied; it was demonstrated that activated carbon amended environments significantly improve methane yield and reduce acidification (Zhang et al. 2020). Noteworthy, potential methane production from FW (480 ± 88 L/kg VS) is *ca.* double that from energy crops; AF methane yield is about 81% (395 ± 84 L / kgVS) of the potential producible methane, and the yield benefits from both co-digestion and FW pretreatments (physical, thermal, and biochemical) (Negri et al. 2020).

4.3.1.4 Biohythane

Biohythane is a mixture of hydrogen (H_2) and methane (CH_4) with hydrogen concentration in the range 10–30% v/v. It can be obtained by mixing H_2, produced in the acidogenic step of the AF, and CH_4, released during the methanogenic steps of the AF, as illustrated in Fig. 4.5. Biohythane overcomes the individual limitations of these gases, respectively the high flammability of H_2 that leads to storage issues and the low flammability CH_4. The flammability trade-off makes biohythane an appropriate and clean fuel with good calorific efficiency and reduces release of contaminants in the atmosphere (David et al. 2019). For these reasons, it can be easily commercialized (Sarkar and Venkata Mohan 2017). Biohythane improves methane engine combustion and further reduces the contaminants released in the atmosphere. Various FW, including distillery spent wash (Pasupuleti and Venkata Mohan 2015) and canteen food waste (Sarkar and Venkata Mohan 2017), were used as feedstock for biohythane production. The possibility to codigest FW and sewage sludge is interesting from both the process and economic points of view (David et al. 2019).

4.3.1.5 Volatile Fatty Acids

VFAs or short-chain (C2-C5) carboxylic acids such as acetic acid, propionic acid, butyric acid, iso-butyric acid, valeric acid and iso-valeric acid are the major coproducts in AF produced in the acidogenic phase of the process, as illustrated in Fig. 4.5. The importance of VFAs in pharmaceuticals, textiles, food and beverages, biofuels, and bioplastic productions made them the focus of interest of many researchers. Their fossilbased production is energy-intensive and has a negative impact on the environment; hence their biological production is gaining immense interest. Since sterile fermentation conditions are not required, FW represents an inexpensive feedstock. The feed composition, inoculum, reactor condition, temperature, organic loading rate, and redox conditions (Dahiya and Joseph 2015) are the key parameter to maneuver in order to maximize their production and to differentiate the type of VFAs.

The concentration and the distribution of VFAs are the result of acidogenic metabolic pathways, which can be tailor-made in the digester for efficient product recovery.

FWs from university canteens, catering, and garbage collection companies are the right candidate for volatile fatty acids production. The best operative conditions for the anaerobic fermenter include a neutral pH range (6.0–7.0), less than 10 days hydraulic retention time, thermophilic temperatures, and an organic loading rate of about 10 kgVS/m^{3}d. This way, VFAs concentration increased from 10% to 25% (Strazzera et al. 2018).

Generation of VFAs was also studied to maximize CH_4 yield in AF of FW (Uçkun Kiran et al. 2014). During the AF, the production of VFA and the synthesis of methane are competing processes because VFA production is majorly limited by the prevalence of methanogenic bacteria in mixed microbial consortia. To enhance the VFA production from food waste, the biocatalyst was exposed to acid-shock, followed by operation under alkaline conditions. These treatments enriched acidogenic spore formers, fatty acid-producing bacteroides, and saccharolytic and proteolytic bacteria, thereby increasing VFA yield (Sarkar and Venkata Mohan 2017).

Thermodynamically, the production of acetic acid by anaerobic acidifying bacteria is simpler; hence for FW that mainly contain carbohydrates, it is expected that acetic acid will be the highest proportion of VFA. Under stress conditions, propionic and butyric acid concentrations are expected to increase.

VFAs can be subsequently used for the production of bioplastics (anoxygenesis), biodiesel (oleaginous metabolism), bioelectricity generation (bioelectrogenesis), and biofertilizers (Venkata Mohan et al. 2016).

Since VFAs are widely applied in the chemical industry, their concentration, separation, and purification are crucial downstream steps for AF of FW. The removal of VFA avoids the inhibition of their production. In-situ recovery of VFA by electrodialysis was studied for this reason and to control the pH, which in turn helps select the microbial community for different purposes. However, it is still difficult to recover one specific acid from the mixed VFAs from AF broth due to the similar physical-chemical characteristics of VFAs; hence their separation and purification remain a hot research need (Zhou et al. 2018).

4.3.1.6 Medium Chain Fatty Acids (C6-C10)

Biological production of medium chain fatty acids (MCFA) proceeds via reverse β-oxidation pathway, studied as a metabolic pathway of *Clostridium kluyveri* (Steinbusch et al. 2011), that enables elongation of VFAs by two carbons per cycle to MCFAs (C6-C10). Reverse β-oxidation requires a carbon source, reducing equivalent and energy from specific electron donors such as ethanol, H_2, and lactate (Steinbusch et al. 2011). The longer carbon chain is related to lower oxygen/carbon ratio and higher hydrophobicity of FW, compared to VFAs. The hydrophobicity of MCFA promotes their separation from the fermentation broth (Gildemyn et al. 2017) and provides higher energy density (Grootscholten et al. 2013). MCFAs are high value products not only for biofuels, but also as antimicrobial, fragrance, and flavoring intermediates (Gildemyn et al. 2017).

4.3.2 *Solventogenesis*

The biochemical production of solvents is called solventogenesis, whose most common products are ethanol, acetone, and butanol as indicated in Fig. 4.6.

The process resembles yeast fermentation of sugars to produce ethanol, but the organisms that carry out the complete solventogenesis are strictly anaerobic (obligate anaerobes). Bacteria from the Class Clostridia (Family Clostridiaceae) are the most used solventogenic agents: the organic acids they produce from several carbon sources during the acidogenesis can be converted to solvents, especially at low pH where acids are undissociated; the switch could also be triggered by nitrogen and phosphate limitation and is interesting because a judicious FW source selection can stimulate solventogenesis, but the exact mechanism of the process is not well understood yet (Kumar et al. 2013).

Some species can also produce isopropanol. Solventogenesis usually produces butanol, acetone, ethanol in a 6:3:1 ratio. *Clostridium acetobutylicum* is the most well-studied since it gives the best results (Kumar et al. 2013).

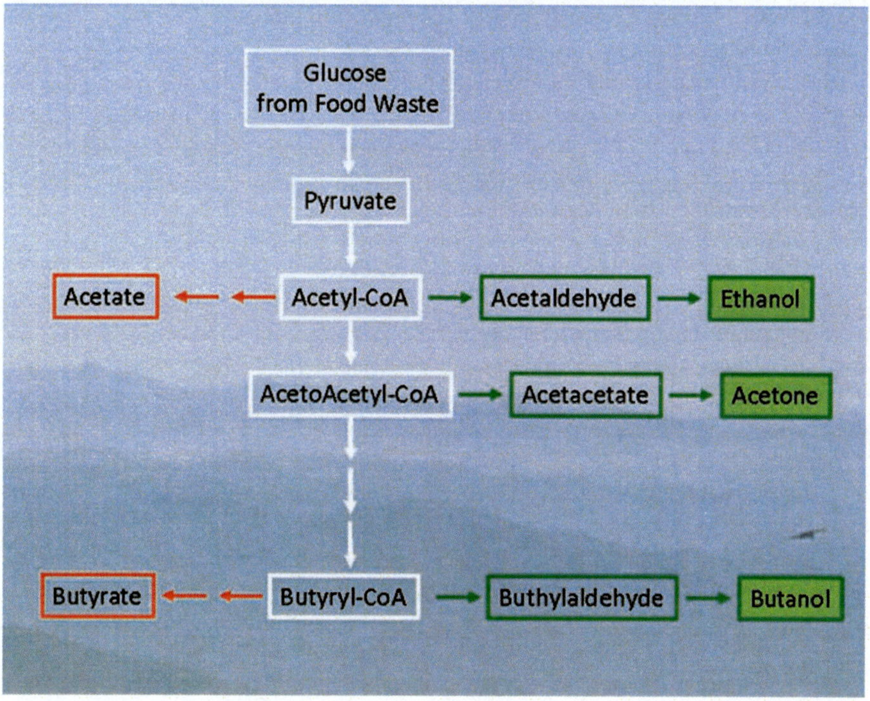

Fig. 4.6 Mechanism of solventogenesis. The usual products of the acid fermentation phase are shown on the left, in red. CoA: Coenzyme A. On the right, the biobased solvents outputs are indicated. Acetyl-CoA and Butyryl-CoA can be restored from acetate and butyrate, respectively, recycling the CoA released in the transformation of AcetoAcetyl-CoA into Acetoacetate

Production of biosolvents from FW using anaerobic microbiomes is a hot research field since they can replace fossil fuels and avoid the dependency on non-renewable sources.

Apart from the product benefits, the high cost is considered a major bottleneck in these biotechnology processes (Wang et al. 2009b).

Biological conversion of FW to biosolvents can decrease the organic pollution load by 75–80% while deriving energy and making the process self-sustainable.

4.3.2.1 Bio-Ethanol

Bio-ethanol can be obtained typically from the yeast fermentation of sugars. Nowadays, worldwide ethanol production demand is more than 100 billion liters. The shift from fossilbased to renewable resources based ethanol production is mandatory. Currently, pilot-scale production of bioethanol relies on the fermentation of sugars from sugarcane and corn as a feedstock, thereby fostering the food versus fuel competition. The use of low-cost agricultural residues viz., wheat straw, rice straw, sugar cane bagasse, and other cellulosic biomass (Humbird et al. 2002) suffers lignin recalcitrance in the conversion of cellulose and hemicelluloses into monosaccharides. By this scenario, FW is a highly eligible feedstock for saccharification and fermentation.

Restaurant waste (corn, potatoes, and pasta) was suitable for the production of ethanol via enzymatic digestion of starch (using α-amylase and glucoamylase) and fermentation of the resulting sugars to ethanol using yeast (Walker et al. 2013).

Bioethanol production from pineapple leaf waste containing 60–85% of holocellulose was reported (Chintagunta et al. 2017). In this case, the simultaneous saccharification and fermentation were carried out by a cellulase cocktail and yeast and resulted in a 7.12% v/v of bioethanol production.

Bioethanol production from potato peeling and mash wastes was evaluated using a co-culture of *Aspergillus niger* and *Saccharomyces cerevisiae*. Under optimized conditions, potato peels and mash wastes produced respectively 6.18% v/v and 9.30% v/v of bioethanol, and the residue obtained after ethanol production could be used as fertilizer (Chintagunta et al. 2016).

Simultaneous hydrolysis and fermentation of unprocessed FW into ethanol using thermophilic anaerobic bacteria in a single reactor was reported, for the first time, by (Dhiman et al. 2017).

The mixing of kitchen waste and waste paper was crucial for cost-effective ethanol production from waste paper (Nishimura et al. 2017) since the kitchen waste represented a carbon and nutrient source as well as an acidity regulator.

4.3.2.2 Bio-Butanol

Bio-butanol (butyl alcohol) is commonly used as a low vapor pressure solvent and a precursor for other compounds. It can also be used as fuel by blending with or replacing gasoline. Biobutanol is non-sensitive to water, and it is characterized by low toxicity and flammability. Starch-rich crops and algae are widely used substrates for bio-butanol production (Stoeberl et al. 2011). The main process for bio-butanol production from sugar-rich FW extract relies on Clostridiaceae bacteria; butanol is coproduced with acetone and ethanol to give ABE (acetone-butanol-ethanol) mixture. Domestic FW, being a rich carbon source, is utilized for ABE production by *Clostridium beijerinckii* P260 (Huang et al. 2015). Ujor et al. first reported the practicability of butanol production using industrial starchy FW (inedible dough, breading's and batter liquids) with batch fermentation carried out with *Clostridium beijerinckii* NCIMB 8052 that produced 0.37 g of ABE per g of industrial starchy FW (Ujor et al. 2014).

Shao and Chen evaluated the potential of *Amorphophallus konjac* waste as a feasible substrate for ABE fermentation by *Clostridium acetobutylicum* ATCC 824. The results indicate that ABE concentration was higher for separate hydrolysis and fermentation steps than simultaneous saccharification and fermentation samples (Shao and Chen 2015).

4.3.3 Biodiesel Production from Renewable Sources: Waste Oil and Oleaginous Metabolism

Chemically, biodiesel is constituted of long-chain fatty acids methyl (or alkyl) esters. Typically, biodiesel is produced from a range of feedstocks, including edible vegetable oils, waste cooking oils, and animal fat. The limited supply of these feedstocks results in the high cost associated with biodiesel production. Hence, renewable resources for the further expansion of the biodiesel industry are needed to overcome this main limitation in biodiesel production.

It is clear that cost-effective lipid sources are mandatory for biodiesel synthesis. Mixed non-edible oils, castor seed oil, and waste fish oil (Fadhil et al. 2017), under optimized conditions, gave 95.20 ± 2.5% w/w of biodiesel with acceptable fuel properties and acceptable overall process economics. Waste pepper seeds without lipid extraction (Lee et al. 2017), waste palm oil (Thushari and Babel 2018), and other waste oils (Hu et al. 2017) proved to be useful for efficient production of biodiesel. Green catalysts (derived from coconut meal residue or an ionic liquid) and parameters optimization were studied (Thushari and Babel 2018) (Hu et al. 2017).

Recently microalgae, yeast, fungi, and bacteria have been explored for lipid production due to their oleaginous metabolism, using FW as feedstock. These lipids can become the primary renewable source of fatty acids for biodiesel production.

Oleaginous microorganisms can accumulate lipids with a short life cycle and lower cultivation area compared to conventional lipid sources. Some eukaryotic yeast strains, such as *Rhodosporidium sp.*, *Rhodotorula sp.* and *Lipomyces sp.* store intracellular lipids up to 70% of their dry biomass when there is a shortage of nitrogen from the culture medium, while the uptake of carbon is not limited and its metabolization continues (Ratledge 2002). *Crptococcus curvatus,* when grown in sweet sorghum bagasse, can synthesize lipids up to 60% on a dry weight basis with 44% of saturated fatty acids (Liang et al. 2012). Lower percentages were reported for fungi as *Aspergillus sp.* (22.1%) (Venkata Subhash and Venkata Mohan 2011). Malonyl-CoA plays a crucial role in fatty acid elongation.

Inexpensive carbon sources include, in chronologic order, molasses (Alvarez et al. 1992), grape must (Buzzini and Martini 2000), radish brine (Malisorn and Suntornsuk 2008), hydrolyzates of agricultural residue (Yu et al. 2011), distillery wastewater, glycerol, rice hulls, wheat straw, waste cooking oil and corncob waste liquor (Venkata Subhash and Venkata Mohan 2011) (Dahiya et al. 2018), food and feed waste (Schneider et al. 2013) cheese whey (Castanha et al. 2014).

Microalgae can utilize both inorganic (carbon dioxide; autotrophic mode) and organic (heterotrophic mode) carbon and are particularly able to accumulate lipids up to 90% of dry weight under specific conditions. Their cultivation results in flue gas CO_2 mitigation (Li et al. 2008). Approximately, microalgae fix 183 tons of carbon dioxide to produce 100 tons of algal biomass (Chisti 2007).

Microalgae yielded higher biomass as well as lipids when grown in the heterotrophic and mixotrophic mode of operations using wastewater as feedstock. CO_2 supplementation to domestic wastewater enhances microalgae lipid accumulation under mixotrophic microenvironment (Prathima Devi and Venkata Mohan 2012). Municipal and food wastewater was used to grow *Chlorella sorokiniana* heterotrophically (Chi et al. 2011). *Schizochytrium mangrovei* and *Chlorella pyrenoidosa* were used to profit from the fungal hydrolysate of canteen waste (rice, noodles, meat, and vegetables) used as culture medium and nutrient source: the algal biomass contained carbohydrates, proteins, lipids, and ω-3 fatty acids (Pleissner et al. 2013). Acidogenic effluent from biohydrogen production is a carbon source that was used for lipid biosynthesis with lipid productivity of 26% (Venkata Mohan and Prathima Devi 2012). *Chlorella ellipsoidea* and *Scenedesmus quadricauda* could profitably grow on effluent from anaerobic digestion of kitchen waste with good lipid productivity (Pei et al. 2017). *Chlamydomonas reinhardtii* grew well with acetate or VFAs under mixotrophic settings (Moon et al. 2013). Short-chain fatty acids provide a promising alternative feedstock for biodiesel production with integrated fed-batch microalgae cultivation biorefinery. VFAs were investigated as organic carbon source during the growth and nutrient stress lipogenic phase in mixotrophic fed-batch microalgae cultivation. Nutrient deprivation granted the highest lipid productivity with acetate: in all cases, palmitic acid was the most abundant fatty acid. Acetate, followed by butyrate and propionate, also gave the highest algal biomass growth (Chandra et al. 2015a). Therefore, VFAs from FW can function as a promising feedstock for biodiesel production within the microalgae biorefinery framework (Fig. 4.7).

4.3.4 *Microbial Electrochemical Technologies*

The synergistic approach of integrating microbial environment with electrochemistry is still in its infancy, but it has recently come into the limelight as a promising futuristic and green strategy to produce energy and renewable feedstock. Microbial

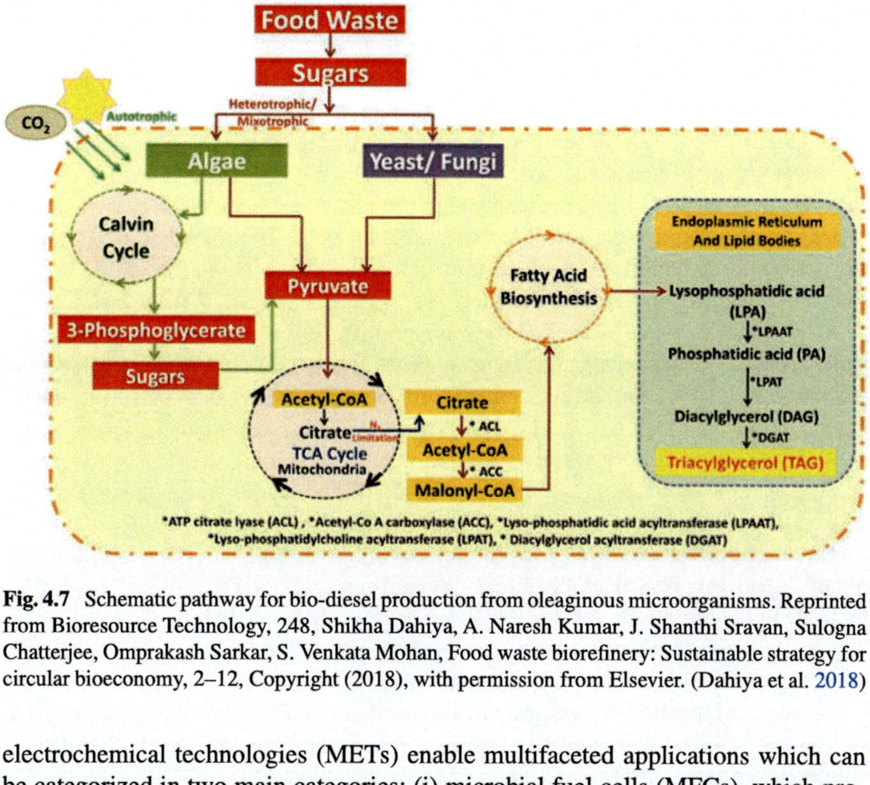

Fig. 4.7 Schematic pathway for bio-diesel production from oleaginous microorganisms. Reprinted from Bioresource Technology, 248, Shikha Dahiya, A. Naresh Kumar, J. Shanthi Sravan, Sulogna Chatterjee, Omprakash Sarkar, S. Venkata Mohan, Food waste biorefinery: Sustainable strategy for circular bioeconomy, 2–12, Copyright (2018), with permission from Elsevier. (Dahiya et al. 2018)

electrochemical technologies (METs) enable multifaceted applications which can be categorized in two main categories: (i) microbial fuel cells (MFCs), which provide power generation from organic substrates (including FW or food chain wastewaters), relying on the spontaneous oxidative reaction of the microorganisms, used as biocatalysts, (ii) eletrofermentation (EF) in which an external potential is supplied to influence the fermentation and microbial metabolism, for the production of hydrogen, target biochemicals or other value added products.

4.3.4.1 Microbial Fuel Cell

Microbes generally carry out their catabolism by available oxidizing substrates; the reducing equivalents (protons and electrons) spontaneously generated during the oxidative reactions are transferred via a series of redox carriers (e. g., NAD^+, FAD) towards terminal electron acceptors; the proton motive force leads to the synthesis of energy-rich phosphate bonds, subsequently used to foster the microbial anabolism (Venkata Mohan et al. 2017). The function of a microbial fuel cell (MFC) is based on harnessing the available electrons by introducing electrodes as intermediary electron acceptors and make them flow from the donors to the acceptors via an external circuit. Bioelectricity is harvested while degrading organic waste and producing specific platform chemicals: microorganisms function as biocatalysts in these fuel cell systems, and the resulting electron flux from metabolism plays a pivotal role in bio-electrogenesis (Venkata Mohan et al. 2014b).

The MFC is composed of the cathode and anode chambers, separated by a proton exchange membrane which enables the transfer of protons to the cathode (Butti et al. 2016). MFC is a sustainable technology that is capable of converting the chemical energy of wastewaters directly to clean bioelectrical energy while treating them (Ishii et al. 2013). The presence of a terminal electron acceptor in the cathode acts as an electron driving force. For example, electrons in the cathode exposed to oxygen combine with it and with protons to form water, or they can be used for the reductive synthesis of valuable products (Venkata Mohan et al. 2014a).

MFC enabled clean bioelectricity production from (i) orange peel biomass by *Enterococcus, Paludibacter* and *Pseudomonas* (Miran et al. 2016); (ii) FW by exoelectrogenic *Geobacter,* and fermentative Bacteroides (Jia et al. 2013); (iii) acidogenic FW leachate (Rikame et al. 2012); (iv) canteen based composite waste (Goud et al. 2011) (iv) distillery wastewaters (Mohanakrishna et al. 2010); (v) food processing plant wastewaters (Oh and Logan 2005), (vi) breweries wastewaters (Dannys et al. 2016).

Under optimized conditions, open-circuit voltage as high as 1.12 V and 90% of organic load (expressed as chemical oxygen demand, COD) removal from FW leachate in the anodic chamber were observed, even if the ion exchange capacity of the ion exchange membrane was reduced due to membrane fouling by deposition of microorganism and cations (Rikame et al. 2012).

Canteen composite FW was ground and used as solid feed for the operation of a solid-state bio-electrofermentation system, which can function on the self-driven bioelectrogenic activity scale (Mohan and Chandrasekhar 2011). The system maximum open-circuit voltage and power density were, respectively, 443 mV and 162.4 mW/m^2, on the 9th day of operation; 72% substrate removal efficiency and simultaneous biohydrogen, bioethanol, and bioelectricity production were obtained at the laboratory scale (Chandrasekhar et al. 2014).

4.3.4.2 Electro-Fermentation

To support their metabolic pathways, all organisms need an electron donor and acceptor. Electro-fermentation (EF) is a novel process that consists of electrochemically controlling microbial fermentative metabolism and rate kinetics with electrodes poised with a certain voltage to anodically subtract or cathodically supply electrons to the microbial cell metabolism.

AF stimulated by the electron flux is an elegant alternative that can be profitably used to overcome the thermodynamic limitations of conventional microbial fermentation and to fine-tune the metabolism towards specific biobased products synthesis. Microbes can exchange electrons with solid conductors through (i) direct transfer from the biofilm to the electrode and (ii) mediated transfer via soluble redox-active molecules (e.g., H_2, flavins, humic acids) which act as electron shuttles for the microbial population. Carbon- and graphene-based electrodes are very common because of their good conductivity, chemical stability, and cheap availability (Sravan et al. 2018). Single or dual-chambered reactors can be used according to the compatibility/incompatibility of the products formed at each electrode.

Anodic and cathodic currents can steer specific microbial metabolic reactions: at the anode, the solid conductor is an electron sink because a positive potential is imposed: it favors oxidative reactions, thereby fostering NADH consumption and creating proton gradients sufficient for ATP generation. At the cathode, the negative imposed potential makes it an electron supply that induces metabolic reductive reactions via the available reducing equivalents, such as chain elongation of carboxylates or alcohols. This way, fermentative routes can be electrically enhanced (Schievano et al. 2016).

Protons produced in the anodic compartment migrate to the cathode via a proton exchange membrane and, there, they are reduced to form H_2 in the presence of electrons arrived from the anode: the small voltage is applied to enable the crossing of endothermic activation energy barrier to form biohydrogen. The modulation of the potential in the EF system can spur different metabolic outputs, such as ethanol, butanol, formate, lactate, propionate, succinate, etc. (Venkata Mohan et al. 2014b).

Electrode introduction in microbial fermentation also has an electrocatalytic effect in increasing the rate of redox reactions and eliminate redox constraints (Shanthi Sravan et al. 2018).

Since most bacteria are negatively charged, imposing a positive potential to the anode accelerates their migration and assists in selective biofilm enrichment for better substrate removal and conversion of the organic sources into valuable platform chemicals (Arunasri et al. 2016). Simultaneously reductive reactions are facilitated at the negative cathode, and syntrophic interactions are at play in bioelectrochemical systems (Venkata Mohan et al. 2014b).

Since EF electrically enhances fermentative routes, it could broaden the possible range of fermentation substrates, including FW, and this would be crucial in biorefineries (Schievano et al. 2016).

A tailor-made platform chemicals production relied on the conversion of composite canteen-based FW to targeted carboxylic acids with coproduction of biogas: in particular, EF enabled the elongation of the chain length and the simultaneous production of biohythane (Shanthi Sravan et al. 2018).

Suspended conductive granular activated carbon can serve to enhance the performance of EF systems to get a huge methane flux (>300 mL CH_4 per g COD) from FW at up to 2.75 V, due to a higher activity of exoelectrogenic bacteria (*Geobacter*) on anodic biofilms and methanogenic activity of *Methanospirillum* in cathodic and activated carbon biofilms (Harb et al. 2020).

Figure 4.8 illustrate the basic principles of EF with various metabolic outputs.

Figure 4.9 illustrates multi-dimensional bioconversion approaches for the valorization of FW to value added biobased products.

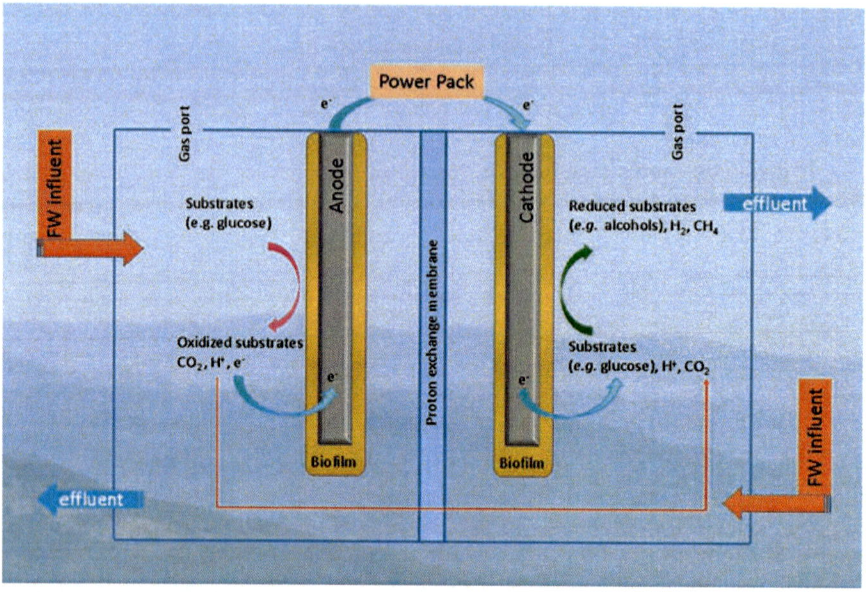

Fig. 4.8 Schematic of electro-fermentation with various metabolic outputs

4.4 Thermal and Thermochemical Methods

Heat-based conversions include combustion, incineration, gasification and subsequent liquefaction, pyrolysis, torrefaction, hydrothermal carbonization, and hydrothermal liquefaction (Tanger et al. 2013), (Zhou et al. 2020). They are particularly appropriate for the conversion of lignocellulosic feedstock; the high moisture content of other FW is challenging, but the wide possibilities offered by thermal conversions also enable their valorization.

Combustion is a highly exothermic complete oxidation of biomass; it is the most common process to produce bioenergy (heat or power) if the moisture content of the waste matrix is low. Complete combustion gives ash, CO_2, water, and heat, but if combustion is not complete, carbon monoxide (CO) and other unburnt substances are present in the flue gas. Incineration is combustion deliberately carried out in order to destroy organic matter.

Gasification is the partial oxidation of biomass with a reduced oxidant level. CO and H_2 are the output products of gasification. Gasification technologies are significant in areas with substantial forestry biomass capacity while they are less important for FW. The product of gasification can be converted into liquid hydrocarbons (liquefaction) via the Fischer–Tropsch process, usually in the presence of metal catalysts, typically at temperatures of 150–300 °C and pressures up to several tens of atmospheres to produce low sulfur liquid fuels or lubrication oil (Peduzzi et al. 2018).

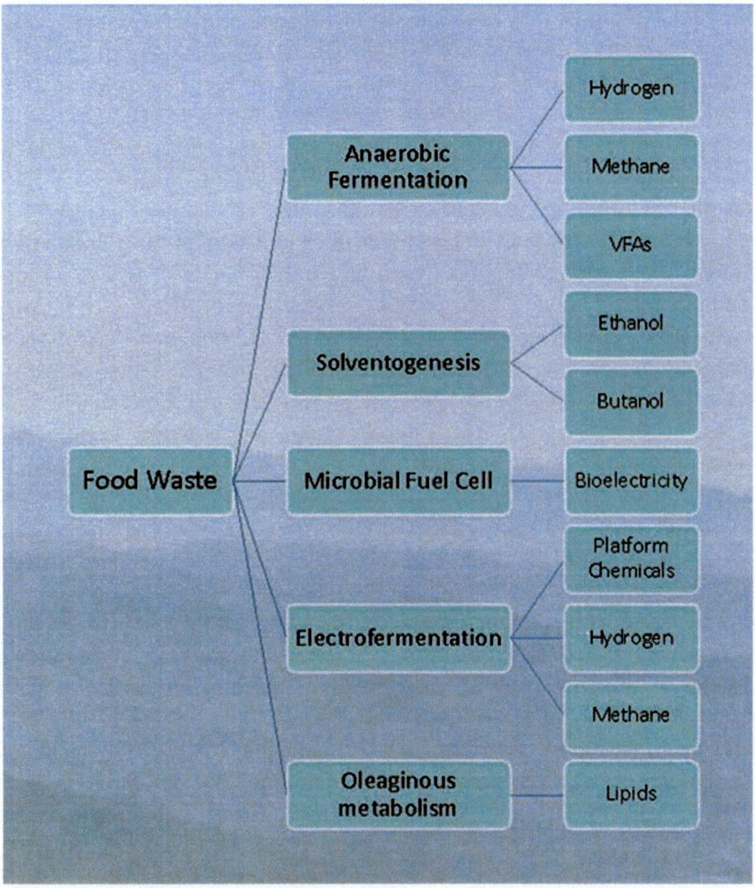

Fig. 4.9 Overview of multi-dimensional bioconversion approaches for valorization of food waste to value added biobased products in circular bioeconomy

Pyrolysis is the thermo-chemical disassembly of biomass at high temperatures (approximately between 500 °C and 800 °C) in the absence of air. Pyrolysis technologies are supposed to be economically efficient for combined heat and power generation only within the biorefinery concept due to the possible cogeneration of energy and other products of higher value (Tanger et al. 2013).

Torrefaction is a gentler form of pyrolysis conducted at lower temperatures, typically between 200 and 320 °C (Peduzzi et al. 2018).

Hydrothermal carbonization is the method of choice for thermal conversion of high moisture content FW into hydrochar; for example, kitchen waste and deionized water (wastewater ratio of 1:1.8) were heated to and kept at 195 °C, 225 °C and 255 °C for 12 h under constant stirring during the whole process. The hydrochar was obtained after cooling and vacuum filtration of the mixture and drying of the solid product at 105 °C for 24 h. It was then used as an additive to regulate the biogas and methane production in the anaerobic digestion of kitchen waste thanks to redox-active moieties on its surface (Zhou et al. 2020).

Torrefaction, slow pyrolysis and hydrothermal carbonization can be all profitably used to produce biochar and hydrochar from bio-mass. Torrefaction is also used as a pre-process for increasing the volumetric energy content of biomass pellets and to avoid microbiological spoilage during storage (Ahmad et al. 2019).

Hydrothermal liquefaction technology for bio-oil production from FW is a thermochemical process that converts wet feedstocks into liquid fuels with subcritical or critical water at temperatures between 250 and 374 °C and pressures from 4 to 22 MPa for a highly reactive environment. Several complex reactions occur in the feedstock, which is usually a complex mixture of carbohydrates, lignin, proteins, and lipids, and water. The conversion can be divided into three main steps: (i) biopolymers are first hydrolyzed into small polar oligomers and monomers thanks to the elevated temperatures; (ii) the harsh temperature and pressure conditions trigger further decomposition reactions, such as dehydration, decarboxylation, and deamination, thereby generating highly reactive species; and (iii) the reactive mixture can give bio-oils or char due to subsequent repolymerization reactions; the process was successfully applied to FW such as pineapple peel, banana peel, and watermelon peel (Chen et al. 2020).

Many of the illustrated techniques are still in their expensive infancy, although they have several potential applications. Efficient thermochemical methods entail standardization of feedstock by the optimization of lignin content and reduction of minerals and moisture content (Tanger et al. 2013).

An interesting study on prominent thermochemical conversion technologies, such as slow pyrolysis, fast pyrolysis, gasification, and hydrothermal liquefaction, and different downstream processing options such as bio-oil upgrading and Fischer–Tropsch conversion for treating poultry litter were recently analyzed. Pyrolysis had lower greenhouse emissions than the other technologies and good economic performance (Bora et al. 2020).

Hybrid approaches based on combined thermochemical–biochemical methods are currently in the limelight of researchers' attention and represent the greatest hope for the exploitation of biomass to produce a broad range of value added products (Shen et al. 2015), but the toxicity of the pyrolytic substrates and the presence of growth inhibitors from raw syngas contaminants are critical challenges which limit the efforts to commercialize hybrid processing.

4.5 Conclusions

The safety of products recovered from FW and the impact (beneficial or not) of recycling them inside the food chain is an open debate.

Scale up is might not be straightforward due to transition of the batch to continuous processes.

It has to be emphasized that if active compounds are obtained in pure form, their safety should be checked similarly to synthetic substances to obtain a market release permission. Avoiding extreme processing conditions (i.e., using non-thermal technologies) and applying safe solvents and materials is highly desirable within the green chemistry framework.

Despite ongoing research towards the development of green procedures free from environmental aggravation and cost, in several cases, integrated eco-compatible technologies are still in their infancy.

Unfortunately, the vast majority of studies have been conducted on a laboratory scale with a specific recovery strategy, on a particular recovery stage, on a specific FW. The cost analysis of a proposed approach to recover a particular value added product from a certain FW, which is necessary to bridge the gap between academia and industry, is not easily feasible. Safety consideration, functionality and organoleptic characteristics of the end product are crucial for its commercialization.

Future research should be oriented towards (i) the extensive testing of new, allegedly "green" processes for cost-effectiveness, and environmental sustainability , (ii) the introduction of novel production strategies to reduce unit operations, which in turn would reduce the associated energy demands and expenses, (iii) the comparative evaluation of various such technologies to ascertain suitability for a given purpose and (iv) the carefully designed scale-up procedures that would pave the way for industrial applications.

Bibliography

2008/98/EC D (2008) Directive 2008/98/EC of the European Parliament and of the Council of 19 November 2008 on waste and repealing certain Directives (Text with EEA relevance). europa.eu

Abejón R, Abejón A, Garea A et al (2016) In silico evaluation of ultrafiltration and nanofiltration membrane cascades for continuous fractionation of protein hydrolysate from tuna processing byproduct. Ind Eng Chem Res 55:7493–7504. https://doi.org/10.1021/acs.iecr.6b01495

Afonso MD, Bórquez R (2002) Review of the treatment of seafood processing wastewaters and recovery of proteins therein by membrane separation processes – prospects of the ultrafiltration of wastewaters from the fish meal industry. Desalination 142:29–45. https://doi.org/10.1016/S0011-9164(01)00423-4

Aghapour Aktij S, Zirehpour A, Mollahosseini A et al (2020) Feasibility of membrane processes for the recovery and purification of bio-based volatile fatty acids: a comprehensive review. J Ind Eng Chem 81:24–40

Aguirre Calvo TR, Perullini M, Santagapita PR (2018) Encapsulation of betacyanins and polyphenols extracted from leaves and stems of beetroot in Ca(II)-alginate beads: a structural study. J Food Eng 235:32–40

Agyeman FO, Tao W (2014) Anaerobic co-digestion of food waste and dairy manure: effects offood waste particle size and organic loading rate. J Environ Manag 133:268–274. https://doi.org/10.1016/j.jenvman.2013.12.016

Ahangari B, Sargolzaei J (2013) Extraction of lipids from spent coffee grounds using organic solvents and supercritical carbon dioxide. J Food Process Preserv 37:1014–1021. https://doi.org/10.1111/j.1745-4549.2012.00757.x

Ahmad FB, Zhang Z, Doherty WOS, O'Hara IM (2019) The outlook of the production of advanced fuels and chemicals from integrated oil palm biomass biorefinery. Renew Sust Energ Rev 109:386–411

Alvarez RM, Rodríguez B, Romano JM et al (1992) Lipid accumulation in Rhodotorula glutinis on sugar cane molasses in single-stage continuous culture. World J Microbiol Biotechnol 8:214–215. https://doi.org/10.1007/BF01195853

Amani T, Nosrati M, Mousav SM, Kermanshahi RK (2011) Study of syntrophic anaerobic digestion of volatile fatty acids using enriched cultures at mesophilic conditions. Int J Environ Sci Technol 8:83–96. https://doi.org/10.1007/BF03326198

Annevelink E, Van Gogh B, Bartels P et al (2017) Vision paper how to achieve resource use efficiency in integrated food and biobased value chains? https://doi.org/10.18174/407896

Aroujalian A, Raisi A (2007) Recovery of volatile aroma components from orange juice by pervaporation. J Memb Sci 303:154–161. https://doi.org/10.1016/j.memsci.2007.07.004

Arunasri K, Annie Modestra J, Yeruva DK et al (2016) Polarized potential and electrode materials implication on electro-fermentative di-hydrogen production: microbial assemblages and hydrogenase gene copy variation. Bioresour Technol 200:691–698. https://doi.org/10.1016/j.biortech.2015.10.091

Bafrani MP (2010) Thesis for the Degree of Doctor of Philosophy Citrus waste biorefinery: process development, Simulation and Economic Analysis

Baiano A, Bevilacqua L, Terracone C et al (2014) Single and interactive effects of process variables on microwave-assisted and conventional extractions of antioxidants from vegetable solid wastes. J Food Eng 120:135–145. https://doi.org/10.1016/j.jfoodeng.2013.07.010

Balti R, Bougatef A, Ali NEH et al (2010) Influence of degree of hydrolysis on functional properties and angiotensin I-converting enzyme-inhibitory activity of protein hydrolysates from cuttlefish (Sepia officinalis) by-products. J Sci Food Agric 90:2006–2014. https://doi.org/10.1002/jsfa.4045

Barba D, Beolchini F, Cifoni D, Veglió F (2001) Whey protein concentrate production in a pilot scale two-stage diafiltration process. Sep Sci Technol 36:587–603. https://doi.org/10.1081/SS-100102948

Barbosa-Pereira L, Pocheville A, Angulo I et al (2013) Fractionation and purification of bioactive compounds obtained from a brewery waste stream. Biomed Res Int 2013:11. https://doi.org/10.1155/2013/408491

Bauer F, Coenen L, Hansen T et al (2017) Technological innovation systems for biorefineries: a review of the literature. Biofuels Bioprod Biorefin 11:534–548. https://doi.org/10.1002/bbb.1767

Berardini N, Knödler M, Schieber A, Carle R (2005) Utilization of mango peels as a source of pectin and polyphenolics. Innov Food Sci Emerg Technol 6:442–452. https://doi.org/10.1016/j.ifset.2005.06.004

Betz M, Kulozik U (2011) Whey protein gels for the entrapment of bioactive anthocyanins from bilberry extract. Int Dairy J 21:703–710. https://doi.org/10.1016/j.idairyj.2011.04.003

Bhattacharjee S, Bhattacharjee C, Datta S (2006) Studies on the fractionation of β-lactoglobulin from casein whey using ultrafiltration and ion-exchange membrane chromatography. J Memb Sci 275:141–150. https://doi.org/10.1016/j.memsci.2005.09.013

Bonfigli M, Godoy E, Reinheimer MA, Scenna NJ (2017) Comparison between conventional and ultrasound-assisted techniques for extraction of anthocyanins from grape pomace. Experimental results and mathematical modeling. J Food Eng 207:56–72. https://doi.org/10.1016/j.jfoodeng.2017.03.011

Boonrod D, Reanma K, Niamsup H (2006) Extraction and physicochemical characteristics of acid-soluble pectin from raw papaya (Carica papaya) peel. Chiang Mai J Sci 33:129–135

Bora RR, Tao Y, Lehmann J et al (2020) Techno-economic feasibility and spatial analysis of thermochemical conversion pathways for regional poultry waste valorization. ACS Sustain Chem Eng 8:5763–5775. https://doi.org/10.1021/acssuschemeng.0c01229

Bosiljkov T, Dujmić F, Cvjetko Bubalo M et al (2017) Natural deep eutectic solvents and ultrasound-assisted extraction: green approaches for extraction of wine lees anthocyanins. Food Bioprod Process 102:195–203. https://doi.org/10.1016/j.fbp.2016.12.005

Bourtoom T, Chinnan MS, Jantawat P, Sanguandeekul R (2009) Recovery and characterization of proteins precipitated from surimi wash-water. LWT – Food Sci Technol 42:599–605. https://doi.org/10.1016/j.lwt.2008.09.001

Boussetta N, Lanoisellé JL, Bedel-Cloutour C, Vorobiev E (2009) Extraction of soluble matter from grape pomace by high voltage electrical discharges for polyphenol recovery: effect of Sulphur dioxide and thermal treatments. J Food Eng 95:192–198. https://doi.org/10.1016/j.jfoodeng.2009.04.030

Bund RK, Pandit AB (2007) Rapid lactose recovery from paneer whey using sonocrystallization: a process optimization. Chem Eng Process Process Intensif 46:846–850. https://doi.org/10.1016/j.cep.2007.05.015

Butti SK, Velvizhi G, Sulonen MLK et al (2016) Microbial electrochemical technologies with the perspective of harnessing bioenergy: maneuvering towards upscaling. Renew Sust Energ Rev 53:462–476

Buzzini P, Martini A (2000) Production of carotenoids by strains of Rhodotorula glutinis cultured in raw materials of agro-industrial origin. Bioresour Technol 71:41–44. https://doi.org/10.1016/S0960-8524(99)00056-5

Caballero AS, Romero-García JM, Castro E, Cardona CA (2020) Supercritical fluid extraction for enhancing polyphenolic compounds production from olive waste extracts. J Chem Technol Biotechnol 95:356–362. https://doi.org/10.1002/jctb.5907

Caderby E, Baumberger S, Hoareau W et al (2013) Sugar cane stillage: a potential source of natural antioxidants. J Agric Food Chem 61:11494–11501. https://doi.org/10.1021/jf4039474

Cahú TB, Santos SD, Mendes A et al (2012) Recovery of protein, chitin, carotenoids and glycosaminoglycans from Pacific white shrimp (Litopenaeus vannamei) processing waste. Process Biochem 47:570–577. https://doi.org/10.1016/j.procbio.2011.12.012

Caldas TW, Mazza KEL, Teles ASC et al (2018) Phenolic compounds recovery from grape skin using conventional and non-conventional extraction methods. Ind Crop Prod 111:86–91. https://doi.org/10.1016/j.indcrop.2017.10.012

Campone L, Celano R, Piccinelli AL et al (2018) Response surface methodology to optimize supercritical carbon dioxide/co-solvent extraction of brown onion skin by-product as source of nutraceutical compounds. Food Chem 269:495–502. https://doi.org/10.1016/j.foodchem.2018.07.042

Cardinali A, Pati S, Minervini F et al (2012) Verbascoside, isoverbascoside, and their derivatives recovered from olive mill wastewater as possible food antioxidants. J Agric Food Chem 60:1822–1829. https://doi.org/10.1021/jf204001p

Casazza AA, Aliakbarian B, Mantegna S et al (2010) Extraction of phenolics from Vitis vinifera wastes using non-conventional techniques. J Food Eng 100:50–55. https://doi.org/10.1016/j.jfoodeng.2010.03.026

Cassano A, Conidi C, Giorno L, Drioli E (2013) Fractionation of olive mill wastewaters by membrane separation techniques. J Hazard Mater 248–249:185–193. https://doi.org/10.1016/j.jhazmat.2013.01.006

Castanha RF, Mariano AP, de Morais LAS et al (2014) Optimization of lipids production by Cryptococcus laurentii 11 using cheese whey with molasses. Brazilian J Microbiol 45:379–387. https://doi.org/10.1590/S1517-83822014000200003

Chandra R, Arora S, Rohit MV, Venkata Mohan S (2015a) Lipid metabolism in response to individual short chain fatty acids during mixotrophic mode of microalgal cultivation: influence on biodiesel saturation and protein profile. Bioresour Technol 188:169–176. https://doi.org/10.1016/j.biortech.2015.01.088

Chandra R, Nikhil GN, Venkata Mohan S (2015b) Single-stage operation of hybrid dark-photo fermentation to enhance biohydrogen production through regulation of system redox condition: evaluation with real-field wastewater. Int J Mol Sci 16:9540–9556. https://doi.org/10.3390/ijms16059540

Chandrasekhar K, Amulya K, Venkata Mohan S (2014) Solid phase bio-electrofermentation of food waste to harvest value-added products associated with waste remediation. Waste Manag 45:57–65. https://doi.org/10.1016/j.wasman.2015.06.001

Chemat F, Zill-E-Huma, Khan MK (2011) Applications of ultrasound in food technology: processing, preservation and extraction. In: Ultrasonics sonochemistry. Elsevier B.V, pp 813–835
Chen YC, Jaczynski J (2007) Protein recovery from rainbow trout (Oncorhynchus mykiss) processing byproducts via isoelectric solubilization/precipitation and its gelation properties as affected by functional additives. J Agric Food Chem 55:9079–9088. https://doi.org/10.1021/jf071992w
Chen Y-C, Tou JC, Jaczynski J (2007) Amino acid, fatty acid, and mineral profiles of materials recovered from rainbow trout (Oncorhynchus mykiss) processing by-products using isoelectric solubilization/precipitation. J Food Sci 72:C527–C535. https://doi.org/10.1111/j.1750-3841.2007.00522.x
Chen WH, Lin YY, Liu HC, Baroutian S (2020) Optimization of food waste hydrothermal liquefaction by a two-step process in association with a double analysis. Energy 199:117438. https://doi.org/10.1016/j.energy.2020.117438
Chi Z, Zheng Y, Jiang A, Chen S (2011) Lipid production by culturing oleaginous yeast and algae with food waste and municipal wastewater in an integrated process. Appl Biochem Biotechnol 165:442–453. https://doi.org/10.1007/s12010-011-9263-6
Chintagunta AD, Jacob S, Banerjee R (2016) Integrated bioethanol and biomanure production from potato waste. Waste Manag 49:320–325. https://doi.org/10.1016/j.wasman.2015.08.010
Chintagunta AD, Ray S, Banerjee R (2017) An integrated bioprocess for bioethanol and biomanure production from pineapple leaf waste. J Clean Prod 165:1508–1516. https://doi.org/10.1016/j.jclepro.2017.07.179
Chisti Y (2007) Biodiesel from microalgae. Biotechnol. Adv. 25:294–306. https://doi.org/10.1016/j.biotechadv.2007.02.001
Chmiel H, Kaschek M, Blöcher C et al (2003) Concepts for the treatment of spent process water in the food and beverage industries. Desalination 152:307–314. https://doi.org/10.1016/S0011-9164(02)01078-0
Coma M, Martinez-Hernandez E, Abeln F et al (2017) Organic waste as a sustainable feedstock for platform chemicals. Faraday Discuss 202:175–195. https://doi.org/10.1039/c7fd00070g
Conidi C, Cassano A, Garcia-Castello E (2014) Valorization of artichoke wastewaters by integrated membrane process. Water Res 48:363–374. https://doi.org/10.1016/j.watres.2013.09.047
Crea R (2002) Method of obtaining a hydroxytyrosol-rich composition from vegetation water, WO/2002/0218310. World Intellectual Property Organization, Crowely
Da Porto C, Natolino A (2017) Supercritical fluid extraction of polyphenols from grape seed (Vitis vinifera): study on process variables and kinetics. J Supercrit Fluids 130:239–245. https://doi.org/10.1016/j.supflu.2017.02.013
da Silva FV, dos Santos RLA, Fujiki TL, et al (2010) Desenvolvimento de sistema de controle para precipitação de bromelina a partir de resíduos de abacaxi. Cienc e Tecnol Aliment 30:1033–1040. doi: https://doi.org/10.1590/S0101-20612010000400031
Dahiya S, Joseph J (2015) High rate biomethanation technology for solid waste management and rapid biogas production: an emphasis on reactor design parameters. Bioresour Technol 188:73–78. https://doi.org/10.1016/j.biortech.2015.01.074
Dahiya S, Sarkar O, Swamy YV, Venkata Mohan S (2015) Acidogenic fermentation of food waste for volatile fatty acid production with co-generation of biohydrogen. Bioresour Technol 182:103–113. https://doi.org/10.1016/j.biortech.2015.01.007
Dahiya S, Kumar AN, Shanthi Sravan J et al (2018) Food waste biorefinery: sustainable strategy for circular bioeconomy. Bioresour Technol 248:2–12
Dahmoune F, Madani K, Jauregi P et al (2013) Fractionation of a red grape marc extract by colloidal gas aphrons. In: Chemical Engineering Transactions. Italian Association of Chemical Engineering – AIDIC, In, pp 1903–1908
Dai K, Wen JL, Wang YL et al (2019) Impacts of medium composition and applied current on recovery of volatile fatty acids during coupling of electrodialysis with an anaerobic digester. J Clean Prod 207:483–489. https://doi.org/10.1016/j.jclepro.2018.10.019
Dang Y-Y, Zhang H, Xiu Z-L (2014) Microwave-assisted aqueous two-phase extraction of phenolics from grape (*Vitis vinifera*) seed. J Chem Technol Biotechnol 89:1576–1581. https://doi.org/10.1002/jctb.4241

Dannys E, Green T, Wettlaufer A et al (2016) Wastewater Treatment with Microbial Fuel Cells: A Design and Feasibility Study for Scale-up in Microbreweries. J Bioprocess Biotech 6:1–6. https://doi.org/10.4172/2155-9821.1000267

David B, Federico B, Cristina C et al (2019) Biohythane production from food wastes. In: Biohydrogen, In, Elsevier, pp 347–368

de Jong E, Higson A, Walsh P, Wellisch M (2012) Value Added Products from Biorefinerie IEA Bioenergy 33

de Torres C, Díaz-Maroto MC, Hermosín-Gutiérrez I, Pérez-Coello MS (2010) Effect of freeze-drying and oven-drying on volatiles and phenolics composition of grape skin. Anal Chim Acta 660:177–182. https://doi.org/10.1016/j.aca.2009.10.005

Deepanraj B, Sivasubramanian V, Jayaraj S (2017) Multi-response optimization of process parameters in biogas production from food waste using Taguchi – Grey relational analysis. Energy Convers Manag 141:429–438. https://doi.org/10.1016/j.enconman.2016.12.013

Dennis C, Aguilera JM, Satin M (2009) Technologies shaping the future. In: Agro-industries for development. CABI Publishing, pp 92–135

Deora NS, Misra NN, Deswal A et al (2013) Ultrasound for improved crystallisation in food processing. Food Eng Rev 5:36–44

Desobry SA, Netto FM, Labuza TP (1997) Comparison of Spray-drying, Drum-drying and Freeze-drying for β-Carotene Encapsulation and Preservation. J Food Sci 62:1158–1162

Dhiman SS, David A, Shrestha N et al (2017) Conversion of raw and untreated disposal into ethanol. Bioresour Technol 244:733–740. https://doi.org/10.1016/j.biortech.2017.07.102

Di Mauro A, Fallico B, Passerini A et al (1999) Recovery of hesperidin from orange peel by concentration of extracts on styrene-divinylbenzene resin. J Agric Food Chem 47:4391–4397. https://doi.org/10.1021/jf990038z

DOE/EERE (2011) Biochemical conversion: using hydrolysis, fermentation, and catalysis to make fuels and chemicals. U.S. Department Of Energy. Office of Energy Efficiency and Renewable

Doyen A, Beaulieu L, Saucier L et al (2011) Impact of ultrafiltration membrane material on peptide separation from a snow crab byproduct hydrolysate by electrodialysis with ultrafiltration membranes. J Agric Food Chem 59:1784–1792. https://doi.org/10.1021/jf103739m

Drosou C, Kyriakopoulou K, Bimpilas A et al (2015) A comparative study on different extraction techniques to recover red grape pomace polyphenols from vinification byproducts. Ind Crop Prod 75:141–149. https://doi.org/10.1016/j.indcrop.2015.05.063

Duba KS, Casazza AA, Ben MH et al (2015) Extraction of polyphenols from grape skins and defatted grape seeds using subcritical water: experiments and modeling. Food Bioprod Process 94:29–38. https://doi.org/10.1016/j.fbp.2015.01.001

Dugmore TIJ, Clark JH, Bustamante J et al (2017) Valorisation of biowastes for the production of green materials using chemical methods. Top Curr Chem 375:46. https://doi.org/10.1007/s41061-017-0133-8

El-Sayed MMH, Chase HA (2011) Trends in whey protein fractionation. Biotechnol Lett 33:1501–1511

Esteban J, Ladero M (2018) Food waste as a source of value-added chemicals and materials: a biorefinery perspective. Int J Food Sci Technol 53:1095–1108. https://doi.org/10.1111/ijfs.13726

Eun JB, Maruf A, Das PR, Nam SH (2020) A review of encapsulation of carotenoids using spray drying and freeze drying. Crit Rev Food Sci Nutr 60:3547–3572. https://doi.org/10.1080/10408398.2019.1698511

European Commission (2015) Sustainable agriculture, forestry and fisheries in the bioeconomy

Fadhil AB, Al-Tikrity ETB, Albadree MA (2017) Biodiesel production from mixed non-edible oils, castor seed oil and waste fish oil. Fuel 210:721–728. https://doi.org/10.1016/j.fuel.2017.09.009

Fang Z, Bhandari B (2010) Encapsulation of polyphenols – a review. Trends Food Sci Technol 21:510–523

Farhat A, Fabiano-Tixier AS, El Maataoui M et al (2011) Microwave steam diffusion for extraction of essential oil from orange peel: kinetic data, extract's global yield and mechanism. Food Chem 125:255–261. https://doi.org/10.1016/j.foodchem.2010.07.110

Ferhat MA, Meklati BY, Smadja J, Chemat F (2006) An improved microwave Clevenger apparatus for distillation of essential oils from orange peel. J Chromatogr A 1112:121–126. https://doi.org/10.1016/j.chroma.2005.12.030

Fernandes P, Cabral JMS (2007) Phytosterols: applications and recovery methods. Bioresour Technol 98:2335–2350. https://doi.org/10.1016/j.biortech.2006.10.006

Fernández CM, Ramos MJ, Pérez Á, Rodríguez JF (2010) Production of biodiesel from winery waste: extraction, refining and transesterification of grape seed oil. Bioresour Technol 101:7019–7024. https://doi.org/10.1016/j.biortech.2010.04.014

Ferri F, Bertin L, Scoma A et al (2011) Recovery of low molecular weight phenols through solid-phase extraction. Chem Eng J 166:994–1001. https://doi.org/10.1016/j.cej.2010.11.090

Fishman ML, Chau HK, Hoagland P, Ayyad K (1999) Characterization of pectin, flash-extracted from orange albedo by microwave heating, under pressure. Carbohydr Res 323:126–138. https://doi.org/10.1016/S0008-6215(99)00244-X

Fuda E, Jauregi P (2006) An insight into the mechanism of protein separation by colloidal gas aphrons (CGA) generated from ionic surfactants. J Chromatogr B Anal Technol Biomed Life Sci 843:317–326. https://doi.org/10.1016/j.jchromb.2006.06.032

Galanakis CM (2012) Recovery of high added-value components from food wastes: conventional, emerging technologies and commercialized applications. Trends Food Sci Technol

Galanakis CM, Tornberg E, Gekas V (2010a) Dietary fiber suspensions from olive mill wastewater as potential fat replacements in meatballs. LWT – Food Sci Technol 43:1018–1025. https://doi.org/10.1016/j.lwt.2009.09.011

Galanakis CM, Tornberg E, Gekas V (2010b) The effect of heat processing on the functional properties of pectin contained in olive mill wastewater. LWT – Food Sci Technol. https://doi.org/10.1016/j.lwt.2010.01.004

Galanakis CM, Tornberg E, Gekas V (2010c) A study of the recovery of the dietary fibres from olive mill wastewater and the gelling ability of the soluble fibre fraction. LWT – Food Sci Technol 43:1009–1017. https://doi.org/10.1016/j.lwt.2010.01.005

Galanakis CM, Tornberg E, Gekas V (2010d) Clarification of high-added value products from olive mill wastewater. J Food Eng 99:190–197. https://doi.org/10.1016/j.jfoodeng.2010.02.018

Galanakis CM, Fountoulis G, Gekas V (2012) Nanofiltration of brackish groundwater by using a polypiperazine membrane. Desalination 286:277–284. https://doi.org/10.1016/j.desal.2011.11.035

Galanakis CM, Markouli E, Gekas V (2013) Recovery and fractionation of different phenolic classes from winery sludge using ultrafiltration. Sep Purif Technol 107:245–251. https://doi.org/10.1016/j.seppur.2013.01.034

Gargiulo V, Lanzetta R, Parrilli M, De Castro C (2009) Structural analysis of chondroitin sulfate from Scyliorhinus canicula: a useful source of this polysaccharide. Glycobiology 19:1485–1491. https://doi.org/10.1093/glycob/cwp123

Gehring CK, Gigliotti JC, Moritz JS et al (2011) Functional and nutritional characteristics of proteins and lipids recovered by isoelectric processing of fish by-products and low-value fish: a review. Food Chem 124:422–431

Ghafoor K, Park J, Choi YH (2010) Optimization of supercritical fluid extraction of bioactive compounds from grape (Vitis labrusca B.) peel by using response surface methodology. Innov Food Sci Emerg Technol 11:485–490. https://doi.org/10.1016/j.ifset.2010.01.013

Gildemyn S, Molitor B, Usack JG et al (2017) Upgrading syngas fermentation effluent using Clostridium kluyveri in a continuous fermentation. Biotechnol Biofuels 10

Girotto F, Lavagnolo MC, Pivato A, Cossu R (2017) Acidogenic fermentation of the organic fraction of municipal solid waste and cheese whey for bio-plastic precursors recovery – effects of process conditions during batch tests. Waste Manag 70:71–80. https://doi.org/10.1016/j.wasman.2017.09.015

González-Centeno MR, Comas-Serra F, Femenia A et al (2015) Effect of power ultrasound application on aqueous extraction of phenolic compounds and antioxidant capacity from grape pomace (Vitis vinifera L.): experimental kinetics and modeling. Ultrason Sonochem 22:506–514. https://doi.org/10.1016/j.ultsonch.2014.05.027

Goud RK, Babu PS, Mohan SV (2011) Canteen based composite food waste as potential anodic fuel for bioelectricity generation in single chambered microbial fuel cell (MFC): bio-electrochemical evaluation under increasing substrate loading condition. Int J Hydrog Energy 36:6210–6218. https://doi.org/10.1016/j.ijhydene.2011.02.056

Grootscholten TIM, Kinsky dal Borgo F, Hamelers HVM, Buisman CJN (2013) Promoting chain elongation in mixed culture acidification reactors by addition of ethanol. Biomass Bioenergy 48:10–16. https://doi.org/10.1016/j.biombioe.2012.11.019

Gupta VK, Mittal A, Malviya A, Mittal J (2009) Adsorption of carmoisine A from wastewater using waste materials-bottom ash and deoiled soya. J Colloid Interface Sci 335:24–33. https://doi.org/10.1016/j.jcis.2009.03.056

Hafid HS, Rahman NA, Md Shah UK, Baharudin AS (2015) Enhanced fermentable sugar production from kitchen waste using various pretreatments. J Environ Manag 156:290–298. https://doi.org/10.1016/j.jenvman.2015.03.045

Hall RM, Mayer DA, Mazzutti S, Ferreira SRS (2018) Simulating large scale SFE applied to recover bioactive compounds from papaya seeds. J Supercrit Fluids 140:302–309. https://doi.org/10.1016/j.supflu.2018.07.013

Harb M, Ermer N, Bounehme Sawaya C, Smith AL (2020) Increased applied voltage in the presence of GAC enhances microbial activity and methane production during anaerobic digestion of food waste. Environ Sci Water Res Technol 6:737–746. https://doi.org/10.1039/c9ew01000a

Hassan SS, Williams GA, Jaiswal AK (2019) Moving towards the second generation of lignocellulosic biorefineries in the EU: drivers, challenges, and opportunities. Renew Sust Energ Rev:590–599

Heebøll-Nielsen A, Justesen SFL, Thomas ORT (2004) Fractionation of whey proteins with high-capacity superparamagnetic ion-exchangers. J Biotechnol 113:247–262. https://doi.org/10.1016/j.jbiotec.2004.06.008

Herculano PN, Moreira KA, Bezerra RP et al (2016) Potential application of waste from castor bean (Ricinus communis L.) for production for xylanase of interest in the industry. 3 Biotech 6. https://doi.org/10.1007/s13205-016-0463-1

Hu S, Li Y, Lou W (2017) Novel efficient procedure for biodiesel synthesis from waste oils with high acid value using 1-sulfobutyl-3-methylimidazolium hydrosulfate ionic liquid as the catalyst. Chinese J Chem Eng 25:1519–1523. https://doi.org/10.1016/j.cjche.2017.03.029

Huang H, Singh V, Qureshi N (2015) Butanol production from food waste: a novel process for producing sustainable energy and reducing environmental pollution. Biotechnol Biofuels 8. https://doi.org/10.1186/s13068-015-0332-x

Humbird D, Davis R, Tao L, et al (2002) Process design and economics for biochemical conversion of lignocellulosic biomass to ethanol: dilute-acid pretreatment and enzymatic hydrolysis of corn stover

Ishii S, Suzuki S, Norden-Krichmar TM et al (2013) Identifying the microbial communities and operational conditions for optimized wastewater treatment in microbial fuel cells. Water Res 47:7120–7130. https://doi.org/10.1016/j.watres.2013.07.048

Iwami K, Hattori M, Yasumi T, Ibuki F (1988) Stability of gliadin-encapsulated unsaturated fatty acids against autoxidation. J Agric Food Chem 36:160–164. https://doi.org/10.1021/jf00079a041

Jafarian Asl P, Niazmand R, Yahyavi F (2020) Extraction of phytosterols and tocopherols from rapeseed oil waste by supercritical CO2 plus CO-solvent: a comparison with conventional solvent extraction. Heliyon 6. https://doi.org/10.1016/j.heliyon.2020.e03592

Jara F, Pilosof AMR (2011) Partitioning of α-lactalbumin and β-lactoglobulin in whey protein concentrate/hydroxypropylmethylcellulose aqueous two-phase systems. Food Hydrocoll 25:374–380. https://doi.org/10.1016/j.foodhyd.2010.07.003

Jarienė E, Danilčenko H, Aleknvičienė P, Kulaitienė J (2008) Expression – extraction of pumpkin oil. In: Experiments in unit operations and processing of foods. Springer, pp 53–61

Jarudilokkul S, Rungphetcharat K, Boonamnuayvitaya V (2004) Protein separation by colloidal gas aphrons using nonionic surfactant. Sep Purif Technol 35:23–29. https://doi.org/10.1016/S1383-5866(03)00111-4

Jia J, Tang Y, Liu B et al (2013) Electricity generation from food wastes and microbial community structure in microbial fuel cells. Bioresour Technol 144:94–99. https://doi.org/10.1016/j.biortech.2013.06.072

Jiang B, Wang J (2013) Method for extracting soy protein, oligosaccharide and isoflavone from soybean wastewater by one-step process. Chinese Pat CN103265614-

Jiang J, Gong C, Wang J et al (2014) Effects of ultrasound pre-treatment on the amount of dissolved organic matter extracted from food waste. Bioresour Technol 155:266–271. https://doi.org/10.1016/j.biortech.2013.12.064

Jiang H, Zhang M, Fang Z et al (2016) Effect of different dielectric drying methods on the physic-chemical properties of a starch-water model system. Food Hydrocoll 52:192–200. https://doi.org/10.1016/j.foodhyd.2015.06.021

Jumah R, Al-Asheh S, Banat F, Al-Zoubi K (2005) Electroosmotic dewatering of tomato paste suspension under AC electric field. Dry Technol 23:1465–1475. https://doi.org/10.1081/DRT-200063524

Kehili M, Kammlott M, Choura S et al (2017) Supercritical CO2 extraction and antioxidant activity of lycopene and β-carotene-enriched oleoresin from tomato (Lycopersicum esculentum L.) peels by-product of a Tunisian industry. Food Bioprod Process 102:340–349. https://doi.org/10.1016/j.fbp.2017.02.002

Kerton FM, Liu Y, Omari KW, Hawboldt K (2013) Green chemistry and the ocean-based biorefinery. Green Chem 15:860–871

Kim S-B, Ji C-I, Woo J-W et al (2012) Simplified purification of chondroitin sulphate from scapular cartilage of shortfin mako shark (Isurus oxyrinchus). Int J Food Sci Technol 47:91–99. https://doi.org/10.1111/j.1365-2621.2011.02811.x

Kitrytė V, Kavaliauskaitė A, Tamkutė L et al (2020) Zero waste biorefining of lingonberry (Vaccinium vitis-idaea L.) pomace into functional ingredients by consecutive high pressure and enzyme assisted extractions with green solvents. Food Chem 322. https://doi.org/10.1016/j.foodchem.2020.126767

Koduvayur Habeebullah SF, Nielsen NS, Jacobsen C (2010) Antioxidant activity of potato peel extracts in a fish-rapeseed oil mixture and in oil-in-water emulsions. JAOCS, J Am Oil Chem Soc 87:1319–1332. https://doi.org/10.1007/s11746-010-1611-0

Koubala BB, Mbome LI, Kansci G et al (2008) Physicochemical properties of pectins from ambarella peels (Spondias cytherea) obtained using different extraction conditions. Food Chem 106:1202–1207. https://doi.org/10.1016/j.foodchem.2007.07.065

Krishna D, Kalamdhad AS (2014) Pre-treatment and anaerobic digestion of food waste for high rate methane production – a review. J Environ Chem Eng 2:1821–1830

Kryževičiute N, Kraujalis P, Venskutonis PR (2016) Optimization of high pressure extraction processes for the separation of raspberry pomace into lipophilic and hydrophilic fractions. J Supercrit Fluids 108:61–68. https://doi.org/10.1016/j.supflu.2015.10.025

Kumar M, Gayen K, Saini S (2013) Role of extracellular cues to trigger the metabolic phase shifting from acidogenesis to solventogenesis in Clostridium acetobutylicum. Bioresour Technol 138:55–62. https://doi.org/10.1016/j.biortech.2013.03.159

Lama-Muñoz A, Rodríguez-Gutiérrez G, Rubio-Senent F et al (2013) A study of the precursors of the natural antioxidant phenol 3,4-dihydroxyphenylglycol in olive oil waste. Food Chem 140:154–160. https://doi.org/10.1016/j.foodchem.2013.02.063

Lee J, Kim J, Ok YS, Kwon EE (2017) Rapid biodiesel synthesis from waste pepper seeds without lipid isolation step. Bioresour Technol 239:17–20. https://doi.org/10.1016/j.biortech.2017.05.011

Li Y, Horsman M, Wu N et al (2008) Biomass production of commercial algae. Life Sci 1:56–60. https://doi.org/10.1021/bp.070371k

Li J, Ban Q, Zhang L, Jha AK (2012a) Syntrophic propionate degradation in anaerobic digestion: a review. Int J Agric Biol 14:843–850

Li S, Zhang X, Andresen JM (2012b) Production of fermentable sugars from enzymatic hydrolysis of pretreated municipal solid waste after autoclave process. Fuel 92:84–88. https://doi.org/10.1016/j.fuel.2011.07.012

Li J, Zu YG, Fu YJ et al (2010) Optimization of microwave-assisted extraction of triterpene saponins from defatted residue of yellow horn (Xanthoceras sorbifolia Bunge.) kernel and evaluation of its antioxidant activity. Innov Food Sci Emerg Technol 11:637–643. https://doi.org/10.1016/j.ifset.2010.06.004

Li Y, Fabiano-Tixier AS, Vian MA, Chemat F (2013) Solvent-free microwave extraction of bioactive compounds provides a tool for green analytical chemistry. TrAC – Trends Anal Chem 47:1–11

Liang Y, Tang T, Siddaramu T et al (2012) Lipid production from sweet sorghum bagasse through yeast fermentation. Renew Energy 40:130–136. https://doi.org/10.1016/j.renene.2011.09.035

Lin CSK, Pfaltzgraff LA, Herrero-Davila L et al (2013) Food waste as a valuable resource for the production of chemicals, materials and fuels. Current situation and global perspective energy. Environ Sci 6:426–464

Lipnizki F (2010) Cross-flow membrane applications in the food industry. In: Membrane technology. Wiley-VCH Verlag GmbH & Co. KGaA, Weinheim, pp 1–24

Liu Y, Whitman WB (2008) Metabolic, phylogenetic, and ecological diversity of the methanogenic archaea. Ann N Y Acad Sci 1125:171–189. https://doi.org/10.1196/annals.1419.019

Liu Y, Shi J, Langrish TAG (2006) Water-based extraction of pectin from flavedo and albedo of orange peels. Chem Eng J 120:203–209. https://doi.org/10.1016/j.cej.2006.02.015

Liu D., Vorobiev D., Savoire R., Lanoiselle JL (2011) Extraction of polyphenols from grape seeds by unconventional methods and extract concentration through polymeric membrane. Food Process Eng a Chang World Proc 11th Int Congr Eng Food 1939–1940

Lo YM, Cao D, Argin-Soysal S et al (2005) Recovery of protein from poultry processing wastewater using membrane ultrafiltration. Bioresour Technol 96:687–698. https://doi.org/10.1016/j.biortech.2004.06.026

Löf D, Schillén K, Nilsson L (2011) Flavonoids: precipitation kinetics and interaction with surfactant micelles. J Food Sci 76:35–39. https://doi.org/10.1111/j.1750-3841.2011.02103.x

Lusquiños F, Comesaña R, Riveiro A et al (2006) Laser cladding from marine wastes for biomedical applications. In: International congress on applications of lasers & electro-optics. Laser Institute of America, p P545

Luque de Castro MD, Priego-Capote F (2007) Ultrasound-assisted crystallization (sonocrystallization). Ultrason Sonochem 14:717–724. https://doi.org/10.1016/j.ultsonch.2006.12.004

Ma J, Zhang L, Mu L et al (2018) Thermally assisted bio-drying of food waste: synergistic enhancement and energetic evaluation. Waste Manag 80:327–338. https://doi.org/10.1016/j.wasman.2018.09.023

Malisorn C, Suntornsuk W (2008) Optimization of β-carotene production by Rhodotorula glutinis DM28 in fermented radish brine. Bioresour Technol 99:2281–2287. https://doi.org/10.1016/j.biortech.2007.05.019

Marques LG, Silveira AM, Freire JT (2006) Freeze-drying characteristics of tropical fruits. Dry Technol 24:457–463. https://doi.org/10.1080/07373930600611919

Martinez GA, Rebecchi S, Decorti D et al (2016) Towards multi-purpose biorefinery platforms for the valorisation of red grape pomace: production of polyphenols, volatile fatty acids, polyhydroxyalkanoates and biogas. Green Chem 18:261–270. https://doi.org/10.1039/c5gc01558h

Martins BC, Rescolino R, Coelho DF et al (2014) Characterization of bromelain from ananas comosus agroindustrial residues purified by ethanol factional precipitation. Chem Eng Trans 37:781–786. https://doi.org/10.3303/CET1437131

Masmoudi M, Besbes S, Chaabouni M et al (2008) Optimization of pectin extraction from lemon by-product with acidified date juice using response surface methodology. Carbohydr Polym 74:185–192. https://doi.org/10.1016/j.carbpol.2008.02.003

McClements DJ, Rao J (2011a) Food-grade nanoemulsions: formulation, fabrication, properties, performance, biological fate, and potential toxicity. Crit Rev Food Sci Nutr 51:285–330. https://doi.org/10.1080/10408398.2011.559558

McClements DJ, Rao J (2011b) Food-grade nanoemulsions: formulation, fabrication, properties, performance, biological fate, and potential toxicity. Crit Rev Food Sci Nutr 51:285–330

McClements DJ, Bai L, Chung C (2017) Recent advances in the utilization of natural emulsifiers to form and stabilize emulsions. Annu Rev Food Sci Technol 8:205–236. https://doi.org/10.1146/annurev-food-030216-030154

McHugh M, Krukonis V (2013) Supercritical fluid extraction: principles and practice

Miran W, Nawaz M, Jang J, Lee DS (2016) Conversion of orange peel waste biomass to bioelectricity using a mediator-less microbial fuel cell. Sci Total Environ 547:197–205. https://doi.org/10.1016/j.scitotenv.2016.01.004

Mohan SV, Chandrasekhar K (2011) Solid phase microbial fuel cell (SMFC) for harnessing bioelectricity from composite food waste fermentation: influence of electrode assembly and buffering capacity. Bioresour Technol 102:7077–7085. https://doi.org/10.1016/j.biortech.2011.04.039

Mohanakrishna G, Venkata Mohan S, Sarma PN (2010) Bio-electrochemical treatment of distillery wastewater in microbial fuel cell facilitating decolorization and desalination along with power generation. J Hazard Mater 177:487–494. https://doi.org/10.1016/j.jhazmat.2009.12.059

Monti L, Donati E, Zambrini AV, Contarini G (2018) Application of membrane technologies to bovine ricotta cheese exhausted whey (scotta). Int Dairy J 85:121–128. https://doi.org/10.1016/j.idairyj.2018.05.007

Moon M, Kim CW, Park WK et al (2013) Mixotrophic growth with acetate or volatile fatty acids maximizes growth and lipid production in Chlamydomonas reinhardtii. Algal Res 2:352–357. https://doi.org/10.1016/j.algal.2013.09.003

Mouratoglou E, Malliou V, Makris DP (2016) Novel glycerol-based natural eutectic mixtures and their efficiency in the ultrasound-assisted extraction of antioxidant polyphenols from agri-food waste biomass. Waste Biomass Valorization 7:1377–1387. https://doi.org/10.1007/s12649-016-9539-8

Murado MA, Fraguas J, Montemayor MI et al (2010) Preparation of highly purified chondroitin sulphate from skate (Raja clavata) cartilage by-products. Process optimization including a new procedure of alkaline hydroalcoholic hydrolysis. Biochem Eng J 49:126–132. https://doi.org/10.1016/j.bej.2009.12.006

Naik M, Rawson A, Rangarajan JM (2020) Radio frequency-assisted extraction of pectin from jackfruit (Artocarpus heterophyllus) peel and its characterization. J Food Process Eng. https://doi.org/10.1111/jfpe.13389

Nasini L, De Luca G, Ricci A et al (2016) Gas emissions during olive mill waste composting under static pile conditions. Int Biodeterior Biodegradation 107:70–76. https://doi.org/10.1016/J.IBIOD.2015.11.001

Nath A, Koris A, Verasztó B, Bhattacharjee C (2015) Selective separation of biomolecules from food waste by magnetic fishing technique – a review. In: Mediterranean microwave symposium. IEEE Computer Society

Nayak A, Bhushan B (2019) An overview of the recent trends on the waste valorization techniques for food wastes. J Environ Manag 233:352–370. https://doi.org/10.1016/j.jenvman.2018.12.041

Negri C, Ricci M, Zilio M et al (2020) Anaerobic digestion of food waste for bio-energy production in China and Southeast Asia: A review. Renew Sustain Energy Rev 133:110138. https://doi.org/10.1016/j.rser.2020.110138

Néel J (1995) Chapter 5 Pervaporation. Membrane Separations Technology. Principles and Applications. Technol 2:143–211. https://doi.org/10.1016/S0927-5193(06)80007-0

Neethirajan S, Jayas DS (2011) Nanotechnology for the food and bioprocessing industries. Food Bioprocess Technol 4:39–47

Ng A, Lecain S, Parker ML et al (1999) Modification of cell-wall polymers of onion waste. III. Effect of extrusion-cooking on cell-wall material of outer fleshy tissues. Carbohydr Polym 39:341–349. https://doi.org/10.1016/S0144-8617(99)00025-9

Ng SK, Plunkett A, Stojceska V et al (2011) Electro-kinetic technology as a low-cost method for dewatering food by-product. Dry Technol 29:1721–1728. https://doi.org/10.1080/07373937.2011.602199

Niphadkar SS, Vetal MD, Rathod VK (2015) Purification and characterization of polyphenol oxidase from waste potato peel by aqueous two-phase extraction. Prep Biochem Biotechnol 45:632–649. https://doi.org/10.1080/10826068.2014.940970

Nishimura H, Tan L, Kira N et al (2017) Production of ethanol from a mixture of waste paper and kitchen waste via a process of successive liquefaction, presaccharification, and simultaneous saccharification and fermentation. Waste Manag 67:86–94. https://doi.org/10.1016/j.wasman.2017.04.030

Ochando-Pulido JM, Corpas-Martínez JR, Vellido-Perez JA, Martinez-Ferez A (2020) Optimization of polymeric nanofiltration performance for olive-oil-washing wastewater phenols recovery and reclamation. Sep Purif Technol 236. https://doi.org/10.1016/j.seppur.2019.116261

Oh SE, Logan BE (2005) Hydrogen and electricity production from a food processing wastewater using fermentation and microbial fuel cell technologies. Water Res 39:4673–4682. https://doi.org/10.1016/j.watres.2005.09.019

Oreopoulou V, Russ W (2007) Utilization of by-products and treatment of waste in the food industry. Springer

Oreopoulou V, Tzia C (2007) Utilization of plant by-products for the recovery of proteins, dietary fibers, antioxidants, and colorants. In: Utilization of by-products and treatment of waste in the food industry. Springer, pp 209–232

Paes J, Dotta R, Barbero GF, Martínez J (2014) Extraction of phenolic compounds and anthocyanins from blueberry (Vaccinium myrtillus L.) residues using supercritical CO2 and pressurized liquids. J Supercrit Fluids 95:8–16. https://doi.org/10.1016/j.supflu.2014.07.025

Paini M, Casazza AA, Aliakbarian B et al (2016) Influence of ethanol/water ratio in ultrasound and high-pressure/high-temperature phenolic compound extraction from Agri-food waste. Int J Food Sci Technol 51:349–358. https://doi.org/10.1111/ijfs.12956

Pal P, Dey P (2013) Process intensification in lactic acid production by three stage membrane integrated hybrid reactor system. Chem Eng Process Process Intensif 64:1–9. https://doi.org/10.1016/j.cep.2012.12.006

Pan XR, Li WW, Huang L et al (2018) Recovery of high-concentration volatile fatty acids from wastewater using an acidogenesis-electrodialysis integrated system. Bioresour Technol 260:61–67. https://doi.org/10.1016/j.biortech.2018.03.083

Panchev IN, Kirtchev NA, Dimitrov DD (2011) Possibilities for application of laser ablation in food technologies. Innov Food Sci Emerg Technol 12:369–374. https://doi.org/10.1016/j.ifset.2011.02.008

Papoutsis K, Grasso S, Menon A et al (2020) Recovery of ergosterol and vitamin D2 from mushroom waste – potential valorization by food and pharmaceutical industries. Trends Food Sci Technol 99:351–366

Pasupuleti SB, Venkata Mohan S (2015) Single-stage fermentation process for high-value biohythane production with the treatment of distillery spent-wash. Bioresour Technol 189:177–185. https://doi.org/10.1016/j.biortech.2015.03.128

Pasupuleti SB, Sarkar O, Venkata Mohan S (2014) Upscaling of biohydrogen production process in semi-pilot scale biofilm reactor: evaluation with food waste at variable organic loads. Int J Hydrog Energy 39:7587–7596. https://doi.org/10.1016/j.ijhydene.2014.02.034

Pasquel Reátegui JL, Machado APDF, Barbero GF et al (2014) Extraction of antioxidant compounds from blackberry (Rubus sp.) bagasse using supercritical CO2 assisted by ultrasound. J Supercrit Fluids 94:223–233. https://doi.org/10.1016/j.supflu.2014.07.019

Patel SR, Murthy ZVP (2010) Optimization of process parameters by Taguchi method in the recovery of lactose from whey using sonocrystallization. Cryst Res Technol 45:747–752. https://doi.org/10.1002/crat.201000139

Patsea M, Stefou I, Grigorakis S, Makris DP (2017) Screening of natural sodium acetate-based low-transition temperature mixtures (LTTMs) for enhanced extraction of antioxidants and pigments from red vinification solid wastes. Environ Process 4:123–135. https://doi.org/10.1007/s40710-016-0205-8

Pavlić B, Đurković AV, Vladić J et al (2016) Extraction of minor compounds (chlorophylls and carotenoids) from yarrow–rose hip mixtures by traditional versus green technique. J Food Process Eng 39:418–424. https://doi.org/10.1111/jfpe.12242

Pavlović MD, Buntić AV, Šiler-Marinković SS, Dimitrijević-Branković SI (2013) Ethanol influenced fast microwave-assisted extraction for natural antioxidants obtaining from spent filter coffee. Sep Purif Technol 118:503–510. https://doi.org/10.1016/j.seppur.2013.07.035

Pedras B, Salema-Oom M, Sá-Nogueira I et al (2017) Valorization of white wine grape pomace through application of subcritical water: analysis of extraction, hydrolysis, and biological activity of the extracts obtained. J Supercrit Fluids 128:138–144. https://doi.org/10.1016/j.supflu.2017.05.020

Peduzzi E, Boissonnet G, Haarlemmer G, Maréchal F (2018) Thermo-economic analysis and multi-objective optimisation of lignocellulosic biomass conversion to Fischer-Tropsch fuels. Sustain Energy Fuels 2:1069–1084. https://doi.org/10.1039/c7se00468k

Pei H, Jiang L, Hou Q, Yu Z (2017) Toward facilitating microalgae cope with effluent from anaerobic digestion of kitchen waste: the art of agricultural phytohormones. Biotechnol Biofuels 10. https://doi.org/10.1186/s13068-017-0759-3

Peng S, Zou L, Zhou W, et al (2019) Encapsulation of lipophilic polyphenols into nanoliposomes using pH-driven method: advantages and disadvantages. J Agric Food Chem. doi: https://doi.org/10.1021/acs.jafc.9b01602

Perera A, Appleton D, Ying LH et al (2012) Large scale purification of geraniin from Nephelium lappaceum rind waste using reverse-phase chromatography. Sep Purif Technol 98:145–149. https://doi.org/10.1016/j.seppur.2012.06.019

Pérez-Serradilla JA, Luque de Castro MD (2011) Microwave-assisted extraction of phenolic compounds from wine lees and spray-drying of the extract. Food Chem 124:1652–1659. https://doi.org/10.1016/j.foodchem.2010.07.046

Périno-Issartier S, Zill-e-Huma A-VM, Chemat F (2011) Solvent free microwave-assisted extraction of antioxidants from sea buckthorn (Hippophae rhamnoides) food by-products. Food Bioprocess Technol 4:1020–1028. https://doi.org/10.1007/s11947-010-0438-x

Pleissner D, Lam WC, Sun Z, Lin CSK (2013) Food waste as nutrient source in heterotrophic microalgae cultivation. Bioresour Technol 137:139–146. https://doi.org/10.1016/j.biortech.2013.03.088

Płotka-Wasylka J, Rutkowska M, Owczarek K et al (2017) Extraction with environmentally friendly solvents. TrAC – Trends Anal. Chem. 91:12–25

Prakash Maran J, Sivakumar V, Thirugnanasambandham K, Sridhar R (2014) Microwave assisted extraction of pectin from waste Citrullus lanatus fruit rinds. Carbohydr Polym 101:786–791. https://doi.org/10.1016/j.carbpol.2013.09.062

Prapakornwiriya N, Diosady LL (2014) Recovery of sinapic acid from the waste effluent of mustard protein isolation by ion exchange chromatography. JAOCS, J Am Oil Chem Soc 91:357–362. https://doi.org/10.1007/s11746-013-2366-1

Prathima Devi M, Venkata Mohan S (2012) CO 2 supplementation to domestic wastewater enhances microalgae lipid accumulation under mixotrophic microenvironment: effect of sparging period and interval. Bioresour Technol 112:116–123. https://doi.org/10.1016/j.biortech.2012.02.095

Quan H, Qiong-Yao Y, Jiang S et al (2011) Structural characterization and antioxidant activities of 2 water-soluble polysaccharide fractions purified from tea (Camellia sinensis) flower. J Food Sci 76. https://doi.org/10.1111/j.1750-3841.2011.02063.x

Quemerais A, Daufin G (1995) Microfiltration performance: physicochemical aspects of whey pretreatment. J Dairy Res 62:269–279. https://doi.org/10.1017/S0022029900030971

Radošević K, Ćurko N, Gaurina Srček V et al (2016) Natural deep eutectic solvents as beneficial extractants for enhancement of plant extracts bioactivity. LWT – Food Sci Technol 73:45–51. https://doi.org/10.1016/j.lwt.2016.05.037

Rahmanian N, Jafari SM, Galanakis CM (2014) Recovery and removal of phenolic compounds from olive mill wastewater. JAOCS, J Am Oil Chem Soc 91:1–18. https://doi.org/10.1007/s11746-013-2350-9

Rajkumar P, Kailappan R, Viswanathan R et al (2007) Foam mat drying of alphonso mango pulp. Dry Technol 25:357–365. https://doi.org/10.1080/07373930601120126

Ramadan MF, Sitohy MZ, Moersel JT (2008) Solvent and enzyme-aided aqueous extraction of goldenberry (Physalis peruviana L.) pomace oil: impact of processing on composition and quality of oil and meal. Eur Food Res Technol 226:1445–1458. https://doi.org/10.1007/s00217-007-0676-y

Ran L, Yang C, Xu M et al (2019) Enhanced aqueous two-phase extraction of proanthocyanidins from grape seeds by using ionic liquids as adjuvants. Sep Purif Technol 226:154–161. https://doi.org/10.1016/j.seppur.2019.05.089

Ratledge C (2002) Regulation of lipid accumulation in oleaginous micro-organisms. Biochemical Society Transactions, In, pp 1047–1050

Ratti C, Kudra T (2006) Drying of foamed biological materials: opportunities and challenges. Drying Technology, In, pp 1101–1108

Reis IAO, Samuel B, Nascimento LAS, et al (2011) Extraction of ascorbic acid using alcohol / phosphate potassium salt – based aqueous two-phase system

Reis IAO, Santos SB, Pereira FDS et al (2014) Extraction and recovery of Rutin from acerola waste using alcohol-salt-based aqueous two-phase systems. Sep Sci Technol 49:656–663. https://doi.org/10.1080/01496395.2013.860461

Rikame SS, Mungray AA, Mungray AK (2012) Electricity generation from acidogenic food waste leachate using dual chamber mediator less microbial fuel cell. Int Biodeterior Biodegrad 75:131–137. https://doi.org/10.1016/j.ibiod.2012.09.006

Roldán E, Sánchez-Moreno C, de Ancos B, Cano MP (2008) Characterisation of onion (Allium cepa L.) by-products as food ingredients with antioxidant and antibrowning properties. Food Chem 108:907–916. https://doi.org/10.1016/j.foodchem.2007.11.058

Rosa A, Era B, Masala C et al (2019) Supercritical CO2 extraction of waste Citrus seeds: chemical composition, nutritional and biological properties of edible fixed oils. Eur J Lipid Sci Technol 121. https://doi.org/10.1002/ejlt.201800502

Saidi S, Saoudi M, Ben Amar R (2018) Valorisation of tuna processing waste biomass: isolation, purification and characterisation of four novel antioxidant peptides from tuna by-product hydrolysate. Environ Sci Pollut Res 25:17383–17392. https://doi.org/10.1007/s11356-018-1809-5

San Román MF, Bringas E, Ibañez R, Ortiz I (2010) Liquid membrane technology: fundamentals and review of its applications. J Chem Technol Biotechnol 85:2–10

Sánchez-García YI, Bhangu SK, Ashokkumar M, Gutiérrez-Méndez N (2018) Sonocrystallization of lactose from whey. Technological Approaches for Novel Applications in Dairy Processing. InTech, In

Santamaría B, Salazar G, Beltrán S, Cabezas JL (2002) Membrane sequences for fractionation of polyphenolic extracts from defatted milled grape seeds. Desalination 148:103–109. https://doi.org/10.1016/S0011-9164(02)00661-6

Santos PH, Baggio Ribeiro DH, Micke GA et al (2019) Extraction of bioactive compounds from feijoa (Acca sellowiana (O. berg) Burret) peel by low and high-pressure techniques. J Supercrit Fluids 145:219–227. https://doi.org/10.1016/j.supflu.2018.12.016

Saravana Pandian P, Sindhanai Selvan S, Subathira A, Saravanan S (2020) Optimization of aqueous two phase extraction of proteins from Litopenaeus vannamei waste by response surface methodology coupled multi-objective genetic algorithm. Chem Prod Process Model:15. https://doi.org/10.1515/cppm-2019-0034

Sarkar A, Kaul P (2014) Evaluation of tomato processing by-products: a comparative study in a pilot scale setup. J Food Process Eng 37:299–307. https://doi.org/10.1111/jfpe.12086

Sarkar O, Venkata Mohan S (2017) Pre-aeration of food waste to augment acidogenic process at higher organic load: valorizing biohydrogen, volatile fatty acids and biohythane. Bioresour Technol 242:68–76. https://doi.org/10.1016/j.biortech.2017.05.053

Sarkar O, Kumar AN, Dahiya S et al (2016) Regulation of acidogenic metabolism towards enhanced short chain fatty acid biosynthesis from waste: metagenomic profiling. RSC Adv 6:18641–18653. https://doi.org/10.1039/c5ra24254a

Schieber A, Stintzing FC, Carle R (2001) By-products of plant food processing as a source of functional compounds – recent developments. Trends Food Sci Technol 12:401–413. https://doi.org/10.1016/S0924-2244(02)00012-2

Schievano A, Pepé Sciarria T, Vanbroekhoven K et al (2016) Electro-fermentation – merging electrochemistry with fermentation in industrial applications. Trends Biotechnol 34:866–878

Schneider T, Rempp T, Graeff-Hönninger S et al (2013) Utilization of soluble starch by oleaginous red yeast rhodotorula glutinis. J Sustain Bioenergy Syst 3:57–63. https://doi.org/10.4236/jsbs.2013.31007

Seidl PR, Goulart AK (2016) Pretreatment processes for lignocellulosic biomass conversion to biofuels and bioproducts. Curr Opin Green Sustain Chem 2:48–53

Seixas FL, Fukuda DL, Turbiani FRB et al (2014) Extraction of pectin from passion fruit peel (Passiflora edulis f.flavicarpa) by microwave-induced heating. Food Hydrocoll 38:186–192. https://doi.org/10.1016/j.foodhyd.2013.12.001

Serre E, Rozoy E, Pedneault K et al (2016) Deacidification of cranberry juice by electrodialysis: impact of membrane types and configurations on acid migration and juice physicochemical characteristics. Sep Purif Technol 163:228–237. https://doi.org/10.1016/j.seppur.2016.02.044

Shanthi Sravan J, Butti SK, Sarkar O et al (2018) Electrofermentation of food waste – regulating acidogenesis towards enhanced volatile fatty acids production. Chem Eng J 334:1709–1718. https://doi.org/10.1016/j.cej.2017.11.005

Shao M, Chen H (2015) Feasibility of acetone-butanol-ethanol (ABE) fermentation from Amorphophallus konjac waste by Clostridium acetobutylicum ATCC 824. Process Biochem 50:1301–1307. https://doi.org/10.1016/j.procbio.2015.05.009

Sharholy M, Ahmad K, Mahmood G, Trivedi RC (2008) Municipal solid waste management in Indian cities – a review. Waste Manag 28:459–467. https://doi.org/10.1016/J.WASMAN.2007.02.008

Sharma HK, Xu C, Qin W (2019) Biological pretreatment of lignocellulosic biomass for biofuels and bioproducts: an overview. Waste Biomass Valorization 10:235–251

Shaw LA, McClements DJ, Decker EA (2007) Spray-dried multilayered emulsions as a delivery method for ω-3 fatty acids into food systems. J Agric Food Chem 55:3112–3119. https://doi.org/10.1021/jf063068s

Shen Y, Jarboe L, Brown R, Wen Z (2015) A thermochemical-biochemical hybrid processing of lignocellulosic biomass for producing fuels and chemicals. Biotechnol Adv 33:1799–1813

Shi M, Zhang Z, Yang Y (2013) Antioxidant and immunoregulatory activity of Ganoderma lucidum polysaccharide (GLP). Carbohydr Polym 95:200–206. https://doi.org/10.1016/j.carbpol.2013.02.081

Silva HD, Cerqueira MA, Souza BWS et al (2011) Nanoemulsions of β-carotene using a high-energy emulsification- evaporation technique. J Food Eng 102:130–135. https://doi.org/10.1016/j.jfoodeng.2010.08.005

Sotiropoulos A, Malamis D, Michailidis P et al (2016) Research on the drying kinetics of household food waste for the development and optimization of domestic waste drying technique. Environ Technol (United Kingdom) 37:929–939. https://doi.org/10.1080/21622515.2015.1092588

Soto ML, Moure A, Domínguez H, Parajó JC (2011) Recovery, concentration and purification of phenolic compounds by adsorption: a review. J Food Eng 105:1–27

Souchon I, Pierre FX, Athes-Dutour V, Marin M (2002) Pervaporation as a deodorization process applied to food industry effluents: recovery and valorisation of aroma compounds from cauliflower blanching water. Desalination 148:79–85. https://doi.org/10.1016/S0011-9164(02)00657-4

Sowbhagya HB, Chitra VN (2010) Enzyme-assisted extraction of flavorings and colorants from plant materials. Crit Rev Food Sci Nutr 50:146–161. https://doi.org/10.1080/10408390802248775

Spigno G, Amendola D, Dahmoune F, Jauregi P (2015) Colloidal gas aphrons based separation process for the purification and fractionation of natural phenolic extracts. Food Bioprod Process 94:434–442. https://doi.org/10.1016/j.fbp.2014.06.002

Sravan JS, Butti SK, Sarkar O, Mohan SV (2018) Electrofermentation: chemicals and fuels. Biomass Biofuels Biochem Microb Electrochem Technol Sustain Platf Fuels Chem Remediat:723–737. https://doi.org/10.1016/B978-0-444-64052-9.00029-7

Steinbusch KJJ, Hamelers HVM, Plugge CM, Buisman CJN (2011) Biological formation of caproate and caprylate from acetate: fuel and chemical production from low grade biomass. Energy Environ Sci 4:216–224. https://doi.org/10.1039/c0ee00282h

Stoeberl M, Werkmeister R, Faulstich M, Russ W (2011) Biobutanol from food wastes – fermentative production, use as biofuel an the influence on the emissions. Proc Food Sci 1:1867–1874. https://doi.org/10.1016/j.profoo.2011.09.274

Strati IF, Oreopoulou V (2011) Effect of extraction parameters on the carotenoid recovery from tomato waste. Int J Food Sci Technol 46:23–29. https://doi.org/10.1111/j.1365-2621.2010.02496.x

Strazzera G, Battista F, Garcia NH et al (2018) Volatile fatty acids production from food wastes for biorefinery platforms: a review. J Environ Manag 226:278–288. https://doi.org/10.1016/j.jenvman.2018.08.039

Subramaniam MB, Blakebrough N, Hashim MA (2007) Clarification of suspensions by colloidal gas aphrons. J Chem Technol Biotechnol 48:41–60. https://doi.org/10.1002/jctb.280480105

Sun RC, Tomkinson J (2002) Characterization of hemicelluloses obtained by classical and ultrasonically assisted extractions from wheat straw. Carbohydr Polym 50:263–271. https://doi.org/10.1016/S0144-8617(02)00037-1

Tahergorabi R, Jaczynski J (2014) Isoelectric solubilization/precipitation as a means to recover protein and lipids from seafood by-products. In: Seafood processing by-products: trends and applications. Springer, New York, pp 101–123

Tahergorabi R, Beamer SK, Matak KE, Jaczynski J (2011a) Effect of isoelectric solubilization/precipitation and titanium dioxide on whitening and texture of proteins recovered from dark chicken-meat processing by-products. LWT – Food Sci Technol 44:896–903. https://doi.org/10.1016/j.lwt.2010.10.018

Tahergorabi R, Beamer SK, Matak KE, Jaczynski J (2011b) Effect of isoelectric solubilization/precipitation and titanium dioxide on whitening and texture of proteins recovered from dark chicken-meat processing by-products. LWT – Food Sci Technol 44:896–903. https://doi.org/10.1016/j.lwt.2010.10.018

Tanger P, Field JL, Jahn CE et al (2013) Biomass for thermochemical conversion: targets and challenges. Front Plant Sci 4

Thushari I, Babel S (2018) Sustainable utilization of waste palm oil and sulfonated carbon catalyst derived from coconut meal residue for biodiesel production. Bioresour Technol 248:199–203. https://doi.org/10.1016/j.biortech.2017.06.106

Tiwari BK (2015) Ultrasound: a clean, green extraction technology. TrAC – Trends Anal Chem 71:100–109

Todd R, Baroutian S (2017) A techno-economic comparison of subcritical water, supercritical CO2 and organic solvent extraction of bioactives from grape marc. J Clean Prod 158:349–358. https://doi.org/10.1016/j.jclepro.2017.05.043

Tsakona S, Galanakis CM, Gekas V (2012) Hydro-Ethanolic mixtures for the recovery of phenols from Mediterranean plant materials. Food Bioprocess Technol 5:1384–1393. https://doi.org/10.1007/s11947-010-0419-0

Tuck CO, Pérez E, Horváth IT et al (2012) Valorization of biomass: deriving more value from waste. Science (80-) 337:695–699. https://doi.org/10.1126/science.1218930

Tzoumaki MV, Moschakis T, Kiosseoglou V, Biliaderis CG (2011) Oil-in-water emulsions stabilized by chitin nanocrystal particles. Food Hydrocoll. https://doi.org/10.1016/j.foodhyd.2011.02.008

Uçkun Kiran E, Trzcinski AP, Ng WJ, Liu Y (2014) Bioconversion of food waste to energy: a review. Fuel 134:389–399

Ujor V, Bharathidasan AK, Cornish K, Ezeji TC (2014) Feasibility of producing butanol from industrial starchy food wastes. Appl Energy 136:590–598. https://doi.org/10.1016/j.apenergy.2014.09.040

Venkata Mohan S (2009) Harnessing of biohydrogen from wastewater treatment using mixed fermentative consortia: process evaluation towards optimization. Int. J. Hydrogen Energy 34:7460–7474

Venkata Mohan S, Sarkar O (2017) Waste to biohydrogen: addressing sustainability with biorefinery. In: Energy engineering. Springer, Singapore, pp 29–37

Venkata Mohan S, Prathima Devi M (2012) Fatty acid rich effluent from acidogenic biohydrogen reactor as substrate for lipid accumulation in heterotrophic microalgae with simultaneous treatment. Bioresour Technol 123:627–635. https://doi.org/10.1016/j.biortech.2012.07.004

Venkata Mohan S, Mohanakrishna G, Goud RK, Sarma PN (2009) Acidogenic fermentation of vegetable based market waste to harness biohydrogen with simultaneous stabilization. Bioresour Technol 100:3061–3068. https://doi.org/10.1016/j.biortech.2008.12.059

Venkata Mohan S, Velvizhi G, Annie Modestra J, Srikanth S (2014a) Microbial fuel cell: critical factors regulating bio-catalyzed electrochemical process and recent advancements. Renew Sust Energ Rev 40:779–797

Venkata Mohan S, Velvizhi G, Vamshi Krishna K, Lenin Babu M (2014b) Microbial catalyzed electrochemical systems: a bio-factory with multi-facet applications. Bioresour Technol 165:355–364

Venkata Mohan S, Nikhil GN, Chiranjeevi P et al (2016) Waste biorefinery models towards sustainable circular bioeconomy: critical review and future perspectives. Bioresour Technol 215:2–12

Venkata Mohan S, Velvizhi G, Chiranjeevi P (2017) Microbial electrochemical platform: biofactory with diverse applications. In: Agarwal A, Agarwal R, Gupta T (eds) Biofuels, green energy and technology. Springer, Singapore, p 35

Venkata Subhash G, Venkata Mohan S (2011) Biodiesel production from isolated oleaginous fungi Aspergillus sp. using corncob waste liquor as a substrate. Bioresour Technol 102:9286–9290. https://doi.org/10.1016/j.biortech.2011.06.084

Vergara-Salinas JR, Vergara M, Altamirano C et al (2015) Characterization of pressurized hot water extracts of grape pomace: chemical and biological antioxidant activity. Food Chem 171:62–69. https://doi.org/10.1016/j.foodchem.2014.08.094

Villeneuve W, Perreault V, Chevallier P et al (2019) Use of cation-coated filtration membranes for demineralization by electrodialysis. Sep Purif Technol 218:70–80. https://doi.org/10.1016/j.seppur.2019.02.032

Vorobiev E, Lebovka N (2010) Enhanced extraction from solid foods and biosuspensions by pulsed electrical energy. Food Eng Rev 2:95–108. https://doi.org/10.1007/s12393-010-9021-5

Walker K, Vadlani P, Madl R et al (2013) Ethanol fermentation from food processing waste. Environ Prog Sustain Energy 32:1280–1283. https://doi.org/10.1002/ep.11700

Wan J, Coventry J, Swiergon P et al (2009) Advances in innovative processing technologies for microbial inactivation and enhancement of food safety – pulsed electric field and low-temperature plasma. Trends Food Sci Technol 20:414–424. https://doi.org/10.1016/j.tifs.2009.01.050

Wang Q, Ying T, Jiang T et al (2009a) Demineralization of soybean oligosaccharides extract from sweet slurry by conventional electrodialysis. J Food Eng 95:410–415. https://doi.org/10.1016/j.jfoodeng.2009.05.024

Wang YH, Li SL, Chen IC, Cheng SS (2009b) Starch hydrolysis characteristics of hydrogen producing sludge in thermophilic hydrogen fermentor fed with kitchen waste. Int J Hydrog Energy 34:7435–7440. https://doi.org/10.1016/j.ijhydene.2009.05.074

Wang X, Hansen C, Allen K (2014) Extraction of anthocyanins from black bean canning wastewater with macroporous resins. J Food Sci 79:E184–E188. https://doi.org/10.1111/1750-3841.12333

Weisz GM, Carle R, Kammerer DR (2013) Sustainable sunflower processing – II. Recovery of phenolic compounds as a by-product of sunflower protein extraction. Innov Food Sci Emerg Technol 17:169–179. https://doi.org/10.1016/j.ifset.2012.09.009

Xu Z, Zhao M, Miao H et al (2014) In situ volatile fatty acids influence biogas generation from kitchen wastes by anaerobic digestion. Bioresour Technol 163:186–192. https://doi.org/10.1016/j.biortech.2014.04.037

Yang EJ, Kim SS, Oh TH et al (2009) Essential oil of citrus fruit waste attenuates LPS-induced nitric oxide production and inhibits the growth of skin pathogens. Int J Agric Biol 11:791–794

Yoshida H (1993) Practical aspects of dewatering enhanced by electro-osmosis. Drying Technology 11(4):784–814

Yoshii H, Soottitantawat A, Liu XD et al (2001) Flavor release from spray-dried maltodextrin/gum Arabic or soy matrices as a function of storage relative humidity. Innov Food Sci Emerg Technol 2:55–61. https://doi.org/10.1016/S1466-8564(01)00019-4

Young SL, Sarda X, Rosenberg M (1993) Microencapsulating properties of whey proteins. 1. Microencapsulation of anhydrous milk fat. J Dairy Sci 76:2868–2877. https://doi.org/10.3168/jds.S0022-0302(93)77625-0

Yu X, Zheng Y, Dorgan KM, Chen S (2011) Oil production by oleaginous yeasts using the hydrolysate from pretreatment of wheat straw with dilute sulfuric acid. Bioresour Technol 102:6134–6140. https://doi.org/10.1016/j.biortech.2011.02.081

Zanello PP, Sforza S, Dossena A et al (2014) Antimicrobial activity of poultry bone and meat trimmings hydrolyzates in low-sodium Turkey food. Food Funct 5:220–228. https://doi.org/10.1039/c3fo60454c

Zeshan, Karthikeyan OP, Visvanathan C (2012) Effect of C/N ratio and ammonia-N accumulation in a pilot-scale thermophilic dry anaerobic digester. Bioresour Technol 113:294–302. https://doi.org/10.1016/j.biortech.2012.02.028

Zhang J, Zhang R, Wang H, Yang K (2020) Direct interspecies electron transfer stimulated by granular activated carbon enhances anaerobic methanation efficiency from typical kitchen waste lipid-rapeseed oil. Sci Total Environ 704:135282. https://doi.org/10.1016/j.scitotenv.2019.135282

Zhao T, Zhou Y, Mao G et al (2013) Extraction, purification and characterisation of chondroitin sulfate in Chinese sturgeon cartilage. J Sci Food Agric 93:1633–1640. https://doi.org/10.1002/jsfa.5937

Zhong K, Lin W, Wang Q, Zhou S (2012) Extraction and radicals scavenging activity of polysaccharides with microwave extraction from mung bean hulls. Int J Biol Macromol 51:612–617. https://doi.org/10.1016/j.ijbiomac.2012.06.032

Zhou M, Yan B, Wong JWC, Zhang Y (2018) Enhanced volatile fatty acids production from anaerobic fermentation of food waste: a mini-review focusing on acidogenic metabolic pathways. Bioresour Technol 248:68–78

Zhou Y, Engler N, Li Y, Nelles M (2020) The influence of hydrothermal operation on the surface properties of kitchen waste-derived hydrochar: biogas upgrading. J Clean Prod 259:121020. https://doi.org/10.1016/j.jclepro.2020.121020

Zill-E-Huma, Abert-Vian M, Elmaataoui M, Chemat F (2011) A novel idea in food extraction field: study of vacuum microwave hydrodiffusion technique for by-products extraction. J Food Eng 105:351–360. https://doi.org/10.1016/j.jfoodeng.2011.02.045

Chapter 5
Biocascading: Platform Molecules, Value Added Chemicals, and Bioactives

Teresa Cecchi

Biorefineries are the physical embodiment of bioeconomy. The idea of the circular economy is based on four crucial pillars (i) products are designed so that consumables are returned to the biosphere without harm after a sequence of uses, (ii) durables are designed to maximize their valuable reuse or up-grade, (iii) renewable energy is used to fuel the process, (iv) waste, generated where the costs of if its reuse and recycling are higher than the value created, potentially ceases to exist because products undergo a cycle of disassembly and reuse.

Initially, the prevailing business model of biorefinery has been inspired by the large scale classical petrochemical refinery, and the most common practice was to focus technologies on energy and fuels. The fuel-related biorefinery concept enjoyed an initial surge in popularity in the mid-2000s; due to fossil fuel-related concerns, biofuels (ethanol and fatty acid methyl esters) were among the first bioproducts to emerge. Anyhow, the solar, wind, and renewable hydroelectric energy are good competitors of energy derived from biomass. As synthetic chemistry needs building blocks to synthesize larger molecules, it is seen as wasteful to take biomass, which is already full of complex molecules, and then burn it to reduce it back down to one of the simplest carbon-containing molecules, that is, carbon dioxide. Moreover, there is a number of performance problems with the fuels themselves due to (i) polyunsaturated fats polymerizing or forming reactive oxygen radicals and (ii) saturated fats, solid at temperatures unsuitable to the majority of winter environments. Furthermore, the objection to using food-grade crops to produce fuels, with malnutrition still being a global issue ("food vs. fuel debate"), contributes to stimulate biorefineries to search for alternative feedstocks and to target different outputs and bioproducts (Dugmore et al. 2017). This represents an imperative and impressive research challenge, as well as a promising topic globally from both an environmental and economic point of view. Actually, the reach of biobased chemicals into consumer articles is continuously expanding. A flexible and multitasking biorefinery, capable of handling feedstock of varying origins, composition and quantities is highly looked-for. A vast gamut of residual streams, which currently find low-value

T. Cecchi, C. De Carolis, *Biobased products from food sector waste*,
https://doi.org/10.1007/978-3-030-63436-0_5

applications based on their calorific content, contain functional molecules that could be used for food and non-food market applications.

It is crucial to have these functional molecules separated in a cost-efficient, energy-efficient, and eco-efficient way to make them available for subsequent use in higher-value applications, which in turn will provide significant employment and income opportunities. It is important to concentrate on supply chain hurdles in proper waste management, including pretreatment and optimization actions, for a sustainable valorization of side streams to produce a wide range of biomaterials (Dupont-Inglis and Borg 2018). If biorefinery outputs aim at food applications, they must also include consumer safety considerations to check that the products will comply with relevant legislation on chemical risk management, toxicity, and safety.

The quantities, qualities, and physical locations of these streams define the optimal location and size of the operational site (centralized large-scale biorefineries or small/medium-scale processing units) that they integrate into.

Active product-driven biorefineries are not very common (Tsagaraki et al. 2017). Nevertheless, according to a 2016 study conducted by the European Commission's Joint Research Centre on EU biobased industry, 50 companies developed 284 putative biobased products (Nattrass et al. 2016) from materials of biological origin (for example, biomasses, algae, crops, marine organisms, and waste); hence their environmental footprint is expected to be limited in comparison to their traditional counterparts.

Figure 5.1 illustrates the projected production of biobased chemicals and materials in Europe 2020/2030 (Hassan et al. 2019).

The untapped market potential of biobased chemicals was pioneeristically emphasized by the Joint Research Centre, the EU Commission's science and knowledge service; its research provides independent scientific advice and support to EU policy (Spekreijse et al. 2019). The top 10 key non-energy and non-traditional biobased chemical product application categories enlist:

- platform chemicals
- solvents
- polymers for plastics
- paints, coatings, inks, and dyes
- surfactants
- cosmetics and personal care products
- adhesives
- lubricants
- plasticizers (and stabilizers for rubber and plastics)
- man-made fibers

They yearly cover about 4.7 Mt that represents 3% of the total EU market. Their overall average annual growth rate is estimated at 3.6%, hence their demand is not expected to grow rapidly mainly because production costs are still higher than for their traditional counterparts (Spekreijse et al. 2019).

Up to date policy interventions aimed at stimulating the untapped potential of the biobased markets via investment incentives (grants, loans, guarantees, etc.), carbon

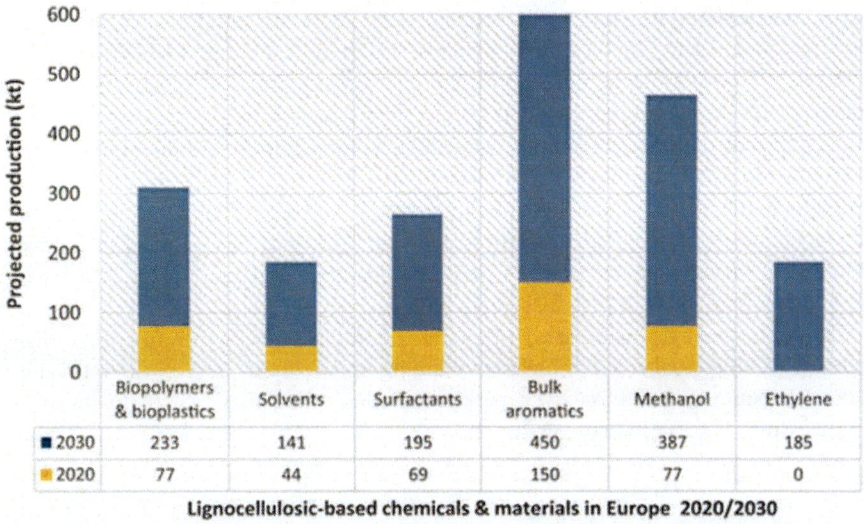

Fig. 5.1 Projected production of biobased chemicals and materials in Europe 2020/2030, (Hassan et al. 2019). Reprinted from Renewable and Sustainable Energy Reviews, Vol 101, Authors: Hassan, Shady S.; Williams, Gwilym A.; Jaiswal, Amit K., Moving towards the second generation of lignocellulosic biorefineries in the EU: Drivers, challenges, and opportunities, Pages No. 590–599, Copyright 2019, with permission from Elsevier

taxes, or making the use of biobased chemicals mandatory in some industries, thus boosting their demand. Highlighting the use of food waste (FW) as feedstock for biorefineries can concur to offer policymakers fresh insights on major market drivers and possible solutions for constraints in the biobased chemical market.

The vast majority of food industry processing wastes, produced in enormous amounts every year, are usually disposed of, with the corresponding economical cost it implies. In the best scenario, they have been used for pet and animal feeding or composting or energy generation. This chapter tries to lay down the promising prospects associated with food waste exploitation within the biorefinery framework. By this scenario, the cascading concept (Annevelink et al. 2017) (European Commission 2015) applied to biomass (biocascading) relies on the physical, chemical, biochemical, and thermochemical technologies illustrated in Chap. 4. Biocascading enables the extraction of quality products, while the biomass is used sequentially as often as possible and finally for energy; the most valuable components are extracted from biomass first according to the hierarchical, smart and efficient use of biomass illustrated in Fig. 4.1 of Chap. 4. (European Commission 2015).

Even the organic fraction of municipal solid waste, often perceived as a challenge for urban agendas, may be an important feedstock for biorefining (Girotto et al. 2017). The energy recovery, composting, and anaerobic digestion or its disposal in a landfill would pre-empt exploiting its full potential for the production of valuable products, achievable in cascading operations, and useful in various market segments.

A chemical building block is a molecule that can be converted to various secondary molecules and, in turn, into a broad range of different downstream uses. Chemical building blocks are also called "platform chemicals" because they form the base for more complex products (Bozell and Petersen 2010).

As regards biobased chemicals, we can have both drop-in and novel molecules. Drop-in biobased chemicals are biobased versions of chemically identical petrochemicals that have established markets. Novel biobased chemicals bear higher financial and technological risks for producers. Still, they can be used to produce molecules with unique and superior properties, such as biodegradability, which cannot be obtained through fossilbased reactions.

In the following, a wide array of chemical building blocks from the three main classes of biological macromolecules (carbohydrates, lipids, and proteins), value-added chemicals, and bioactives will be detailed to emphasize the prospects of money-spinning FW management.

5.1 Carbohydrates

Of the three main classes of biological macromolecules, carbohydrates are the most hydrophilic; they are the primary energy source. Chemically, they are polar hydrates of carbon, prone to hydrogen bonding. In biomass, they can be found as simple sugars (monosaccharides), dimers (disaccharides), or polymers (polysaccharides). Glucose is the most abundant monosaccharide. It is known as α- or β-glucose according to the hydroxyl axial or equatorial position on carbon 1 in the ring. It possesses six carbon atoms, it is classed as hexose, and other hexose sugars are all isomers to glucose. Pentose sugars consist of five-membered rings.

Sugars from FW can be used in fermentations to produce biohydrogen, bioethanol and biogas, as illustrated in Chap. 3 and Chap. 4. As sugars are a primary energy source for many non-photosynthetic cells, fermentation to ethanol (for fuel or alcoholic beverages) has been the first mean of sugar conversion. Many recent advances in sugar valorization have utilized fermentation to target other molecules involved in the fermentation pathway, particularly platform molecules and simple sugars can also be used in microbial fuel cells, as illustrated in Sect. 4.3.4.1. In the following, we will focus on non-energy biobased chemicals that can be obtained from sugars.

A final selection of 12 building blocks from a list of more than 300 candidates was put forward in 2004 (Werpy and Petersen 2004). Criteria for inclusion were: marketability, properties, performance, prior industry experience, the technical complexity of the synthesis pathways, and the ability to compete directly against existing chemicals derived from petroleum. These twelve crucial building blocks are midway chemicals that can be converted to a number of high-value biobased compounds and new families of useful molecules, due to their multiple functional groups. While biological transformations account for the majority of routes from feedstocks to building blocks, chemical transformations predominate in converting building blocks to secondary chemicals or families of derivatives and intermediates. The twelve sugar-based building blocks have acidic, alcohol, and ester functional groups. They are diacids, such as succinic, fumaric, malic acid, and 2,5-furan dicarboxylic acid. Other acids, namely 3-hydroxy propionic acid, aspartic acid, glucaric acid, glutamic acid, itaconic acid, levulinic acid, a cyclic ester, 3-hydroxybutyrolactone, and polyols, such as glycerol, sorbitol, and xylitol/arabinitol complete the building blocks list (Werpy and Petersen 2004). They are categorized in Fig. 5.2, which details the chemical or biochemical pathway to the specific building block from sugars.

Figure 5.3 illustrates some of the possible transformation pathways of the twelve sugar-based building blocks to derivatives and their potential use and commercial significance.

In 2010 an updated evaluation of technology advances provided a revised list of biobased product opportunities from carbohydrates and an overview of the technology developments that could realize commercial success by properly integrating biofuels with biobased product outputs (Bozell and Petersen 2010). Criteria for inclusion were:

1. The presence of extensive recent literature.

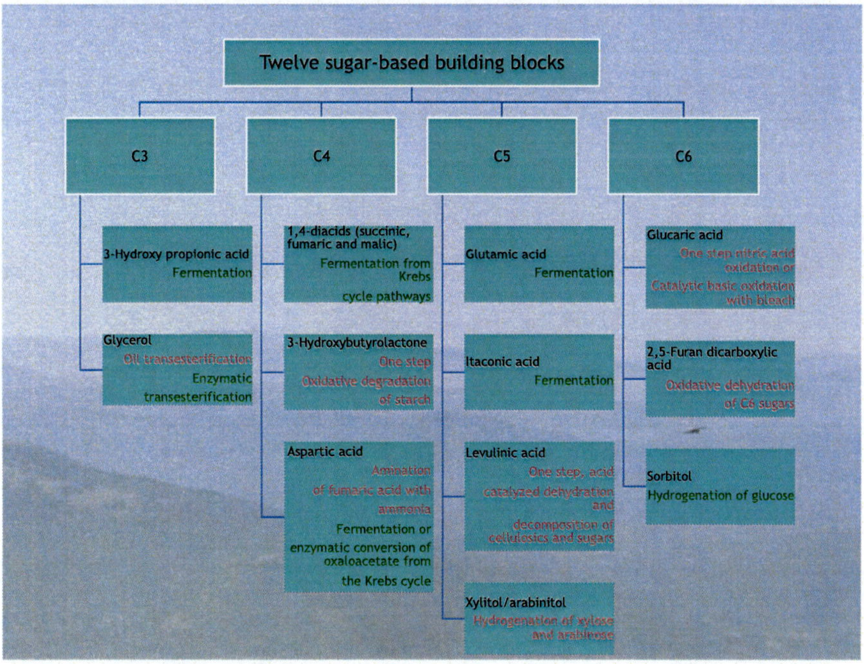

Fig. 5.2 Twelve sugar-based building block chemicals. The pathway to the specific building block from sugars can be a biotransformation (in green) or a chemical transformation (in magenta) (Werpy and Petersen 2004)

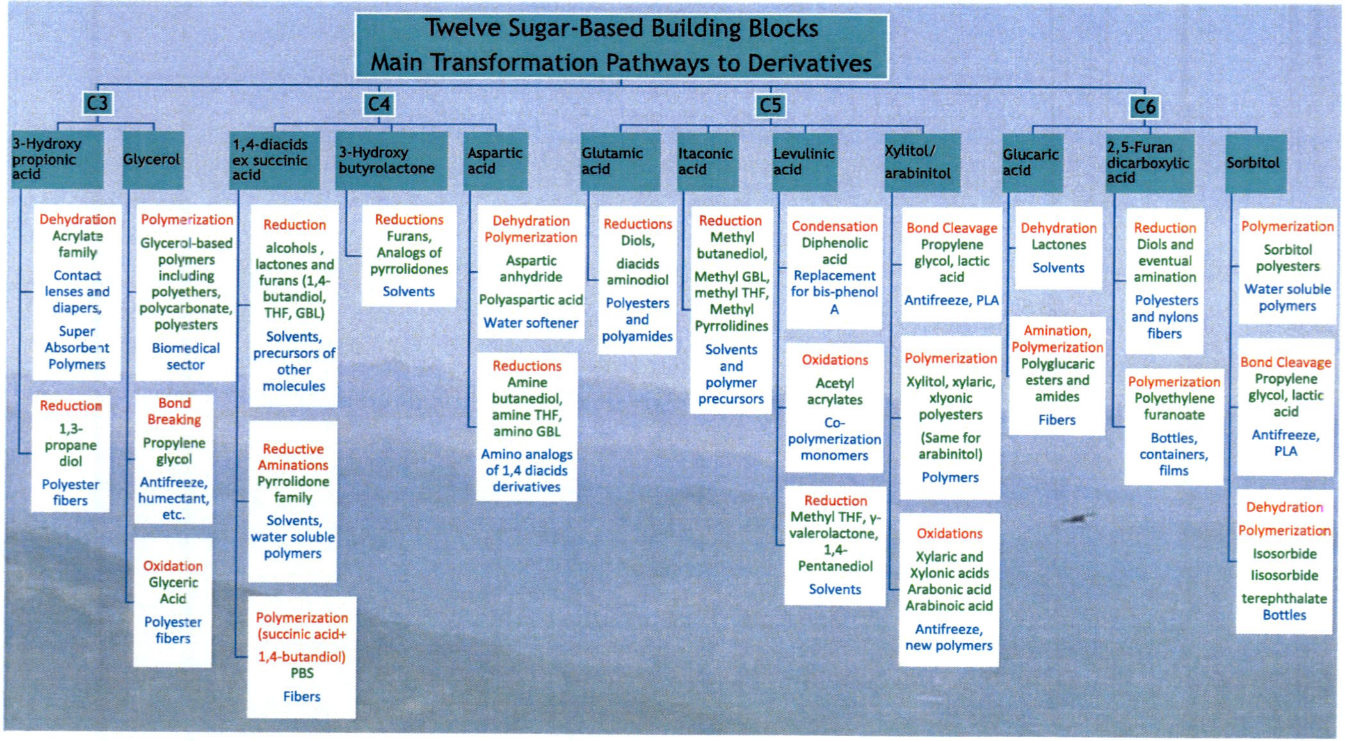

Fig. 5.3 Main transformation pathways to derivatives of the twelve sugar-based building blocks (Werpy and Petersen 2004). Reactions in red, derivatives in green, potential use of derivatives in blue. Abbreviations: tetrahydrofuran (THF), γ- butyrolactone (GBL), polybutylene succinate (PBS), polylactic acid (PLA)

2. Multiple product applicability.
3. The technology provides direct substitutes for existing petrochemicals.
4. Applicability to high volume product.
5. The biobased compound has potential as a platform chemical.
6. Industrial scale-up is underway.
7. The biobased compound is an existing commercial product, and new uses are forecast.
8. The biobased compound is a primary building block.
9. Commercial biobased production is well established.

Fumaric, malic, aspartic, glutamic, itaconic, glucaric acids, and 3-hydroxybutyrolactone were excluded from the revised list. In contrast, ethanol, lactic acid, biohydrocarbons (e.g.isoprene), furans such as furfural and 5-hydroxymethylfurfural are new entries (Bozell and Petersen 2010). The inclusion of furfural and 5-hydroxymethylfurfural stemmed from their feasible preparation from C5 and C6 saccharide-based renewable resources and their versatile chemistry. Hydrogenation of furfural provides both furfuryl alcohol and tetrahydrofurfuryl alcohol, which are nonhazardous useful chemical intermediates often used as herbicide adjuvants. Furfuryl alcohol is a precursor to a variety of resins for high-tech applications. Furfuryl alcohol can be transformed into levulinic acid, which is a starting feedstock with downstream products via δ-valerolactone. Furfural decarbonylisation would result in the production of furan and tetrahydrofuran, which are important commodity chemicals (EU 2015). Furfural and 5-hydroxymethylfurfural can be transformed into useful monomers for polymer production: for example, the conversion of furfural to adipic acid and hexamethylene diamine would enable the production of biobased Nylon 66 (John et al. 2019).

Terpenes were also included in the revised list. They are an important class of secondary metabolites that consist of multiple isoprene units (C_5H_8). According to the number of multiple isoprene units, they are categorized as hemiterpene (one isoprene unit), monoterpene (2 units), sesquiterpene (3), diterpene (4) and so on. Terpenoids are modified terpenes with added, missing, or shifted methyl and oxygenated functional groups allowing added reactivity. They all comprise a wide range of cycloaliphatic and hydrocarbon chain structures with repeating isoprenyl double bonds, easy to functionalize. The inclusion of terpenes, mainly recognized as odorous chemicals, is due to the fact that their reactivity can be very rich and thermoplastics can be obtained from their polymerization; although they have been investigated since the early twentieth century, to date, few terpenes and terpenoids have been studied as polymers building blocks (John et al. 2019). Still, developments in terpenoid chemistry, spurred by a growing interest in biobased polymers, are expected.

Ethanol and lactic acid were also included since they were widely recognized as crucial for bio diesel and polylactic acid production (described in Chap. 6), respectively (John et al. 2019).

In 2015 the various feedstock pretreatment and downstream process options for sugars (most fermentation-based) were mapped and revisited (EU 2015). In order to assess the development status of these technologies, a key metric, the Nasa TRL (Technology Readiness Level) scale was introduced in the hermeneutics of EU funded projects by the research and innovation programme Horizon 2020 (Horizon 2020 WORK PROGRAMME 2018–2020 General Annexes 2017). The development level of a specific technology starts from the basic principle research (TRL1), continues to pilot scale testing and technology validation in relevant environment (TRL 5), and ends up with the mass deployment of a fully commercialized technology and market outreach with the actual system proven (TRL 9).

The minimum timeframe of at least ten years is expected in regions where policy is supportive to successfully progress from having an established pilot (TRL5) to reach commissioning of a first commercial plant (TRL8) (EU 2015).

Table 5.1 illustrates promising biobased chemical targets as assessed in 2004, 2010, and 2015.

As indicated in Table 5.1, in 2015, a total of 10 products (acrylic acid, adipic acid, 1,4-butanediol (BDO), farnesene, 2,5- Furandicarboxylic acid (FDCA), isobutene, polyhydroxyalkanoates (PHAs), polyethylene (PE), polylactic acid (PLA) and succinic acid) were down-selected from a larger list. Inclusion criteria were a minimum TRL 5, the presence of market players already involved in the pathway, and notable potential market size, among others.

It is important to emphasize that omission of a specific compound in Table 5.1 does not mean it is without merit because the biorefining industry is in a state of rapid change and expansion. Both inclusions and exclusions are never definitive. For these reasons, in the following, we will illustrate the potential of FW to provide some important carbohydrate-based compounds, value-added chemicals, bioactives, and chemical building blocks. Biobased polymers will be dealt with in Chap. 6.

5.1.1 *Starch and Lignocellulosic Biomass*

Starch consists of amylose, an unbranched linear chain of α-glucose monomers linked by α-1,4 glycosidic bonds, and amylopectin, a branched chain due to *ca.* one α-1,6 glycosidic linkage per 30 α-1,4 glycosidic bonds. Many kinds of FW are starch-rich. Starch was effectively transformed into small monomeric sugars via a mixture of α-amylase, β-amylase, and glucoamylase enzymes (Dahiya et al. 2018).

Starch can be recovered from wastewater produced during potato washing for chips, snacks, or fries, thereby avoiding starch sedimentation in drainage pipelines (Catarino et al. 2007) and also from plantain peel wastes (Hernández-Carmona et al. 2017).

Fruit processing for juice and jams production generates skin, pulp, seed waste, and pomace, which are rich in value-added compounds like starch. Compared to starch from conventional sources such as cereal and tubers, fruit waste starches have unique features such as gelatinization temperature and enthalpy and lower breakdown viscosity due to their varied morphologies and structures (Kringel et al. 2020;

Table 5.1 Comparison of top sugar-based building block in the last two decades

C number	2004	2010	2015
	Werpy and Petersen (2004)	Bozell and Petersen (2010)	EU (2015)
C2	–	Ethanol	–
C3	3-Hydroxy-propionic acid	Hydroxypropionic acid/aldehyde	–
	Glycerol	Glycerol and derivatives	–
	–	Lactic acid	–
	–	–	Acrylic acid
C4	1,4-dicarboxylic acids (succinic, fumaric and malic)	Succinic acid	Succinic acid
	3-Hydroxybutyrolactone	–	–
	Aspartic acid	–	–
	–	–	1,4-butanediol (BDO)
	–	–	Iso-butene
C5	Glutamic acid	–	–
	Itaconic acid	–	–
	Levulinic acid	Levulinic acid	–
	Xylitol/Arabinitol	Xylitol	–
	–	Biohydrocarbons (from isoprene)	Farnesene (from isoprene)
C6	Glucaric acid	–	–
	2,5-furan dicarboxylic acid	Furanics (C6 and C5)	2,5- Furandicarboxylic acid (FDCA)
	Sorbitol	Sorbitol	–
	–	–	Adipic acid
Polymers	–	–	Polyethylene (PE)
	–	–	Polyhydroxyalkanoates (PHAs)
	–	–	Polylactic acid (PLA)

Schieber 2017). Brewery effluents can also be a source of soluble starch for algae cultivation (Amenorfenyo et al. 2019).

Lignocellulosic biomass, whose structure is described in Chap. 4 (Sec 4.2.1.1) and graphically shown in Fig. 4.2, contains variable levels of cellulose in association with tough substrates such as hemicellulose, pectin, and lignin and other composites. These components were regarded as ballast substances from vegetal waste streams but can be an impressive source of biobased chemicals.

Lignin is the broadest distributed aromatic biopolymer (Key and Bozell 2016). It is a cell-wall component bonding cells together in the woody stems, providing them with rigidity and impact resistance. Noteworthy, its fundamental structure remains unknown since it varies according to its botanical and pedo-climatological origin.

Still, the three main aromatic subunits found in this amorphous three-dimensional polymer network are depicted in Fig. 5.4.

Unfortunately, a very small percentage of overall lignin is used for depolymerization into smaller, simpler aromatic molecules suitable for chemical modification and/or subsequent polymerization (vanillin, phenols derivatives, cresols, ferulic and coumaric acids, etc.). The remaining lignin primarily serves as a (bio)fuel (de Oliveira et al. 2015).

Hemicellulose is an amorphous polysaccharide. It is formed through biosynthetic routes different from that of cellulose but, like cellulose, most hemicelluloses are supporting material in the cell wall. Hemicellulose polymers, at variance with cellulose, may be branched; it is embedded in plant cell walls and binds with pectin to cellulose to form a network of cross-linked fibers that keeps the cell wall rigidity. Assorted kinds of hemicelluloses (mannans, xylans, glucuronoxylan, arabinoxylan, glucomannan, and xyloglucan, among others) are known according to the diverse sugars which constitute short chains of less than 3000 monomer units: besides glucose, the five-carbon sugars xylose and arabinose, the six-carbon sugars mannose and galactose, the six-carbon deoxy sugar rhamnose, the acidified form of regular sugars such as glucuronic and galacturonic acids can be present. Xylose is usually the sugar monomer present in the largest amount in hardwood, while mannose is very abundant in softwood (Ebringerová et al. 2005). Hemicellulose can be depolymerized into pentose units (mainly xylose) that can further be transformed into 2-furfural, as indicated in Fig. 5.5. 2-Furfural can, in turn, be transformed into 2-furanmethanol and 2-furancarboxylic acid.

Cellulose is the most abundant biopolymer, and it is a structural material in plants. Unlike hemicellulose, cellulose consists of longer chains derived exclusively from glucose: 7,000–15,000 monomers are present in each polymer of cellulose (Gibson 2012). Cellulose differs from starch since the different bond angle

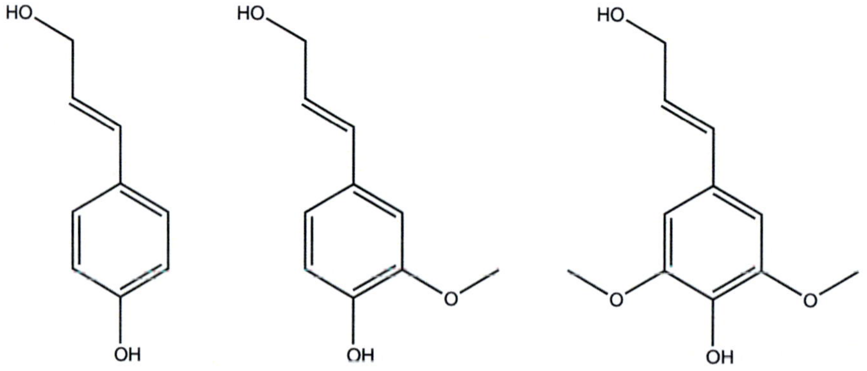

p-Coumaryl alcohol Coniferyl alcohol Synapil alcohol

Fig. 5.4 Main aromatic subunits found in lignin

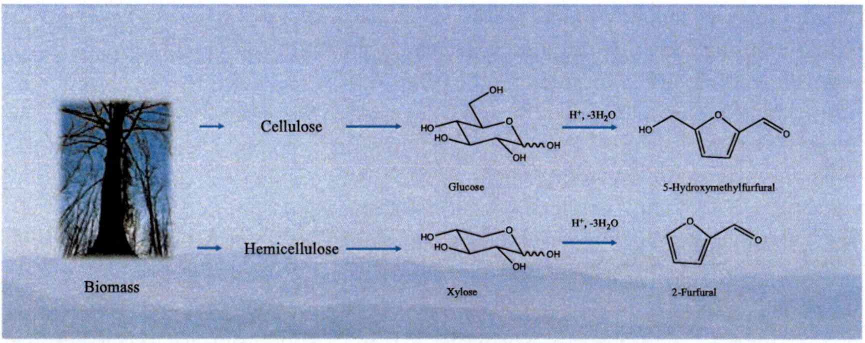

Fig. 5.5 Production of furfural and 5-hydroxymethylfurfural from biomass

between β-glucose monomers, due to the β-1,4 glycosidic linkage, is key to very different resulting properties: for example, it is insoluble in water, indigestible by many animals (including humans), and more resistant to breakdown without specific enzymes: this makes recycling of the material relatively easy. Pretreatment of this kind of lignocellulosic biomass to release fermentable sugars is a major hurdle for full valorization, and it was treated in Chap. 4. Microbial fermentation of simple sugars to alcohol or acids in the absence of oxygen or anaerobic digestion by which microorganisms break down biomass microorganisms in the absence of oxygen to form biogas are both possible biotechnological routes (Hassan et al. 2019). Efficient thermochemical conversion processes include gasification, which is the partial oxidation of biomass in the presence of reduced oxidant level, at variance with combustion, which features the complete oxidation of biomass. Pyrolysis is performed at elevated temperatures (500 °C - 800 °C) in the absence of air, while torrefaction, a milder form of pyrolysis, is conducted at lower temperatures (200 °C - 320 °C). Combined thermochemical–biochemical methods are actively under investigation even if the toxicity of the crude pyrolytic substrates and the presence of growth inhibitors from raw syngas contaminants are critical challenges (Hassan et al. 2019).

Considerable efforts have been devoted to the depolymerization of cellulose to obtain 5-hydroxymethyl furfural (5-HMF), as indicated in Fig. 5.5. Furfural from hemicellulose and 5-HMF from cellulose are promising biomass-derived furanic building blocks (Bozell and Petersen 2010). 5-HMF is synthesized from glucose and fructose via dehydration under acidic conditions. The reaction may be performed either in water, ionic liquids, or organic solvents, in particular polar aprotic solvents. After the hydrolysis of cellulose towards glucose catalyzed by acids in an aqueous medium, glucose is dehydrated to 5-HMF in a dimethyl sulfoxide /H_2O mixture with the aid of various metal-based catalysts (Kougioumtzis et al. 2018). 5-HMF is a versatile platform molecule. It can be easily transformed into 2-methyltetrahydrofuran (a biobased potential substitute of the fossilbased tetrahydrofuran), 2,5-dimethylfuran (a biofuel), levulinic acid (a keto acid that can be obtained from rehydration of 5-HMF, featured in Table 5.1 because it is one of the twelve biobased building blocks), 5-hydroxy-4-keto-2-pentenoic acid (a highly eligible monomer for polymer production due to the acid and alcohol groups in the same molecule), and 2,5-Furan dicarboxylic acid. This di-acid is a promising platform chemical since 2004 (see Table 5.1) because it is a putative substitute of terephthalic acid, a crucial building block of polyethylene terephthalate, and the product of its polymerization, polyethylene furanoate (PEF), was shown to be successful in replacing PET in many applications and in commercial production for plastic bottles (Dugmore et al. 2017).

The catalytic degradation of cellulose from vegetable waste over a solid Brønsted acid, Amberlyst 36, in an aqueous solution with/without polar aprotic solvent enabled the synthesis of levulinic acid. Even if the levulinic acid yield is lower than that obtained with homogeneous Brønsted acids, the solid Brønsted acid avoids corrosion of the reactors, difficult separation, and energy-intensive processes, which are the major hurdles of using homogeneous Brønsted acids (Cheng and Yang 2016; Chen et al. 2017).

5.1.2 *Glucose and Other Simple Sugars*

Considering the wide range of sugars-based building blocks (illustrated in Figs. 5.2 and 5.3 and detailed in Table 5.1) and the eclectic FW sources of sugars, the various possibilities of their mining and recovery from FW is a major challenge, taking into account that the kinds of sugars may vary for FW composition and origin (Dahiya et al. 2018; Li et al. 2012). The correct pretreatment is crucial to minimize the formation of sugar degraded compounds and inhibitory byproducts and to maximize sugar yields, thereby increasing the economic viability of the process.

Solubilisation and extraction technologies to get sugars from respective biowaste, such as pressing, ultra-filtration, and hot acid extraction, have all been experienced. Still increasingly, free sugars in mixtures have also been studied for direct fermentation. Such a biorefinery concept would be adaptable to many different types of FW including mixed FW (Dugmore et al. 2017). Physical, chemical, physico-chemical, biological, and enzymatic pretreatment methods, or a combination of these strategies, can be used to get target sugars. Dilute hydrochloric acid and sulfuric acid pretreatments are promising methods for the valorization of carbohydrate rich FW, especially in combination with an enzymatic method to control the formation of sugar degraded compounds such as furfurals and hydroxyl methyl furfurals. Ultrasonic treatment of FW increased the amount of reducing sugars dissolved out of FW (Jiang et al. 2014).

Glucose and fermentable sugars can be obtained from the hydrolysis of starch and cellulose recovered from FW. For example, *Actinobacillus succinogenes* fermented glucose, and fructose from fruit and vegetable waste hydrolysate as the sole feedstock to produce succinic acid with high yield (Dessie et al. 2018). High fructose syrup is usually obtained by enzymatic saccharification of starch to glucose, followed by enzymatic isomerization of glucose to fructose. It would be crucial to produce high fructose syrup from FW. The possibility of getting it from the conversion of beverage waste, which is a significant sewage stream, was actually studied (Haque et al. 2017). The steps involved (i) enzymatic hydrolysis with pectinase and sucrase at 50 °C and pH of 4.0–4.5; (ii) an activated carbon treatment for the decolorisation of the mixture; (iii) cation and anion exchange chromatography to remove 98% of ions; (iv) a ligand exchange chromatography for fructose separation. A total of 47.5% of sugars in the beverage waste hydrolysate were recovered and valorized as high fructose syrup via a green process.

Among cheap sources of simple sugars, we find fruit and vegetable waste. The residue of pigmented orange pulp wash was stabilized by removal of enzymes and microorganisms via ultra-filtration; then 80% of the water was eliminated by a reverse osmosis treatment to obtain a purified transparent liquid that is a sugar concentrate containing about 250 g/L of sugars (glucose, fructose and sucrose), as well as 9 g/L of citric acid and 1 g/L of pectin. It could be easily used as a natural sweetener in food and beverage industries (Scordino et al. 2007) but it can also be used for subsequent chemical and biochemical transformations. Sugars for the production of 5-hydroxymethylfurfural were obtained from chicory waste (Stökle and Kruse 2019) but also from cassava (Ona et al. 2019). Vegetables have one of the highest wastage rates at retail and consumer levels and a combined thermic-enzymatic hydrolysis maximized the concentration of fermentable sugars in the final broth (Díaz et al. 2017).

A very high yield of glucose was obtained from kitchen waste (0.79 g /g kitchen waste) after a combined pretreatment by 1.5% HCl and glucoamylase (Hafid et al. 2015). Kumar et al. obtained glucose from potato peel waste under microwave irradiation using silicotungstic acid as a catalyst for the chemical hydrolysis of potato peel starch: the yield after 15 min of microwave irradiation was 59% (Kumar et al. 2016).

Brewer's spent grain holds high potential in obtaining fermentable sugars through enzymatic hydrolysis, and sonication (Luft et al. 2019) or microwave-assisted hydrothermal pretreatment technology (López-Linares et al. 2019).

Whey, a significant dairy effluent, is an important source of lactose. Its valorization is detailed in Chap. 2 and Chap. 4. We hereby recall that whey permeate from ultrafiltration containing lactose was hydrolyzed using enzymes to release fermentable sugars and then integrated into wheat fermentation as a co-substrate to partially replace process water for ethanol production, using *Saccharomyces cerevisiae* (Parashar et al. 2016).

Pure lactose can be recovered from whey via nanofiltration of the ultrafiltration permeate: in fact, it represents the nanofiltration retentate. The subsequent valorization of this dairy sugar may involve the use of crystallization, purification, hydrolysis (to get glucose and galactose), and transformation into various valuable derivatives and nutraceuticals (see below) (Smithers 2008) (Nath et al. 2016).

A high-value rare sugar, D-tagatose, is a naturally occurring monosaccharide as sweet as sucrose but with 38% fewer calories. Low quantities of this natural sweetener are found in fruits and dairy products. Kim et al. (Kim et al. 2017) purified L-arabinose isomerase from *Paenibacillus polymyxa* to convert onion juice residue to D-tagatose. The yield of 9.9% was acceptable considering the high value of D-tagatose and the low value of the onion agro-residue.

Large amounts of coffee residue waste are daily produced worldwide, since coffee is one of the most consumed beverages. Their disposal poses environmental problems due to their caffeine, polyphenols, and tannins content. To recover D-mannose, a sugar monomer widely used as a nutrient supplement, coffee residue waste was pretreated with ethanol and enzymatically hydrolyzed to produce sugars. A subsequent selective fermentation of glucose and galactose to ethanol, retained D-mannose in the fermented broth. After decolorization and pervaporation steps, D-mannose was finally recovered (Nguyen et al. 2017).

5.1.3 Pectin

Pectin is a complex polysaccharide and a soluble dietary fiber; it is a bioactive compound found in the primary cell walls and in the middle lamellas between plant cells to bind them together (Mohnen 2008). It is the major adhesive material between cells (Willats et al. 2001). It has a galacturonic acid backbone. The average chain length, the degree of branching in the 'hairy' regions, the content of specific sugars, galacturonic acid, and uronic acid, and molecular weight distribution depend on the biomass sources and impact the overall properties and, in particular, the gelling properties of the pectin (Dugmore et al. 2017). Pectin is actually the jelly-like matrix that binds plant cells together and in which cellulose fibrils are embedded.

Pectin extraction is a highly lucrative valorization method within the fruit waste biorefinery system due to its high concentrations within fruit and its numerous applications in food processing industry: it is an efficient gelling and thickening agent for confectionery, jams, jellies, and fruit juices or it may serve as fat substitute in meat products (Galanakis et al. 2010). Pectin is also utilized in the medicine, pharmaceutical, and cosmetic industry as a binding and carrier agent.

For the manufacture of pectin, FW such as cocoa husks (Mollea et al. 2008) papaya peels (Boonrod et al. 2006), citrus waste (Bafrani 2010), and passion fruit peel and rind flours (Kulkarni and Vijayanand 2010) (Canteri et al. 2010), are often used. Mango waste, provided pectin extraction yields of up to 21% (Sirisakulwat et al. 2010) and simultaneous recovery of polyphenols was also possible using a styrene-divinylbenzene resin (Berardini et al. 2005).

Pectin sources also include apple peel (Shalini and Gupta 2010), peach pomace (Galanakis 2012), pigmented orange pulp wash, treated to recover also sugars and citric acid, (Scordino et al. 2007), green beans, carrots, leek and celeriac waste streams (Christiaens et al. 2015), chokeberry, black currant, pear, cherry, carrot, and apple pomace (Nawirska and Kwaśniewska 2005). Pectins structure from different FW streams differs considerably (Christiaens et al. 2015). Green labeled pectic oligosaccharides and pectin were extracted, using proteases and cellulases able to act, respectively, on proteins and cellulose present in cell walls, from different plant byproducts, such as chicory roots, citrus peel, cauliflower florets and leaves, endive, and sugar beet pulps (Zykwinska et al. 2008).

An uncommon use of pectin was recently reported by Grassino et al. (Grassino et al. 2016). They demonstrated that the pectin extracted from tomato peel serves as an efficient and safe corrosion inhibitor for tin, even at very low concentrations. The maximum inhibition efficiency was reported as 71%. The isolation of pectin from a cheap and abundant renewable resource, such as tomato waste, turns the waste disposal problem of the canning factory into the opportunity to produce a value-added tin corrosion inhibitor, according to a waste-to-wealth strategy. The substitution of ordinary toxic corrosion inhibitors by pectins in the canning industry is noteworthy within the circular economy approach.

For extraction of pectin from plant tissues, the most frequently used technique makes use of heat refluxing extraction with acidified water followed by pectin precipitation, usually with ethanol; treatment with endo-xylanase and other enzymes were also successfully tested. The mineral acids traditionally employed in the extraction of pectins conflict with the principles of green chemistry due to their hazardous nature. Moreover, despite the high yield generally obtained, the long extraction times at high temperature leads to thermal degradation of the pectin. Green, less harsh extraction methods include using ammonium oxalate/oxalic acid and then precipitation with ethanol, or ultrasound- and microwave-assisted extraction techniques, yielding non-degraded pectin in high yields (Grassino et al. 2016; Dugmore et al. 2017) (Rodríguez Robledo and Isabel Castro Vázquez 2020).

5.1.4 Xanthan Gum

Xanthan gum is an anionic polyelectrolyte and a microbial exopolysaccharide produced by the plant pathogen *Xanthomonas campestris,* a species of bacteria that use glucose or sucrose as the sole carbon source. It comprises complex repeating units of pentasaccharides (two glucose, two mannose and one glucuronic acid residue), along with a pyruvic acid unit.

Xanthan gum serves as a thickening and gelling agent, emulsifier, lubricant, binder, and stabilizer; it is a potential substitute for agar, which is significantly used in microbiological studies. In food industries, it is appreciated as a gelling agent, thickener or stabilizer; it is an important commodity for use also in paper, paint, ceramics, oil, cosmetic, and pharmaceutical industries thanks to its temperature and pH stability, pseudoplasticity, and water solubility (Jose et al. 2015).

Utilization of low cost and unconventional substrate for the production of xanthan gum is expected to reduce the cost for its fermentative output and to make the process economically viable: whey, citrus wastes, corn steep liquor, glycerol, chicken feathers, molasses and glucose syrup, and olive oil wastewaters were studied as cheaper carbon sources (Murad et al. 2019).

Sapodilla (*Manilkara zapota*) peel, mango (*Mangifera indica*), custard apple (*Annona reticulate*) unused pulp, muskmelon (*Cucumis melo*) peel, sweet potato

(*Ipomoea batatas*) peel, cauliflower (*Brassica oleracea* var. *botrytis*) leaves and stalks were screened in exploratory research as potential substrates for xanthan gum production; *Manilkara zapota* peels provided high yielding and cost-effective media for Xanthan gum production by *Xanthomonas campestris* (Jose et al. 2015).

Li et al. (Li et al. 2016; Li et al. 2017) used kitchen waste, instead of the expensive glucose or sucrose, as a sole substrate using *Xanthomonas campestris* LRELP-1. The maximal production of xanthan gum of 11.3 g/L was obtained using kitchen waste hydrolysate. The study revealed a low-cost strategy for the production of xanthan gum as well as an effective strategy for kitchen waste management.

5.2 Lipids

At variance with carbohydrates, lipids are the most hydrophobic of the three main classes of biological macromolecules. They are used in biological systems for energy storage and as water repellents. In triglycerides, three fatty acids are bound to one molecule of glycerol via ester bonds. In phospholipids, similar to triglycerides, one fatty acid is replaced by a phosphate group, this way, they are the polar lipids that constitute the cell membranes, responsible for creating an hydrophobic barrier between the aqueous environment inside and outside the cell.

Saturated fatty acids are unbranched carboxylic acids with at least 12 carbon atoms; in the presence of one or more double bonds, the fatty acid is unsaturated or polyunsaturated (double bonds are not conjugated and in *cis* position).

Lipids are particularly important for biodiesel production (see also Sect. 3.5.2, and Sect. 4.3.3) according to the reaction scheme illustrated in Fig. 5.6 that details the transesterification of lipid from FW with methanol in the presence of chemical and enzymatic catalysts. Biodiesel is a promising substitute for the fossilbased diesel fuels. It is clear that the demand for biodiesel is expected to rise over the coming years. Since the world tackles a reported food shortage, commonly used edible oils from food crops represent the least eligible feedstock in this context. Researchers are currently engaged in the upcycling of FW for biodiesel production, which is a key step to reduce FW problem and also meet the energy demand: the cost of biodiesel production is high and the investigation of low cost feedstocks that are economically feasible is mandatory. FW contains 5–30% of lipids, and it is obviously one the cheapest feedstock in this respect. Anyhow, isolation of lipids is required since FW also contains carbohydrates, protein, and minerals (Pleissner et al. 2014) (Pleissner et al. 2013). Solvent extraction effectively isolates lipids, but separation and recycling of harmful organic solvents will increase the cost of biodiesel production. Hexane is a common choice, but it is listed among the substances of very high concern, under the criteria set up by REACH (Registration, Evaluation, Authorisation, and Restriction of Chemicals) (The European Parliament and the Council of the European Union 2006). A solventless lipid isolation from FW was obtained after fungal hydrolysis and centrifugation from the resultant hydrolysate (Pleissner et al. 2014) (Pleissner et al. 2013) (Karmee and Lin 2014).

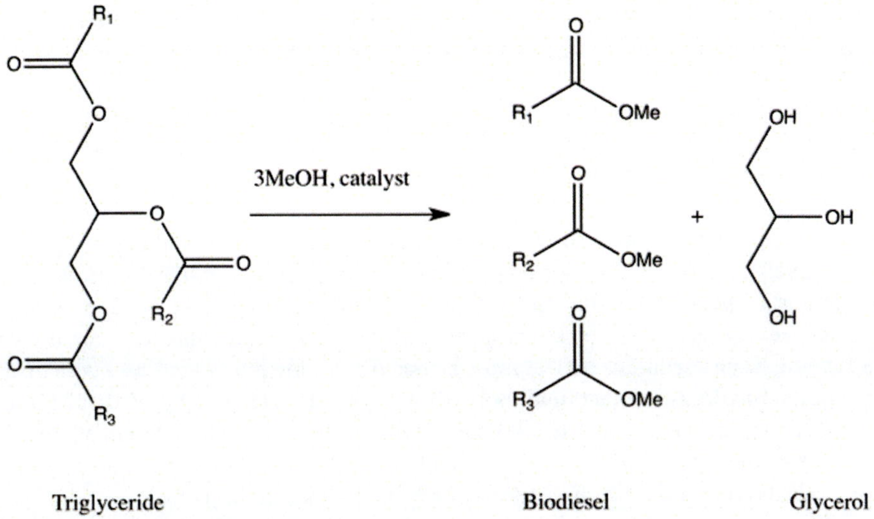

Fig. 5.6 Transesterification of lipid from food waste with methanol in the presence of chemical and enzymatic catalysts for biodiesel production

An innovative FW-based biorefinery concept aimed at the production of fatty acids and lactic acid from mixed restaurant food and bakery wastes within a 'zero-waste' approach. The treatment included the hydrolysis of food and bakery wastes using extracellular fungal enzymes in submerged fermentation. By hydrolysis, 0.27 g glucose, 4.7 mg free amino nitrogen, and 1.9 mg phosphate were recovered per gram dry waste material. After hydrolysis, the nutrient-rich hydrolysate was used as medium in fed-batch cultures of the heterotrophic microalga *Chlorella pyrenoidosa*. The extraction of lipids from algal biomass and from lipid-rich solids remaining after the hydrolysis resulted in a saturated and unsaturated fatty acid-rich feedstock. Defatted waste-derived solids and algal biomass were furtherly used in lactic acid production using *Bacillus coagulans* (Pleissner et al. 2015).

Supercritical CO_2 extraction, whose principles are described in Sect. 4.2.3.4, is an attractive green alternative to hexane extraction. It was used to extract lipids from FW such as wheat straw, corn leaves, and rice bran (Dugmore et al. 2017). Subcritical water extraction, similarly described in Sect. 4.2.3.2, can also be profitably used for lipid extraction.

For biodiesel production, lipid from FW was transformed into fatty acid methyl esters (FAME) with methanol, using a base and *Candida antarctica* lipase-B as catalyst. The base (KOH) catalyzed transesterification reaction gave an excellent 100% FAME yield at 60 °C and 1:10 M ratio of lipid to methanol in 2 hours. Lipase yielded a 90% FAME conversion at 40 °C in 24 hours (Karmee et al. 2015). The kitchen waste was dried to improve the efficiency of the lipid extraction in the soxhlet apparatus using methanol as solvent. Fatty acids (caproic acid, lauric acid, mystric acid, palmitic acid, stearic acid and oleic acid) were transesterified to produce biodiesel. Concentrations of FAMEs were gas chromatographically determined and biodiesel properties were found satisfactory, compared with various standards (Barik et al. 2018). Since glycerol is a by-product of the biodiesel synthesis, according to the biorefinery philosophy, it is perhaps inevitable that this crude glycerol should be further utilized. It proved to be suitable for the synthesis of polyhydroxyalkanoates (PHA) by microorganisms, which are natural PHA producers (*Bacillus megaterium*) with yields as high as 62%, similar to those obtained using sugarcane bagasse and whey (Naranjo et al. 2013). In a comprehensive biorefinery structure, Kachrimanidou et al. complemented the production of biodiesel with that of poly 3-hydroxybutyrate and poly (3-hydroxybutyrate-co-3-hydroxyvalerate) from a mixture of crude glycerol and levulinic acid from the lignocellulosic sunflower meal (the leftover residue from oil extraction), thereby valorizing all byproducts from biodiesel production (Kachrimanidou et al. 2014).

Crude glycerol can be exploited for the production of single cell oil (microbial lipid) and citric acid, an important nutrient for fermentation of many products, by *Yarrowia lipolytica* strains (Papanikolaou and Aggelis 2009). *Lactobacillus brevis* N1E9.3.3 yielded 0.89 g of 1,3-propanediol per g of glycerol (Vivek et al. 2016).

Glycerol provides access to a range of monomers for subsequent polymerizations. Glycerol based polymers, with variable architectures available, including polyethers, polycarbonates, and polyesters, have important biomedical applications as carriers for drug delivery, antibacterials, and for tissue repair (Zhang and Grinstaff 2014). Glycerol can also be obtained from sugars, but this route is not attractive from the economic point of view (EU 2015).

The lipids extracted from FW can be used not only for biodiesel production; base catalyzed hydrolysis of triglycerides is known as the saponification reaction, crucial for surfactant industries. Lipids can also be used as greases and lubricants to reduce friction between moving surfaces. Autoxidation of double bonds to form reactive epoxides, prone to nucleophilic addition, is the first step in the cross-linking needed to create resins with a broad range of useful properties such as high strength, excellent corrosion, weather resistance, and electrical insulation (Dugmore et al. 2017).

Lipids are also important for the synthesis of wax esters, which find application for the production of candles, lubricants, food as well as cosmetics. Three oleaginous yeasts, namely *Cryptococcus curvatus*, *Rhodosporidium toruloides* and *Lipomyces starkeyi,* were used to produce microbial oils from FW and confectionary waste streams, cheese whey, and wine lees. Microbial oils (see Sect. 4.3.3) were converted into wax esters via enzymatic catalysis using a lipase catalyzed synthesis. Transesterification of microbial oils to behenyl and cetyl esters was more efficient than the transesterification of control palm oil, and the conversion yields were up to 87.3% and 69.1%, respectively. Enzymes were efficiently reused during microbial transesterification. Wax esters from microbial oils are comparable to natural waxes, and they have potential applications in cosmetics, chemical, and food industries (Papadaki et al. 2017).

Waste lipid usefulness for biobased resins production will be dealt with in Sect. 6.6.

5.3 Protein, Peptides, Amino Acids

Of the three groups of macronutrients in biomass, proteins are possibly the most assorted and complex. The monomers of proteins are amino acids bonded through amide bonds (known as peptide bonds in protein chemistry between the amine and acid groups. Each amino acid is characterized by R groups, which are aliphatic chains, eventually comprising other acidic, basic, and aromatic functional groups, providing a range of polarities and hydrophobicities. The sequence of the amino acids is the primary structure and it determines the nature of intramolecular interactions. These define the secondary and tertiary structures, the 3D shape, and nature of the protein. The wide range of protein structures and characteristics makes protein processing a very challenging task, as there is no "universal" and established approach.

Proteins are particularly important because they can be broken down into their constituent amino acids, which are useful chiral platform molecules (natural amino acids occur exclusively in the L-isomer with the exception of the achiral glycine). They are useful in asymmetric synthesis either in introducing a chiral center in the molecule or as a temporary moiety to create diastereomers and thus facilitate their separation especially necessary for drug development whereby one enantiomer may be inactive or even harmful.

The nature of the FW obviously influences the type of protein that can be valorized. The estimate of the quality and quantity of proteins in diverse FWs enables their ranking. For this reason, Prandi et al. studied 39 FW streams by Kjeldahl analysis, amino acid analysis, protein analysis, and racemization degree for assessing the potentialities and the limitations of many waste streams for further valorization as a source of proteins. For each waste stream the specific nitrogen-to-protein conversion factor was obtained. All wastes of dairy and brewery origin, rapeseed press cake, sea buckthorn spent pulp, leek leaves, parsley waste, pumpkin kernel cake, and mushroom waste have relevant and abundant proteins; since there are streams with valuable proteins, but in too little amount and other streams rich in proteins of low nutritional value, the amino acid analysis is useful to indicate possible use for supplementing specific amino acids (Prandi et al. 2019).

Academic and industrial research focused on some specific protein source.

Potatoes waste contains precious protein. Ammonium sulfate precipitation led to the highest protein yield and to the recovery of protein isolates enriched in patatin. Protease inhibitors from potatoes have been proven to be appetite suppressants with trypsin inhibiting activities. They have a molecular weight of around 21 kDa hence they have scope to be separated via ultrafiltration. They were precipitated by $FeCl_3$ and $MnCl_2$ (Waglay et al. 2014) (Dugmore et al. 2017).

Cheese processing whey, whose valorization was deeply detailed in Chap. 2 and Chap. 4, is a precious source of oligopeptides and proteins which can be used in supplements and soft drinks (Madureira et al. 2010). Whey protein containing all 20 amino acids is the focus of interest of those who want to enhance muscle growth, including athletes and sportspeople familiar with protein powders in gyms and sports shops. Success in weight loss, anticancer activity, tissue regenerating, and antiinflammatory properties were also reported. Emerging, non-thermal extraction and isolation technologies make the applications cost-effective (Smithers 2008; Dugmore et al. 2017).

Collagen, the most abundant protein in animal organisms, provides the structural integrity for body tissue. It is a particularly abundant by-product from seafood production (Pal and Suresh 2016) or slaughterhouse residues. Collagen can have nutraceutical and bio-medical applications, and it has been studied for anticancer drug delivery purposes (Baiano 2014).

Keratin, an insoluble protein that makes up the exterior parts of animals such as hair, nails, claws, feathers, etc., is of interest for its structural properties (Chojnacka et al. 2011). Waste feathers generated in the poultry industry are an excellent source of the harder, insoluble, mechanically strong keratin types. These properties are due to the high amounts of cysteine able to provide some "built-in" cross-linking via disulfide bonds. Tanabe et al. studied the polymerization of films from keratin; glycerol proved to be the best additional cross-linking agent required to get the highest tensile strength of the films; films mixed with chitosan also demonstrated superior mechanical properties to keratin alone, hence these biowastes are an interesting opportunity for an integrated biorefinery by using feedstocks from different sources (Tanabe et al. 2004). Poole et al. studied the use of chicken feather keratin for the production of environmentally sustainable protein fibers, comparing them to other natural (e.g., cotton and silk) and synthetic (e.g., nylon and polypropylene) fibers and demonstrated the ability of keratin fibers to form threads without the need of additional cross-linking agents. The availability of feather keratin in up to five million tonnes annually is rewarding from the feasibility point of view (Poole et al. 2009). The difficulty in solubilizing the keratin suggests the need for a green extraction method to be able to claim polymers and fibers from keratin as "green" materials.

Protein hydrolyzates from fish byproducts have also been proposed as seafood flavors for soups or surimi (Kristinsson and Rasco 2000).

Sub-critical water hydrolysis proved to be a green and profitable way for recovering high added-value protein hydrolysates in the form of peptides or amino acids from both animal and vegetable wastes. These recovered biomolecules are precious in fields such as biotechnology and food industry. Key parameters as reaction time, temperature, amino acid degradation, and kinetic constants were crucial for maximizing the recovery. These hydrolysates are greatly appreciated for their functional properties and are used as nutraceuticals and antioxidants (Marcet et al. 2016).

5.4 Organic Acids

Organic acids are weak acids. Most of them are can be obtained from sugars and are included in the sugar-based building blocks list in its original (Werpy and Petersen 2004) or revised (Bozell and Petersen 2010; EU 2015) versions.

They are particularly interesting because they can be transformed to a variety of derivatives with a wide range of functional groups, as illustrated in Fig. 5.3 for those organic acids included in the original list (Werpy and Petersen 2004).

5.4.1 Volatile Organic Acids (VFA)

VFAs or short-chain (C2-C5) carboxylic acids such as acetic acid, propionic acid, butyric acid, iso-butyric acid, valeric acid and iso-valeric acid are the major co-products in anaerobic fermentation. VFA production from waste-stream is gaining momentum due to increasing market demand as well as its cost-effective and environmentally friendly approach; VFAs produced from waste streams are mainly acetic, propionic, and butyric acid (Atasoy et al. 2018). They find important applications in pharmaceuticals, textiles, food and beverages, and polymers sectors. VFAs can be subsequently used for the production of bioplastics (anoxygenesis), biodiesel (oleaginous metabolism), bioelectricity generation (bioelectrogenesis), and biofertilizers (Venkata Mohan et al. 2016). Their fossilbased production is energy-intensive and has a negative impact on the environment; hence their biological production via anaerobic fermentation of sugar substrates by both thermophilic and mesophilic anaerobic bacteria is gaining immense interest (See Sect. 4.3.1 for the description of the anaerobic fermentation) (Sarkar and Venkata Mohan 2017).

Since sterile fermentation conditions are not required, FW represents an inexpensive feedstock. Parameters such as pH, temperature, organic loading rate, retention time, and kind of substrate influence VFA production.

Acetic acid is the main VFA since its production by anaerobic acidifying bacteria is thermodynamically favored. It is used for the production of vinyl acetate, which represents an important monomer in polymers, adhesives, dyes. Acetic acid is also used as a solvent and for the production of other chemicals. Acetic acid bacteria are cataloged in ten genera in the family *Acetobacteriaceae.* The development of anaerobic acidifying bacteria stains more tolerant to acetic acid could increase the end concentration of the product. Pal and Nayak (Pal and Nayak 2016) developed an economically viable approach for the concomitant production of acetic acid and recovery of whey protein from whey via a multistage membrane integrated hybrid reactor system. The eco-friendly and cost effective process resulted in the continuous output of 98% pure acetic acid.

Vinegar, commonly used for food preservation, is a mixture of acetic acid and water usually produced by wine microbial fermentation; actually, acetic acid is a food additive commonly used to regulate acidity. Roda et al. (Roda et al. 2017) reported vinegar production from pineapple peels enzymatically saccharified and fermented with *Saccharomyces cerevisiae* for 7–10 days under aerobic conditions at an incubation temperature of 25 °C. Acetic acid was obtained from the further fermentation of this alcohol medium by *Acetobacter aceti,* for 30 days at 32 °C to get a concentration of 5% acetic acid clear vinegar. Pineapple waste is not the only alternative sustainable feedstock for the production of vinegar, as olive oil press-mill wastewater was transformed into olive vinegar with a high percentage of phenolic antioxidants indicating its nutraceutical potential: sugar addition and inoculum of selected yeast were necessary (De Leonardis et al. 2018).

Propionic acid is a well-known preservative and food additive; it also has broad industrial applications, including bactericides, fungicides, herbicides, pharmaceuticals, perfumes, and emulsifying agents. Its esters are used in the food industry as aroma additive; it is an animal feed supplement and a fishing bait additive (Atasoy et al. 2018). The expensive petrochemical synthetic route of this metabolite can be profitably replaced by its biotechnological production, using cheap and easily available waste biomass for anaerobic digestion. Under stress conditions, propionic and butyric acid concentrations are expected to increase compared to acetic acid. Propionibacteria produce propionic acid under anaerobic conditions. Research on propionic acid production mainly focuses on increasing yield and productivity inducing bacteria acid-tolerance since propionic acid inhibits cell growth by disturbing the pH gradient across cell membrane even at low concentrations (Koutinas et al. 2014).

For example, the knockout of the acetate kinase gene of a genetically modified *Propionibacterium acidipropionici* (a Gram-positive, anaerobic bacterium) reduced acetate production, increased propionic acid concentration and stimulated acid tolerance (Zhang and Yang 2009). Crude glycerol, which is the main by-product of biodiesel production, was evaluated as a cheaper carbon source than glucose and proved to enable good productivity compared to a substrate such as sugarcane molasses (Koutinas et al. 2014).

Apple pomace was a carbon source in the process of propionic-acetic fermentation via wild strain *Propionibacterium freudenreichii* T82 bacteria. After 120 hours of cultivation, propionic acid biosynthesis reached its maximum (1.771 g/L) while the conentration of the acetic acid was 7.049 g/L (Piwowarek et al. 2016). Genetically modified bacteria or immobilized cells are the focus of interest of many researchers as regards the specific industrial biotechnological production of propionic acid (Koutinas et al. 2014).

The efficient recovery of high-quality short-chain fatty acids (including propionic acid) from wine vinasse and waste activated sludge mixtures was realized. An exceptional short-chain fatty acids promotion was due to key enzymes and diversity and abundance of fermentative bacteria, which resulted in the rapid hydrolysis and acidification of bioavailable substrates. Furthermore, the dewaterability of fermentation residues was facilitated by wine vinasses as skeleton builders (Luo et al. 2019).

Butyric acid is a food and feed additive, it is used as a solvent, and it is an important chemical intermediate. Its esters are used in the food industry as aroma additives. Butyric acid was the prevalent acid in FW anaerobic fermentation under acidic conditions, using inocula based on anaerobic activated sludge at pH 6.0 after 20 days of fermentation. VFA composition analysis showed that butyric acid concentration was followed by acetate acid and propionic acid (Wang et al. 2014). Production of VFA by anaerobic digestion of Cassava wastewaters was also successfully reported: 3400 mg/L of VFA were obtained with a predominance of acetic (63%) and butyric acid (22%), followed by propionic acid (12%) (Hasan et al. 2015).

In an anaerobic digestion process, VFA production can be enhanced by (i) shortening the reaction time to prevent the final methanogenesis; (ii) adjusting pH above 8.0 or below 6.0 to impair the growth of methanogens; (iii) adding methanogenic inhibitor (Atasoy et al. 2018). For example, in order to limit the prevalence of methanogenic bacteria and to enhance VFA production from FW, the biocatalyst was exposed to acid-shock, which stimulated the production of redox mediators, followed by operation under alkaline conditions. The presence of *Clostridium autoethanogenum* and *Propionibacterium freudenreichii* resulted in the effective utilization of complex carbohydrates, facilitating acidification (Sarkar et al. 2016). Oxygen-assisted pre-aeration before hydrolysis of FW was also exploited to assist in the suppression of methanogenic activity and resulted in enhanced acidogenic product formation (10% more VFA production than the control) (Sarkar and Venkata Mohan 2017).

Electrofermentation (see Sect. 4.3.4.2) was used to fine-tune metabolites production and to overcome the thermodynamic limitations of microbial fermentation, thereby controlling the metabolic routes towards specific biobased products. Electrofermentation of the composite canteen-based FW gave fatty gave acid profiles with a higher amount of acetic acid, followed by butyric acid and propionic acid with co-generation of biogas. Fermentation stimulated by the electron flux enabled the elongation of the chain length to enhance the conversion of conventional substrates to targeted carboxylic acids. This emerging technology could open new frontiers in waste biorefinery (Shanthi Sravan et al. 2018).

5.4.2 1,4-Dicarboxylic Acids (Fumaric, Succinic, Malic Acid)

The family of 4-carbon diacids shares similar biochemical sugar-related production paths. They have the potential to be key building blocks for deriving both commodity and specialty chemicals, and they are among the twelve sugar-based building blocks (Werpy and Petersen 2004) even if even only succinic acid is included in the revised list in 2010 (Bozell and Petersen 2010) and 2015 (EU 2015).

Fumaric acid is one of the intermediates in the citric acid cycle. It has a fruit-like taste and it has a role as a food acidity regulator; it finds applications also in pharmaceuticals as well as in the preparation of polyester resins and polyhydric alcohols, and as a mordant for dyes. The demand is increasing each year. Mostly the fumaric acid is fossilbased. Interestingly, different waste biomass can be used as a source for fumaric acid production. For example, solid-state fermentation of apple industry waste biomass by *Rhizopus oryzae* 1526 (Das et al. 2015) yields ca. 52 g of fumaric acid per kg of apple pomace substrate; it was clear that small size fungal pellets favored more fumaric acid production than large-sized fungal pellets.

Succinic acid is a four carbon, linear, saturated di-carboxylic acid. Its name comes from the Latin word *succinum* that means amber, where it was first detected. Succinic acid is commonly used as an acidity regulator and flavor agent in the food and beverage industry and in dietary supplements. Its applications include high-value niche uses such as personal care products, pharmaceuticals, cosmetics, and large volume applications such as bio-polymers and solvents. It is also useful to produce commodity chemicals that are currently petroleum-derived, such as 1,4-butanediol (used for the synthesis of γ-butyrolactone), or, in the presence of phosphoric acid and high temperature, of its dehydration product, namely tetrahydrofuran (EU 2015).

Succinic acid has predominantly been produced by energy-intensive fossilbased chemical strategies from butane; petrochemical succinic acid production costs are increasing, making its bioproduction more and more competitive. Various microorganisms have been screened and tested for the biotechnological production of succinic acid with good yields (Bechthold et al. 2008). Most of the natural succinic acid producers are bacteria isolated from the bovine rumen. *Anaerobiospirillum succiniciproducens, Actinobacillus succinogenes, Mannheimia succiniciproducens, Basfia succiniciproducens* are the most widely used bacterial strain to get biobased succinic acid that represents one of the currently fastest-growing markets. Nowadays biobased succinic acid could easily overtake fossilbased alternatives due to the level and breadth of industry activity in the product due to price competitiveness, lower environmental impact, and ease of production (EU 2015). The upcycling of FW for its fermentative production by *Actinobacillus succinogenes* is a viable alternative synthetic approach. Various FW types were used as substrates. Citrus peel pretreated with dilute acid and enzymatically saccharified (Patsalou et al. 2017) gave 0.70 g of succinic acid per gram of total sugars consumed. Oil palm empty fruit bunch, an abundant agricultural residue available in Malaysia (Akhtar and Idris 2017), was enzymatically saccharified and the fermentation resulted in 0.47 g of succinic acid per g of substrate. Diverse fruit and vegetable wastes were hydrolyzed; the fungal hydrolysis by *Aspergillus niger* and *Rhizopus oryzae* produced glucose and fructose; *Actinobacillus succinogenes* used fruit and vegetable wastes hydrolysate as the sole feedstock to make succinic acid with high yield (Dessie et al. 2018).

Cake and pastry hydrolysates, mixed with magnesium carbonate were used as feedstock in *Actinobacillus succinogenes* fermentation, and a cation-exchange resin-based process was subsequently used to recover the succinic acid from fermentation broth with high crystal purity (Zhang et al. 2013).

Waste bread served as a nutrient rich feedstock for the fermentative succinic acid production by *Actinobacillus succinogenes*. Amylolytic and proteolytic enzymes from *Aspergillus awamori* and *Aspergillus oryzae* hydrolysed a bread suspension to generate over 100 g/L glucose and 490 mg/L free amino nitrogen. *A. succinogenes* fermentations of the hydrolysate led to the production of 0.55 g succinic acid per gram of bread (Leung et al. 2012).

Anaerobiospirillum succiniciproducens efficiently produces succinic acid (Bozell and Petersen 2010) and batch and continuous cultivation of *Anaerobiospirillum succiniciproducens* for the fermentation of nontreated whey lactose was characterized by high productivity (Lee et al. 2000).

A techno-economic study for pilot-scale fermentative production of succinic acid by *Actinobacilus succinogenes* demonstrated that the conversion of 1 ton/day of bakery waste into succinic acid resulted in a return on investment, payback period, and internal rate of return of the project of 12.8%, 7.2 years and 15.3% respectively, thereby indicating that the fermentative succinic acid production from bakery waste via a biorefinery approach was feasible and economically attractive (Lam et al. 2014).

Mixed food waste from restaurants was used to produce a nutrient-complete FW hydrolysate, using crude enzymes from *Aspergillus awamori* and *Aspergillus oryzae*. An overall yield of 0.224 g succinic acid/g substrate, using FW hydrolysate and recombinant *Escherichia coli,* demonstrated the enormous potential of mixed FW as renewable resource in the biotechnological production of biobased chemicals and materials (Sun et al. 2014).

Succinic acid market is expected to strongly expand in the bioplastic sector and particularly in the production of polybutylene succinate. In Sect. 6.4 the importance of this platform chemical, which has considerable potential economics and environmental meaning for the bioplastics sector, will be detailed.

Fossil based malic acid is produced from maleic anhydride, which in turn is produced from vapor-phase oxidation of butane. Malic acid production from sugars relies on the fermentative conversion from fumaric to malic acid. Malic acid can be cyclized to form hydroxysuccinic anhydride, which, via reduction, gives the 3-hydroxybutryolactone, which represents, along with its precursor malic acid, a platform chemical (Werpy and Petersen 2004) as illustrated in Figs. 5.2 and 5.3. Citrus waste is a possible source of malic acid (Satari and Karimi 2018).

5.4.3 Lactic Acid

Lactic acid, according to the IUPAC nomenclature, is 2-hydroxy-propionic acid; it exists in two optically active enantiomers, the L(+)- and the D(−)-lactic acid (respectively L-LA and D-LA). Lactic acid is well-recognized building blocks integrated into the revised list of sugar-based platform chemicals in 2010 (Bozell and Petersen 2010); it is an important organic acid for the production of several downstream chemicals in food, pharmaceutical, and cosmetic industries. Lactic acid esterification gives lactate esters of interest as new "green" solvents; its catalytic reduction leads to propylene glycol, which can be further dehydrated to give propylene oxide. The direct dehydration of lactic acid gives acrylic acid, even if in low yield (Bozell and Petersen 2010).

Its primary use is for the production of polylactate (PLA), as detailed in Sect. 6.2, a polymer that exhibits performance properties similar to polystyrene, a storage resistance to fatty foods equivalent to polyethylenterephthalate, excellent barrier properties for flavors and aromas, and good heat sealability. Nowadays, LA used for PLA production is mainly derived from renewable edible crops such as sugar beet and corn (Castro-Aguirre et al. 2016). Lactic acid is among the established biobased products that already dominate global production (EU 2015). The eco-friendly fermentative strategy was stimulated by environmental concerns about petroleum-based synthesis.

Commercial LA is usually and selectively produced by the bacterial fermentation of carbohydrates by homofermentative organisms belonging to the genus *Lactobacillus* and *Bacillus*. *L. delbrueckii, L. amylophilus, L. bavaricus, L. casei, L. maltaromicus,* and *L. salivarius* predominantly produce the L(+)-isomer. Strains such as *L. lactis, L. jensenii,* and *L. acidophilus* generate either the D(−)-isomer or mixtures of both enantiomeric LA. Most strains are anaerobic and convert pyruvic acid, the end-product of the Embden-Meyerhof pathway, to lactate by either of the two enzymes, L- or D-lactate dehydrogenase; pyruvate is the electron acceptor of NADH (nicotinamide adenine dinucleotide) oxidation (Singhvi et al. 2018).

Commercial fermentation, usually conducted in a batch process, takes a few days to complete but continuous fermentations proved to be feasible and advantageous (Olszewska-Widdrat et al. 2020).

During a typical LA fermentation, LA production results in a drop in pH that has an inhibitory effect on the metabolic activity of LA-producing strains. To avoid this drawback, a neutralizing agent such as calcium carbonate or hydroxide is used in industrial processes to produce calcium lactate, which does not inhibit further LA synthesis in the fermentation broth. Filtration removes the cell biomass and other insolubles, then, after evaporation and recrystallization, corrosive sulfuric acid is used to liberate crude LA from calcium lactate, generating a large amount of calcium sulfate (gypsum) as solid waste. A filtration step separates gypsum; LA purification, via esterification and subsequent hydrolysis, is one of the most expensive stages of the production process and a reactive distillation system was advantageous for lactic acid purification (Komesu et al. 2015).

The process is not only costly but also complicated; by this scenario, the use of acid-tolerant strains, obtained by genome shuffling through protoplast fusion, gene deletion, exploitation of knockdown libraries and adaptive evolution to acidic pH, would avoid the need for neutralizers, thereby simplifying the downstream process, and lowering the overall cost of the fermentative LA production; another strategy to circumvent the toxicity of lactic acid towards microbial cells was in situ LA recovery (Singhvi et al. 2018).

A derivative of thermotolerant *Bacillus subtilis* (strain 168) that grows at 50 °C was metabolically engineered to enable fermentation of glucose at 48 °C in mineral salts medium to D-lactate. This platform organism could minimize process costs in support of PLA industry (Awasthi et al. 2018). *Lactobacillus delbrueckii* gave 90% yield based on glucose fed; engineered yeast species, such as *Pichia stipitis* were able to ferment xylose from all lignocellulosic waste to lactate (Bozell and Petersen 2010).

To limit the land use share for bioplastics, a wide range of high sugar-containing FW such as whey, coffee mucilage, corn cobs, corn stalks, rice bran, barley, wheat bran and brewer's spent grains (Venus 2006; Venus and Richter 2006), kitchen and canteen waste (Tashiro et al. 2013; Ohkouchi and Inoue 2006) (Wang et al. 2010), pineapple syrup and grape invertase, as substrate and enzyme (Ueno et al. 2003), wheat bran (Germec et al. 2019), soybean vinasse (Karp et al. 2011), *Curcuma longa* biomass (Nguyen et al. 2013), carbohydrates, starchy and lignocellulosic biomasses free from food-related concerns, glycerol, microalgae, whey (Abdel-Rahman et al. 2013; Vasala et al. 2005) (Vishnu et al. 2006), fruit and vegetable wastes (Wu et al. 2015), and yoghurt waste (Alonso et al. 2010) have been explored as alternative feedstock for fermentative LA production.

Dairy waste are particularly interesting since dairy industries generate high amounts of whey from milk processing for various manufactured products. Whey disposal is a major pollution problem (Abdel-Rahman et al. 2013). Whey is a valuable raw material (see Sect. 2.8) for biotechnological lactic acid production, because it comprises the fermentable sugar, lactose and water-soluble vitamins, mineral salts, and other essential nutrients for microbial growth. Theoretically, 1 mol of the disaccharide lactose, consisting of 1 mol of glucose and 1 mol of galactose, yields 4 mol of lactic acid through a homofermentative pathway (Abdel-Rahman et al. 2013). To obtain lactic acid from whey, Vasala et al. (Vasala et al. 2005) used *Lactobacillus salivarius* an ideal organism for lactic acid fermentation in high-salt high lactose conditions, typical of whey. A scalable fermentation protocol to obtain a whey-based food disinfectant (Martin-Diana et al. 2006) with the highest lactic acid concentration and the lowest lactose levels made use of a specific mesophilic-lactic acid bacteria starter mix. This fermented whey represents an effective and environmentally-safe alternative to the use of chlorine as a disinfectant for the food processing industry (Santos et al. 2019).

The market for yogurt has also grown rapidly over the past few years. Consequently, rejected, damaged, or expired yoghurts create huge amounts of waste products. Yoghurt is usually sweetened with additional sugars, such as sucrose and glucose, or supplemented with fruits syrups; bioconversion of total sugars resulted in higher lactic acid production than cheese whey containing fewer sugars, in pH controlled batch fermentation after 34 h, with a hierarchical mode consumption by *Lactobacillus casei* in the order glucose > sucrose > lactose (Alonso et al. 2010).

Kitchen waste is another important substrate for lactic acid production. Defatted hydrolysate of mixed restaurant FW and bakery waste (obtained making use of *Aspergillus awamori* and *Aspergillus oryzae)* were successfully fermented to lactic acid by *Bacillus coagulans* (Pleissner et al. 2015). *Lactobacillus manihotivorans* LMG18011 proved to be useful for L-lactic acid production from kitchen waste (Ohkouchi and Inoue 2006). A bacterial consortium from marine-animal-resource compost led to accumulation of butyric acid at low temperatures, whereas at 50 °C, 34.5 g/L of L-lactic acid was produced with 90% lactic acid selectivity and 100% optical purity. Bacteria present in the original compost diminished rapidly and *Bacillus coagulans* strains became dominant at temperatures above 45 °C. The achievement of 100% optical purity of L-lactic acid using a bacterial consortium is noteworthy from the industrial point of view (Tashiro et al. 2013). Zhang et al. (Zhang et al. 2017) optimized by statistical design the synergistic lactic acid production from school canteen waste and waste activated sludge by interactive control of pH and incubation temperature. Optimum pH for lactic acid production decreased with increasing temperature. Nguyen et al. (Nguyen et al. 2013) used *Lactobacillus coryneformis* and *Lactobacillus paracasei* for the simultaneous saccharification co-fermentation of waste *Curcuma longa* biomass. Under optimized nitrogen sources and concentrations, enzyme compositions, and raw material concentrations, 97.13 g/L and 91.61 g/L of L- and D-lactic acid were obtained.

L-LA is the preferred building block for food and pharmaceutical applications since elevated levels of D-LA are harmful to humans; hence it would be desirable to selectively produce this enantiomer via a specific strain of lactic acid bacteria (Reddy et al. 2008) as *S. thermophilus* and *L. Bulgaricus* (Tang et al. 2013). Bio-production of optically pure L-LA from restaurant FW at ambient temperature relied on the regulation of key enzyme activity by sewage sludge supplementation and intermittent alkaline fermentation. The production of optically pure L-LA acid was achieved from FW at ambient temperature with a yield of 0.52 g/g of total chemical oxygen demand of the substrate (Li et al. 2015). *Streptococcus thermophilus* and *Lactobacillus bulgaricus* selectively produced L-LA from furfural residues and corn kernels with hydrolyzed yeast as cheap nutrients (Tang et al. 2013). Soybean vinasse was another useful substrate for optically pure L-LA production also at pilot scale (Karp et al. 2011).

5.4.4 Gluconic and Glucaric Acids

Gluconic (2,3,4,5,6-pentahydroxyhexanoic acid) and glucaric (2,3,4,5-tetrahydroxy hexanedioic acid) acids are oxidative products of glucose. The latter is included in the original twelve sugar based building blocks (Werpy and Petersen 2004) but not in the revised lists (Bozell and Petersen 2010; EU 2015).

Gluconic acid occurs naturally in fruit, honey, and wine and finds application in different fields like food, pharmaceutical, textile, and leather industries. It is an acidity regulator, while its anion, gluconate, is used in many pharmaceutical products. The use of FW as feedstock may lower its production cost that prevents its extensive use. Sharma et al. (Sharma et al. 2008) used *Aspergillus niger* ARNU-4 in a solid-state fermentation of sugarcane molasses, incorporating tea waste as a novel solid support. After process parameters optimization, maximum gluconic acid yield (76.3 g/L) was observed with 70% moisture level and an incubation temperature of 30 °C. The addition of 0.5% of yeast extract increased the yield to 82.2 g/L.

Glucaric acid is obtained from catalytic oxidation of glucose (Lee et al. 2019). Its current market is minor, but it has huge potential as a replacement for phosphates in detergents since phosphates are blamed for stimulating algae growth and aquatic eutrophisation. Glucaric acid is also appreciated for corrosion inhibition properties (de Jong et al. 2012). It is a precursor of adipic acid, used as monomer for the production of nylon and polyurethane (EU 2015). Even if marketed glucaric acid is presently chemically produced its biothecnological production from carbohydrate-rich substrates is currently possible, via engineered cells of *Escherichia coli*, *Saccharomyces cerevisiae* and *Pichia pastoris* (Chen et al. 2020).

5.4.5 Acrylic Acid

Acrylic acid is systematically named prop-2-enoic acid. It is a corrosive liquid with a characteristic acrid smell. Acrylic acid and its esters readily polymerize by reacting at their double bond, forming homopolymers (polyacrylic acid) or copolymers with other monomers (e.g., acrylamides, acrylonitrile, vinyl, styrene, and butadiene), used in the manufacture of coatings, adhesives, diapers, textiles, resins, personal care products, synthetic rubbers, and paints. Conventional petrochemical acrylic acid is produced via the oxidation of propylene. Biobased acrylic can be obtained, for example, from dehydration of FW sugar-derived lactic acid (EU 2015).

5.4.6 Adipic Acid

Adipic acid is a C6 straight-chain dicarboxylic acid used as monomer for the production of nylon and polyurethane and to manufacture plasticizers and lubricant compounds. Food grade adipic acid is appreciated as a gelling aid, an acidulant, and a leavening agent. Adipic acid is usually produced from various petrochemical feedstocks, such as cyclohexane. Biobased production of adipic acid relies on hydrogenation of FW glucose derived glucaric acids (EU 2015).

5.4.7 Citric Acid

About 70% of the overall citric acid production is used in the food industry as an acidifying as well as a flavor-enhancing agent in food and beverages. It is commonly used also in chemical, cosmetics and pharmaceutical applications like acidulation, chelation, emulsification, preservation, and as a plasticizer (Soccol and Vandenberghe 2003).

The use of FW as an economically viable and cost-effective substrate for citric acid fermentative production can tackle the increase in demand of citric acid.

Molasses, starch, and sucrose-based FW are the most eligible substrates, yet a wide gamut of fruit and vegetable wastes, and other agroindustrial residues such as coffee husk and cassava bagasse proved to be suitable for a cost-effective citric acid production.

An adequate supply of carbon, nitrogen, and phosphorus and a pH of 2.0 is required for optimum production of citric acid; oxygen supply enhanced yields of citric acid and decreased fermentation times (Nayak and Bhushan 2019). Numerous microorganisms such as *Aspergillus sp.* and yeasts like *Candida tropicalis*, *C. oleophila*, *C. guillermondii* etc. can be used but the filamentous fungi *Aspergillus niger* are the most widely used during the fermentation process.

Aspergillus niger was demonstrated to be able to produce citric acid using various feedstock such as banana peel (Karthikeyan and Sivakumar 2010), fruits wastes (Kumar et al. 2003), pineapple peel (Sashi Prabha and Rangaiah 2014), and apple pomace ultrafiltration sludge (Dhillon et al. 2011). In this case, response surface methodology was used to optimize various process parameters and maximize citric acid production (44.9 g/100 g dry substrate after 144 h of fermentation).

The citric acid bioproduction ability of *Aspergillus niger* was also successfully tested on brewery spent liquid supplemented with 40% (w/w) apple pomace ultrafiltration sludge (Dhillon et al. 2012). Recently, grape processing industry side stream (Schieber 2017; Galanakis 2017) and the residue of pigmented orange pulp wash (Scordino et al. 2007), have been investigated as substrates for citric acid production.

5.5 Phosphate

Phosphorus is essential for all living organisms. At the current rate of consumption, many countries will face a crisis as their phosphorus reserves are depleted. Phosphorus is historically used as a mineral fertilizer in agriculture; nowadays, the demand for phosphorus is increasing since it is used in a wide variety of products like pharmaceuticals, flame retardants, and chemicals. Due to the rising global population and the global trend towards phosphorus-intensive diets, the demand for phosphorus will increase by 50–100% by 2050 (Cordell 2010). Mineral reserves of phosphorus are limited and unevenly distributed; hence countries need to prioritize securing a supply of phosphorus from alternative sources to sustain their food production. On a local level, urbanization can easily result in excess of phosphorus in surface water where it can cause euthrophization and toxic algal blooms; hence it should be recovered from wastewaters and watersheds; anyhow the food industry is a real urban mine since animal byproducts (e.g., bones) and other FW are rich sources of phosphorus (EU 2014).

Harvesting phosphorus during FW recycling can be cost-effective since FW is the solid waste fraction containing the most phosphorus per kilogram. In a feasibility study concerning phosphorus recovery from household solid organic waste in Sweden, three forms of recycling FW were compared. The highest yield was obtained when phosphorus was harvested from anaerobically digested FW (Lu 2014) compared to compost; recovery after incineration of municipal solid waste is less favorable due to heavy metal contamination that poses concerns for the potential application to agricultural land. Another study confirmed the potential of anaerobic digestion of restaurant FW (used as a substrate) not only for biogas recovery but also for phosphorus recovery. The substrate to inoculum ratio was studied to maximize phosphorus recovery in an anaerobic batch digester at mesophilic (35 ± 1 °C) condition and pH = 6.0 for 15 days (Selaman and Wid 2019).

It is, however, clear that decreasing FW is about 80 times more efficient than recycling FW in reducing phosphorus demand (Vaccari et al. 2019).

5.6 Nutraceuticals: Antioxidants and Other Bioactives

The inception of the nutraceutical concept is due to increasing health consciousness. The portmanteau word "nutraceutical" was introduced in 1989 by De Felice and indicates "a food (or part of a food) that provides medical or health benefits, including the prevention and/or treatment of a disease" (Kalra 2003). Nutraceuticals are nutritive pharmaceuticals found in foods or parts of foods that are supposed to provide health and/or medical benefits, including prevention, protection, and treatment of a disease (Brower 1998) with no side effects. For example, they are claimed to function as antimicrobial, antidiabetic, antihypertensive, anticoagulant, anticancer, or hypo-cholesteraemic agents.

Nutraceuticals, functional, or health foods have gained wide interest during the last two decades. Their health-promoting effects are likely due to a complex mix of biochemical synergistic interactions that resulted in some evidence-based clinical success (Volpe and Sotis 2015; Borghi and Cicero 2017; Vonder Haar et al. 2016; Barrios et al. 2017; Cicero and Colletti 2015; Jain et al. 2018; Mostacci et al. 2017; Pagano et al. 2018; Pérez-Sánchez et al. 2018; Di Hsu and Cheng 2018; Cicero et al. 2018; Venkatakrishnan et al. 2019; Cicero et al. 2019). The functional food industry is gaining momentum and is seen more than from a research perspective. Polyphenols, alkaloids, carotenoids, prebiotics, phytosterols, tannins, fatty acids, terpenoids, saponins, and soluble and insoluble dietary fibers are recognized as potential health promoting compounds (Bergamin et al. 2019).

Molecules endowed with antioxidant capacity are probably the most recognized nutraceuticals. Oxidative stress is caused by a discrepancy between the production of reactive oxygen species (ROS) and the ability to neutralize them. Antioxidants limit the damage of ROS preventing their reaction with vulnerable substrates, thereby preventing oxidation of molecules. They may act via electron transfer (ET) reactions or Hydrogen Atom Transfer reactions (Cecchi et al. 2019). Hence, these compounds play an important role as preservatives in foods and cosmetics. Antioxidants also reduce the risk of certain human diseases. Synthetic antioxidants may not be as effective as natural antioxidants since they lack synergistic effects with the food matrix (Cecchi et al. 2010); it follows that extraction of antioxidants from FW is noteworthy.

Some FW, like peels, seeds etc., are excellent sources for the production of antioxidants. A wide variety of fruit waste including peach (Plazzotta et al. 2020), strawberry, sour-cherry, black currant, red currant, blackberry, and raspberry (Krisch et al.

2009), citrus fruit (Matharu et al. 2016; Fernández-López et al. 2004; Scordino et al. 2007), clementine (Pfukwa et al. 2019), mango (Schieber 2017; Abdalla et al. 2007; Ajila et al. 2010), banana (Schieber 2017; Maneerat et al. 2017), pineapple (Schieber 2017; Upadhyay et al. 2012), and tropical fruits (Da Silva et al. 2014), among others (Krisch et al. 2009; Ayala-Zavala et al. 2010; Górecka et al. 2010; Rohm et al. 2015; Peschel et al. 2006) (Balasundram et al. 2006; Sharma et al. 2020) can be exploited to recover precious antioxidants. Antioxidants can also be extracted from vegetable wastes, such as broccoli (Dominguez-Perles et al. 2011; Aires et al. 2017), chicory (Lante et al. 2011), carrot (Schieber 2017), turnip greens (Chihoub et al. 2019), thistle (Drouet et al. 2019; Chihoub et al. 2019), lettuce (Plazzotta et al. 2018; Lafarga et al. 2019), radicchio (Perotto et al. 2018), artichokes (Pagano et al. 2018), cucumbers (Chen et al. 2019), beans (Yang et al. 2019), green beans (Aires et al. 2017), pumpkin (Hataminia et al. 2018; Kulczyński et al. 2020), zucchini (Lafarga et al. 2019), and coffee waste (Murthy and Madhava Naidu 2012).

In the following, we will detail some specific important bioactives from specific FW.

Amado et al. (Amado et al. 2014) optimized, via response surface strategy, antioxidant extraction from potato peel waste. The optimum conditions of extraction were 34 min of extraction time, the temperature of 89.9 °C and ethanol concentration of 71.2% and 38.6% for extraction of phenolics and flavonoids, respectively. Potato peel antioxidants can effectively limit oxidation of soybean oil, lipid and protein of minced horse mackerel, and radiation-processed lamb meat (Mohdaly et al. 2010; Sabeena Farvin et al. 2012; Kanatt et al. 2005; Okuno et al. 2002).

Barba et al. (Barba et al. 2016) reviewed green approaches for extraction of antioxidant bioactive compounds from winery wastes and byproducts like grapes stalk, grapes marc, grapes seeds, etc. Emerging technologies detailed in Chap. 4 can replace conventional ones with a number of advantages, such as reduced processing time and energy, the absence of toxic solvents, and increased selectivity and yields. The characteristics of the matrix being processed, the needed selectivity, and the value of the extract enable a judicious choice of the best extraction technique; the scale-up of non-conventional technologies is not straightforward, due to equipment and installation costs. Grape seeds was a rich source of vitamin E and, in general, winery byproducts are an excellent source of antioxidants such as polyphenols including catechin, epicatechin, and proanthocyanidins (flavanols), gallic acid (phenolic acid), resveratrol (stilbene), enocyanin (anthocyanin), procyanidins (Schieber 2017; Galanakis 2017; Makris 2018; Luo et al. 2019).

Olive pomace is a rich source of antioxidant phenolic compounds, and their utilization has many environmental benefits. (Wang et al. 2017) extracted them via an eco-friendly ultrasound-assisted enzymatic hydrolysis of olive waste. The phenolic extract obtained with a yield of 4%, after a treatment time of 40 min at 55 °C and pH 5.75, can be used as a food additive enhancing antioxidant properties in fatty food, with better economic benefits than synthetic additives.

Polyphenols and carotenoids from a gamut of fruit byproducts could be profitably used as a natural food or beverage preservatives able to extend the shelf-life of the product by delaying the formation of off-flavors and rancidity, thanks to their antioxidant activity (Oreopoulou and Tzia 2007).

Lycopene, a potent antioxidant red pigment, was extracted from tomato processing waste using an acetone-hexane mixture with a yield of 3.47–4.03 mg/100 g of processing waste and a recovery of 65.22–75.75% (Poojary and Passamonti 2015).

Quercetin is an antioxidant flavonoid commonly found in fruits and vegetables, and it is useful as a supplement in foods and beverages. Quercetin protects normal cells of chemotherapy and radiotherapy side effects; hence quercetin may have a key role in anticancer treatment. Onion skin waste from processed onions can be valorized since it is a precious source of quercetin and its glycoside, which inhibited the urease and xanthine oxidase activity in vitro and showed potent antioxidant activity in vitro scavenging assays (Nile et al. 2017). Quercetin extraction increased by 1.61-fold after the onion skin waste was enzymatically saccharified with a cocktail of cellulase, pectinase, and xylanase. A novel nano-matrix was used for the easy separation and purification of quercetin via its adsorption. The recovery yield of quercetin was approximately 75% after treatment with ethylenediaminetetraacetic acid (Choi et al. 2015); other important flavonoids such as myricetin, luteolin, and kaempferol can also be obtained from onion waste (Corell et al. 2018).

Papaya seed proved to be a good source of bioactive and health-protective benzyl isothiocyanate; it can be profitably recovered via large scale supercritical fluid extraction (Hall et al. 2018). Astaxanthin is a naturally occurring xanthophyll carotenoid that displays different biological activities, including antioxidant, proapoptotic, cardioprotective, anti-hypersensitive, and anticancer effects in preclinical studies. It is found in nature primarily in marine organisms such as microalgae, salmon, trout, and crustaceans (Faraone et al. 2020). It can be obtained as co-product in the recovery of chitosan from a very cheap and abundant source: shrimp wastes. It can be extracted by several chemical strategies and green approaches like microbial fermentation or enzymatic extraction (Prameela et al. 2017).

Harnedy and FitzGerald detailed several bioactive peptides, proteins and amino acids from marine processing waste and fish for the production of a range of nutraceuticals. Fish, shellfish, mollusks, and crustaceans wastes contain significant levels of high quality proteins (10–23% (w/w)), which represent substrates for biofunctional peptide mining. Fish protein hydrolysates such as fermented sauces and pastes used as staples in Southeast Asia, Scandinavian and Eskimo cultures, can be obtained from marine-derived waste; they display antioxidant, antihypertensive, anticoagulant, antimicrobial, ACE inhibitory, and calcium-binding activities; they have potential as an appetite suppressant and HIV-1 protease inhibitors (Harnedy and FitzGerald 2012). Marine processing waste is a source of endogenous bioactive peptides and bioactive proteins, which were reported to have antimicrobial, antifungal, antiviral, antiprotozoal, antiproliferative, calcium-binding, cytotoxic, opioid, antidiabetic, anticoagulant, hemolytic, agglutinating, and antioxidant activity (Harnedy and FitzGerald 2012). The antioxidant capacity of eumelanin extracted from Cephalopod ink can add value to this abundant residue of cuttlefish and squid processing (Cecchi et al. 2019)

Apart from antioxidants, other target bioactives are the focus of academic and market interest.

Docosahexaenoic acid (DHA) is an omega-3 essential polyunsaturated fatty acid with positive impacts on hypertension, diabetics, myocardial infarction, and cancer. Deficiencies of DHA are related to some diseases. It is usually extracted from fish oil. Quilodran et al. (Quilodrán et al. 2010) studied the potential of residual mash from brewery by-product and liquid residues from potato chip processing factory as a substrate for the production of DHA by *Thraustochytriidae sp.* AS4-A1. The strain productivity with supplementation of yeast extract, B-complex vitamins, and mono-sodium glutamate reached 540 mg/L/ day. The DHA production was observed during the growth period, and astaxanthin was co-produced. It follows that brewery by-product represents a cost-effective and excellent nutrient source for the production of DHA. Dietary fiber and fats rich in the essential linoleic acid can be recovered from both pea pod and broad bean pod (Mateos-Aparicio et al. 2010).

Lactose-derived nutraceuticals can be synthesized through different chemical and biochemical reactions (hydrolysis, transgalactosylation, oxidation, reduction, isomerization, and hydrogenolysis), considering raw whey or isolated lactose as feedstock. For example, lactulose is a nonabsorbable disaccharide with prebiotic properties; lactitol is a sugar alcohol and a bulking agent for sugar-free product; both lactulose and lactitol are useful for treating patients with chronic constipation; among other lactose-derived nutraceuticals, lactobionic acid, produced from the oxidation of lactose, combines a sweet taste and a pH reducing effects; it finds important application in skincare products, but it is not used as a food ingredient yet. Galacto-oligosaccharides find application in food and pharmaceutical industry as prebiotics to promote gut health; moreover they increase calcium and magnesium absorption; similarly, lactosucrose is a potential prebiotic oligosaccharide (Nath et al. 2016).

Strictly pharmaceutical compounds can also be obtained from FW. Recovery of bromelain, which is a group of proteolytic enzymes with important anti-inflammatory, antithrombotic, and fibrinolytic properties (da Silva et al. 2010) can be recovered from pineapple (*Ananas comosus*) waste with 98% of the total enzymatic activity (Martins et al. 2014). Chondroitin sulfate, a structural component of cartilage widely used for the treatment of osteoarthritis, was obtained from scapular cartilage of Shortfin mako shark by enzymatic hydrolysis and fractional precipitation using isopropyl alcohol (Kim et al. 2012), from waste of *Scyliorhinus canicula*, (Gargiulo et al. 2009), and skate cartilage byproducts (Murado et al. 2010).

5.7 Enzymes

The industrial use of enzymes is growing every day in detergent, paper (bleaching), textiles manufacturing, bioethanol production, food and beverage processing (fruit juice extraction, vegetable oil extraction, alcoholic beverages, tea, and coffee fermentation).

Their use versus traditional chemical processes enables costs and waste reduction and avoids harsh pressure or temperature conditions (Singh et al. 2016). They have tremendous industrial applications because of their high specificity and environmental friendliness.

Solid-state fermentation, carried out in the absence of water, and submerged fermentation are the main processes used to produce enzymes; solid-state fermentation has become outstanding due to its higher yield and its simpler and more versatile equipment (Kriaa and Kammoun 2016). The multi-faceted application of enzymes has led academia to develop technologies for their production from cheaper substrates as agroindustrial residues. Wastes from the fruit and vegetable sectors are renowned as one of the best waste sources of commercially relevant enzymes (Balasundram et al. 2006) (Sharma et al. 2020).

Unfortunately, the biorefinery of food wastes for enzyme recovery on an industrial scale is not operative yet; promising studies on a laboratory scale focus on the immobilization and purification of enzymes via a single step (Nayak and Bhushan 2019).

5.7.1 Amylase

Amylases are very important enzymes that degrade starch to smaller carbohydrates units like glucose, maltose, and maltotriose. They find industrial applications in paper, textile, food, detergent, and ethanol production. Commercial carbon and nitrogen sources are commonly used for the production of amylases. *Bacillus*, *Aspergillus*, *Rhizopus* and *Rhizobial* isolates are crucial sources of industrial amylases (Unakal et al. 2012). Agroresidues or food and kitchen waste are a cost-effective substrate for amylase production by *Chryseobacterium* and *Bacillus* species (Hasan et al. 2017). Banana peel (Radha Krishna et al. 2012) enabled the production of amylase by *Aspergillus niger* NCIM 616 via solid-state fermentation, a promising strategy when compared to submerged fermentation. Supplementation of mineral salts to the medium improved amylase production. Potatoes peel was another important carbon source for amylase production via the thermophilic bacterium *Bacillus subtilis* K-18 (KX881940) (Mushtaq et al. 2017). *Bacillus subtilis* similarly produced α-amylase from the wastes of banana (Unakal et al. 2012).

Starch-based cereal substrates, such as wheat bran, are easily colonized by the strains of *Bacillus sp.* and *Aspergillus sp.* for the optimal production of α-amylase (Nayak and Bhushan 2019).

Breweries spent grain characterized by high protein content produced amylase enzyme under solid-state fermentation by *Aspergillus oryzae* (Francis et al. 2003).

5.7.2 Cellulase

Cellulases are complex enzymes; their sequential action results in the complete hydrolysis of cellulose; hence they are crucial for the lignocellulosic biomass waste bioconversion. Cellulases are actually an enzyme complex composed of three major groups of enzymes, each with specific activities: (i) endoglucanases (endocellulase) randomly cleave the internal bonds of the amorphous region, releasing oligosaccharides; (ii) exoglucanases (exocellulase), including cellobiohydrolases, are responsible for the hydrolysis of terminal reducing and non-reducing groups; they remove two to four units from the ends of the chains resulting in tetrasaccharides or disaccharides, such as cellobiose; (iii) β-glucosidase hydrolyze the exocellulase product into monosaccharides, namely glucose.

Cellulase is made use of in different industries like biofuel, paper and pulp, textile, detergent, food, and feed industry. Free cellulases can be synthesized by fungi or bacteria. The former dominate commercial applications due to their high levels of expression and secretion (Carvalho et al. 2012).

Cellulase production, by growing *Aspergillus niger* NRRL3 on solid-state fermentation, made use of soybean hulls and waste paper as supports (Julia et al. 2016), increasing the volumetric productivity at shorter times, with a positive impact on overall cellulase complex production economics. The activities of endoglucanase, exoglucanase, and glucosidase (5914.29 U/L, 4551.19 U/L, and 984.01 U/L, respectively) were, correspondingly, 4 times, 9.5 times, and 1.7 times higher than waste paper alone at the same fermentation time.

A wide gamut of agricultural and kitchen waste residues like corn cobs, carrot peelings, wheat bran and straw, orange and potato peelings, pineapple peelings sawdust, rice husk moistened with water were alkali pretreated and then used for the production of the complete cellulase system by *Aspergillus niger* NS-2, in solid-state fermentation without any additional nutritional sources. Maximum production was obtained after 96 h of incubation. Wheat bran gave the highest cellulase yield (Bansal et al. 2012). Similarly, potato peels were a proper substrate for the production of cellulolytic enzymes (Carvalho et al. 2012). Peel pea waste was used as a carbon source for cellulase production under solid-state cultivation by *Tricoderma reesei* (Verma et al. 2011).

5.7.3 *Xylanase*

Xylanases degrade the linear plant polysaccharide xylan, one kind of hemicellulose, into xylose, thus breaking down the β-1,4 backbone of hemicellulose. The hydrolysis of hemicellulose generates several isomers of pentoses and hexoses.

Xylanase is important in the food, feed, juices, paper, and pulp industries. Cellulase-free xylanases are important in pulp biobleaching as alternatives to the use of toxic chlorinated compounds.

Grape pomace left out after juice extraction from grapes, is neither suitable as animal food nor for fertilizer applications since the high phenol content poses concerns for zootechnical uses and may inhibit seed germination. The feasibility of the production of xylanases by *Aspergillus awamori* from grape pomace was studied (Botella et al. 2007) (Díaz et al. 2012): additional carbon source and manipulation of the initial moisture content of grape pomace are important factors in enzyme production. It is rewarding to observe that these enzymes, obtainable from the grape industry byproducts, are commonly used just in the clarification processes in wine cellars.

5.7.4 *Pectinase*

Pectinase breaks down pectin, another polysaccharide found in plant cell walls. Pectic enzymes, including pectolyase, pectozyme, and the most used polygalacturonase, are commonly used in processes involving the degradation of plant materials, such as the extraction of fruit juice and wine production; pectinase is also used to clear the wine or the juice because the presence of pectin causes slight cloudiness. Pectinases are also used during the retting of plant material. Chelating agents or acid pretreatment enhance the effect of the enzyme and improve the separation of the cellulosic fiber from the woody stem. Pectinases are crucial enzymes in food industries for the production of pectic oligosaccharides, DNA extraction from plants, and degumming of fibers. In order to meet the increasing demands, it is essential to develop strategies for their cost-effective production.

Pectinases can be obtained from fungi that produce them to damage the middle lamella and insert fungal hyphae in plants in order to extract nutrients. Pectinase is denatured at high pH and temperature.

The upcycling of several agroindustrial residues, as well as fruits and vegetable wastes, would result in cost-effective production of pectinases. These residues are ideal substrates for pectinase production. Ahmed et al. (Ahmed et al. 2016) used the soluble carbohydrates of the citrus waste peel as the sole carbon source for pectinase production by *Aspergillus niger* and submerged fermentation. The maximum enzyme yield was obtained on the fifth day of fermentation. Uzuner and Cekmecelioglu (Uzuner and Cekmecelioglu 2015) got pectinase by *Bacillus subtilis* using the hazelnut shell hydrolysate as a suitable low-cost medium for submerged fermentation; the statistical design of experiments enabled the optimization of the polygalacturonase activity (5.60 U/mL) with an incubation time of 72 h, pH of 7.0, incubation temperature of 30 °C, yeast extract concentration of 0.5% w/v and 0.02% w/v of KH_2PO_4.

5.7.5 *Protease*

Proteases are hydrolytic enzymes that catalyze the hydrolysis of the peptide bond and the degradation of proteins. Proteases are widely distributed in plants, animals, and mainly in microbes. They are particularly useful in medicine, food, and detergent industries (Kandasamy et al. 2016).

Waste bread pieces are feasible raw materials for protease production by *Aspergillus awamori* in a packed-bed reactor; this way, a major FW in many countries is transformed into an ideal substrate for solid-state fermentation. The highest protease activity after a stepwise parameter optimization strategy was 83.2 U/g of bread with an incubation time of 144 h. This smart upcycling of waste bread economically and ecologically compares to its anaerobic digestion for methane production (Melikoglu et al. 2013; Melikoglu et al. 2015). Cofee pulp waste and corncobs (Kandasamy et al. 2016), soy fiber and a mixture of cow hair and sludge (Marín et al. 2018) were also exploited to produce protease through solid-state fermentation.

5.8 Activated Carbon Adsorbent and Biochar

Activated carbon adsorbent are progressively used in various industrial purification and separation fields. In particular, wastewater remediation is a hot topic due to increased public awareness about its adverse effect on the ecosystem and the limited resource availability of clean water. Anyhow, adsorption is not only useful for wastewater remediation but also for the improvement of drinking water quality. Activated carbons are among the most successful adsorbent materials. They have an exceptional adsorption capacity for the majority of organic and inorganic pollutants, e.g., dyes, heavy metals, pharmaceuticals, and phenols. The use of activated carbon for wastewater treatment is effective, but the high cost of fossilbased activated carbon impairs the widespread use of the technology (Bansal and Goyal 2005). By this scenario, lignocellulosic waste and byproducts/residues from the agroindustrial sector hold the promise to be a cheaper and sustainable alternative for the synthesis of quality activated carbons (Nayak et al. 2018), (Nayak et al. 2017) with a large surface area, and a vast gamut of surface functional groups (carboxyl, carbonyl, phenol, quinone, lactone, etc.) bound to the edges of the graphite-like layers. The rich surface chemistry increases the affinity of activated carbons towards different water-based pollutants (Alslaibi et al. 2013). Various FW like bran and husk (rice, wheat), shell (groundnut, coconut, hazelnut, walnut, peanut), stalks (cotton, grape, sunflower), pomace (grape, olive, apple), cow dung, sawdust, soybean hulls, corn cob, banana, orange peels, sugar beet pulp, etc. have demonstrated their suitability as substrates for the production of waste-based activated carbon for water treatment; heavy metals were the most studied target adsorbates (Nayak and Bhushan 2019), (Wang et al. 2018). The adsorption capacity of various FW based activated carbon for cadmium, copper, chromium, lead, mercury, nickel, and selenium laden wastewaters ranged from 7 to 2000 mg/g (Nayak and Bhushan 2019). Alkali modified rice husk activated carbon was almost twofold more effective than the unmodified rice husk carbon for the removal of Cd^{2+} from wastewater (Ye et al. 2010). Saygili et al. (Sayǵili et al. 2015) reported the conversion of grape industrial processing waste into high-quality mesoporous carbon sorbent. The optimum conditions gave an activated carbon with a carbon content of 81.76%, a yield of 44.13%, a surface area of 1455 m^2/g, a total pore volume of 2.318 cm^3/g, and an average pore diameter of 6.81 nm. It effectively removes cationic and anionic dyes, such as Methylene blue and Metanil yellow, from aqueous solution. The adsorptive property was well described by the Langmuir isotherm model. The adsorption monolayer adsorption capacity was found to be higher than that of commercially available carbonaceous materials. Microorganisms adsorption onto activated carbon was also studied to alter the adsorption properties of activated carbon. The pH of the treated wastewater plays a definite role in affecting the nature of the binding and the pollutant removal efficiency (Nayak and Bhushan 2019) (Ye et al. 2010) (Wang et al. 2018). Thermochemical conversion methods dealt with in Chap. 4 (Sect. 4.4) are involved in the synthesis of bioadsorbents. The method of synthesis categorized as 'physical' involves initial heating the low-cost precursors at lower temperatures (400~700 °C) to produce chars (carbonization) followed by heating at higher temperatures (800~1000 °C) in the presence of an activating agent (usually steam or CO_2). The carbon atoms on the surface of the carbonized microcrystal react with the activating gas and make the pores expand while the surface area increases, according to the endothermic reactions (1), (2): $C+H_2O \rightarrow H_2+CO$ (1) $C+CO_2 \rightarrow 2CO$ (2). Impurity removal, crushing, refining, and drying complete the process. In the synthetic process classified as "chemical", the organic waste is mixed with the activator (basic solutions, mineral-based and organic-based acid solutions, organic compounds, and oxidizing agents), which erodes the cellulose in the raw material; then activated carbon is made by carbonization, activation, rinsing and drying. Physical activation and chemical activation can also be mixed. In general, organic waste is treated with an activator first, and then it is treated by a physical method at high temperatures with the activating gas. Microwave energy can replace the traditional heating method.

Biochar is the pyrogenic carbon that can be produced from FW when the organic materials undergo thermal decomposition (pyrolysis) in an environment in the absence of oxygen at low temperatures, low heating rates, and with a very short residence time. The obtained FW Biochar was profitably used as a soil amendment since it improves soil fertility, it sequesters carbon from the environment in the soil for hundreds of years, and it is able to suppress soil-borne plant pathogens by providing soil aeration and favoring beneficial soil microorganisms. Biochar is a coal-substitute and a gas adsorbent (Ahmad et al. 2019) (De Corato et al. 2018). It is useful in the development of polymer composite, and also in the improvement of the mechanical properties of green admixtures with cement mortar, thereby sequestering high volume carbon in civil infrastructure (Gupta et al. 2018).

Corn stalk was used to prepare a biochar-based catalyst for the delignification of corn stalk and its cleaner conversion into 5-hydroxymethylfurfural, a well-known furanic sugar-based platform chemical (see Table 5.1) (Liu et al. 2018).

Carbonaceous materials from agricultural residues (biochar, charcoal, activated carbon, and their magnetic forms, coal fly ash) were used to immobilize TiO_2 for greener photocatalytic technologies (Colmenares et al. 2016). Cost-effective and environmentally friendly bioadsorbents regeneration methods are the focus of academic interest.

5.9 Chitin and Chitosan

Chitin, the most abundant natural amino polysaccharide, is a structural biopolymer found in arthropods exoskeletons (e.g., crustaceans and insects) but also available in mushrooms and yeasts. Seafood waste is chitinaceous in nature. Actually, the shell of crabs, lobsters, shrimps, and prawns all contain chitin from which chitosan can be extracted (Galanakis 2012; Ravindran and Jaiswal 2016; Yan and Chen 2015; Prameela et al. 2017; Kumar et al. 2018). Approximately 18 Tg of chitin-containing waste is produced annually (Tsang et al. 2019); dried shrimps and crabs waste have about 50% of chitin on a dried basis (Yan and Chen 2015).

The utilization of chitin is impaired by the fact that both its α–/β-forms are insoluble in a majority of solvents. Its common disposal (ocean dumping, incineration, landfilling) leads to environmental pollution, eutrophication, and bio-fouling.

Chitin can be completely solubilized in acidic conditions when it is converted into its deacetylated product, chitosan (Franca et al. 2011). The obtained chitosan is a polymer of N-acetylglucosamine and glucosamine units. The alkaline N-deacetylation of chitin converts acetyl glucosamine to glucosamine, thereby restoring the primary amino functional group and significantly modifying the physical, chemical, and biological properties of chtin (Rinaudo 2006; Hirano et al. 1989). The degree of deacetylation can be described as an actual percentage conversion of acetyl glucosamine to glucosamine. Chitosan derivatives are different from original chitin. The crystalline structure is rigid due to high hydrogen bonding, but the amino groups of the chitosan are soluble in mildly acidic aqueous solutions. However, they are insoluble in alkaline media and water (Pillai et al. 2009).

Chitin and its most important derivative, chitosan, are the focus of interest of extensive worldwide academic and industrial research to alter the surface characteristics of these biopolymers, thereby allowing use in further diversified applications. In this respect, chitosan is more versatile than chitin due to its additional free amino moieties. Chitin only features hydroxyl groups for modification. Chitin, being a highly insoluble and chemically rather unreactive material, has far fewer direct applications but can give interesting compounds if appropriately modified (Pillai et al. 2009). A plethora of oxygen- and nitrogen-containing chemicals are accessible from chitin and/or chitosan: acetic, lactic, formic, oxalic, glycolic, and levulinic acids, levoglucosenone, 5-HMF, hydroxyethyl-2-amino-2-deoxyhexopyranoside, hydroxyethyl-2-acetamido-2-deoxyhexopyranoside, 3-acetamido-5-acetylfuran and proximicin A (alkaloid with cytostatic effects in various human tumour cell lines) (Sadiq et al. 2018), proximicin B and C (natural antibiotics), pyrrole, and dihydrofuropyridine are potentially synthesizable (Wang et al. 2013; Pierson et al. 2014) (Gao et al. 2016; Hülsey 2018). A wide range of chemical modifications, including acylation, alkylation, Schiff base formation, phthaloylation, silylation, tosylation, quaternary salt formation, sulfation and thiolation, phosphorylation, methylation, azylation, nitration, sulphonation, xanthation, N-succinylation, O-acetylation, H-bonding, etc. have been explored (Jayakumar et al. 2008; Jayakumar et al. 2010b; Jayakumar et al. 2010a; Yong et al. 2014; Zhang and Yan 2016).

Chitin and chitosan polymers can be developed into different forms such as gels, films, nanofibers, nanofibrils, membranes, scaffolds, beads, sponges, microparticles, and nanoparticles. They have excellent properties such as biodegradability, bio-compatibility, nontoxicity, and high adsorption capacity (Bobbink et al. 2015) (Anitha et al. 2014; Pillai et al. 2009).

Chitosan can be produced by using chemical, fermentative, or enzymatic processes. The chemical conversion of chitin into chitosan is a very competitive process (Gómez-Ríos et al. 2017) based on three fundamental steps: demineralization (by the reaction of HCl with calcium carbonate), deprotonation, and deacetylation (by using NaOH at different concentrations). Further treatments can be added if pigments should be extracted; the extraction can be microwave-assisted (El Knidri et al. 2016; Hülsey 2018). Short (around 5 min) pretreatment of prawn shell waste in glycerol, a co-product of biodiesel production, facilitates at 200 ° C the subsequent demineralization and deproteinization steps using the environmentally friendly citric acid. The resulting chitosan has a higher purity and a higher crystallinity as compared to the harsher conventional process (Hülsey 2018).

Chiosan is nontoxic, biodegradable, bioadhesive, bioactive, and biocompatible and finds applications in the medical sector for, e.g., tissue engineering and wound dressings (Jiménez-Gómez and Cecilia 2020). It exhibits antibacterial activity against *Enterococcus faecalis*, *Escherichia coli*, *Staphylococcus aureus*, and antimycotic activity against *Candida albicans* (Hussein et al. 2013).

Chitosan is also used in the environmental sector, in wastewater treatment, due to its inhibitory effect on fecal coliform bacteria, *Vibrio* spp., and staphylococci, and in the purification of drinking water (Al-Manhel et al. 2018).

Quaternized chitosan derivatives are interesting due to their high antimicrobial activity and mucoadhesiveness; at variance with chitosan, soluble only in acidic solutions, these derivatives are soluble also at neutral and basic pH due to the insertion of quaternary permanent cationic moieties on the polysaccharide backbone (Freitas et al. 2020) Chitosan/halloysite nanotubes bionanocomposites were demonstrated to be cytocompatible; hence they can be used as scaffold materials in tissue engineering (Liu et al. 2012). Crosslinked carboxymethylchitosan-g-poly(acrylic acid) copolymer is a superabsorbent material with a very high swelling ratio (Chen and Tan 2006).

A promising approache to modify chitin and chitosan is the chemically- and radiation-initiated graft copolymerization of various vinyl monomers onto the trunk polymers. Limited cases of polycondensation and oxidative coupling are the non-vinyl graft copolymerization methods. Ring-opening graft copolymerization is also possible. The polymers selected for grafting onto chitin/chitosan include poly(2-alkyl oxazolines), poly(ethylene glycol)s, block polyethers, poly(ethylene imine)s, poly(2-hydroxyalkanoate)s, polyurethanes, poly (dimethylsiloxane)s, and dendrimer-like hyperbranched polymers (Zohuriaan-Mehr 2005). Graft copolymerization of chitosan via atom transfer radical polymerization proved to be a promising strategy (Thakur and Thakur 2014). Chitosan derivatives have important antibacterial, chelating, and complexation properties (Kobayashi et al. 2017).

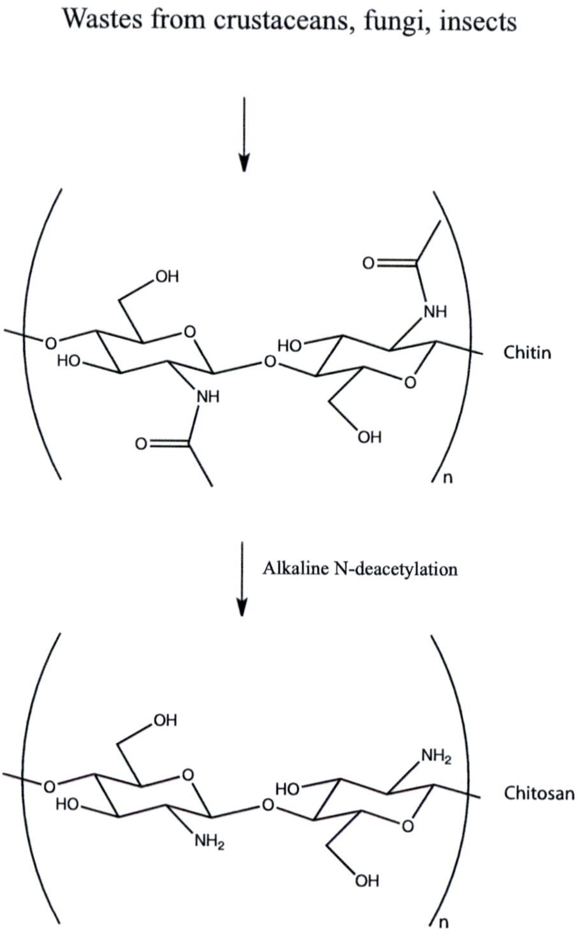

Fig. 5.7 Schematic of conversion from chitin to chitosan and their natural resources

The utilization of shrimp shell waste for the production of chitosan is well known. The Techno-economic analysis demonstrated that the process is cost-competitive (Gómez-Ríos et al. 2017). However, other possible chitin sources are increasingly studied (Hülsey 2018).

Figure 5.7 presents a schematic diagram of chitin to chitosan conversion.

5.10 pH Indicator Films

Food technology researchers are engaged with the development of intelligent packing for better preservation and increased shelf life of food products. In this context, a pH indicator film that is nontoxic and produces a response to pH change represents a sensor that provides visual information to the customers concerning food deterioration. Luchese et al. (Luchese et al. 2017) developed a pH indicator film embedding blueberry agro-waste to starch-based films. Anthocyanins from the blueberry by-product of fruit processing industry change their color in an acidic or basic environment. The pH indicator films were evaluated with buffers having different pH. The results indicate that blueberry powder acts as a potential pH indicator for intelligent food packing as well as for sensible food deterioration. It can be easily anticipated that blackberry, red grape, and red cabbage waste can be used for the same reason.

Chitosan-based smart films were produced using blueberry and blackberry pomace extracts as active agents up to 4% w/v. They showed excellent antioxidant potential that was not impaired by film production. The film permeability to oxygen and mechanical properties were not altered significantly, while water vapor permeability slightly decreased, and the film stiffness increased. Clear color changes occurred with changing pH from 2 to 10: films with blueberry changed from rose to blue-green and with blackberry from red to dark violet. The most significant change was obtained just in the pH range from 4 to 7, which is important for the detection of food spoilage in actual foodstuff. Films with blackberry pomace extract showed the highest antioxidant capacity since pure blackberry pomace extract was a better antioxidant than the blueberry one (Kurek et al. 2018).

A pH-sensitive indicator, based on a PVA/chitosan film doped with anthocyanins from red cabbage is another example of intelligent food packaging (Pereira et al. 2015).

5.11 Pigments

The use of synthetic coloring agents in food was reported to be linked to an increase in cancer rate and other health problems (Dilrukshi et al. 2019). It follows that the use of safe and natural coloring agents is of utmost importance and, actually, their demand increased during the last decades. Biotechnological pigment production using microorganisms is a safe strategy but not often economically. Agroindustrial residues serve as a cost-effective substrate for the fermentative production of microbial pigments. Eligible carbon, nitrogen, and mineral source for the production of pigments are molasses, corn steep liquor, bran, whey, etc. Fermentation-derived food grade pigments from cheaply available agroindustrial residues can make the process cost-effective and environmentally friendly (Panesar et al. 2015). Carotenes are unsaturated isoprene derivatives used in feed, pharmaceutical, and food industries, usually obtained from *Blakeslea trispora* using an expensive synthetic medium. Their importance is linked to cardio-protection and cancer prevention. Nanou and Roukas (Nanou and Roukas 2016) developed a fermentation strategy using waste cooking oil, supplemented with corn steep liquor and butylated hydroxytoluene, for the production of carotenoids, such as β-carotene (74.2%), γ-carotene (23.2%), and lycopene (2.6%), from *Blakeslea trispora*. Carotenoids production (up to 49.3 ± 0.2 mg/g dry biomass) was stimulated by oxidative stress induced by hydroperoxides of waste cooking oil. Cheng and Yang (Cheng and Yang 2016) evaluated carotenoid production by *Rhodotorula mucilaginosa* using waste health drink, ketchup and molasses as the substrate.

Similarly, *Chryseobacterium artocarpi* CECT 8497 proved to be efficient for the production of yellowish-orange pigment that can be used as a coloring agent from agroindustrial waste (Aruldass et al. 2016). A statistical design approach enabled the optimization of pigment production (152 mg/L) from liquid pineapple waste. Pineapple waste gave a threefold higher production compared to the nutrient broth. The pigment can be used in soap manufacture.

Monascus purpureus is able to produce pigmented fungal metabolites, known as azaphilones, which find wide applications in the food industry as food colorants, with grape waste as a growth substrate. Since several factors affect pigment production, like carbon and nitrogen source, agitation, aeration, etc., a chemometric approach was used to maximize the output (Silveira et al. 2008).

5.12 Single-Cell Protein (SCP) and Fungal Biomass

In our hungry world, one billion people do not have access to a diet that provides enough protein and energy. Animal meat, the main source of protein, causes severe impacts on the environment (Souza Filho et al. 2018), and vegan-protein concentrate is appreciated by vegan consumers.

Single-cell proteins are edible microbial proteins that can be obtained from unicellular microorganisms, such as algae, yeasts, fungi, or bacteria. They are envisioned as a substitute for protein-rich foods for human consumption or as animal feeds.

The metabolic pathways of microorganisms enable the conversion of low protein organic substrates into high protein biomass, up to 60–82% of dry cell weight. Besides high protein content, SCP also contains fats, carbohydrates, nucleic acids, amino acids, vitamins and minerals. Adequate aeration and supplies of carbon, nitrogen, and phosphorus are essential to support the growth of microorganisms. SCP production relies on agricultural waste (starch, molasses, fruit, and vegetable wastes) as nutrient sources or autotrophic growth.

Saccharomyces sp., Candida utilis, Endomycopsis fibuligera, and Pichia burtonii are able to exploit FW as a proficient source for the production of SCP (Sharma et al. 2020). *Saccharomyces cerevisiae* and *Kluyveromyces marxianus* were grown on food industry waste via solid-state fermentation. The protein, nutrient minerals content, as well as volatile aroma compounds in fermented products were analyzed. The highest protein and fat were obtained from substrate fermented by *Kluyveromyces marxianus* and can be used for livestock feed enrichment (Aggelopoulos et al. 2014).

Filamentous fungi have the ability to grow on a wide range of substrates of different origins and produce high-quality biomass. Its high protein and fat levels make it an expedient material that can be used for animal/fish feed or even for human consumption (if edible fungi are used). The biomass production from dairy waste was studied for two food-grade fungi traditionally used for the production of oriental fermented foods and enzymes, namely, *Aspergillus oryzae* and *Neurospora intermedia*; ethanol and glycerol were obtained as co-products. The edible fungi-based biorefineries feedstock included cheese whey, milk, yoghurt, cream, and fermented cream; conversion of lactose resulted in 0.48 g of protein-rich fungal biomass and 0.06 g of ethanol per gram of waste (Mahboubi et al. 2017).

Waste bread proved useful for the cultivation of food-grade edible strains of filamentous fungi, *Neurospora intermedia*, *Aspergillus oryzae*, belonging to ascomycetes and *Mucor indicus*, *Rhizopus oryzae*, belonging to zygomycetes group. The fungal biomasses is rich in protein and can be used as animal or fish feed component, and ethanol is, again, a co-product (Nair et al. 2017); hence the integration of the process at running ethanol plants might be the most advantageous strategy.

The amino acids profile of SCP from *Kluyveromyces marxianus* var. *marxianus* (ATCC 8554) growing in whey as substrate was studied. The crude protein content in *K. marxianus* was 42.19%. The profile of amino acids obtained suggests the potential use of this unicellular protein, taking into account the international reference patterns of FAO (Páez et al. 2008).

Olive mill wastes, which poses a concern from the environmental point of view, was used as a supplement (0–90%) for the cultivation of different strains of oyster mushrooms. The waste supplement only influenced the color of fruiting bodies, but total phenolic content, antioxidant activities were similar to that of control (wheat straw) (Ruiz-Rodriguez et al. 2010).

Mixed culture of *Kluyveromyces marxianus* and *Candida krusei* were used for SCP production and efficient (80.2%) organic load removal from the whey; extreme operating conditions (high temperature 40 °C and low pH 3.5) during batch and continuous aerobic fermentation minimized contamination (Yadav et al. 2014).

Vegan-mycoprotein concentrate from the pea-processing industry by-product was attained using edible filamentous fungal strains of Ascomycota and Zygomycota phyla. *Aspergillus oryzae* among the tested fungi, gave the best protein yield of 0.26 g per g of pea-processing by-product (Souza Filho et al. 2018).

Defatted rice bran was profitably transformed into SCP, soluble and insoluble dietary fiber, and minerals by a freshwater green microalga *Chlorella sorokiniana* (Pruksasri et al. 2019).

Solid-state fermentation using onion waste as substrate resulted in the production of fruiting bodies of commercially important edible mushrooms such as *Pleurotus sajor-caju*, with a yield of 45.73%; this result is important this mushroom is claimed to have interesting health benefits (Pereira et al. 2017).

Bibliography

Abdalla AEM, Darwish SM, Ayad EHE, El-Hamahmy RM (2007) Egyptian mango by-product 1. Compositional quality of mango seed kernel. Food Chem 103:1134–1140. https://doi.org/10.1016/j.foodchem.2006.10.017

Abdel-Rahman MA, Tashiro Y, Sonomoto K (2013) Recent advances in lactic acid production by microbial fermentation processes. Biotechnol Adv 31:877–902

Aggelopoulos T, Katsieris K, Bekatorou A et al (2014) Solid state fermentation of food waste mixtures for single cell protein, aroma volatiles and fat production. Food Chem 145:710–716. https://doi.org/10.1016/j.foodchem.2013.07.105

Ahmad FB, Zhang Z, Doherty WOS, O'Hara IM (2019) The outlook of the production of advanced fuels and chemicals from integrated oil palm biomass biorefinery. Renew Sust Energ Rev 109:386–411

Ahmed I, Zia MA, Hussain MA et al (2016) Bioprocessing of citrus waste peel for induced pectinase production by *Aspergillus niger* ; its purification and characterization. J Radiat Res Appl Sci 9:148–154. https://doi.org/10.1016/j.jrras.2015.11.003

Aires A, Carvalho R, Saavedra MJ (2017) Reuse potential of vegetable wastes (broccoli, green bean and tomato) for the recovery of antioxidant phenolic acids and flavonoids. Int J Food Sci Technol 52:98–107. https://doi.org/10.1111/ijfs.13256

Ajila CM, Aalami M, Leelavathi K, Rao UJSP (2010) Mango peel powder: a potential source of antioxidant and dietary fiber in macaroni preparations. Innov Food Sci Emerg Technol 11:219–224. https://doi.org/10.1016/j.ifset.2009.10.004

Akhtar J, Idris A (2017) Oil palm empty fruit bunches a promising substrate for succinic acid production via simultaneous saccharification and fermentation. Renew Energy 114:917–923. https://doi.org/10.1016/j.renene.2017.07.113

Al-Manhel AJ, Al-Hilphy ARS, Niamah AK (2018) Extraction of chitosan, characterisation and its use for water purification. J Saudi Soc Agric Sci 17:186–190. https://doi.org/10.1016/j.jssas.2016.04.001

Alonso S, Herrero M, Rendueles M, Díaz M (2010) Residual yoghurt whey for lactic acid production. Biomass Bioenergy 34:931–938. https://doi.org/10.1016/j.biombioe.2010.01.041

Alslaibi TM, Abustan I, Ahmad MA, Foul AA (2013) A review: production of activated carbon from agricultural byproducts via conventional and microwave heating. J Chem Technol Biotechnol 88:1183–1190. https://doi.org/10.1002/jctb.4028

Amado IR, Franco D, Sánchez M et al (2014) Optimisation of antioxidant extraction from Solanum tuberosum potato peel waste by surface response methodology. Food Chem 165:290–299. https://doi.org/10.1016/j.foodchem.2014.05.103

Amenorfenyo DK, Huang X, Zhang Y et al (2019) Microalgae brewery wastewater treatment: potentials, benefits and the challenges. Int J Environ Res Public Health 16. https://doi.org/10.3390/ijerph16111910

Anitha A, Sowmya S, Kumar PTS et al (2014) Chitin and chitosan in selected biomedical applications. Prog Polym Sci 39:1644–1667

Annevelink E, Van Gogh B, Bartels P et al (2017) Vision paper how to achieve resource use efficiency in integrated food and biobased value chains? https://doi.org/10.18174/407896

Aruldass CA, Aziz A, Venil CK et al (2016) Utilization of agro-industrial waste for the production of yellowish-orange pigment from Chryseobacterium artocarpi CECT 8497. Int Biodeterior Biodegrad 113:342–349. https://doi.org/10.1016/j.ibiod.2016.01.024

Atasoy M, Owusu-Agyeman I, Plaza E, Cetecioglu Z (2018) Bio-based volatile fatty acid production and recovery from waste streams: current status and future challenges. Bioresour Technol 268:773–786

Awasthi D, Wang L, Rhee MS et al (2018) Metabolic engineering of Bacillus subtilis for production of D-lactic acid. Biotechnol Bioeng 115:453–463. https://doi.org/10.1002/bit.26472

Ayala-Zavala JF, Rosas-Domínguez C, Vega-Vega V, González-Aguilar GA (2010) Antioxidant enrichment and antimicrobial protection of fresh-cut fruits using their own byproducts: looking for integral exploitation. J Food Sci 75. https://doi.org/10.1111/j.1750-3841.2010.01792.x

Bafrani MP (2010) THESIS FOR THE DEGREE OF DOCTOR OF PHILOSOPHY Citrus waste biorefinery: process development, Simulation and Economic Analysis

Baiano A (2014) Recovery of biomolecules from food wastes — a review. Molecules 19:14821–14842. https://doi.org/10.3390/molecules190914821

Balasundram N, Sundram K, Samman S (2006) Phenolic compounds in plants and Agri-industrial by-products: antioxidant activity, occurrence, and potential uses. Food Chem 99:191–203. https://doi.org/10.1016/j.foodchem.2005.07.042

Bansal RC, Goyal M (2005) Activated carbon adsorption. CRC Press. ISBN 9780367578077

Bansal N, Tewari R, Soni R, Soni SK (2012) Production of cellulases from Aspergillus niger NS-2 in solid state fermentation on agricultural and kitchen waste residues. Waste Manag 32:1341–1346. https://doi.org/10.1016/j.wasman.2012.03.006

Barba FJ, Zhu Z, Koubaa M et al (2016) Green alternative methods for the extraction of antioxidant bioactive compounds from winery wastes and by-products: a review. Trends Food Sci Technol 49:96–109

Barik S, Paul KK, Priyadarshi D (2018) Utilization of kitchen food waste for biodiesel production. In: IOP conference series: earth and environmental science

Barrios V, Escobar C, Cicero AFG et al (2017) A nutraceutical approach (Armolipid plus) to reduce total and LDL cholesterol in individuals with mild to moderate dyslipidemia: review of the clinical evidence. Atheroscler Suppl 24:1–15. https://doi.org/10.1016/j.atherosclerosissup.2016.10.003

Bechthold I, Bretz K, Kabasci S et al (2008) Succinic acid: a new platform chemical for biobased polymers from renewable resources. Chem Eng Technol 31:647–654

Berardini N, Knödler M, Schieber A, Carle R (2005) Utilization of mango peels as a source of pectin and polyphenolics. Innov Food Sci Emerg Technol 6:442–452. https://doi.org/10.1016/j.ifset.2005.06.004

Bergamin A, Mantzioris E, Cross G et al (2019) Nutraceuticals: reviewing their role in chronic disease prevention and management. Pharmaceut Med

Bobbink FD, Zhang J, Pierson Y et al (2015) Conversion of chitin derived N-acetyl-d-glucosamine (NAG) into polyols over transition metal catalysts and hydrogen in water. Green Chem 17:1024–1031. https://doi.org/10.1039/c4gc01631a

Boonrod D, Reanma K, Niamsup H (2006) Extraction and physicochemical characteristics of acid-soluble pectin from raw papaya (Carica papaya) Peel. Chiang Mai J Sci 33:129–135

Borghi C, Cicero AFG (2017) Nutraceuticals with a clinically detectable blood pressure-lowering effect: a review of available randomized clinical trials and their meta-analyses. Br J Clin Pharmacol 83:163–171

Botella C, Diaz A, de Ory I et al (2007) Xylanase and pectinase production by Aspergillus awamori on grape pomace in solid state fermentation. Process Biochem 42:98–101. https://doi.org/10.1016/j.procbio.2006.06.025

Bozell JJ, Petersen GR (2010) Technology development for the production of biobased products from biorefinery carbohydrates-the US Department of Energy's "top 10" revisited. https://doi.org/10.1039/b922014c

Brower V (1998) Nutraceuticals: poised for a healthy slice of the healthcare market? Nat Biotechnol 16:728–730

Canteri MHG, Scheer AP, Wosiacki G et al (2010) A comparative study of pectin extracted from passion fruit rind flours. J Polym Environ 18:593–599. https://doi.org/10.1007/s10924-010-0206-z

Carvalho T, Santos D, Paulo D et al (2012) Optimisation of solid state fermentation of potato peel for the production of cellulolytic enzymes. Food Chem 133:1299–1304. https://doi.org/10.1016/j.foodchem.2011.11.115

Castro-Aguirre E, Iñiguez-Franco F, Samsudin H et al (2016) Poly(lactic acid)—mass production, processing, industrial applications, and end of life. Adv Drug Deliv Rev 107:333–366

Catarino J, Mendonça E, Picado A et al (2007) Getting value from wastewater: by-products recovery in a potato chips industry. J Clean Prod 15:927–931. https://doi.org/10.1016/j.jclepro.2005.12.003

Cecchi T, Passamonti P, Cecchi P (2010) Optimization of the measurement of Italian monocultivar extra virgin olive oil antioxidant power via the Briggs-Rauscher reaction. Food Anal Methods 3:1–6. https://doi.org/10.1007/s12161-009-9071-6

Cecchi T, Pezzella A, Di Mauro E et al (2019) On the antioxidant activity of eumelanin biopigments: a quantitative comparison between free radical scavenging and redox properties. Nat Prod Res:1–9. https://doi.org/10.1080/14786419.2018.1542391

Chen Y, Tan HM (2006) Crosslinked carboxymethylchitosan-g-poly(acrylic acid) copolymer as a novel superabsorbent polymer. Carbohydr Res 341:887–896. https://doi.org/10.1016/j.carres.2006.01.027

Chen SS, Yu IKM, Tsang DCW et al (2017) Valorization of cellulosic food waste into levulinic acid catalyzed by heterogeneous BrØnsted acids: temperature and solvent effects. Chem Eng J 327:328–335. https://doi.org/10.1016/j.cej.2017.06.108

Chen S, Huang H, Huang G (2019) Extraction, derivatization and antioxidant activity of cucumber polysaccharide. Int J Biol Macromol 140:1047–1053. https://doi.org/10.1016/j.ijbiomac.2019.08.203

Chen LZ, Huang SL, Hou J, et al (2020) Cell-based and cell-free biocatalysis for the production of d-glucaric acid. Biotechnol. Biofuels 13:1–11

Cheng YT, Yang CF (2016) Using strain Rhodotorula mucilaginosa to produce carotenoids using food wastes. J Taiwan Inst Chem Eng 61:270–275. https://doi.org/10.1016/j.jtice.2015.12.027

Chihoub W, Dias MI, Barros L et al (2019) Valorisation of the green waste parts from turnip, radish and wild cardoon: nutritional value, phenolic profile and bioactivity evaluation. https://doi.org/10.1016/j.foodres.2019.108651

Choi IS, Cho EJ, Moon JH, Bae HJ (2015) Onion skin waste as a valorization resource for the by-products quercetin and biosugar. Food Chem 188:537–542. https://doi.org/10.1016/j.foodchem.2015.05.028

Chojnacka K, Górecka H, Michalak I, Górecki H (2011) A review: valorization of keratinous materials. Waste Biomass Valorization 2:317–321

Christiaens S, Uwibambe D, Uyttebroek M et al (2015) Pectin characterisation in vegetable waste streams: a starting point for waste valorisation in the food industry. LWT – Food Sci Technol 61:275–282. https://doi.org/10.1016/j.lwt.2014.12.054

Cicero AFG, Colletti A (2015) Nutraceuticals and blood pressure control: results from clinical trials and meta-analyses. High Blood Press Cardiovasc Prev 22:203–213. https://doi.org/10.1007/s40292-015-0081-8

Cicero AFG, Colletti A, Bellentani S (2018) Nutraceutical approach to non-alcoholic fatty liver disease (NAFLD): the available clinical evidence. Nutrients 10

Cicero AFG, Fogacci F, Bove M et al (2019) Three-arm, placebo-controlled, randomized clinical trial evaluating the metabolic effect of a combined nutraceutical containing a bergamot standardized flavonoid extract in dyslipidemic overweight subjects. Phyther Res. https://doi.org/10.1002/ptr.6402

Colmenares JC, Varma RS, Lisowski P (2016) Sustainable hybrid photocatalysts: Titania immobilized on carbon materials derived from renewable and biodegradable resources. Green Chem 18:5736–5750. https://doi.org/10.1039/c6gc02477g

Cordell D (2010) The story of phosphorus sustainability implications of global phosphorus scarcity for food security. Linköping Studies in Arts and Science:509. Thesis

Corell L, Armenta S, Esteve-Turrillas FA, de la Guardia M (2018) Flavonoid determination in onion, chili and leek by hard cap espresso extraction and liquid chromatography with diode array detection. Microchem J 140:74–79. https://doi.org/10.1016/j.microc.2018.04.014

da Silva FV, dos Santos RL, Fujiki TL et al (2010) Desenvolvimento de sistema de controle para precipitação de bromelina a partir de resíduos de abacaxi. Cienc e Tecnol Aliment 30:1033–1040. https://doi.org/10.1590/S0101-20612010000400031

Da Silva LMR, De Figueiredo EAT, Ricardo NMPS et al (2014) Quantification of bioactive compounds in pulps and by-products of tropical fruits from Brazil. Food Chem 143:398–404. https://doi.org/10.1016/j.foodchem.2013.08.001

Dahiya S, Kumar AN, Shanthi Sravan J et al (2018) Food waste biorefinery: sustainable strategy for circular bioeconomy. Bioresour Technol

Das RK, Brar SK, Verma M (2015) A fermentative approach towards optimizing directed biosynthesis of fumaric acid by Rhizopus oryzae 1526 utilizing apple industry waste biomass. Fungal Biol 119:1279–1290. https://doi.org/10.1016/j.funbio.2015.10.001

De Corato U, De Bari I, Viola E, Pugliese M (2018) Assessing the main opportunities of integrated biorefining from agro-bioenergy co/by-products and agroindustrial residues into high-value added products associated to some emerging markets: a review. Renew Sust Energ Rev 88:326–346. https://doi.org/10.1016/J.RSER.2018.02.041

de Jong E, Higson A, Walsh P, Wellisch M (2012) Value Added Products from Biorefinerie IEA Bioenergy 33

De Leonardis A, Macciola V, Iorizzo M et al (2018) Effective assay for olive vinegar production from olive oil mill wastewaters. Food Chem 240:437–440. https://doi.org/10.1016/j.foodchem.2017.07.159

de Oliveira F, Ramires EC, Frollini E, Belgacem MN (2015) Lignopolyurethanic materials based on oxypropylated sodium lignosulfonate and castor oil blends. Ind Crop Prod 72:77–86. https://doi.org/10.1016/j.indcrop.2015.01.023

Dessie W, Zhang W, Xin F et al (2018) Succinic acid production from fruit and vegetable wastes hydrolyzed by on-site enzyme mixtures through solid state fermentation. Bioresour Technol 247:1177–1180. https://doi.org/10.1016/j.biortech.2017.08.171

Dhillon GS, Brar SK, Verma M, Tyagi RD (2011) Apple pomace ultrafiltration sludge - a novel substrate for fungal bioproduction of citric acid: optimisation studies. Food Chem 128:864–871. https://doi.org/10.1016/j.foodchem.2011.03.107

Dhillon GS, Brar SK, Verma M (2012) Biotechnological potential of industrial wastes for economical citric acid bioproduction by Aspergillus Niger through submerged fermentation. Int J Food Sci Technol 47:542–548. https://doi.org/10.1111/j.1365-2621.2011.02875.x

Di Hsu K, Cheng KC (2018) From nutraceutical to clinical trial: frontiers in Ganoderma development. Appl Microbiol Biotechnol 102:9037–9051

Díaz AB, De Ory I, Caro I, Blandino A (2012) Enhance hydrolytic enzymes production by Aspergillus awamori on supplemented grape pomace. Food Bioprod Process 90:72–78. https://doi.org/10.1016/j.fbp.2010.12.003

Díaz AI, Laca A, Laca A, Díaz M (2017) Treatment of supermarket vegetable wastes to be used as alternative substrates in bioprocesses. Waste Manag 67:59–66. https://doi.org/10.1016/j.wasman.2017.05.018

Dilrukshi PGT, Munasinghe H, Buddhika A et al (2019) Identification of synthetic food Colours in selected confectioneries and beverages in Jaffna District, Sri Lanka. J Food Qual 2019:8. https://doi.org/10.1155/2019/7453169

Dominguez-Perles R, Moreno DA, Carvajal M, Garcia-Viguera C (2011) Composition and antioxidant capacity of a novel beverage produced with green tea and minimally-processed byproducts of broccoli. Innov Food Sci Emerg Technol 12:361–368. https://doi.org/10.1016/j.ifset.2011.04.005

Drouet S, Leclerc EA, Garros L et al (2019) A green ultrasound-assisted extraction optimization of the natural antioxidant and anti-aging flavonolignans from milk thistle silybum marianum (L.) gaertn. Fruits for cosmetic applications. Antioxidants 8. https://doi.org/10.3390/antiox8080304

Dugmore TIJ, Clark JH, Bustamante J et al (2017) Valorisation of biowastes for the production of green materials using chemical methods. Top Curr Chem 375:46. https://doi.org/10.1007/s41061-017-0133-8

Dupont-Inglis J, Borg A (2018) Destination bioeconomy – the path towards a smarter, more sustainable future. New Biotechnol 40:140–143

Ebringerová A, Hromádková Z, Heinze T (2005) Hemicellulose. Adv Polym Sci 186:1–67

El Knidri H, El Khalfaouy R, Laajeb A et al (2016) Eco-friendly extraction and characterization of chitin and chitosan from the shrimp shell waste via microwave irradiation. Process Saf Environ Prot 104:395–405. https://doi.org/10.1016/j.psep.2016.09.020

EU (2014) Phosphorus DG ENTR Introduction Paper

EU (2015) From the Sugar Platform to biofuels and biochemicals. Final Rep Eur Comm Dir Energy 183. doi: contract No. ENER/C2/423–2012/SI2.673791

European Commission (2015) Sustainable Agriculture, Forestry and Fisheries in the Bioeconomy

Faraone I, Sinisgalli C, Ostuni A et al (2020) Astaxanthin anticancer effects are mediated through multiple molecular mechanisms: a systematic review. Pharmacol Res 155

Fernández-López J, Fernández-Ginés JM, Aleson-Carbonell L et al (2004) Application of functional citrus by-products to meat products. Trends Food Sci Technol 15:176–185. https://doi.org/10.1016/j.tifs.2003.08.007

Franca EF, Freitas LCG, Lins RD (2011) Chitosan molecular structure as a function of N-acetylation. Biopolymers 95:448–460. https://doi.org/10.1002/bip.21602

Francis F, Sabu A, Nampoothiri KM et al (2003) Use of response surface methodology for optimizing process parameters for the production of α-amylase by Aspergillus oryzae. Biochem Eng J 15:107–115. https://doi.org/10.1016/S1369-703X(02)00192-4

Freitas ED, Moura CF, Kerwald J, Beppu MM (2020) An overview of current knowledge on the properties, synthesis and applications of quaternary chitosan derivatives. Polymers (Basel) 12:1–41. https://doi.org/10.3390/polym12122878

Galanakis CM (2012) Recovery of high added-value components from food wastes: conventional, emerging technologies and commercialized applications. Trends Food Sci Technol 26:68–87

Galanakis CM (2017) Handbook of grape processing by-products: sustainable solutions

Galanakis CM, Tornberg E, Gekas V (2010) Dietary fiber suspensions from olive mill wastewater as potential fat replacements in meatballs. LWT – Food Sci Technol 43:1018–1025. https://doi.org/10.1016/j.lwt.2009.09.011

Gao X, Chen X, Zhang J et al (2016) Transformation of chitin and waste shrimp shells into acetic acid and pyrrole. ACS Sustain Chem Eng 4:3912–3920. https://doi.org/10.1021/acssuschemeng.6b00767

Gargiulo V, Lanzetta R, Parrilli M, De Castro C (2009) Structural analysis of chondroitin sulfate from Scyliorhinus canicula: a useful source of this polysaccharide. Glycobiology 19:1485–1491. https://doi.org/10.1093/glycob/cwp123

Germec M, Ozcan A, Turhan I (2019) Bioconversion of wheat bran into high value-added products and modelling of fermentations. Ind Crops Prod:139. https://doi.org/10.1016/j.indcrop.2019.111565

Gibson LJ (2012) The hierarchical structure and mechanics of plant materials. J R Soc Interface 9:2749–2766. https://doi.org/10.1098/rsif.2012.0341

Girotto F, Lavagnolo MC, Pivato A, Cossu R (2017) Acidogenic fermentation of the organic fraction of municipal solid waste and cheese whey for bio-plastic precursors recovery – effects of process conditions during batch tests. Waste Manag 70:71–80. https://doi.org/10.1016/j.wasman.2017.09.015

Gómez-Ríos D, Barrera-Zapata R, Ríos-Estepa R (2017) Comparison of process technologies for chitosan production from shrimp shell waste: a techno-economic approach using Aspen plus ®. Food Bioprod Process 103:49–57. https://doi.org/10.1016/j.fbp.2017.02.010

Górecka D, Dziedzic K, Pachołek B, Górecka M (2010) Raspberry pomace as a potential Fiber source for cookies enrichment. Acta Sci Pol Technol Aliment 9:451–462

Grassino AN, Halambek J, Djaković S et al (2016) Utilization of tomato peel waste from canning factory as a potential source for pectin production and application as tin corrosion inhibitor. Food Hydrocoll 52:265–274. https://doi.org/10.1016/j.foodhyd.2015.06.020

Gupta S, Kua HW, Koh HJ (2018) Application of biochar from food and wood waste as green admixture for cement mortar. Sci Total Environ 619–620:419–435. https://doi.org/10.1016/j.scitotenv.2017.11.044

Hafid HS, Rahman NA, Md Shah UK, Baharudin AS (2015) Enhanced fermentable sugar production from kitchen waste using various pretreatments. J Environ Manag 156:290–298. https://doi.org/10.1016/j.jenvman.2015.03.045

Hall RM, Mayer DA, Mazzutti S, Ferreira SRS (2018) Simulating large scale SFE applied to recover bioactive compounds from papaya seeds. J Supercrit Fluids 140:302–309. https://doi.org/10.1016/j.supflu.2018.07.013

Haque MA, Yang X, Ong KL et al (2017) Bioconversion of beverage waste to high fructose syrup as a value-added product. Food Bioprod Process 105:179–187. https://doi.org/10.1016/j.fbp.2017.07.007

Harnedy PA, FitzGerald RJ (2012) Bioactive peptides from marine processing waste and shellfish: a review. J Funct Foods 4:6–24

Hasan SDM, Giongo C, Fiorese ML et al (2015) Volatile fatty acids production from anaerobic treatment of cassava waste water: effect of temperature and alkalinity. Environ Technol 36:2637–2646. https://doi.org/10.1080/09593330.2015.1041426

Hasan MM, Marzan LW, Hosna A et al (2017) Optimization of some fermentation conditions for the production of extracellular amylases by using Chryseobacterium and Bacillus isolates from organic kitchen wastes. J Genet Eng Biotechnol 15:59–68. https://doi.org/10.1016/j.jgeb.2017.02.009

Hassan SS, Williams GA, Jaiswal AK (2019) Moving towards the second generation of lignocellulosic biorefineries in the EU: drivers, challenges, and opportunities. Renew Sust Energ Rev 101:590–599

Hataminia F, Farhadian N, Karimi M, Ebrahimi M (2018) A novel method for squalene extraction from pumpkin seed oil using magnetic nanoparticles and exploring the inhibition effect of extracted squalene on angiogenesis property. J Taiwan Inst Chem Eng 91:1–9. https://doi.org/10.1016/j.jtice.2018.05.017

Hernández-Carmona F, Morales-Matos Y, Lambis-Miranda H, Pasqualino J (2017) Starch extraction potential from plantain peel wastes. J Environ Chem Eng 5:4980–4985. https://doi.org/10.1016/j.jece.2017.09.034

Hirano S, Tsuchida H, Nagao N (1989) N-acetylation in chitosan and the rate of its enzymic hydrolysis. Biomaterials 10:574–576. https://doi.org/10.1016/0142-9612(89)90066-5

Horizon 2020 WORK PROGRAMME 2018–2020 General Annexes E from P 19-CDC (2017) Technology Readiness Levels Introduction

Hülsey MJ (2018) Shell biorefinery: a comprehensive introduction. Green Energy Environ 3:318–327

Hussein MHM, El-Hady MF, Shehata HAH et al (2013) Preparation of some eco-friendly corrosion inhibitors having antibacterial activity from sea food waste. J Surfactant Deterg 16:233–242. https://doi.org/10.1007/s11743-012-1395-3

Jain S, Buttar HS, Chintameneni M, Kaur G (2018) Prevention of cardiovascular diseases with anti-inflammatory and anti- oxidant nutraceuticals and herbal products: an overview of pre-clinical and clinical studies. Recent Patents Inflamm Allergy Drug Discov 12:145–157. https://doi.org/10.2174/1872213x12666180815144803

Jayakumar R, Selvamurugan N, Nair SV et al (2008) Preparative methods of phosphorylated chitin and chitosan-an overview. Int J Biol Macromol 43:221–225

Jayakumar R, Menon D, Manzoor K et al (2010a) Biomedical applications of chitin and chitosan based nanomaterials - a short review. Carbohydr Polym 82:227–232

Jayakumar R, Prabaharan M, Nair SV, Tamura H (2010b) Novel chitin and chitosan nanofibers in biomedical applications. Biotechnol Adv 28:142–150

Jiang J, Gong C, Wang J et al (2014) Effects of ultrasound pre-treatment on the amount of dissolved organic matter extracted from food waste. Bioresour Technol 155:266–271. https://doi.org/10.1016/j.biortech.2013.12.064

Jiménez-Gómez CP, Cecilia JA (2020) Chitosan: A Natural Biopolymer with a Wide and Varied Range of Applications. Molecules 25. https://doi.org/10.3390/molecules25173980

John G, Nagarajan S, Kumar P et al (2019) Natural monomers : a mine for functional and sustainable materials – occurrence, chemical modification and polymerization. Prog Polym Sci 92:158–209

Jose A, Prasun B, Vivien M, et al (2015) Utility of xanthan gum produced from agro food waste. Proc India Int Sci festival- young Sci meet 1–21

Julia BM, Belén AM, Georgina B, Beatriz F (2016) Potential use of soybean hulls and waste paper as supports in SSF for cellulase production by Aspergillus Niger. Biocatal Agric Biotechnol 6:1–8. https://doi.org/10.1016/j.bcab.2016.02.003

Kachrimanidou V, Kopsahelis N, Papanikolaou S et al (2014) Sunflower-based biorefinery: poly(3-hydroxybutyrate) and poly(3-hydroxybutyrate-co-3-hydroxyvalerate) production from crude glycerol, sunflower meal and levulinic acid. Bioresour Technol 172:121–130. https://doi.org/10.1016/j.biortech.2014.08.044

Kalra EK (2003) Nutraceutical - definition and introduction

Kanatt SR, Chander R, Radhakrishna P, Sharma A (2005) Potato peel extract - a natural antioxidant for retarding lipid peroxidation in radiation processed lamb meat. J Agric Food Chem 53:1499–1504. https://doi.org/10.1021/jf048270e

Kandasamy S, Muthusamy G, Balakrishnan S et al (2016) Optimization of protease production from surface-modified coffee pulp waste and corncobs using Bacillus sp. by SSF. 3 Biotech 6:1–11. https://doi.org/10.1007/s13205-016-0481-z

Karmee SK, Lin CSK (2014) Lipids from food waste as feedstock for biodiesel production: case Hong Kong. Lipid Technol 26:206–209. https://doi.org/10.1002/lite.201400044

Karmee SK, Linardi D, Lee J, Lin CSK (2015) Conversion of lipid from food waste to biodiesel. Waste Manag 41:169–173. https://doi.org/10.1016/j.wasman.2015.03.025

Karp SG, Igashiyama AH, Siqueira PF et al (2011) Application of the biorefinery concept to produce l-lactic acid from the soybean vinasse at laboratory and pilot scale. Bioresour Technol 102:1765–1772. https://doi.org/10.1016/j.biortech.2010.08.102

Karthikeyan A, Sivakumar N (2010) Citric acid production by Koji fermentation using banana peel as a novel substrate. Bioresour Technol 101:5552–5556. https://doi.org/10.1016/j.biortech.2010.02.063

Key RE, Bozell JJ (2016) Progress toward lignin valorization via selective catalytic technologies and the tailoring of biosynthetic pathways. ACS Sustain Chem Eng 4:5123–5135

Kim S-B, Ji C-I, Woo J-W et al (2012) Simplified purification of chondroitin sulphate from scapular cartilage of shortfin mako shark (Isurus oxyrinchus). Int J Food Sci Technol 47:91–99. https://doi.org/10.1111/j.1365-2621.2011.02811.x

Kim HM, Song Y, Wi SG, Bae HJ (2017) Production of D-tagatose and bioethanol from onion waste by an intergrating bioprocess. J Biotechnol 260:84–90. https://doi.org/10.1016/j.jbiotec.2017.09.013

Kobayashi H, Techikawara K, Fukuoka A (2017) Hydrolytic hydrogenation of chitin to amino sugar alcohol. Green Chem 19:3350–3356. https://doi.org/10.1039/c7gc01063j

Komesu A, Martinez PFM, Lunelli BH et al (2015) Lactic acid purification by reactive distillation system using design of experiments. Chem Eng Process Process Intensif 95:26–30. https://doi.org/10.1016/j.cep.2015.05.005

Kougioumtzis MA, Marianou A, Atsonios K et al (2018) Production of 5-HMF from cellulosic biomass: experimental results and integrated process simulation. Waste Biomass Valorization 9:2433–2445. https://doi.org/10.1007/s12649-018-0267-0

Koutinas AA, Vlysidis A, Pleissner D et al (2014) Valorization of industrial waste and by-product streams via fermentation for the production of chemicals and biopolymers. Chem Soc Rev 43:2587. https://doi.org/10.1039/c3cs60293a

Kriaa M, Kammoun R (2016) Producing *Aspergillus tubingensis* CTM507 glucose oxidase by solid state fermentation versus submerged fermentation: process optimization and enzyme stability by an intermediary metabolite in relation with diauxic growth. J Chem Technol Biotechnol 91:1540–1550. https://doi.org/10.1002/jctb.4753

Kringel DH, Dias ARG, da Zavareze ER, Gandra EA (2020) Fruit wastes as promising sources of starch: extraction, properties, and applications. Starch/Staerke 72

Krisch J, Galgoczy L, Vágvölgyi C, et al (2009) Antimicrobial and antioxidant potential of waste products remaining after juice pressing

Kristinsson HG, Rasco BA (2000) Fish protein hydrolysates: production, biochemical, and functional properties. Crit Rev Food Sci Nutr. https://doi.org/10.1080/10408690091189266

Kulczyński B, Gramza-Michałowska A, Królczyk JB (2020) Optimization of extraction conditions for the antioxidant potential of different pumpkin varieties (Cucurbita Maxima). Sustain 12. https://doi.org/10.3390/su12041305

Kulkarni SG, Vijayanand P (2010) Effect of extraction conditions on the quality characteristics of pectin from passion fruit peel (Passiflora edulis f. flavicarpa L.). LWT – Food Sci Technol 43:1026–1031. https://doi.org/10.1016/j.lwt.2009.11.006

Kumar D, Jain VK, Shanker G, Srivastava A (2003) Utilisation of fruits waste for citric acid production by solid state fermentation. Process Biochem 38:1725–1729. https://doi.org/10.1016/S0032-9592(02)00253-4

Kumar VB, Pulidindi IN, Gedanken A (2016) Glucose production from potato peel waste under microwave irradiation. J Mol Catal A Chem 417:163–167. https://doi.org/10.1016/j.molcata.2016.03.025

Kumar A, Kumar D, George N et al (2018) A process for complete biodegradation of shrimp waste by a novel marine isolate Paenibacillus sp. AD with simultaneous production of chitinase and chitin oligosaccharides. Int J Biol Macromol 109:263–272. https://doi.org/10.1016/j.ijbiomac.2017.12.024

Kurek M, Garofulić IE, Bakić MT et al (2018) Development and evaluation of a novel antioxidant and pH indicator film based on chitosan and food waste sources of antioxidants. Food Hydrocoll 84:238–246. https://doi.org/10.1016/j.foodhyd.2018.05.050

Lafarga T, Rodríguez-Roque MJ, Bobo G et al (2019) Effect of ultrasound processing on the bioaccessibility of phenolic compounds and antioxidant capacity of selected vegetables. Food Sci Biotechnol 28:1713–1721. https://doi.org/10.1007/s10068-019-00618-4

Lam KF, Leung CCJ, Lei HM, Lin CSK (2014) Economic feasibility of a pilot-scale fermentative succinic acid production from bakery wastes. Food Bioprod Process 92:282–290. https://doi.org/10.1016/j.fbp.2013.09.001

Lante A, Nardi T, Zocca F et al (2011) Evaluation of red chicory extract as a natural antioxidant by pure lipid oxidation and yeast oxidative stress response as model systems. J Agric Food Chem 59:5318–5324. https://doi.org/10.1021/jf2003317

Lee PC, Lee WG, Kwon S et al (2000) Batch and continuous cultivation of Anaerobiospirillum succiniciproducens for the production of succinic acid from whey. Appl Microbiol Biotechnol 54:23–27. https://doi.org/10.1007/s002530000331

Lee Y, Andrew Lin K-Y, Kwon EE, Lee J (2019) Renewable routes to monomeric precursors of nylon 66 and nylon 6 from food waste. J Clean Prod. https://doi.org/10.1016/j.jclepro.2019.04.194

Leung CCJ, Cheung ASY, Zhang AYZ et al (2012) Utilisation of waste bread for fermentative succinic acid production. Biochem Eng J 65:10–15. https://doi.org/10.1016/j.bej.2012.03.010

Li S, Zhang X, Andresen JM (2012) Production of fermentable sugars from enzymatic hydrolysis of pretreated municipal solid waste after autoclave process. Fuel 92:84–88. https://doi.org/10.1016/j.fuel.2011.07.012

Li X, Chen Y, Zhao S et al (2015) Efficient production of optically pure l-lactic acid from food waste at ambient temperature by regulating key enzyme activity. Water Res 70:148–157. https://doi.org/10.1016/j.watres.2014.11.049

Li P, Li T, Zeng Y et al (2016) Biosynthesis of xanthan gum by Xanthomonas campestris LRELP-1 using kitchen waste as the sole substrate. Carbohydr Polym 151:684–691. https://doi.org/10.1016/j.carbpol.2016.06.017

Li P, Zeng Y, Xie Y et al (2017) Effect of pretreatment on the enzymatic hydrolysis of kitchen waste for xanthan production. Bioresour Technol 223:84–90. https://doi.org/10.1016/j.biortech.2016.10.035

Liu M, Zhang Y, Wu C et al (2012) Chitosan/halloysite nanotubes bionanocomposites: structure, mechanical properties and biocompatibility. Int J Biol Macromol 51:566–575. https://doi.org/10.1016/j.ijbiomac.2012.06.022

Liu L, Yang X, Hou Q et al (2018) Corn stalk conversion into 5-hydroxymethylfurfural by modified biochar catalysis in a multi-functional solvent. J Clean Prod 187:380–389. https://doi.org/10.1016/j.jclepro.2018.03.234

López-Linares JC, García-Cubero MT, Lucas S et al (2019) Microwave assisted hydrothermal as greener pretreatment of brewer's spent grains for biobutanol production. Chem Eng J:1045–1055. https://doi.org/10.1016/j.cej.2019.03.032

Lu X (2014) Feasibility study: phosphorus recovery from household solid organic waste. Master of Science Thesis

Luchese CL, Sperotto N, Spada JC, Tessaro IC (2017) Effect of blueberry agro-industrial waste addition to corn starch-based films for the production of a pH-indicator film. Int J Biol Macromol 104:11–18. https://doi.org/10.1016/j.ijbiomac.2017.05.149

Luft L, Confortin TC, Todero I et al (2019) Ultrasound technology applied to enhance enzymatic hydrolysis of Brewer's spent grain and its potential for production of fermentable sugars. Waste Biomass Valorization 10:2157–2164. https://doi.org/10.1007/s12649-018-0233-x

Luo J, Zhu Y, Song A et al (2019) Efficient short-chain fatty acids recovery from anaerobic fermentation of wine vinasse and waste activated sludge and the underlying mechanisms. Biochem Eng J 145:18–26. https://doi.org/10.1016/j.bej.2019.02.010

Madureira AR, Tavares T, Gomes AMP et al (2010) Invited review: physiological properties of bioactive peptides obtained from whey proteins. J Dairy Sci 93:437–455. https://doi.org/10.3168/jds.2009-2566

Mahboubi A, Ferreira JA, Taherzadeh MJ, Lennartsson PR (2017) Value-added products from dairy waste using edible fungi. Waste Manag 59:518–525. https://doi.org/10.1016/j.wasman.2016.11.017

Makris DP (2018) Green extraction processes for the efficient recovery of bioactive polyphenols from wine industry solid wastes – recent progress. Curr Opin Green Sustain Chem 13:50–55

Maneerat N, Tangsuphoom N, Nitithamyong A (2017) Effect of extraction condition on properties of pectin from banana peels and its function as fat replacer in salad cream. J Food Sci Technol 54:386–397. https://doi.org/10.1007/s13197-016-2475-6

Marcet I, Álvarez C, Paredes B, Díaz M (2016) The use of sub-critical water hydrolysis for the recovery of peptides and free amino acids from food processing wastes. Review of sources and main parameters. Waste Manag 49:364–371

Marín M, Artola A, Sánchez A (2018) Production of proteases from organic wastes by solid-state fermentation: downstream and zero waste strategies. 3 Biotech 8:205. https://doi.org/10.1007/s13205-018-1226-y

Martin-Diana AB, Rico D, Frias J et al (2006) Whey permeate as a bio-preservative for shelf life maintenance of fresh-cut vegetables. Innov Food Sci Emerg Technol 7:112–123. https://doi.org/10.1016/j.ifset.2005.08.002

Martins BC, Rescolino R, Coelho DF et al (2014) Characterization of bromelain from ananas comosus agroindustrial residues purified by ethanol factional precipitation. Chem Eng Trans 37:781–786. https://doi.org/10.3303/CET1437131

Mateos-Aparicio I, Redondo-Cuenca A, Villanueva-Suárez MJ et al (2010) Pea pod, broad bean pod and okara, potential sources of functional compounds. LWT – Food Sci Technol 43:1467–1470. https://doi.org/10.1016/j.lwt.2010.05.008

Matharu AS, de Melo EM, Houghton JA (2016) Opportunity for high value-added chemicals from food supply chain wastes. Bioresour Technol 215:123–130

Melikoglu M, Lin CSK, Webb C (2013) Stepwise optimisation of enzyme production in solid state fermentation of waste bread pieces. Food Bioprod Process 91:638–646. https://doi.org/10.1016/j.fbp.2013.04.008

Melikoglu M, Lin CSK, Webb C (2015) Solid state fermentation of waste bread pieces by Aspergillus awamori: Analysing the effects of airflow rate on enzyme production in packed bed bioreactors. Food Bioprod Process 95:63–75. https://doi.org/10.1016/j.fbp.2015.03.011

Mohdaly AAA, Sarhan MA, Smetanska I, Mahmoud A (2010) Antioxidant properties of various solvent extracts of potato peel, sugar beet pulp and sesame cake. J Sci Food Agric 90:218–226. https://doi.org/10.1002/jsfa.3796

Mohnen D (2008) Pectin structure and biosynthesis. Curr Opin Plant Biol 11:266–277

Mollea C, Chiampo F, Conti R (2008) Extraction and characterization of pectins from cocoa husks: a preliminary study. Food Chem 107:1353–1356. https://doi.org/10.1016/j.foodchem.2007.09.006

Mostacci B, Liguori R, Cicero AF (2017) Nutraceutical approach to peripheral neuropathies: evidence from clinical trials. Curr Drug Metab 18. https://doi.org/10.2174/1389200218666171031145419

Murad HA, Abo-Elkhair AG, Azzaz HH (2019) Production of xanthan gum from nontraditional substrates with perspective of the unique properties and wide industrial applications. JSMC Microbiol 1:6

Murado MA, Fraguas J, Montemayor MI et al (2010) Preparation of highly purified chondroitin sulphate from skate (Raja clavata) cartilage by-products. Process optimization including a new procedure of alkaline hydroalcoholic hydrolysis. Biochem Eng J 49:126–132. https://doi.org/10.1016/j.bej.2009.12.006

Murthy PS, Madhava Naidu M (2012) Sustainable management of coffee industry by-products and value addition - a review. Resour Conserv Recycl 66:45–58

Mushtaq Q, Irfan M, Tabssum F, Iqbal Qazi J (2017) Potato peels: a potential food waste for amylase production. J Food Process Eng 40. https://doi.org/10.1111/jfpe.12512

Nair RB, Eh-Hser Nay T, Lennartsson PR, Taherzadeh MJ (2017) Waste bread valorization using edible filamentous Fungi. In: EUBCE 2017 – 25th European biomass conference and exhibition, Stockholm Sweden, June 12-15, 2017

Nanou K, Roukas T (2016) Waste cooking oil: a new substrate for carotene production by Blakeslea trispora in submerged fermentation. Bioresour Technol 203:198–203. https://doi.org/10.1016/j.biortech.2015.12.053

Naranjo JM, Posada JA, Higuita JC, Cardona CA (2013) Valorization of glycerol through the production of biopolymers: the PHB case using Bacillus megaterium. Bioresour Technol 133:38–44. https://doi.org/10.1016/j.biortech.2013.01.129

Nath A, Verasztó B, Basak S et al (2016) Synthesis of lactose-derived nutraceuticals from dairy waste whey—a review. Food Bioprocess Technol 9:16–48

Nattrass L, Biggs C, Bauen A, Parisi C, Rodríguez-Cerezo E G-BM (2016) The EU bio-based industry - Publications Office of the EU. https://publications.europa.eu/en/publication-detail/-/publication/e7174c29-ec15-11e5-8a81-01aa75ed71a1/language-en. Accessed 26 Jul 2019

Nawirska A, Kwaśniewska M (2005) Dietary fibre fractions from fruit and vegetable processing waste. Food Chem 91:221–225. https://doi.org/10.1016/j.foodchem.2003.10.005

Nayak A, Bhushan B (2019) An overview of the recent trends on the waste valorization techniques for food wastes. J Environ Manag 233:352–370. https://doi.org/10.1016/j.jenvman.2018.12.041

Nayak A, Bhushan B, Gupta V, Sharma P (2017) Chemically activated carbon from lignocellulosic wastes for heavy metal wastewater remediation: effect of activation conditions. J Colloid Interface Sci 493:228–240. https://doi.org/10.1016/j.jcis.2017.01.031

Nayak A, Bhushan B, Rodriguez-Turienzo L (2018) Recovery of polyphenols onto porous carbons developed from exhausted grape pomace: a sustainable approach for the treatment of wine wastewaters. Water Res 145:741–756. https://doi.org/10.1016/j.watres.2018.09.017

Nguyen CM, Kim JS, Nguyen TN et al (2013) Production of l- and d-lactic acid from waste Curcuma longa biomass through simultaneous saccharification and cofermentation. Bioresour Technol 146:35–43. https://doi.org/10.1016/j.biortech.2013.07.035

Nguyen QA, Cho E, Trinh LTP et al (2017) Development of an integrated process to produce D-mannose and bioethanol from coffee residue waste. Bioresour Technol 244:1039–1048. https://doi.org/10.1016/j.biortech.2017.07.169

Nile SH, Nile AS, Keum YS, Sharma K (2017) Utilization of quercetin and quercetin glycosides from onion (Allium cepa L.) solid waste as an antioxidant, urease and xanthine oxidase inhibitors. Food Chem 235:119–126. https://doi.org/10.1016/j.foodchem.2017.05.043

Ohkouchi Y, Inoue Y (2006) Direct production of l(+)-lactic acid from starch and food wastes using Lactobacillus manihotivorans LMG18011. Bioresour Technol 97:1554–1562. https://doi.org/10.1016/j.biortech.2005.06.004

Okuno S, Yoshinaga M, Nakatani M et al (2002) Extraction of antioxidants in Sweetpotato waste powder with supercritical carbon dioxide. Food Sci Technol Res 8:154–157. https://doi.org/10.3136/fstr.8.154

Olszewska-Widdrat A, Alexandri M, López-Gómez JP, et al (2020) Batch and continuous lactic acid fermentation based on a multi-substrate approach. Microorganisms 8:1–14. https://doi.org/10.3390/microorganisms8071084

Ona JI, Halling PJ, Ballesteros M (2019) Enzyme hydrolysis of cassava peels: treatment by amylolytic and cellulolytic enzymes. Biocatal Biotransformation 37:77–85. https://doi.org/10.1080/10242422.2018.1551376

Oreopoulou V, Tzia C (2007) Utilization of plant by-products for the recovery of proteins, dietary fibers, antioxidants, and colorants. In: Utilization of by-products and treatment of waste in the food industry. Springer, pp 209–232

Páez G, Jiménez E, Mármol Z et al (2008) Perfil de aminoácidos de la proteína unicelular de :Kluyveromyces marxianus var. marxianus. Interciencia 33:297–300

Pagano E, Romano B, Izzo AA, Borrelli F (2018) The clinical efficacy of curcumin-containing nutraceuticals: an overview of systematic reviews. Pharmacol Res 134:79–91

Pal P, Nayak J (2016) Development and analysis of a sustainable technology in manufacturing acetic acid and whey protein from waste cheese whey. J Clean Prod 112:59–70. https://doi.org/10.1016/j.jclepro.2015.07.085

Pal GK, Suresh PV (2016) Sustainable valorisation of seafood by-products: recovery of collagen and development of collagen-based novel functional food ingredients. Innov Food Sci Emerg Technol 37:201–215. https://doi.org/10.1016/j.ifset.2016.03.015

Panesar R, Kaur S, Panesar PS (2015) Production of microbial pigments utilizing agro-industrial waste: a review. Curr Opin Food Sci 1:70–76

Papadaki A, Mallouchos A, Efthymiou MN et al (2017) Production of wax esters via microbial oil synthesis from food industry waste and by-product streams. Bioresour Technol 245:274–282. https://doi.org/10.1016/j.biortech.2017.08.004

Papanikolaou S, Aggelis G (2009) Biotechnological valorization of biodiesel derived glycerol waste through production of single cell oil and citric acid by Yarrowia lipolytica. Lipid Technol 21:83–87. https://doi.org/10.1002/lite.200900017

Parashar A, Jin Y, Mason B et al (2016) Incorporation of whey permeate, a dairy effluent, in ethanol fermentation to provide a zero waste solution for the dairy industry. J Dairy Sci 99:1859–1867. https://doi.org/10.3168/jds.2015-10059

Patsalou M, Menikea KK, Makri E et al (2017) Development of a citrus peel-based biorefinery strategy for the production of succinic acid. J Clean Prod 166:706–716. https://doi.org/10.1016/j.jclepro.2017.08.039

Pereira VA, de Arruda INQ, Stefani R (2015) Active chitosan/PVA films with anthocyanins from Brassica oleraceae (red cabbage) as time-temperature indicators for application in intelligent food packaging. Food Hydrocoll 43:180–188. https://doi.org/10.1016/j.foodhyd.2014.05.014

Pereira GS, Cipriani M, Wisbeck E et al (2017) Onion juice waste for production of Pleurotus sajor-caju and pectinases. Food Bioprod Process 106:11–18. https://doi.org/10.1016/j.fbp.2017.08.006

Pérez-Sánchez A, Barrajón-Catalán E, Herranz-López M, Micol V (2018) Nutraceuticals for skin care: a comprehensive review of human clinical studies. Nutrients 10

Perotto G, Ceseracciu L, Simonutti R et al (2018) Bioplastics from vegetable waste: via an eco-friendly water-based process. Green Chem 20:894–902. https://doi.org/10.1039/c7gc03368k

Peschel W, Sánchez-Rabaneda F, Diekmann W et al (2006) An industrial approach in the search of natural antioxidants from vegetable and fruit wastes. Food Chem 97:137–150. https://doi.org/10.1016/j.foodchem.2005.03.033

Pfukwa TM, Fawole OA, Manley M et al (2019) Food preservative capabilities of grape (Vitis vinifera) and clementine mandarin (Citrus reticulata) by-products extracts in South Africa. Sustain 11. https://doi.org/10.3390/su11061746

Pierson Y, Chen X, Bobbink FD et al (2014) Acid-catalyzed chitin liquefaction in ethylene glycol. ACS Sustain Chem Eng 2:2081–2089. https://doi.org/10.1021/sc500334b

Pillai CKS, Paul W, Sharma CP (2009) Chitin and chitosan polymers: chemistry, solubility and fiber formation. Prog Polym Sci 34:641–678

Piwowarek K, Lipińska E, Hać-Szymańczuk E (2016) Possibility of using apple pomaces in the process of propionic-acetic fermentation. Electron J Biotechnol 23:1–6. https://doi.org/10.1016/j.ejbt.2016.07.004

Plazzotta S, Calligaris S, Manzocco L (2018) Innovative bioaerogel-like materials from fresh-cut salad waste via supercritical-CO2-drying. Innov Food Sci Emerg Technol 47:485–492. https://doi.org/10.1016/j.ifset.2018.04.022

Plazzotta S, Ibarz R, Manzocco L, Martín-Belloso O (2020) Optimizing the antioxidant biocompound recovery from peach waste extraction assisted by ultrasounds or microwaves. Ultrason Sonochem 63. https://doi.org/10.1016/j.ultsonch.2019.104954

Pleissner D, Lam WC, Sun Z, Lin CSK (2013) Food waste as nutrient source in heterotrophic microalgae cultivation. Bioresour Technol 137:139–146. https://doi.org/10.1016/j.biortech.2013.03.088

Pleissner D, Kwan TH, Lin CSK (2014) Fungal hydrolysis in submerged fermentation for food waste treatment and fermentation feedstock preparation. Bioresour Technol 158:48–54. https://doi.org/10.1016/j.biortech.2014.01.139

Pleissner D, Lau KY, Schneider R et al (2015) Fatty acid feedstock preparation and lactic acid production as integrated processes in mixed restaurant food and bakery wastes treatment. Food Res Int 73:52–61. https://doi.org/10.1016/j.foodres.2014.11.048

Poojary MM, Passamonti P (2015) Extraction of lycopene from tomato processing waste: kinetics and modelling. Food Chem 173:943–950. https://doi.org/10.1016/j.foodchem.2014.10.127

Poole AJ, Church JS, Huson MG (2009) Environmentally sustainable fibers from regenerated protein. Biomacromolecules 10:1–8

Prameela K, Venkatesh K, Immandi SB et al (2017) Next generation nutraceutical from shrimp waste: the convergence of applications with extraction methods. Food Chem 237:121–132

Prandi B, Faccini A, Lambertini F et al (2019) Food wastes from agrifood industry as possible sources of proteins: a detailed molecular view on the composition of the nitrogen fraction, amino acid profile and racemisation degree of 39 food waste streams. Food Chem 286:567–575. https://doi.org/10.1016/j.foodchem.2019.01.166

Pruksasri S, Wollinger KK, Novalin S (2019) Transformation of rice bran into single-cell protein, extracted protein, soluble and insoluble dietary fiber, and minerals. J Sci Food Agric 99:5044–5049. https://doi.org/10.1002/jsfa.9747

Quilodrán B, Hinzpeter I, Hormazabal E et al (2010) Docosahexaenoic acid (C22:6n-3, DHA) and astaxanthin production by Thraustochytriidae sp. AS4-A1 a native strain with high similitude to Ulkenia sp.: evaluation of liquid residues from food industry AS nutrient sources. Enzym Microb Technol 47:24–30. https://doi.org/10.1016/j.enzmictec.2010.04.002

Radha Krishna P, Srivastava AK, Ramaswamy NK et al (2012) Banana Peel as substrate for α-amylase production using Aspergillus Niger NCIM 616 and process optimization. Indian J Biotechnol 11:314–319

Ravindran R, Jaiswal AK (2016) Exploitation of food industry waste for high-value products. Trends Biotechnol 34:58–69. https://doi.org/10.1016/J.TIBTECH.2015.10.008

Reddy G, Altaf M, Naveena BJ et al (2008) Amylolytic bacterial lactic acid fermentation - a review. Biotechnol Adv 26:22–34

Rinaudo M (2006) Chitin and chitosan: properties and applications. Prog Polym Sci 31:603–632

Roda A, Lucini L, Torchio F et al (2017) Metabolite profiling and volatiles of pineapple wine and vinegar obtained from pineapple waste. Food Chem 229:734–742. https://doi.org/10.1016/j.foodchem.2017.02.111

Rodríguez Robledo V, Isabel Castro Vázquez L (2020) Pectin - Extraction, Purification, Characterization and Applications. In: Pectins - Extraction, Purification, Characterization and Applications. IntechOpen

Rohm H, Brennan C, Turner C et al (2015) Adding value to fruit processing waste: innovative ways to incorporate fibers from berry pomace in baked and extruded cereal-based foods—a SUSFOOD project. Foods 4:690–697. https://doi.org/10.3390/foods4040690

Ruiz-Rodriguez A, Soler-Rivas C, Polonia I, Wichers HJ (2010) Effect of olive mill waste (OMW) supplementation to oyster mushrooms substrates on the cultivation parameters and fruiting bodies quality. Int Biodeterior Biodegrad 64:638–645. https://doi.org/10.1016/j.ibiod.2010.07.003

Sabeena Farvin KH, Grejsen HD, Jacobsen C (2012) Potato peel extract as a natural antioxidant in chilled storage of minced horse mackerel (Trachurus trachurus): effect on lipid and protein oxidation. Food Chem 131:843–851. https://doi.org/10.1016/j.foodchem.2011.09.056

Sadiq AD, Chen X, Yan N, Sperry J (2018) Towards the Shell biorefinery: sustainable synthesis of the anticancer alkaloid Proximicin a from chitin. ChemSusChem 11:532–535. https://doi.org/10.1002/cssc.201702356

Santos MIS, Fradinho P, Martins S et al (2019) A novel way for whey: cheese whey fermentation produces an effective and environmentally-safe alternative to chlorine. Appl Sci 9. https://doi.org/10.3390/app9142800

Sarkar O, Venkata Mohan S (2017) Pre-aeration of food waste to augment acidogenic process at higher organic load: valorizing biohydrogen, volatile fatty acids and biohythane. Bioresour Technol 242:68–76. https://doi.org/10.1016/j.biortech.2017.05.053

Sarkar O, Kumar AN, Dahiya S et al (2016) Regulation of acidogenic metabolism towards enhanced short chain fatty acid biosynthesis from waste: metagenomic profiling. RSC Adv 6:18641–18653. https://doi.org/10.1039/c5ra24254a

Sashi Prabha M, Rangaiah GS (2014) Citric acid production using Ananas comosus and its waste with the effect of alcohols. Int J Curr Microbiol Appl Sci 3:747–754

Satari B, Karimi K (2018) Citrus processing wastes: environmental impacts, recent advances, and future perspectives in total valorization. Resour Conserv Recycl 129:153–167. https://doi.org/10.1016/J.RESCONREC.2017.10.032

Sayılı H, Güzel F, Önal Y (2015) Conversion of grape industrial processing waste to activated carbon sorbent and its performance in cationic and anionic dyes adsorption. J Clean Prod 93:84–93. https://doi.org/10.1016/j.jclepro.2015.01.009

Schieber A (2017) Side streams of plant food processing as a source of valuable compounds: selected examples. Annu Rev Food Sci Technol 8:97–112. https://doi.org/10.1146/annurev-food-030216-030135

Scordino M, Di Mauro A, Passerini A, Maccarone E (2007) Highly purified sugar concentrate from a residue of citrus pigments recovery process. LWT - Food Sci Technol 40:713–721. https://doi.org/10.1016/j.lwt.2006.03.007

Selaman R, Wid N (2019) Effects of substrate to inoculum ratio on phosphorus recovery from different composition of food waste using anaerobic batch digestion. J Phys Conf Ser 1358:012030. https://doi.org/10.1088/1742-6596/1358/1/012030

Shalini R, Gupta DK (2010) Utilization of pomace from apple processing industries: a review. J Food Sci Technol

Shanthi Sravan J, Butti SK, Sarkar O et al (2018) Electrofermentation of food waste – regulating acidogenesis towards enhanced volatile fatty acids production. Chem Eng J 334:1709–1718. https://doi.org/10.1016/j.cej.2017.11.005

Sharma A, Vivekanand V, Singh RP (2008) Solid-state fermentation for gluconic acid production from sugarcane molasses by Aspergillus Niger ARNU-4 employing tea waste as the novel solid support. Bioresour Technol 99:3444–3450. https://doi.org/10.1016/j.biortech.2007.08.006

Sharma P, Gaur VK, Kim SH, Pandey A (2020) Microbial strategies for bio-transforming food waste into resources. Bioresour Technol 299:122580. https://doi.org/10.1016/j.biortech.2019.122580

Silveira ST, Daroit DJ, Brandelli A (2008) Pigment production by Monascus purpureus in grape waste using factorial design. LWT - Food Sci Technol 41:170–174. https://doi.org/10.1016/j.lwt.2007.01.013

Singh R, Kumar M, Mittal A, Mehta PK (2016) Microbial enzymes: industrial progress in 21st century. 3 Biotech 6

Singhvi M, Zendo T, Sonomoto K (2018) Free lactic acid production under acidic conditions by lactic acid bacteria strains: challenges and future prospects. Appl Microbiol Biotechnol 102:5911–5924

Sirisakulwat S, Sruamsiri P, Carle R, Neidhart S (2010) Resistance of industrial mango peel waste to pectin degradation prior to by-product drying. Int J Food Sci Technol 45:1647–1658. https://doi.org/10.1111/j.1365-2621.2010.02314.x

Smithers GW (2008) Whey and whey proteins-From "gutter-to-gold". Int Dairy J 18:695–704

Soccol CR, Vandenberghe LPS (2003) Overview of applied solid-state fermentation in Brazil. Biochem Eng J 13:205–218. https://doi.org/10.1016/S1369-703X(02)00133-X

Souza Filho PF, Nair RB, Andersson D et al (2018) Vegan-mycoprotein concentrate from pea-processing industry byproduct using edible filamentous fungi. Fungal Biol Biotechnol 5:5. https://doi.org/10.1186/s40694-018-0050-9

Spekreijse J, Lammens T, Parisi C, et al (2019) Insights into the European market for bio-based chemicals

Stökle K, Kruse A (2019) Extraction of sugars from forced chicory roots. Biomass Convers Biorefinery 9:699–708. https://doi.org/10.1007/s13399-019-00374-9

Sun Z, Li M, Qi Q et al (2014) Mixed food waste as renewable feedstock in succinic acid fermentation. Appl Biochem Biotechnol 174:1822–1833. https://doi.org/10.1007/s12010-014-1169-7

Tanabe T, Okitsu N, Yamauchi K (2004) Fabrication and characterization of chemically cross-linked keratin films. Mater Sci Eng C 24:441–446. https://doi.org/10.1016/j.msec.2003.11.004

Tang Y, Bu L, He J, Jiang J (2013) L(+)-lactic acid production from furfural residues and corn kernels with treated yeast as nutrients. Eur Food Res Technol 236:365–371. https://doi.org/10.1007/s00217-012-1865-x

Tashiro Y, Matsumoto H, Miyamoto H et al (2013) A novel production process for optically pure l-lactic acid from kitchen refuse using a bacterial consortium at high temperatures. Bioresour Technol 146:672–681. https://doi.org/10.1016/j.biortech.2013.07.102

Thakur VK, Thakur MK (2014) Recent advances in graft copolymerization and applications of chitosan: a review. ACS Sustain Chem Eng 2:2637–2652. https://doi.org/10.1021/sc500634p

The European Parliament and the Council of the European Union (2006) REACH Regulation. Off J Eur Union:219–219. https://doi.org/10.4324/9781315270326-156

Tsagaraki E, Karachaliou E, Delioglanis IEK (2017) D2.1 bio-based products and applications potential. Bioways

Tsang YF, Kumar V, Samadar P et al (2019) Production of bioplastic through food waste valorization. Environ Int 127:625–644. https://doi.org/10.1016/j.envint.2019.03.076

Ueno T, Ozawa Y, Ishikawa M et al (2003) Lactic acid production using two food processing wastes, canned pineapple syrup and grape invertase, as substrate and enzyme. Biotechnol Lett 25:573–577. https://doi.org/10.1023/A:1022888832278

Unakal C, Kallur RI, Kaliwal BB (2012) Production of α-amylase using banana waste by Bacillus subtilis under solid state fermentation. Eur J Exp Biol 2:1044–1052

Upadhyay A, Chompoo J, Araki N, Tawata S (2012) Antioxidant, antimicrobial, 15-LOX, and AGEs inhibitions by pineapple stem waste. J Food Sci 77. https://doi.org/10.1111/j.1750-3841.2011.02437.x

Uzuner S, Cekmecelioglu D (2015) Enhanced pectinase production by optimizing fermentation conditions of Bacillus subtilis growing on hazelnut shell hydrolyzate. J Mol Catal B Enzym 113:62–67. https://doi.org/10.1016/j.molcatb.2015.01.003

Vaccari DA, Powers SM, Liu X (2019) Demand-driven model for global phosphate rock suggests paths for phosphorus sustainability. Environ Sci Technol 53:10417–10425. https://doi.org/10.1021/acs.est.9b02464

Vasala A, Panula J, Neubauer P (2005) Efficient lactic acid production from high salt containing dairy by-products by Lactobacillus salivarius ssp. salicinius with pre-treatment by proteolytic microorganisms. J Biotechnol 117:421–431. https://doi.org/10.1016/j.jbiotec.2005.02.010

Venkata Mohan S, Nikhil GN, Chiranjeevi P et al (2016) Waste biorefinery models towards sustainable circular bioeconomy: critical review and future perspectives. Bioresour Technol 215:2–12

Venkatakrishnan K, Chiu HF, Wang CK (2019) Extensive review of popular functional foods and nutraceuticals against obesity and its related complications with a special focus on randomized clinical trials. Food Funct 10:2313–2329

Venus J (2006) Utilization of renewables for lactic acid fermentation. Biotechnol J 1:1428–1432. https://doi.org/10.1002/biot.200600180

Venus J, Richter K (2006) Production of lactic acid from barley: strain selection, phenotypic and medium optimization. Eng Life Sci 6:492–500. https://doi.org/10.1002/elsc.200520136

Verma N, Bansal MC, Kumar V (2011) Pea peel waste: a lignocellulosic waste and its utility in cellulase production BY Trichoderma reesei under solid state cultivation. Bioresources 6:1505–1519

Vishnu C, Naveena BJ, Altaf M et al (2006) Amylopullulanase - a novel enzyme of L. amylophilus GV6 in direct fermentation of starch to L(+) lactic acid. Enzym Microb Technol 38:545–550. https://doi.org/10.1016/j.enzmictec.2005.07.010

Vivek N, Pandey A, Binod P (2016) Biological valorization of pure and crude glycerol into 1,3-propanediol using a novel isolate Lactobacillus brevis N1E9.3.3. Bioresour Technol 213:222–230. https://doi.org/10.1016/j.biortech.2016.02.020

Volpe R, Sotis G (2015) Nutraceuticals: definition and epidemiological rationale for their use in clinical practice. High Blood Press Cardiovasc Prev 22:199–201. https://doi.org/10.1007/s40292-015-0092-5

Vonder Haar C, Peterson TC, Martens KM, Hoane MR (2016) Vitamins and nutrients as primary treatments in experimental brain injury: clinical implications for nutraceutical therapies. Brain Res 1640:114–129

Waglay A, Karboune S, Alli I (2014) Potato protein isolates: recovery and characterization of their properties. Food Chem 142:373–382. https://doi.org/10.1016/j.foodchem.2013.07.060

Wang M, Xu Z, Qiu T et al (2010) Kinetics of lactic acid fermentation on food waste by lactobacillus bulgaricus. In: Advanced materials research, In, pp 1235–1238

Wang Y, Pedersen CM, Deng T et al (2013) Direct conversion of chitin biomass to 5-hydroxymethylfurfural in concentrated ZnCl2 aqueous solution. Bioresour Technol 143:384–390. https://doi.org/10.1016/j.biortech.2013.06.024

Wang K, Yin J, Shen D, Li N (2014) Anaerobic digestion of food waste for volatile fatty acids (VFAs) production with different types of inoculum: effect of pH. Bioresour Technol 161:395–401. https://doi.org/10.1016/j.biortech.2014.03.088

Wang Z, Wang C, Zhang C, Li W (2017) Ultrasound-assisted enzyme catalyzed hydrolysis of olive waste and recovery of antioxidant phenolic compounds. Innov Food Sci Emerg Technol 44:224–234. https://doi.org/10.1016/j.ifset.2017.02.013

Wang C, Chen P, Li Z (2018) Progress in preparation and application of organic waste based activated carbon. In: IOP Conference Series: Materials Science and Engineering

Werpy T, Petersen G (2004) Top value added chemicals from biomass volume I. US NREL medium: ED; size: 76 pp. pages. https://doi.org/10.2172/15008859

Willats WGT, Mccartney L, Mackie W, Knox JP (2001) Pectin: cell biology and prospects for functional analysis. Plant Mol Biol 47:9–27

Wu Y, Ma H, Zheng M, Wang K (2015) Lactic acid production from acidogenic fermentation of fruit and vegetable wastes. Bioresour Technol 191:53–58. https://doi.org/10.1016/j.biortech.2015.04.100

Yadav JSS, Bezawada J, Ajila CM et al (2014) Mixed culture of Kluyveromyces marxianus and Candida krusei for single-cell protein production and organic load removal from whey. Bioresour Technol 164:119–127. https://doi.org/10.1016/j.biortech.2014.04.069

Yan N, Chen X (2015) Sustainability: Don't waste seafood waste. Nature 524:155–157

Yang QQ, Gan RY, Ge YY et al (2019) Ultrasonic treatment increases extraction rate of common bean (Phaseolus vulgaris l.) antioxidants. Antioxidants 8. https://doi.org/10.3390/antiox8040083

Ye H, Zhu Q, Du D (2010) Adsorptive removal of cd(II) from aqueous solution using natural and modified rice husk. Bioresour Technol 101:5175–5179. https://doi.org/10.1016/j.biortech.2010.02.027

Yong SK, Shrivastava M, Srivastava P et al (2014) Environmental applications of chitosan and its derivatives. Rev Environ Contam Toxicol 233:1–43. https://doi.org/10.1007/978-3-319-10479-9_1

Zhang H, Grinstaff MW (2014) Recent advances in glycerol polymers: chemistry and biomedical applications. Macromol Rapid Commun 35:1906–1924. https://doi.org/10.1002/marc.201400389

Zhang J, Yan N (2016) Formic acid-mediated liquefaction of chitin. Green Chem 18:5050–5058. https://doi.org/10.1039/c6gc01053a

Zhang A, Yang S-T (2009) Engineering *Propionibacterium acidipropionici* for enhanced propionic acid tolerance and fermentation. Biotechnol Bioeng 104:n/a-n/a. https://doi.org/10.1002/bit.22437

Zhang AYZ, Sun Z, Leung CCJ et al (2013) Valorisation of bakery waste for succinic acid production. Green Chem 15:690–695. https://doi.org/10.1039/c2gc36518a

Zhang W, Li X, Zhang T et al (2017) High-rate lactic acid production from food waste and waste activated sludge via interactive control of pH adjustment and fermentation temperature. Chem Eng J 328:197–206. https://doi.org/10.1016/j.cej.2017.06.174

Zohuriaan-Mehr MJ (2005) Advances in chitin and chitosan modification through graft copolymerization: a comprehensive review. Iran Polym J (English Ed) 14:235–265

Zykwinska A, Boiffard MH, Kontkanen H et al (2008) Extraction of green labeled pectins and pectic oligosaccharides from plant byproducts. J Agric Food Chem 56:8926–8935. https://doi.org/10.1021/jf801705a

Chapter 6
Biobased Polymers from Food Waste Feedstock and Their Synthesis

Teresa Cecchi

6.1 Introduction

Plastics are mixtures of organic polymers of high molecular mass and other additives. The etymology of the word "Polymer" derives from two Greek words, 'Poly', which means many (numerous), and 'Meros', which means units (parts). Monomers are the chemical building blocks of polymers. A monomer can be reacted together with other monomers in a process called polymerization; monomers have the capacity to chemically bond to two other monomers or, in some cases, to more than two other monomers in cross-linked polymers.

Polymerization converts monomer molecules into a polymer. The process can be categorized into two important groups, namely step-growth and addition polymerizations. A step-growth polymerization is typical of monomers with functional groups such as –OH, $-NH_2$, –COOH, –COCl. Their reactions generate small molecular byproducts such as H_2O and NH_3 during the build-up of the polymer. In the case of water, the polymerization is called polycondensation. In contrast, addition polymerization involves the activation/addition of unsaturated monomers with a free radical or ionic initiator, yielding a much higher molecular weight polymer.

Polymerizations can be carried out in one of four different ways: in bulk, in a solvent, in suspension, and in emulsion. Reaction initiators are usually necessary. In bulk polymerization, only monomers provide the liquid portion of the reactor contents, whereas, in solution processes, the addition of solvents can control viscosity and temperature. Suspension polymerization counteracts the heat problem generated by the exothermic reaction by suspending droplets of water-insoluble monomer in an aqueous phase. These droplets are obtained by vigorous agitation of the system and are in size range from 0.01 to 0.5 cm diameter. Emulsion polymerization differs from the suspension method: the particles in the system are much smaller, from 0.05 to 5μm diameter, and the initiator is in the aqueous phase rather than in the monomer droplets. Besides monomers and an initiator, for emulsion polymerization prerequisite components are a dispersing medium, usually water, and an

T. Cecchi, C. De Carolis, *Biobased products from food sector waste*,
https://doi.org/10.1007/978-3-030-63436-0_6

emulsifying agent. The emulsifying agents are surfactants; their amphipathic nature ensures the stability of the initial monomer emulsion and of the polymer dispersion formed. Common surfactants used in emulsion polymerization comprise fatty acids, sodium lauryl sulfate, and α-olefin sulfonate.

The product of the polymerization reaction is the polymer, which is a long-chain molecule composed of a large number of repeating units covalently bonded. The repeating unit is a part of the macromolecule whose repetition would produce the complete polymer chain (except for the end-groups). The repeating unit may comprise one or more kinds of monomeric units. If a polymer is derived from only one species of monomer, it is called homopolymer; on the converse, copolymers derive from more than one species of monomer (Cowie 2007).

If the monomer is asymmetric, it can be distinguishable from its mirror image. In this case, it is called chiral. The chirality of the resulting polymer represents the relative stereochemistry of adjacent chiral centers within the macromolecule. An important characteristic of chiral polymers is tacticity, that is, "the orderliness of the succession of configurational repeating units in the main chain of a regular macromolecule, a regular oligomer molecule, a regular block, or a regular chain" according to the IUPAC definition (Jenkins et al. 1996). In isotactic polymers all the substituents are located on the same side of the macromolecular backbone. They are usually semicrystalline macromolecules and often form a helix configuration. In syndiotactic polymers, the substituents have alternate positions along the chain. In atactic polymers, the random nature of substituents along the chain generally results in amorphous macromolecules; hence tacticity may also be inferred from the melting temperature.

Plastic polymers can also be classified as thermosetting and thermoplastic matrices. The thermoset materials have crosslinking structures, which prevent the remelting, hence they show dimensional stability and resistance to high temperatures, but cannot be recycled, reshaped, or remolded. This is the case of poly(vinyl ester), phenol-formaldehyde resins, and epoxy resins. Thermoplastic polymers are not crosslinked and can be easily recycled or remolded because they become soft when heated, even if this can be a drawback. Polyolefines and polyesters are common thermoplastics. Elastomers are rubber-like elastic polymers that can be stretched extensively beyond their original length and can retract rapidly to their original dimensions: weak binding forces enable the polymer stretching, and a few crosslinks among chains permit the polymer to retract to the original shape. Thread forming polymers are called fibers: they are characterized by strong molecular forces, which lead to strong interactions of the chains and thus impart the crystalline nature (Cowie 2007).

Polymer science and technology aim to control macromolecular synthesis as regards uniform length, branching, molecular weight distribution, composition, topology, functionality, stereospecificity etc.

Fossilbased polymers are one of the major stories of the past century. The attractive qualities of fossilbased plastics lead us, around the world, to a voracious appetite and over-consumption of plastic goods. They have become a vital asset for humanity but also the crucial emblem of waste, pollution, and ecotoxicity. The use of disposable objects is increasingly perceived by consumers as a negatively valued practice.

There are many facets to the production and use of plastics of petrochemical origin. They are durable, lightweight, low cost, water-resistant, strong and tough, chemically inert, corrosion resistant, and thermal/electrical insulators. Anyhow, these advantages turn into disadvantages from the environmental point of view, since fossilbased plastics are non-renewable goods with a high "carbon footprint" They are also very slow to degrade. They may pose public health concerns because they release toxic fumes if burnt, their recycling is expensive and not always straightforward, they are not heat resistant, and they may become brittle at low temperatures. The rapid emergence of bioplastics is one of the major stories of the last decades because consumers ask for environmentally-sustainable products, there are increased restrictions for the use of polymers with high "carbon footprint" and boosted attention to sequestration of CO_2; hence industries are stimulated to focus on the development of new biobased feedstocks.

In the event of a resource crunch, the shifting toward a biobased material policy will be crucial for a sustainable future of polymer science and technology.

The economic and environmental needs claim the development of sustainable and renewable biobased polymers to address the challenge of finite fossil resources in the polymers market (Mirabella et al. 2014).

Bioplastics or biobased polymers or bio-polymers are terms that describe several different classes of materials, depending on the background of authors. From an environmental viewpoint, the renewability of feedstocks and degradability of the polymers are the most critical factors.

According to European Bioplastics, plastic material is defined as a bioplastic if it is either biobased, biodegradable, or both (European Bioplastics 2018). Anyhow there are biobased polymers that are not plastic at all.

The nomenclature ambiguity in this field of study is addressed by the European standard CEN/TR 15932:2010 (Plastics – Recommendation for terminology and characterization of biopolymers and bioplastics) that takes into consideration the current use of biopolymer-related terms in various settings (Imre et al. 2019).

The European Standard names the polymeric materials as follows:

- Biobased polymers: they are polymers with constitutional units that are totally or in part derived from biomass. Polymers synthesized by living organisms and synthetic biobased polymers, made of monomers that derive from renewable resources, belong to this category.
- Biodegradable polymers: they are polymers that can be biodegraded according to the relevant standards, such as the European Standard EN 13432:2000, ISO 17088:2012, and the US Standard ASTM D6400, discussed in Chap. 9 (Sec. 9.5). Biodegradability biochemically depends on the structure of the polymer chain; it does not depend on the origin of the raw materials. It has to be emphasized that, in fact, all plastics materials are theoretically biodegradable; since fossilbased plastics degrade at a slow rate in the natural environment, because the vast majority of microorganisms did not develop the ability to use them as carbon source,

they are considered non-biodegradable. According to the European Society for biomaterials, biodegradation is the process in which biological agents (microbes and enzymes) play a dominant role in degradation (Williams 1986).

- Biocompatible polymers: they are polymers compatible with human or animal tissues and suitable for medical use because they do not harm the body or its metabolism while fulfilling the expected function. Actually, for implantable systems, there are stringent requirements: (i) a benign and not excessively hydrophobic material is required; (ii) the material should exhibit long-term stability in a wet, oxygen- and enzyme-rich environment and should biodegrade within a specific time, which can be different for different applications; (iii) very low toxicity of degradation products or remnants of catalysts is needed. The biocompatibility of the material with cells should be tested before in vivo studies are conducted.

It follows that the generic term 'biopolymer' covers a rather wide range of polymeric materials. They are a whole family of materials with different properties and applications.

Figure 6.1 illustrates a global polymer classification according to renewability and biodegradability coordinates and lists the most significant examples in each category. The horizontal axis shows the biodegradability of plastic, whereas the vertical axis features the different origins of raw materials. This categorizes plastics into four groups.

Drop-in bioplastics are the "bio-similar" copy of the petrochemical plastics made from biomass instead of fossil oil. Polyolefins, polyamides, polyurethanes, and polyethylene terephthalate are examples of drop-in bioplastics if the producer has replaced the fossilbased chemicals with biobased chemicals while using the same machinery and equipment for their production.

The drop-in bioplastics are high-priced compared to the original fossilbased plastics because of the economy of scale and smaller production and processing capacity; moreover, research and development costs are higher, and biomass cannot be transported through pipelines but in trucks. On the converse, the end-of-life options are similar as regards biodegradability. The carbon footprint for drop-in bioplastic is lower because the CO_2 that is released when the material is returned to the biosphere had been photosynthetically captured by the plant during its growth, and no additional CO_2 is brought into the atmosphere, at variance with fossilbased plastic analogs.

Generally speaking, while the price of conventional plastics oscillates with oil prices, the prices of bioplastics are higher but more stable than conventional plastics. Bioplastic costs are expected to decline as economies of scale are achieved, and logistics are fully developed. Product design may also promote price reductions. For example, PLA is stiffer than PS and less plastic is necessary to make products with comparable rigidity (Zhao et al. 2020). Furthermore, labor costs entailed in recycling plastic objects can be reduced by using biodegradable versions.

Polymer processability, mainly governed by rheological properties, such as extensional viscosities and melt strength, depends on polymer molecular weight,

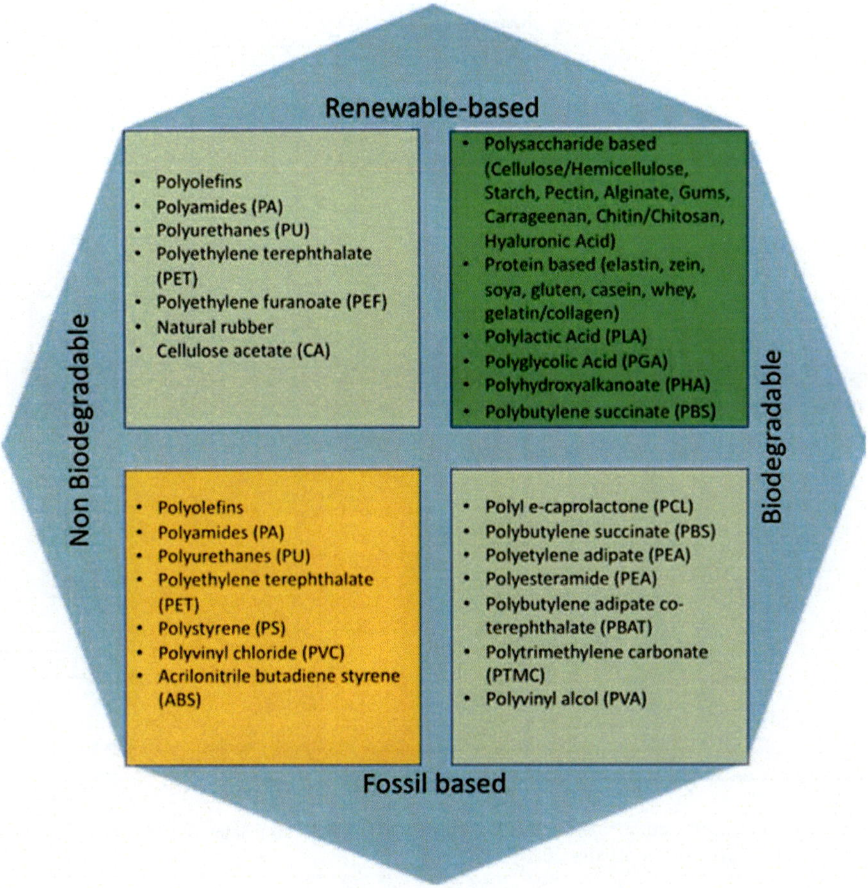

Fig. 6.1 Polymer classification according to renewability and biodegradability coordinate; abbreviations are detailed

molecular weight distribution, degree of crystallinity, and chain structure. Most biodegradable bioplastics have worse processability (via a variety of thermal techniques) than fossilbased plastics (Zhao et al. 2020).

Currently, biobased polymers production was 4.2 million tonnes, which represents about 1% of fossilbased polymers in 2020 (Skoczinski et al. 2021), with a market boom due to sophisticated biopolymers and novel applications.

Their production capacity was projected to increase from around 2.11 million tonnes in 2020 to about 2.87 million tonnes in 2025, as illustrated in Fig. 6.2. In 2020 their compound annual growth rate, 8%, is more than double the overall growth of polymers (Skoczinski et al. 2021).

The most important market segment is packaging, followed by consumers goods, textiles, and agriculture/horticulture, but biopolymers are increasingly used in many other sectors such as automotive and transport, building, and construction, coating

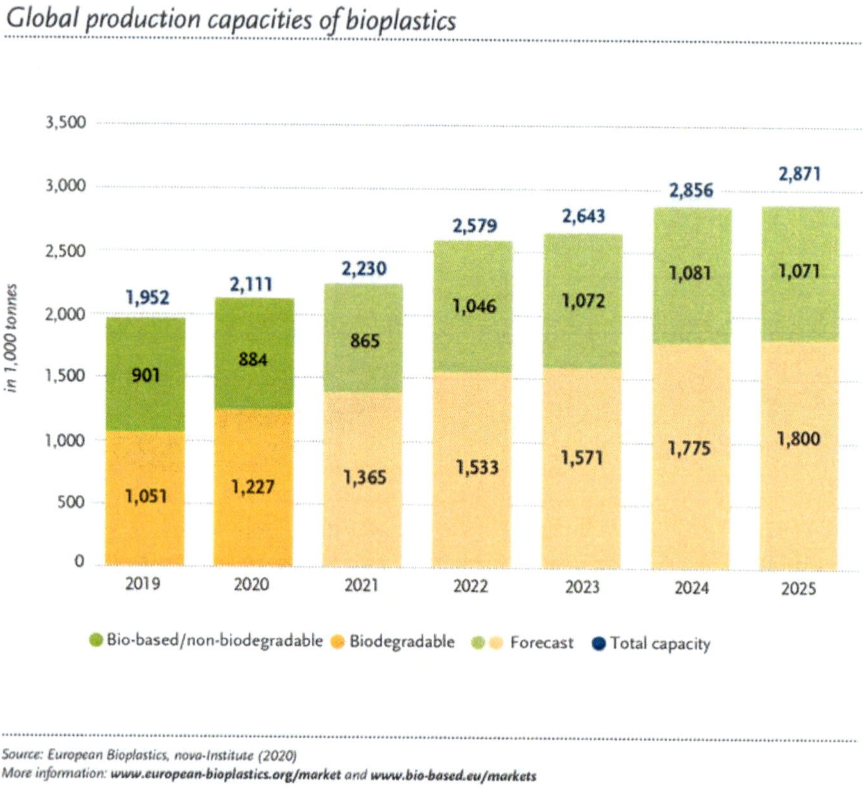

Fig. 6.2 Global production capacities of bioplastics 2019–2025. Reprinted with permission from the Author. (European Bioplastics 2020)

and adhesives, electrics and electronics, and others (European Bioplastics 2020), as detailed in Fig. 6.3.

The application of biodegradable biopolymers is possible and advantageous mainly in the packaging, agriculture, and the medical sector. In short service life applications, biodegradability is of utmost importance. For long-term applications, however, biodegradability might be disadvantageous.

In order to be selected as potential monomers for bioplastic production, platform chemicals need to overcome the gap between laboratory/pilot scale and commercial/industrial scale using suitable and profitable production strategies, including microbial strain development to convert renewable resources into value-added products.

Virgin bioplastic feedstocks raise the issue as to whether they have a negative impact on food supply. European Bioplastics claims there is no competition between the renewable feedstock for food, feed, and the production of bioplastics, and actually, the land used to grow the renewable feedstock for the bioplastics production was less than 0.02% of the global agricultural area (European Bioplastics 2020). Anyhow, the use of food waste (FW) instead of edible crops is an opportunity and a

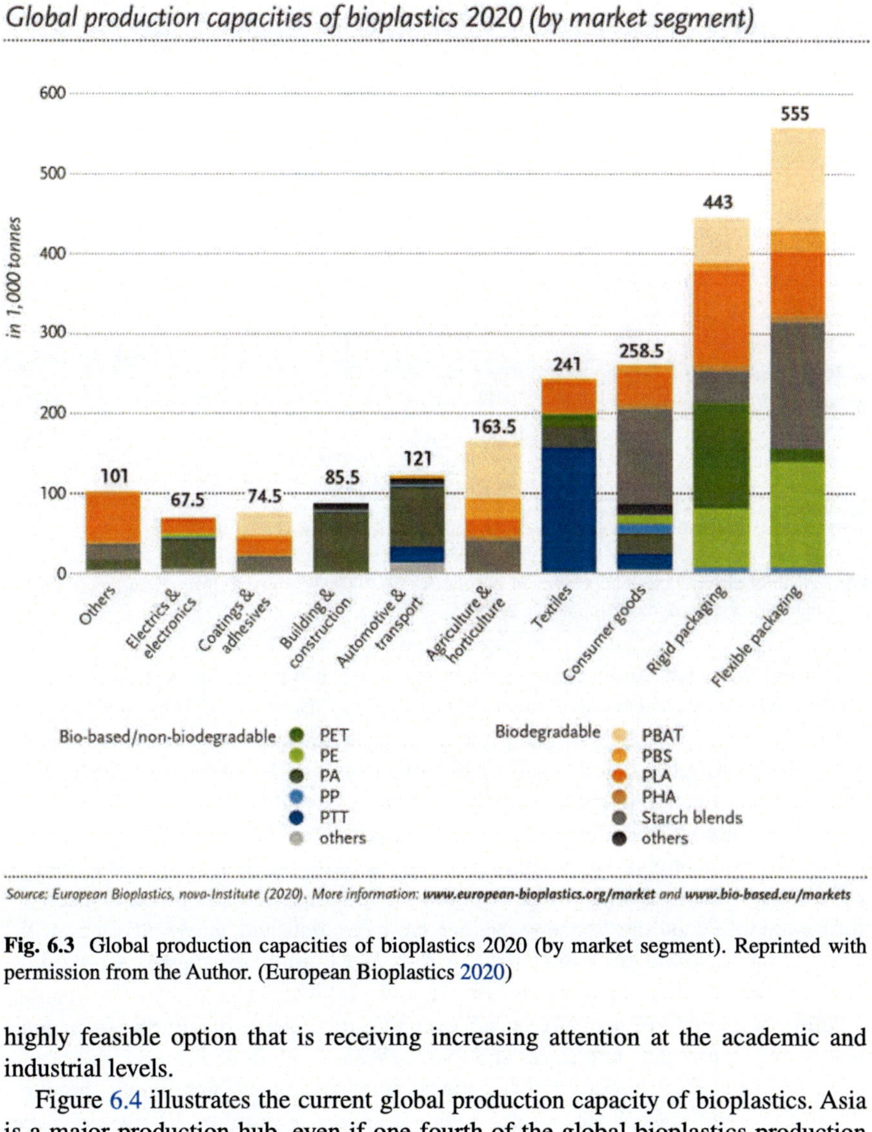

Fig. 6.3 Global production capacities of bioplastics 2020 (by market segment). Reprinted with permission from the Author. (European Bioplastics 2020)

highly feasible option that is receiving increasing attention at the academic and industrial levels.

Figure 6.4 illustrates the current global production capacity of bioplastics. Asia is a major production hub, even if one-fourth of the global bioplastics production capacity is located in Europe that ranks highest in research and development and is the industry's largest market worldwide.

Biopolymers can be present in the FW. This is the case of polysaccharides found in plants (such as starch, cellulose, agar, hemicelluloses, pectin, alginate, konjac, carrageenan, gum) or in animals (chitin/chitosan, hyaluronic acid). Protein of vegetal (zein, soy, gluten) and animal origin (casein, whey protein, collagen, gelatin, elastin, silk, fibroin) are other biopolymers. They can also be synthesized by microorganisms such as fungi (pullulan, elsinan, scleroglucan) and bacterial species (glucans, xanthan gum, polygalactosamine, curdlan, gellan, dextrans, and polyhydroxyalkanoates (PHA)).

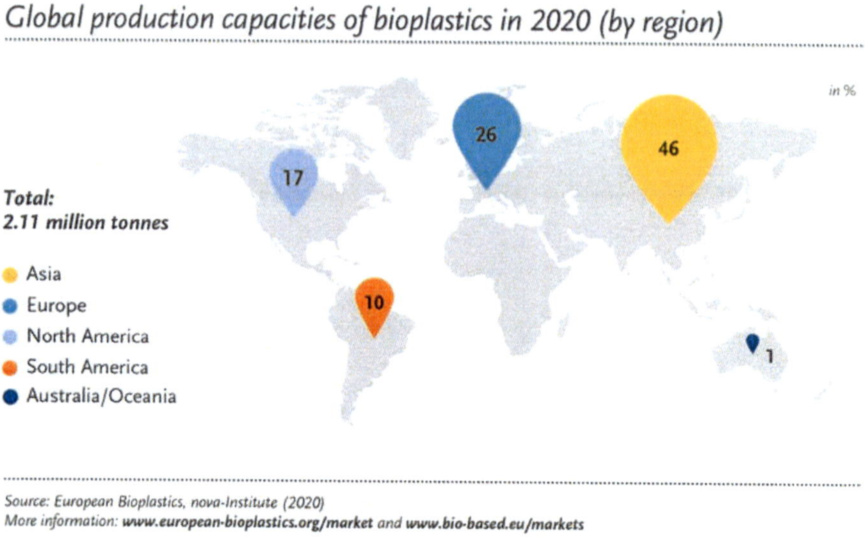

Fig. 6.4 Global production capacities of bioplastics 2020 (by region). Reprinted with permission from the Author. (European Bioplastics 2020)

Natural polymers such as polysaccharides and chitin can be converted into marketable derivatives via bulk or surface chemical modification (Cunha and Gandini 2010) that usually aims at reducing the polymer hydrophilicity (Sanchez-Vazquez et al. 2013). Similarly, cellulose-based bioplastics can be obtained via physico-chemical modification of cellulose. High-molecular-weight cellulose is highly crystalline and almost insoluble in aqueous media (Sandhya et al. 2013). Surface chemical modification gives ethers, esters, and acetals, which are the most common derivatives of the hydroxyl groups in cellulose structures. Plastic properties can be introduced into cellulose fibers by mechanical treatments and, above all, by esterification of the hydroxyl group with the acid to confer fluidity, resistance, and durability comparable to those of synthetic polymers (Abdul Khalil et al. 2016). Conversion of cellulose into polyols can also be achieved through liquefaction during the manufacture of polyurethane polyesters and foams (Wang et al. 2008a).

Apart from PHAs, directly synthesized by microorganisms, biopolymers production from a gamut of FW that ranges from agroindustrial residues to organic waste from household level, relies on polymerization of simple biomolecules accessible either through substrate fermentation by microbes, including the genetically modified ones, or by chemical processing.

Figure 6.5 illustrates the current global production capacities of bioplastics, detailed by material type. More than half the produced bioplastics are biodegradable. Today, there is a bioplastic substitute for nearly every fossil plastic material with the additional advantages of a reduced carbon footprint and additional waste management options (van den Oever and Molenveld 2017).

PLA, PCL, and PGA are hydrophobic aliphatic polyesters based on hydroxyalkanoic acids, particularly important for the biomedical sector because of their

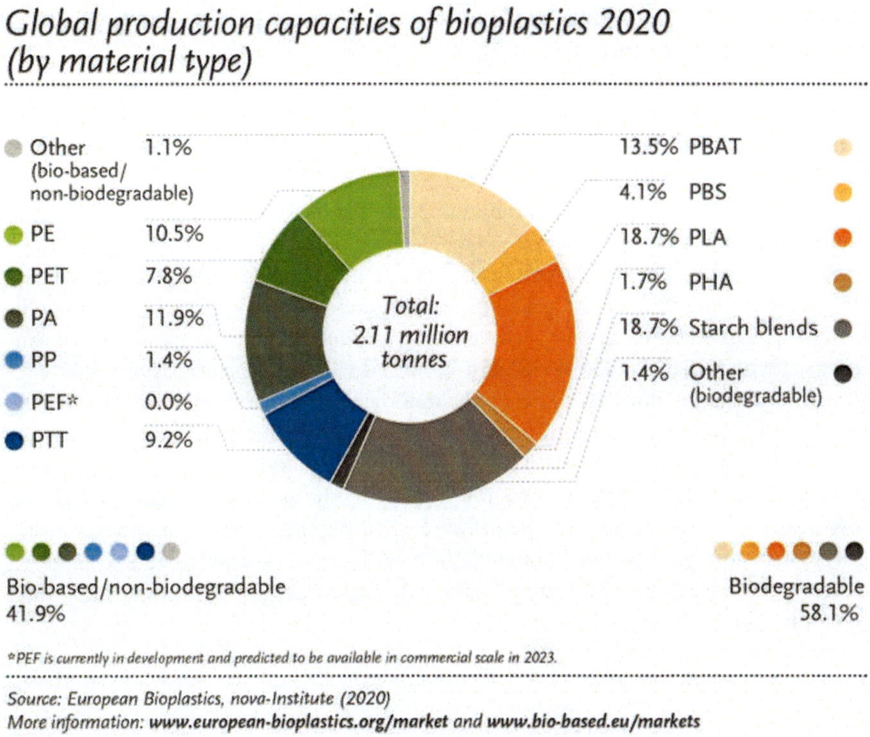

Fig. 6.5 Global production capacities of bioplastics 2020 (by material type). Reprinted with permission from the Author. (European Bioplastics 2020)

biocompatibility and low immunogenicity; they are biodegradable by enzymes and/or hydrolysis under physiological conditions. Their good mechanical properties enable their use as sutures, implants for bone fixation, drug delivery vehicles, and tissue engineering scaffolds (Oh 2011; Jacobson et al. 2008). Biodegradable polymers like PLA, PHA, PBS, PGA, PCL, etc. are high value- added products for sustained release systems, pesticides, fertilizers, packaging, paper coatings, and compost bags.

PLA and PHAs rank highest in the field of research and development and show the fastest rates of market development. Production capacities of PLA show high growth rates due to its excellent barrier properties. High-performance PLA grades are an excellent replacement for PS, PP, and ABS (acrylonitrile butadiene styrene) in challenging applications. Increased volumes of PLA production are needed since PLA was sold out in 2019. Biobased PE and PP are breaking ground, while biobases polyamides, PBAT, PHAs, and casein polymers expansion is envisioned (Skoczinski et al. 2021).

PHAs production has been in development for a while, but their commercial production capacities are projected to increase almost tenfold in the next five years due to their wide gamut of physical and mechanical properties and consumer awareness about the importance of sustainable and ecofriendly bioplastics.

As regards non-biodegradable bioplastics such as biobased PE and PET, new capacities are expected in the coming years. Biobased PP entered the market in 2019. As regards PET, the focus has shifted to the development of PEF (polyethylene furanoate). It is predicted to enter the market in 2023 as a biobased PET alternative with superior barrier and thermal properties for the packaging sector (European Bioplastics 2020).

Biobased polymers can be primarily manufactured using three principal routes: (i) partial modification of natural biobased polymers (e.g. starch, cellulose, and chitin), (ii) polymerization of biobased monomer produced by fermentation or conventional chemistry (e.g., PLA, PBS, and biobased PE) (iii) bacterial synthesis (e.g., PHA). This chapter does not focus on the modification of natural polymers into their marketable derivatives but introduces four among the most interesting biodegradable thermoplastic biopolymers, viz., PLA, PHA, PBS, and PGA with a focus on the history of their development, their properties, synthesis, commercial market, and low-cost FW-based feedstock identified for production. Biobased thermosetting polymers, copolymerization, and blending will be shortly presented. Chapter 8 details the mechanical testing and characterization of bioplastics and biocomposites

6.2 Polylactic Acid (PLA)

PLA is the most popular bioplastic, and its production capacity will increase significantly in the next years.

PLA has been studied for over a century. In the early 1800s, PLA was discovered when Pelouze condensed lactic acid by distillation of water. This polycondensation process produces low yield, low purity, and low-molecular-weight PLA and the cyclic dimer of lactic acid, namely, 3,6-dimethyl-1,4-dioxane-2,5-dione, known as lactide (Benninga 1990).

Nearly a century later, Wallace Carothers found that heating lactide in vacuum conditions produced PLA (Carothers and Arvin 1929). This method, namely ring-opening polymerization (ROP), was later patented by DuPont in 1954 (Auras et al. 2010a). This process is difficult on an industrial scale due to the high cost of purification, limititing it to the production of medical products, such as sutures, implants, and drug carriers.

Since the 1970s, PLA has been approved by the US food and drug administration (FDA) for food and pharmaceutical applications since the PLA depolymerization products are not toxic (Garlotta 2001).

In the 1990s, a commercially viable lactide ROP was developed to polymerize high-molecular-weight PLA (Auras et al. 2010a). Since then, PLA has become the emblem of biopolymers, and it currently attracts an exponentially increasing research attention.

6.2.1 FWs as Feedstocks

Lactic acid (LA) is the monomer of PLA. Lactic acid is one of the most promising value-added building blocks that can be derived from sugars, and it is a midway chemical for the production of various commodity and specialty chemicals (Bozell and Petersen 2010).

The IUPAC name is 2-hydroxy-propionic acid, it exists in two optically active enantiomers, the L(+)- and the D(−)-lactic acid (respectively L-LA and D-LA). Lactic acid can be either prepared chemically or biotechnologically. Anyhow, the fermentative biotechnological LA production is usually preferred because of the low substrate costs, moderate operating temperatures, and lower energy consumption.

During a typical LA fermentation, LA production results in a drop in pH that has an inhibitory effect on the metabolic activity of LA-producing strains. To avoid this drawback, a neutralizing agent such as calcium carbonate or hydroxide is used in industrial processes to produce calcium lactate in the fermentation broth. This salt does not inhibit further LA synthesis. Filtration removes the cell biomass and other insolubles, and then, after evaporation and recrystallization, corrosive sulfuric acid is used to liberate crude LA from calcium lactate, generating a large amount of calcium sulfate (gypsum) as solid waste. A filtration step separates gypsum; LA purification via esterification and subsequent hydrolysis is one of the most expensive stages of the production process and a reactive distillation system was advantageous for lactic acid purification (Komesu et al. 2015). The process is costly and complicated; by this scenario, the use of acid-tolerant strains, obtained by genome shuffling through protoplast fusion, gene deletion, exploitation of knockdown libraries and adaptive evolution to acidic pH, would avoid the need for neutralizers, thereby simplifying the downstream process, and lowering the overall cost of the fermentative LA production; another strategy to circumvent the toxicity of lactic acid towards microbial cells was in situ LA recovery (Singhvi et al. 2018).

Commercial LA is usually and selectively produced by the bacterial fermentation of carbohydrates by homofermentative organisms belonging to the genus *Lactobacillus* and *Bacillus*. *L. delbrueckii, L. amylophilus, L. bavaricus, L. casei, L. maltaromicus,* and *L. salivarius* predominantly produce the L(+)-isomer. Strains such as *L. lactis, L. jensenii,* and *L. acidophilus* generate either the D(–)-isomer or mixtures of both LA enantiomers. Interestingly, at variance with the chemical synthesis, only one of the two enantiomers is produced if a specific strain of lactic acid bacteria is chosen (Reddy et al. 2008). Most strains are anaerobic and convert pyruvic acid, the end-product of the Embden-Meyerhof pathway, to lactate by either of the two enzymes, L- or D-lactate dehydrogenase; pyruvate is the electron acceptor of NADH (nicotinamide adenine dinucleotide) oxidation (Singhvi et al. 2018).

Bio-production of optically pure L-LA from FW at ambient temperature relied on the regulation of key enzyme activity by sewage sludge supplementation and intermittent alkaline fermentation. Production of optically pure L-LA acid was achieved from FW at ambient temperature with a yield of 0.52 g/g of total chemical oxygen demand of the substrate (Li et al. 2015). *Streptococcus thermophilus* and *Lactobacillus bulgaricus* selectively produced L-LA from furfural residues and corn kernels with hydrolyzed yeast as cheap nutrients (Tang et al. 2013).

L-LA is the preferred building block for food and pharmaceutical application of PLA since elevated levels of D-LA are harmful to humans. The presence of a single optical lactic acid is preferable because different optical lactic acids can affect the properties of PLA, such as the rate of crystallization, the extent of crystallization, the melting point, mechanical strength, and degradability.

Most simple sugars can be fermented, but the expensive food-based sugars make their use scarcely feasible from the economic point of view. Nowadays, LA used for PLA production is mainly derived from renewable edible crops such as sugar beet and corn (Castro-Aguirre et al. 2016). To limit the land use share for bioplastics, high sugar-containing FW such as whey, coffee mucilage, corn cobs, corn stalks, rice bran, barley, wheat bran and brewer's spent grains, (Venus 2006; Venus and Richter 2006), kitchen and canteen waste (Tashiro et al. 2013; Ohkouchi and Inoue 2006) (Wang et al. 2010), pineapple syrup and grape invertase as substrate and enzyme, respectively (Ueno et al. 2003), wheat bran (Germec et al. 2019), soybean vinasse (Karp et al. 2011), *Curcuma longa* biomass (Nguyen et al. 2013), carbohydrates, starchy and lignocellulosic biomasses free from food-related concerns, glycerol, microalge, whey (Abdel-Rahman et al. 2013; Vasala et al. 2005; Vishnu et al. 2006), fruit and vegetable wastes (Wu et al. 2015) and yoghurt waste (Alonso et al. 2010) have been explored as alternative feedstocks for fermentative LA production

The production of lactic acid from these and other FWs was more extensively dealt with in Chap. 5 (Sect. 5.4.3).

A techno-economic assessment investigated the technical feasibility, profitability and extent of investment risk for LA, lactide and PLA production using FW powder as the raw material in a plant. Sensitivity analysis demonstrated that the prices of lactic acid, lactide, and PLA, as well as the marketability of byproducts (animal feed), were crucial factors for the profitability in the plant (Kwan et al. 2018).

6.2.2 Synthesis

The production of PLA can be conducted by different synthetic routes, viz. direct condensation polymerization, azeotropic dehydrative polycondensation, and polymerization through lactide formation and ROP, as illustrated in Fig. 6.6.

Since the lactic acid monomer has both –OH and –COOH groups, the reaction can take place directly by self-condensation/dehydration equilibrium. Condensation of lactic acid is usually performed in bulk by distillation of condensation water, while vacuum and temperature are progressively increased (Mehta et al. 2005).

This preparation strategy also involves ring-chain equilibrium involving the depolymerization of PLA into the dimer of lactic acid, i.e., lactide. The polymer obtained by this process has a low molecular weight because it is hard to remove water from the highly viscous reaction mixture completely. At a certain point, the

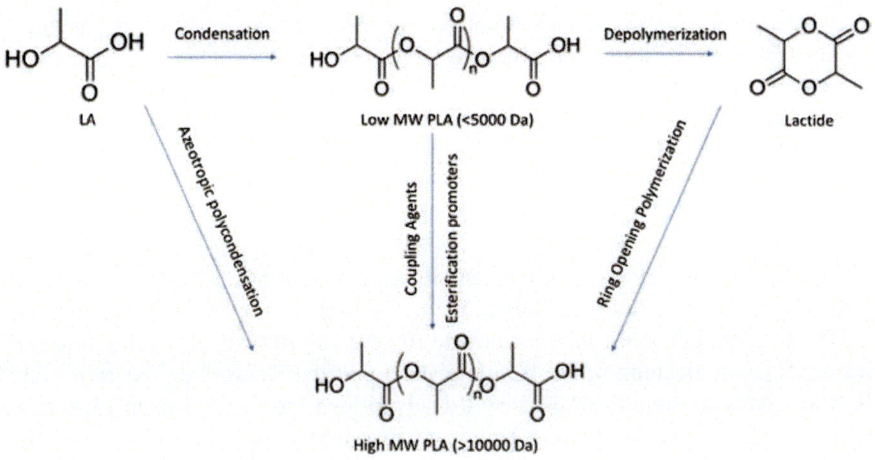

Fig. 6.6 PLA preparation technologies

polymerization and depolymerization rates are the same. A reaction temperature lower than 200 °C limits the generation of the entropically favored lactide. The low molecular weight polymer is brittle and almost unusable. External coupling agents, such as carbonyl diimidazole, bis(trichloromethyl) carbonate, and dicyclohexylcarbodiimide, are employed to increase the chain length promoting esterification between low molecular weight PLA molecules obtained in the initial condensation. These reagents produce reaction byproducts that must be removed; for example, dicyclohexylcarbodiimide forms insoluble dicyclohexylurea, which must be filtered in the purification step. Moreover, the increased number of reaction steps makes the procedure more expensive. Alternatively, reactions involving chain-extending agents are cheaper, requiring low amounts of chain-extending agents. Chain extenders are usually bifunctional low molecular weight chemicals. At variance with coupling agents, they do not bind different PLA chains but they can increase the molecular weight of polymers in a fast reaction without a separate purification step via their reaction with the hydroxyl or carboxyl end groups. Some possible chain-extending agents are isocyanates, acid chlorides, anhydrides, epoxides, thiirane, and oxazoline; isocyanates are toxic chain extending agents (Li and Yang 2006). Chain extending agents are not biodegradable or bioabsorbable and are not compatible with medical applications because of possible severe in vivo complications (Hartmann 1998).

PLA can alternatively be produced using the azeotropic dehydrative polycondensation approach. This method brought a breakthrough in increasing the molecular weight of PLA. Solvents with high boiling point are used to remove the dissociated water by means of the so-called azeotropic distillation technique. A general procedure for this route consists of reduced pressure distillation of lactic acid for 2–3 h at 130 °C to remove the majority of the condensation water. Then, the catalyst and azeotropic solvent, such as toluene, are added. The water formed during the polymerization is removed as water/solvent azeotrope. Then the water/solvent vapor condenses to give a liquid that separates in two phases. The upper one consists of the solvent, that is reintroduced in the reaction mixture. The lower phase comprises water, which is eliminated. The polymer can then be isolated as is or dissolved and precipitated for further purification. This polymerization technique requires significant amounts of catalysts. They remain as impurities in the polymer and may initiate unwanted degradation during subsequent processing work at elevated temperatures, uncontrolled polymer hydrolysis, and toxicity in medical applications (Garlotta 2001). The use of solvents increases the PLA price and makes the processes ecologically unattractive since it is hard to remove solvent totally from the end product.

The third and most exploited synthetic route is the ROP of the cyclic dimer of lactic acid. The opening of the lactide gives a unit that is acyclic. The controlled ROP of lactide is significant because it leads to high molecular weight PLA with specific and desirable properties. Due to the two asymmetric carbon atoms in the molecule, lactide exists in three different stereoisomers: DD-lactide, LL-lactide enantiomers, and *meso*-lactide, as illustrated in Fig. 6.7; the DD-lactide and LL-lactide enantiomers can form a 1: 1 racemic stereocomplex (rac-lactide), which melts at 126–127 °C (Hartmann 1998), higher than the pure isomers, which melt at 97 °C, while *meso*-lactide melts at 52°C (Ehsani et al. 2014). DD-lactide and LL-lactide are also known, respectively, as D-lactide and L-lactide and, according to the Cahn–Ingold–Prelog sequence rules, they sequentially correspond to (R-R)-Lactide and (S-S)-Lactide. The use of high-temperatures induces racemization.

Lactide is currently obtained by the depolymerization of low molecular-weight PLA (1000–5000 MW), obtained condensing lactic acid, with or without catalysts, at 110–180 °C and removing the water of condensation under vacuum 25 mm Hg under reduced pressure and high temperature. This prepolymer is then subjected to temperatures of 180–215 °C at 0.1–15 mm Hg to form and distill off the crude

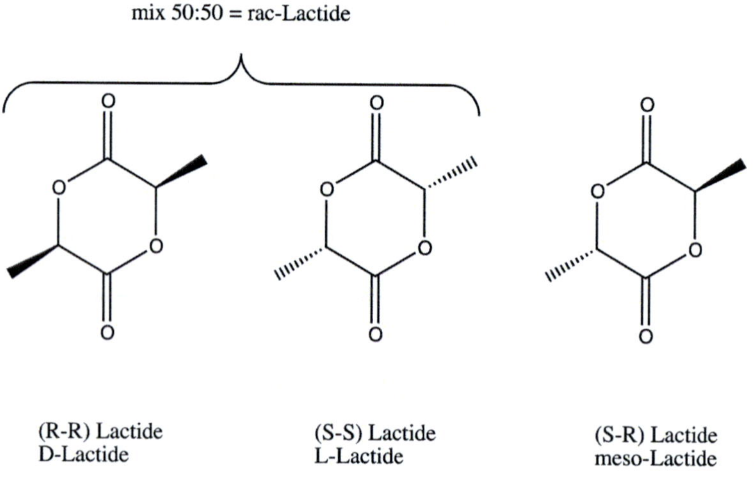

Fig. 6.7 Lactide stereoisomers

lactide and remove water and acid impurities. Purification can be performed via recrystallization or distillation, even if the small difference in boiling temperatures of LL- or DD-lactide and meso-lactide makes the latter procedure quite labor-intensive. Furthermore, for the highest stereochemical purity, recrystallization is needed. The ROP, depending on the type of catalyst used, can follow three different pathways: cationic, anionic, or coordination/insertion. Both cationic and anionic ROP are usually done in solvent systems, and are susceptible to racemization and transesterification. The coordination/insertion ROP is usually performed using tin(II) 2-ethylhexanoate, $Sn(Oct)_2$, as a catalyst. Although $Sn(Oct)_2$ has been accepted as a food additive by the U.S. FDA, the toxicity associated with most tin compounds is a considerable drawback in the case of biomedical applications (Albertsson and Varma 2003; Kowalski et al. 2005). In fact, a medical PLA-based product within a tissue would releases continuously tin during the course of biosorption of PLA, potentially leading to critical tin concentrations with adverse effects on enzymes, receptors, and the whole organisms due to cyto- and genotoxicity (Yamada et al. 2008; Tanzi et al. 1994); the ecotoxicity (Mehta et al. 2005; Hege and Schiller 2014) of tin has also to be taken into account since PLA disposable products are increasingly used. The development of safe catalysts for biopolymers synthesis is of great interest. It follows that tin replacement strategies are of utmost importance. Concerns about catalyst toxicity prompted the academic study of the biocatalytic ROP of lactide using mild reaction requirements and recyclable enzymes such as *Candida antarctica* lipase B immobilized onto chitin and chitosan (Omay and Guvenilir 2013). Cerium trichloride heptahydrate is a nontoxic and easy to handle Lewis acid catalyst in organic reactivity; a new and efficient lactic acid polymerization by multimetallic cerium complexes yielded a PLA suitable for biomedical applications, in high yields up to 95% and a molecular weight ranging from 9000 to 145000 g/mol (Pastore et al. 2021). Iron-based catalysts were also successfully explored to avoid the tin-based catalysts toxicity (Hege and Schiller 2014).

6.2.3 *Properties and Uses*

PLA possesses high density, elongation, and tear resistance similar to PET (Lunt 1998) and PS (Zhao et al. 2020). It follows that the potential of PLA as an alternative material to conventional plastics is high if properly manufactured, since it can be lightweight with low processing temperature, no environmental pollution, and good 3D printability.

If the polymer molecular weight is not very high (<50,000 Da), PLA is amorphous, with a melting point of 130–150 °C and a glass transition temperature (see Sec. 9.2) around 45 °C (Rivero et al. 2016).

On the converse, high molecular weight PLA (>50,000 Da) has a melting temperature of 170–180 °C, a glass transition temperature of 58–65 °C, and a tensile modulus 2.7–16 GPa (Rivero et al. 2016; Södergård and Stolt 2002; Middleton and

Tipton 2000). Annealing and nucleation have been used to enhance the mechanical properties of PLA polymers (Simmons et al. 2019). Exemplary physical properties of PLA are detailed in Table 6.1 ; they strongly depend on the molecular weight of the polymer and its tacticity.

Lactic acid, which is the building block of PLA, is chiral; hence its polymerization may lead to isotactic, syndiotactic, and atactic/heterotactic PLA primary structures. L-LA and D-LA lead to two enantiomeric forms PLAs, namely, poly(L-LA) (PLLA) and poly(D-LA) (PDLA), respectively, while their mixture leads to poly(D-, L-LA) (PDLLA).

Isotactic PLLA homopolymer is a semicrystalline material with a melting temperature of about 170–180 °C (Husseinsyah and Zakaria 2011). On the converse, the atactic and heterotactic stereoblocks, with a random arrangement of L- and D-units, lead to amorphous and low-quality PLA. PLA has quickly developed into a competitive material, but the control of crystallinity is a bottleneck in extended utilization. A specific stereocomplex of PDLA and PLLA has a higher glass transition temperature, with increased crystallinity and mechanical strength (Lv et al. 2018); hence, although present day PLA is composed of mainly L-LA, an L- and D- LA strereocomplex copolymer is expected to improve the quality of PLA and expand its use (Awasthi et al. 2018) due to an enhanced melting point to over 200 °C (Rivero et al. 2016): advanced therapeutic delivery carriers and tissue engineering devices, as well as stabilization of colloidal systems in microparticles, micelles, and hydrogels, could be attained (Li et al. 2016a). An environment-friendly polyhydroxy compound, pentaerythritol, was also used as a crystallization promoter to form the PDLA/PLLA regular stereocomplex; the temperature stability is maximized when a 1:1 blend is used (Chen et al. 2019), but even at lower concentrations there is still a substantial improvement of thermal stability (Zhang et al. 2019).

The use of an aryl amide as a nucleating agent improved the crystallization rate of PLLA/PDLA blend and increased the fraction of the stereocomplex (Xie et al. 2016). Other strategies to facilitate the stereocomplex crystallization, thereby increasing the heat resistance, include copolymerization, flow-induced crystallization, and the use of a polyhydroxy compound, such as pentaerythritol, to stimulate nucleation (Chen et al. 2019).

Polylactic acid can profitably be processed like most thermoplastics into fiber (for example, using conventional melt spinning processes) and film (Zhang et al. 2019).

PLA is soluble in dioxane, acetonitrile, chloroform, methylene chloride, 1,1,2-trichloroethane, and dichloroacetic acid but it is not soluble in water, alcohols, and alkanes. Crystalline PLLA is not soluble in polar solvents, such as acetone, ethyl acetate, or tetrahydrofuran (Farah et al. 2016).

PLA is a thermoplastic aliphatic polyester that can be attacked by microorganisms but it also undergoes simple hydrolytic cleavage of the ester moieties of the polymer backbone with only little or even no assistance from enzymes resulting in the formation of non-harmful and not toxic lactic acid monomers (Pawar et al. 2014), even if the potential toxicity of additives and catalysts has to be considered. Actinobacteria belonging to the family *Pseudonocardiaceae*, and some members of the family *Micromonosporaceae*, *Streptomycetaceae*, *Streptosporangiaceae*, and *Thermomonosporaceae* can degrade PLA; PLA composting and anaerobic digestion are feasible while PLA is less degraded in the environment (Folino et al. 2020). Heterogeneous PLA was effectively composted with garden waste (Ghorpade et al. 2001). PLA can also be depolymerized to its monomer by thermal depolymerization or hydrolysis with boiling water or steam (Madhavan Nampoothiri et al. 2010).

Due to the biodegradability, not toxic nature of degradation products, and mechanical properties, PLA has been used extensively for medical application (Hamad et al. 2015) such as tissue engineering, drug delivery systems, implants, sutures, anchors, screws, scaffolds, bone grafting, reattachment of ligaments, tendons (Auras et al. 2010b).

It has been used in oral, orthopedic, auricular, and craniofacial augmentations in plastic surgery. Degradation via hydrolysis into lactic acid, which is then incorporated into the Krebs cycle, results in the resorption of the material (Athanasiou et al. 1996).

PLA is a crucial biobased polyesters for food packaging that is one of its major applications because of its superior transparency, ease of processing, and availability in the market. In this context, the major drawback of PLA is its brittleness and low toughness (Auras et al. 2004). According to IUPAC, a plasticizer is a substance or material incorporated in a material that increases its flexibility, workability, or distensibility. The addition of plasticizers increases the flexibility of PLA and makes it more useful in the packaging sector (Madhavan Nampoothiri et al. 2010). Phthalates are commonly used, but health concerns about them stimulated research about the possibility to use low molecular weight, not volatile molecules such as polyols (namely: glycerol, ethylene glycol, diethylene glycol, triethylene glycol and sorbitol, mannitol and xylitol), glucose monoesters and partial fatty acid esters, and triethyl citrate as nontoxic plasticizers. To avoid the use of phthalates to mitigate the brittleness of PLA, oligomers of lactic acid (with molar mass around 1000 Da) are blended with commercial PLA to improve ductility, obtaining transparent and ductile materials able for films manufacturing. These oligomers are proposed as innovative and fully compatible and biodegradable plasticizers for PLA (Burgos et al. 2014) (Madhavan Nampoothiri et al. 2010).

As a food contact substance, PLA is one of the most attractive biopolymers with many short-term or disposable applications, such as disposable cutlery (plates, cups, lids, and drinking straws), upholstery, disposable garments, personal care products, covers, and diapers (Farah et al. 2016).

PLA is becoming one of the most popular alternatives to traditional petroleum-based plastics in automotive applications; designed plasticizers and impact modifiers confer adequate levels of ductility while maintaining rigidity and strength, making PLA an interesting competitor of mineral-filled polypropylene for automotive parts possibly subjected to serious loading and unfavorable conditions (Notta-Cuvier et al. 2014).

PLA is used as a feedstock material for 3D printers (Rajpurohit and Dave 2019). Pure PLLA is used as a long-lasting volume enhancer in dermatologic surgery (Mazzuco and Sadick 2016).

PLA-based polymeric drug delivery systems were used for custom-made 3D printing of pharmaceutical products for controlled drug release (Water et al. 2015). 3D printing of PLA biomedical devices was realized using patient-specific anatomical data (Farah et al. 2016).

Durable applications of PLA have been significantly impaired by its inherent brittleness and limited thermal stability. These points, therefore, represent the main key-parameters to be improved for its industrial implementation.

Blending or copolymerization of PLA with another polymer are used to overcome these limitations, as illustrated in Sect. 6.5.

6.3 Polyhydroxyalkanote (PHA)

PHAs are linear polyesters of hydroxyacids synthesized by a wide range of bacteria through bacterial fermentation (Reis et al. 2011). Polyhydroxybutyrate was the first member of PHA to be discovered in 1926 (Li et al. 2016b). PHAs, particularly, are natural storage polymers of many bacterial species similar to polyethylene and polypropylene. PHAs synthesis is stimulated by unbalanced growth as part of a survival mechanism of the microbes (Hassan et al. 2013b). PHAs are storage biopolymers that accumulate in bacteria cytoplasm within granules of 0.2–0.7μm in diameter, especially if bacteria are grown in a media that is limited in a nutrient essential for growth (typically nitrogen or phosphorus) or under stressful conditions. If the carbon source is abundant, bacteria convert the extracellular carbon into these intracellular storage biopolymers that are energy and carbon reserve materials (Venkateswar Reddy and Venkata Mohan 2012). PHAs are then degraded and used for growth when the limiting nutrient is available and the starvation conditions for nitrogen and/or phosphorus are over (Reis et al. 2011).

Today there are more than 150 classes of PHAs, but the presence of refractive intracellular bodies due to these biopolymers was already reported by Beijerinck in 1888. The French microbiologist Maurice Lemoigne in 1925 determined that the formula of those granular inclusions in *Bacillus megaterium* was $(C_4H_6O_2)n$ (Lemoigne 1926). Researches and patents for the production of PHAs actually started only in the '70s (Albuquerque and Malafaia 2018).

The general molecular structure of PHAs is presented in Fig. 6.8. Depending on the carbon numbers in the 3-hydroxyalkanoate units, more than 150 different PHAs monomers are known, which renders them the largest group of natural polyesters.

Among others, poly(3-hydroxybutyrate) (PHB), poly(3-hydroxyvalerate) (PHV), poly(3-hydroxybutyrate-co-3-hydroxyhexanoate) (PHBH), and poly(3-hydroxybutyrate-co-3-hydroxyvalerate) (PHBV) can be easily obtained from fermentation reactions (Sudesh et al. 2000).

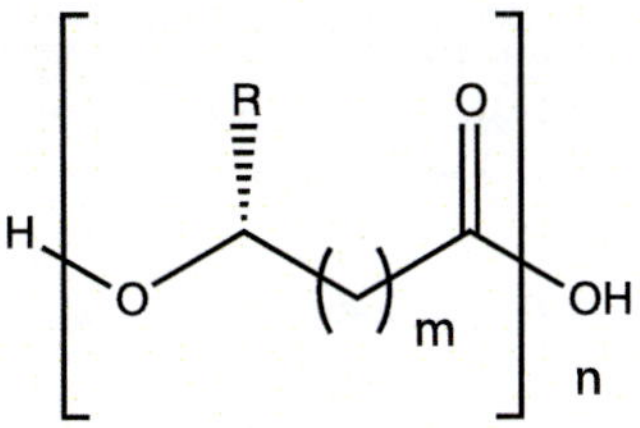

Fig. 6.8 PHA structure. Poly(3-hydroxybutyrate): R = methyl, m=1; Poly(3-hydroxyvalerate): R = ethyl, m=1; Poly(3-hydroxyhexanoate): R = propyl, m=1; Poly(3-hydroxyoctanoate): R = pentyl, m=1; Poly(3-hydroxydecanoate): R = heptyl, m=1; Poly(3-hydroxydodecanoate): R = nonyl, m=1

Over the past decade, enormous efforts have been put into the valorization of FW into PHAs, yet, technological advances to convert FW into PHAs are needed to profit from their commercialization with economic viability. The simplest and most common type of PHA is PHB.

PHAs have become one of the key drivers of the biodegradable polymers market after a long development stage; their production capacities continue to grow in the next years (Briassoulis and Giannoulis 2018).

Although 250 types of natural PHAs producers have been identified, including deep-sea bacteria (Numata et al. 2013), only a few prokaryotic microorganisms have been exploited for the commercial production of PHAs. *Alcaligenes latus, Bacillus megaterium, Cupriavidus necator, and Pseudomonas oleovorans*, are the most effective PHAs producers (Reddy et al. 2003). Marine bacteria, such as *Halomonas hydrothermalis, H. campaniensis LS21*, manifested a huge potential for bioplastic production since they do not need freshwater and can grow in FW-like mixed substrates (Tsang et al. 2019; Yue et al. 2014).

PHAs accumulating prokaryotic microorganisms can be broadly subdivided into two groups according to the chain length of the monomers they produce. Short-chain-length PHAs (SCL-PHAs) and medium-chain-length PHAs (MCL-PHAs) have monomers ranging from 3 to 5 carbons or from 6 to 16 carbons, respectively, in length.

6.3.1 FWs as Feedstocks

PHAs production cost is significantly higher than the capital cost of equipment and strongly influences the final PHAs price (Van Wegen et al. 1998). The current industrial expense for PHAs production (due to complicated bioprocessing, low conversion of carbon substrates, poor growth of microorganisms, difficult downstream separation, and high temperature fermenters sterilization) is the major hurdle for viable marketability of PHAs (Koller et al. 2005; Chen et al. 2020). Unadulterated cultures and mixed cultures are different cultivation strategies. Compared with pure culture technology, mixed culture costs less on sterilization. Genetic engineering is crucial to develop strains that are capable of efficiently producing PHAs from affordable renewable resources and contamination resistant bacteria, especially extremophiles, may push PHAs productions towards cost competitiveness and reproducible molecular weights, structures, and thermo- mechanical properties (Chen et al. 2020).

For these reasons, the substitution of virgin feedstock with FW represents a turning point because utilization of cheap FW as a carbon source seems a promising strategy. Wastewater from palm oil, olive oil, and hydrolysates of starch (e.g., corn and tapioca), cellulose, and hemicellulose are good putative feedstocks, viable from the economic point of view (Reis et al. 2011). Other food wastes used for PHAs production were banana residue (Naranjo et al. 2014), spent coffee grounds (Obruca et al. 2014a; Obruca et al. 2014b) distillery spent wash (Amulya et al. 2014), and margarine waste (Morais et al. 2014), dairy waste (RamKumar Pandian et al. 2010), whey (Amaro et al. 2019; Berwig et al. 2016), paddy straw (Sandhya et al. 2013), and other FW (Venkateswar Reddy and Venkata Mohan 2012).

Follonier et al. evaluated the potential of sugars and fatty acids derived from nine different fruit pomace and waste frying oil for the low-cost production of medium chain length PHAs by *Pseudomonas resinovorans* and waste frying oil as a resource for the bio-production of medium chain length polyhydroxyalkanoates. Apricot pomace was characterized by the lowest level of inhibitors, but the maximum PHAs production (21.3 g/L) was observed with grape pomace (Follonier et al. 2014).

Un-utilizable and toxic molasses spent wash was the feedstock for the production of PHB; a 52% removal of chemical oxidation demand was concomitant to a PHB accumulation of 28% (Khardenavis et al. 2009).

Several research groups focused on the bioplastic production cost analysis. Three case studies of biorefineries provide information on techno-economical possibilities and cost analysis of the PHAs production. Its economic viability is contingent on several factors (e.g., type, cost, and availability of waste generated in the selected area, processing and transport cost of the biomass, and general industrial expenses). The first case study focused on palm oil waste; the milling activity results in 50–70 tons of biomass waste per ha of palm plantations (Tsang et al. 2019). The waste biomass can be further utilized for the generation of biofuels for combined heat and power plants, ethanol and methane production, or composting, briquette and medium-density fiberboard construction, and pulp and paper manufacture (Chiew and Shimada 2013). Anyhow, non-food sugars produced from palm fronds (i.e., around 42.8 wt% of glucan, 12.5 wt% of xylan, and 2.3 wt% of arabinan) can be used for economical production of the most frequently studied and characterized PHA, that is PHB (Zahari et al. 2015) (Tsang et al. 2019). Its cost can be drastically reduced compared to using commercial glucose by using the renewable sugars contained in palm frond.

In the second case, the rejected and wasted proportions of total banana crops, which can be more than 30%, were used as a precious source for glucose, ethanol, and PHB production in a biorefinery (Naranjo et al. 2014; Quinaya and Alzate 2014). The study demonstrated that if the PHAs production is embedded into a multiproduct biorefinery, the economic and environmental availability of the process through energy and mass integration strategies is improved: the former can reduce up to 30.6% the global energy requirements of the process, and the latter allows a 35% in water savings. The optimal use of energy and water resources decreases both the production cost and the negative environmental impacts (Naranjo et al. 2014).

In the third case, techno-economic and environmental analysis for the production of ethanol, PHB, and electricity from sugarcane bagasse demonstrated that the use of a biorefinery could reduce the selling price of PHB by 50%, making it competitive with typical fossil plastic (Moncada et al. 2013).

6.3.2 *Biosynthesis*

PHA is biosynthesized via glycosylation or β-oxidation pathway. For example, the genes for the synthesis of PHB reside within an operon that includes a β-ketothiolase, an acetoacetyl- coenzyme A reductase, which synthesize the monomeric substrate 3-hydroxybutyryl CoA, and a PHA synthase, which polymerizes this monomer to form PHAs (Pohlmann et al. 2006). With respect to the synthesis of PHAs by the β-oxidation pathway, 2-enoyl-CoA hydratase and 3-ketoacyl-acyl carrier protein reductase are known to be responsible for converting 3-enoyl-CoA and 3-ketoacyl-CoA into (R)-3-hydroxyacyl-CoA (Fiedler et al. 2002). The PHA synthase plays a crucial role in the synthesis of high-performance PHAs because its substrate specificity and specific activity affect the monomeric compositions of the resulting PHA (Rehm 2003).

The biosynthesis process of PHA practically goes through as substrate preparation, enzymatic hydrolysis, PHA-accumulating fermentation, and PHA extraction and drying from the PHA-rich biomass (Serafim et al. 2008).

An initial acidogenic fermentation of the feedstock often proved to be very important for the PHA synthesis. The initial acidogenic fermentation is needed to make the waste suitable for PHAs production; in fact, waste carbohydrates would be stored as glycogen and not PHA; on the converse, VFA obtained in the acidogenic fermentation of carbohydrates are readily convertible into PHAs (Reis et al. 2011). A three-stage biotechnological process made use of waste-based feedstocks, and comprised an initial acidogenic fermentation of the feedstock, the second phase of selection of PHA-storing bacterial biomass, and the last phase in which PHA was stored in batch conditions (Reis et al. 2011). Integration of acidogenic fermentation with PHA production increased the biopolymer yield (Venkateswar Reddy and Venkata Mohan 2012: Amulya et al. 2015) due to the ready availability of VFA. Zhang et al. (2014) highlighted that the consumption of even-numbered VFAs or odd-numbered VFAs was correlated, respectively, with the PHB or PHV synthesis. Interconnections of biotechnological pathways enable the co-production of biofuels and bio-products, thereby maximizing FW utilization (Venkateswar Reddy et al. 2014).

6.3.3 *Properties and Uses*

PHAs have great potential as a substitute for traditional plastics since they are strong, tough, and moisture resistant; their performance regarding heat resistance and barrier properties may mimic that of PVC and PET even if they are more rigid and less thermally stable than many fossilbased plastics (Zhao et al. 2020); their characteristics are tunable by the monomeric composition, selection of microbial production strain, substrates, process parameters during production, and post-synthetic processing. They are acknowledged as better oxygen barrier (than PE and PET), better water vapor barrier (than PP), and good fat/odor barrier. Their biodegradability and rubbery-like properties, their thermoplastic properties (similar to those of the fossilbased plastics) as well as their peculiar properties have prompted their usage in various fields. They represent biological alternatives for diverse technomers of petrochemical origin. Their hydrolytic degradation products are nontoxic and demonstrate their potential for applications in the medical sector. Similarly to PLA, PHAs can be biodegraded via composting and under anaerobic conditions: in the first case, the process lasts less than four months, while only two weeks are needed for the anaerobic biodegradation; biodegradability is less than and 10% and 50% in aquatic and soil environments, respectively, after one year (Folino et al. 2020).

PHA can be processed with usual methods such as injection molding, extrusion, thermoforming, film blowing, etc. For example, PHA can be transformed into injection-molded items such as film and sheet, fibers, laminates, and coated articles. There is a broad range of PHA-based marketable products for a variety of applications, including food packaging, disposable bottles, and water-resistant film, paints, adhesives, nonwoven fabrics, synthetic paper products, waxes, foams. They are important for agroindustrial applications (carriers and matrices for controlled release of nutrients, fertilizers, and pesticides), therapeutic applications (controlled release of active pharmaceutical ingredients), as well as for medical and pharmaceutical applications (bone, nerve, cardiovascular, and cartilage tissue engineering, 3D constructs, implants, chiral substrates for drug synthesis and drug delivery) (Van Wegen et al. 1998; Ahmed et al. 2018; Chen and Patel 2012; Liang and Qi 2014; Koch and Mihalyi 2018).

Unfortunately, the applications of PHAs are restricted due to several undesirable physical properties.

Due to poor PHA mechanical properties and the high costs for their microbial production, insufficient production volume, relatively slow crystallization rates, narrow thermal processing window, high hydrophobicity, and high brittleness, PHA exploitation was mostly limited to academic research (Chen et al. 2013), and the utilization of PHAs as direct substitutes for synthetic plastics was a major challenge.

For example, large crystals in PHB cause poor mechanical properties; its thermal processing window is narrow because the melting and decomposition temperatures are similar, and this renders PHB susceptible to thermal degradation. Modification of PHA properties by side-chain adjustment does not tackle the optimization of the properties efficiently, hence blending or copolymerization was also explored (Li et al. 2016b).

Copolymerization, blending, and composite making have been explored in order to modify PHA properties; in this context, the reduction of hydrophobicity and the increase of biodegradability are the focus of interest of biomedical application research (Samui and Kanai 2019). Cost competitiveness and reproducible molecular weights, structures, and thermo- mechanical properties can be obtained by genetically engineered strains and contamination resistant bacteria (Chen et al. 2020).

6.4 Polybutylene Succinate (PBS)

Succinic acid and 1,4-butanediol are building blocks for synthesizing PBS (Su et al. 2019), whose structure is illustrated in Fig. 6.9.

PBS is a well-known biobased thermoplastic aliphatic polyester with interesting thermo-mechanical properties comparable to polypropylene, the second largest polymer (~55million t/year worldwide); thus the potential application area is enormous (EU 2015) also taking into account its proven biodegradability and compostability (according to DIN EN 13432). However, PBS would hardly be a polypropylene competitor due to the cheapness of the latter (EU 2015). PBS is used for blending with starch polymers to improve properties.

In the 1930s, Wallace Hume Carothers, capitalized on previous work by Lourenço (1863), Davidoff (1886), and Voländer (1894) about the esterification of succinic acid and ethylene glycol and begun a more systematic study of succinic acid based polyesters. The elimination of water on a continuous distillation process leads to polymers with molar masses significantly higher than what was previously synthesized (Carothers and Arvin 1929). The study was prompted by the need to find a synthetic alternative to natural silk fiber; since the properties of the final products

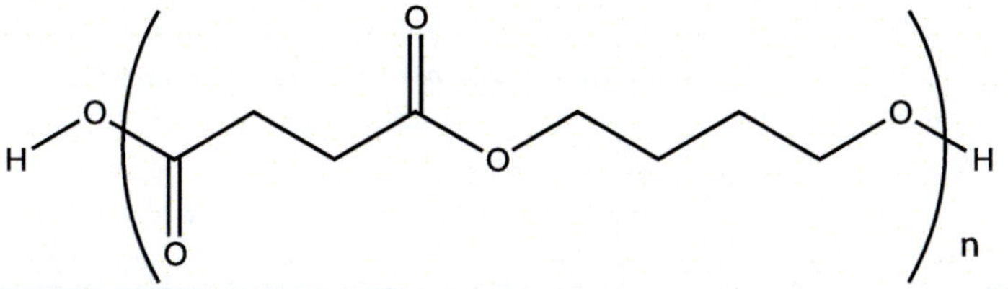

Fig. 6.9 PBS structure

did not show the expected qualities, Carothers focused on polyamides and invented with his colleague Julian Hill Nylon 6,6.

At the beginning of the 1990s, after being forgotten for more than 40 years (Flory and Leutner 1952), these polymers came into the spotlight due to the increasing demand for biodegradable and biobased polymers. PBS is commercially available since 1993 (Gigli et al. 2016).

6.4.1 FW as Feedstocks

For a complete biobased PBS, its biobased building blocks, namely succinic acid and 1,4-butanediol, are required.

Succinic acid was recognized as a key sugar-based building block; according to the U.S. Department of Energy, it is a platform molecule with high-value niche applications such as personal care products and food additives, but also large volume applications such as bio-polyesters, polyurethanes, resins, and coatings. Moreover, it can be used as the precursor of various commodity chemicals, including 1,4-butanediol, which is the other building block of PBS, but also tetrahydrofuran and γ-butyrolactone, which have a wide range of applications ranging from detergents to pharmaceuticals (Werpy and Petersen 2004; Bozell and Petersen 2010; EU 2015). Succinic acid has mostly been produced by chemical processes from fossil feedstocks, via hydrogenation of maleic anhydride, but the biotechnological production of succinic acid from renewable biomass by anaerobic bacteria came into the limelight because succinic acid is an intermediate of the tricarboxylic acid cycle; it is a dicarboxylic acid produced as a fermentation product of anaerobic metabolism (Gottschalk 1986). Various FW have been explored as succinic acid feedstock, and several microorganisms have been screened and tested for the biotechnological production of succinic acid with good yields, as discussed in Sect. 5.4.2 (Bechthold et al. 2008).

1,4-Butanediol is a colorless, viscous liquid industrially used as a solvent in industrial cleaners and in the manufacture of valued polymers such as polybutylene terephthalate, synthetic polyether polyurea fibers, and polyurethanes. It was first commercially manufactured in 1930 through the reaction of formaldehyde with acetylene and subsequent hydrogenation (Reppe process), and this process still accounts for about 42% of global production. Anyhow, in the petrochemical industry, 1,4-Butanediol can be produced in various ways from maleic anhydride, propylene oxide, and butadiene (EU 2015). Biobased 1,4-Butanediol is regarded as a direct drop-in chemical. Its production can either take place via direct fermentation of sugars or via the hydrogenation of biobased succinic acid. The biobased 1,4-butanediol manufacture process is not performed worldwide on a large scale yet, but its production is increasingly supplemented by biotechnological processes; recently, a genetically modified *Escherichia coli* strain was used under microaerobic conditions to convert glucose into 1,4-Butanediol (Pooth et al. 2020). Metabolic engineered bacterial synthesis represents an emerging and advantageous technology, alternative over the succinic acid hydrogenation route, with a lower risk of market fragmentation, thanks to the use of more abundant and low-cost feedstock, such as wheat straw (Forte et al. 2016). Although a complete sugar-based PBS can be made, nowadays, it is mostly fossilbased; oil price fluctuations and consumers' inclination for biobased products will probably influence future PBS production (Zhao et al. 2020).

6.4.2 *Synthesis*

A two steps esterification of succinic acid with 1,4-butanediol is the most common way to produce PBS. First, an excess of the diol is condensed with the diacid to form PBS oligomers with the elimination of water. Secondly, these oligomers are trans-esterified, in the presence of titanium-, zirconium-, tin- or germanium-based organometallic or metal-oxide catalysts, under vacuum, to form a high molar mass polymer. Titanium(IV) butoxide is the most common catalyst (Su et al. 2019). PBS can be produced via melt polycondensation at moderate costs.

Diisocyanates, typically hexamethylene diisocyanate, are used as chain extenders to obtain high molecular weight PBS, reducing brittleness and increasing elongation at break (Gigli et al. 2012b).

The enzymatic synthesis of PBS made use of lipase from *Candida Antarctica* (Azim et al. 2006). The ability of lipases to catalyze the direct polycondensation of a linear aliphatic diacid and a diol increases with increasing the monomer chain length. The diffusion of this greener route was hampered by enzyme leaching and inactivation, and the use of solvents to avoid polymer precipitation (Gigli et al. 2016). Further optimization of the process is likely to reduce biobased PBS price.

6.4.3 *Properties and Uses*

PBS is a semicrystalline polymer with high crystallization ability (35–45%) (Soccio et al. 2008), and its melting temperature (115 °C) (Gigli et al. 2012b; Xu and Guo 2010) is important for applications that require a high temperature range. Noteworthy, its mechanical performance mimics that of PP and PE (Zhao et al. 2020). Due to a glass transition temperature much lower than room temperature, PBS has a wide processing/temperature window (Fabbri et al. 2014).

Its wide temperature window for thermoplastic processing makes the resin appropriate for extrusion, injection molding, thermoforming, and film blowing; however, its stiffness and melt viscosity for processing are often insufficient for various end-use applications (Gigli et al. 2016).

PBS commercialization is mainly devoted to biodegradable food packaging (tea cups, plates and bowls, coffee capsules, food service ware, and catering products) and compostable bags, and other disposable items; PBS is also used in agriculture and fishing sectors for the production of mulching films and fishing nets, respectively; nonwoven textiles, diapers, and foams in PBS were also produced. PBS is also significant in the automotive sector (wood-PBS composites, composites with natural fibers), in electronics and other consumer goods applications. PBS and its copolymers are particularly expedient in biomedicine emerging topics, such as controlled drug release by microspheres or nanocarriers and tissue engineering via films and scaffolds (produced by salt leaching, electrospinning, or extrusion) but only a few examples of in vivo studies regarding PBS and PBS-based systems are present in the literature; hence more exploration on its biocompatibility is needed to validate the possibility to employ this polymer in biomedicine (Gigli et al. 2016).

Materials must fulfill strict requirements to be employed as biomaterials regarding biocompatibility, mechanical properties, and biodegradation rate (Vert 2009). The quite slow hydrolysis rate under physiological conditions (due to the high crystallinity and hydrophobicity of this polymer (Soccio et al. 2012)) and low flexibility of PBS could hinder its uses for certain applications; therefore, blending and/or copolymerization have been used to fine tune surface wettability, degree of crystallinity, mechanical properties and biodegradation rate; chitosan and hydroxyapatite, are important fillers in this respect (Gigli et al. 2016). Polybutylene Succinate Adipate (PBSA) is one of the most promising PBS copolymers.

As regards PBS biodegradation, *Amycolatopsis* sp. HT-6 and several thermophilic actinomycetes such as *Microbispora rosea*, *Excellospora japonica*, and *E. viridilutea* can degrade PBS (Tokiwa et al. 2009). PBS can be biodegraded by composting (90% in three months), but the biodegradation in anaerobic (landfill), soil, and aquatic environments is not easy (Folino et al. 2020). PBS ionomers with improved hydrodegradability containing sulfonated succinate units have been synthesized and characterized (Bautista et al. 2015).

Table 6.1 illustrates exemplary physical properties of PBS, PBSA, PLA, PHA (explained in Chaps. 8 and 9) compared to fossilbased polymers: it has to be emphasized that reported data may be different from other literature data for the same physical properties of the same material because of the difference in molecular weight, production processes, and so on.

6.5 Polyglycolic Acid (PGA)

Polyglycolic acid is the simplest polyester; it is a thermoplastic and biodegradable linear, aliphatic biopolymer. Similarly to PLA, it can be prepared starting from glycolic acid by means of polycondensation or ROP. PGA has been known since 1954, and since it is a tough fiber-forming polymer, in 1962, it was used to develop the first synthetic absorbable suture (Gilding and Reed 1979).

Table 6.1 Exemplary physical properties of PBS, PBSA, PLA, PHA (explained in Chaps. 8 and 9) compared to fossilbased polymers (Xu and Guo 2010; Weise et al. 2018; Farah et al. 2016; Su et al. 2019 ; Rivero et al. 2016; Johansson et al. 2012; Kim et al. 2005; Totaro et al. 2016; Bugnicourt et al. 2014; Chaiwutthinan et al. 2013; Avérous and Pollet 2012; Ojijo et al. 2012b; Sasa et al. 2011; Gigli et al. 2016; Numata et al. 2013; Bugnicourt et al. 2014; Zhao et al. 2020; Södergård and Stolt Stolt 2002; Middleton and Tipton 2000)

	PBS	PBSA	PLA	PHA	HDPE	LDPE	PS	PP
Glass transition temperature (Celsius)	−32	−45	55–65	2	−120	−120 to −40	105	−5
Melting point (Celsius)	114–115	83.1−96	140–180	160–175	129	110	Amorphous	163–186
Heat distortion temperature (HDT-B, Celsius)	97	69	55	–	82	49	95	110
Tensile strength (MPa)	26.5–34	19	50–70	15–40	24–32	10–12	24–60	33–34.5
Tensile modulus (GPa)	0.71	0.16	2.7–16	1–2	0.5	0.3	3.4	1.4
Elongation at break (%)	8–560	807	4–7	1–17	150–700	300–500	3–4	100–415
Flexural strength (MPa)	36.5	–	48–110	–	21	6	80	49
Notched Izod impact (J/m)	300	>400	29	22–35	40	>400	19–150	20
Degree of crystallinity (%)	34–45	20–30	0–40	40–70	69	49	0	56-60
Density (g/cc)	1.22	1.23	1.25	1.17–1.25	0.95	0.92	1.05	0.91–0.94

6.5.1 FWs as Feedstocks

The monomer of PGA is glycolic acid (GA); it is a platform chemical that is currently produced by petrochemical synthesis, in a process where formaldehyde is carbonylated, but its biotechnological and sustainable production from renewable resources is receiving substantial interest.

GA is formed from the glyoxylate shunt route. D-glucose, D-xylose, ethanol, and acetate have been experienced as substrates by using hosts *E. coli*, *C. glutamicum*, *S. cerevisiae,* or *K. lactis*. Genetic modifications aim to enhance the flux from isocitrate to GA via accumulation of isocitrate (through deletion of genes encoding isocitrate dehydrogenase and malate synthase) or overexpression of isocitrate lyase and glyoxylate reductase encoding genes (Salusjärvi et al. 2019).

It is important to design metabolic pathways that make use of naturally producing glycolic acid microorganisms (chemolithotrophic iron-and sulfur-oxidizing bacteria, *Alcaligenes sp.* ECU0401, and a variety of yeast and acetic acid bacteria) and enable utilization of several carbon sources at the same time from various renewable feedstocks as FW streams rich in sugars and starch, lignocellulosic waste, and even gaseous substrates (CO_2, CO and H_2) (Salusjärvi et al. 2019; Jha and Kumar 2019).

6.5.2 Synthesis

Low molecular weight PGA was first synthesized in 1932 by Carothers (Carothers et al. 1932). In the 1950s, DuPont patented a method for synthesizing high-molecular-weight PGA (Lowe 1954). It was known as a tough fiber-forming polymer. Nowadays, it is an extremely expensive, high value-added product; it follows that technology to permit inexpensive mass production is highly desirable.

High-molecular-weight PGA is commonly synthesized via a ROP from highly purified glycolide (GL) in the bulk condition since polycondensation starting from glycolic acid would only give oligomers, as shown in Fig. 6.10. Unfortunately, the viscosity of the generated polymer increases with increasing polymerization time as the chain length increases during the reaction. Agitation difficulties impair the efficiency of the method, which needs to be manufactured in batches using small-scale apparatuses. A process to obtain high-molecular-weight PGA continuously from the intermediate GL, with high levels of purity, was experienced in a commercial production plant at a lower cost than previously possible using a temperature range

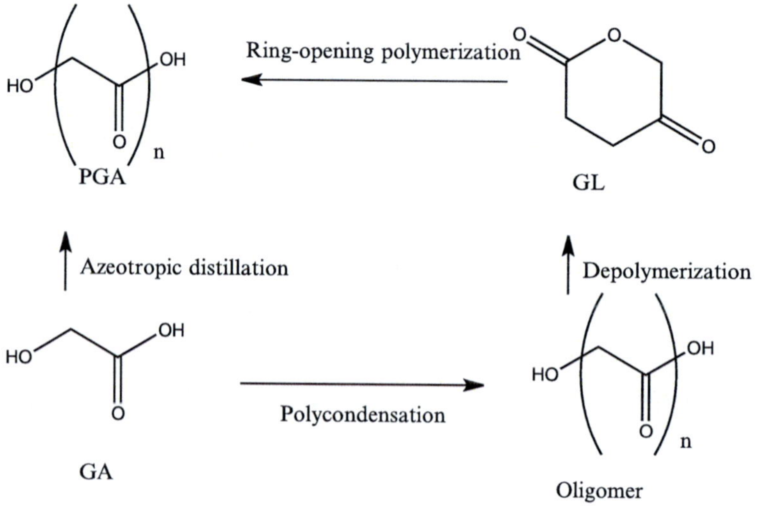

Fig. 6.10 PGA preparation technologies

between the melting point of the GL (~90 °C) and the melting point of high-molecular-weight PGA (~220 °C). This way, the polymerization is firstly induced in the molten state, then the polymer is precipitated as a solid, and polymerization is promoted in the solid state so that high-molecular-weight PGA is generated (Yamane et al. 2014).

The ROP was considered restrictive because of the high cost of the raw material (glycolide) and the associated high energy consumption. The direct synthesis of PGA via an azeotropic distillation (Fig. 6.10) parallels that for the PLA production; it enables tunable molecular weights, better solubility in organic solvents, and faster degradation than reference PGA (Sanko et al. 2019).

The cationic alternating copolymerization of formaldehyde (from trioxane) and carbon monoxide (CO), a sustainable C1 feedstock obtainable from biomethanol or biogas, constitutes an inexpensive and efficient pathway for the synthesis of PGA, by-passing the standard route involving glycolide. PGA was successfully synthesized with yields up to 92% from trioxane and CO, with triflic acid initiator at 170 °C over 3 days. The chemico-physical characterization of the polymer via ^{1}H NMR, ^{13}C NMR, and FT-IR spectra, is consistent with that of the commercial one, prepared from the ROP of glycolide. Molecular weight analysis suggested the formation of oligomers, which can be converted into high molecular weight PGA *via*: (i) their post-polymerization polycondensation, catalyzed by $Zn(OAc)_2 \cdot 2H_2O$; (ii) inclusion of glycerol as a branching agent during the C1 copolymerization (Göktürk et al. 2015).

6.5.3 Properties and Uses

The glass transition temperature of polyglycolide is between 35 and 40 °C and its melting point is in the range of 225–230 °C. The high degree of crystallinity, around 45–55%, results in its insolubility in water and almost all common organic solvents (Budak et al. 2020). It offers high mechanical strength and high gas-barrier performance. The ester linkage in the PGA backbone provides hydrolytic instability. The degradation process is erosive: the polymer is converted back to its monomer glycolic acid due to water diffusion firstly into the amorphous (non-crystalline) regions and then into the crystalline regions of the polymer matrix, cleaving the ester bonds. Under physiological conditions, PGA is hydrolyzed, and it is also broken down by certain enzymes, especially those with esterase activity; hence biodegradation is faster in vivo than in vitro. The glycolic acid monomers can enter the tricarboxylic acid cycle, or they can be excreted by urine (Wada 2007).

PGA finds important applications in the biomedical practice because of its expedient mechanical, biocompatible, and biodegradable properties (Budak et al. 2020). Its hydrolytic instability is a drawback for many applications, but currently, polyglycolide and its copolymers with lactic acid, ε-caprolactone, and trimethylene carbonate are widely used for the synthesis of absorbable sutures, and are being evaluated for other applications in the biomedical field. The bioresorption /

degradation process in the body into water-soluble absorbable monomers lasts a few weeks depending on the PGA molecular weight; it has been explored for implantable medical devices (anastomosis rings, pins, rods, plates, and screws) and tissue engineering or controlled drug delivery. PGA nerve conduits proved to be an effective artificial biodegradable peripheral nerve regeneration-inducing tube (Fujimaki et al. 2019). Production of high molecular weight PGA is needed to achieve sufficient mechanical stability for biomedical applications (Sanko et al. 2019) and for food packaging applications; in this case, tin (II) based catalyst needed to give high molecular weight PGA should rather be removed from the produced PGA since tin (II) compounds are known to be toxic (Budak et al. 2020). Copolymerization and blending are often used for a fine tuning of PGA properties, as described below. The use of ultra-strong and biodegradable PGA in shale gas and oil exploration is of great interest (Yamane et al. 2014).

6.6 Biobased Thermosetting Polymers

6.6.1 *Phenolic Resins*

Phenolic resins are synthesized from phenol (or substituted phenols) with formaldehyde. They were the first commercial synthetic plastics (Bakelite), widely used for the production of molded products, including billiard balls, coatings, and adhesives, and the first circuit boards.

There are many variations in both production and input materials that are used to produce a wide variety of resins for special purposes.

Biobased phenols from FW have been studied for fully biobased phenolic resins. Cashew nut shell liquid, an agricultural by-product of cashew *(Anacardium occidentale)* processing, is abundantly available in tropical countries. It represents one of the major and economical resources of naturally occurring phenols. Cashew nut shell liquid is a reddish-brown viscous fluid containing four major components, namely cardanol, cardol, anacardic acid and 2-methylcardol (Voirin et al. 2014). They are phenolic lipids, illustrated in Fig. 6.11, with varying degree of unsaturation in C15 alkyl side chain at the meta position.

Cardanol can react with formaldehyde to form novolac resin (using excess phenol and an acid or metal salt catalyst) or resole resins (using excess formaldehyde and a basic catalyst) with improved flexibility compared with conventional analogs (Quirino et al. 2014) due to the side chains. They also impart a hydrophobic nature to the polymer, making it resistant to moisture and weathering.

Chemical modifications of cardanol can be made at the reactive phenolic group, or on the mono/di/tri unsaturated side chain in the meta position; π-π stacking and functionalization of the aromatic ring have also been studied and a wide gamut of cardanol-based amphiphiles and other functionalized monomers are reported for the synthesis of fine chemicals and polymers (John et al. 2019).

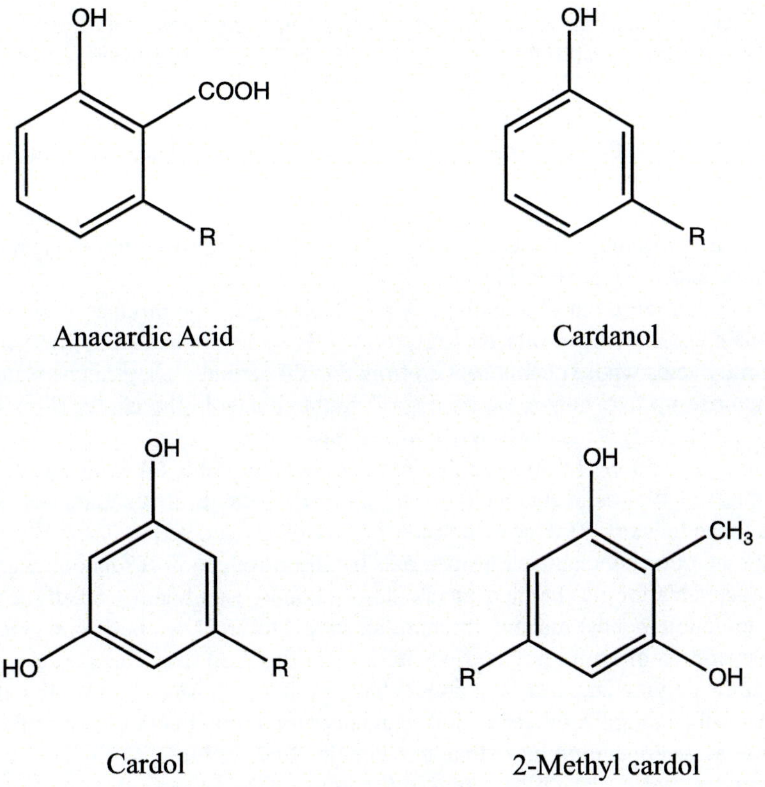

Fig. 6.11 Phenolic compounds present in cashew nut shell liquid

Lignin is a by-product derived from the wood pulp in the paper industry, but it is also contained in many agro-industry wastes. Up to 98% of overall lignin is utilized as a (bio)fuel. Due to its phenolic moieties (Laurichesse and Avérous 2014), lignin is a good candidate as another natural substitute for phenolic building block for phenolic based resins with innovative macromolecular architectures (Kouisni et al. 2011).

Tannin compounds extracted from vegetal sources can also be considered renewable building blocks in the design of phenolic resins (Barbosa et al. 2010).

6.6.2 *Epoxy and Polyurethane Resins*

Epoxy resin is a thermosetting material whose synthesis relies on the high reactivity of the epoxy moiety to many functional groups such as carboxyl, amino, and hydroxyl groups. Polyepoxide networks represent a major class of thermosetting polymers. Owing to their singular mechanical and electrical properties, chemical resistance, and minimal shrinkage after curing, they are extensively used as coatings, in electronic materials and integrated circuit boards, in adhesives and structural carbon fiber adhesives, as in structural applications as well.

Most epoxy resins are industrially manufactured from bisphenol A via the glycidylation route with epichlorohydrin to produce the diglycidyl ether of bisphenol A. Bisphenol A is considered an endocrine disruptor; its toxicity, due to significant hormonal activity, and legal issues stimulate an increasing number of studies on its replacement with natural sources of epoxy monomers both from the academic and industrial sides (Ng et al. 2017).

Waste oils represent a potential low-cost material for the production of resins. The main constituents of oils are triglycerides. They are formed from glycerol and three fatty acids whose chains contain from 8 to 24 carbons. They can be saturated or comprise up to 5 non-conjugated C=C double bonds in the cis- configuration. Longer chain fatty acids characterize marine fats.

The composition of fatty acid differentiates one vegetable oil from another; for example, ricinoleic acid characterizes castor oil, a vegetable oil of significant importance to the industry. The most common fatty acids are detailed in Fig. 6.12.

Oils are quite nonreactive raw materials but the introduction of functional groups at sites suitable for the chemical attack (double bonds, ester bonds, and allylic positions to double bonds) make them more reactive. Different synthetic strategies for the preparation of useful polymeric materials starting from waste oils are possible. The direct polymerization of the double bonds suffers the lack of active functional groups, but heating the oil above 300 °C induces the formation of conjugated diene (if fatty acids containing more than one double bond are present); the presence of dienophiles makes the process proceeds through Diels-Alder addition (Petrović 2010); anyhow this kind of triglyceride-based materials is not very useful for structural applications.

For this reason, triglycerides are often chemically modified and then polymerized. Vegetable oils can be converted into useful polymerizable monomers by epoxidation with a peroxy acid (Prilezhaev-epoxidation), catalytic epoxidation via acidic ion exchange resin, metal-catalyzed epoxidation, or chemo-enzymatic epoxidation. The latter epoxidation method is safe, environmentally friendly, and has a high conversion rate of epoxidation (90%) (Tan and Chow 2010).

Soybean oil, fish, and castor oil can be epoxidized at the C=C double bonds. Figure 6.13 illustrates epoxidized linolein, one of the major components of epoxidized soybean oil.

Epoxidized oils as such are used as plasticizers and hydrochloric acid scavengers in PVC items such as cling film for wrapping foods and toys, among others.

Anyhow, the importance of epoxidized vegetable oils (soybean and linseed oils) stems from the fact that they are among the few biobased epoxy building blocks that reach industrial-scale production and can replace the glycidyl ether of bisphenol A, which is the main epoxy monomer.

The possibility of using epoxidized vegetable oils as sustainable substitutes of epoxy monomers for biobased epoxy thermosetting polymers (Petrović 2010) has

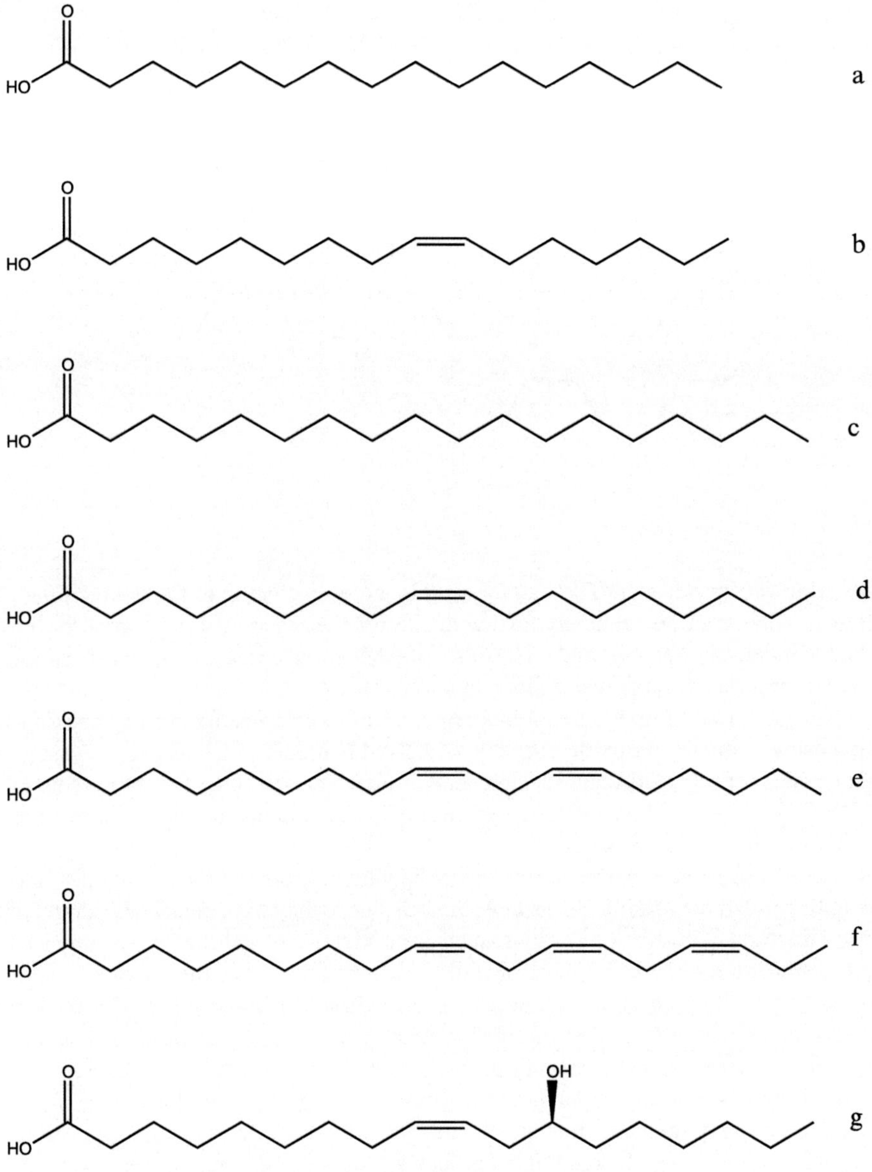

Fig. 6.12 Common fatty acid found in vegetable oils. (**a**) palmitic (C16:0); (**b**) palmitoleic (C16:1); (**c**) stearic (C18:0); (**d**) oleic (C18:1), (**e**) linoleic (C18:2); (**f**) linolenic acids (C18:3), (**g**) ricinoleic acid (C18:1, OH)

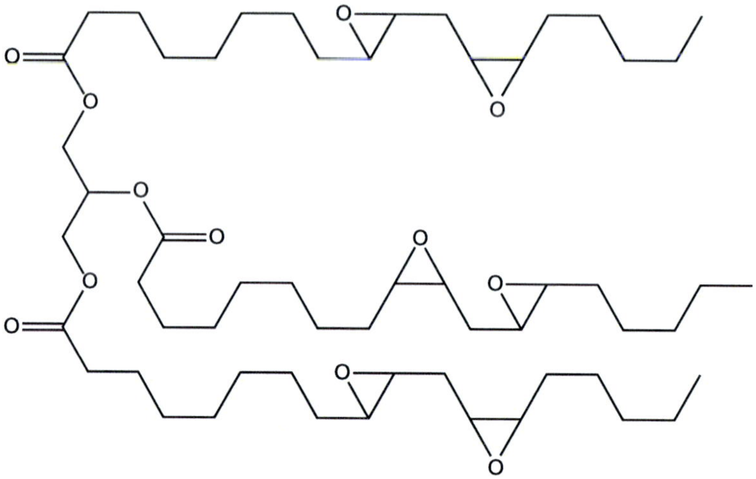

Fig. 6.13 Epoxidized linolein

been known for decades. The oxirane ring is more reactive than the double bond, thus providing an energetically favorable site for the reaction. Epoxidation of the double bonds of waste oils and on the *meta* alkyl chain of cardanol was also used for the preparation of epoxy resins (John et al. 2019).

Novel biobased epoxy resins were prepared from waste vegetable oil by epoxidation using hydrogen peroxide in acetic acid. Acid hardeners such as citric acid, tartaric acids, sebacic acids, and succinic anhydride were used. Thermal stability up to 350 °C, biodegradability, and biocompatibility based on blood clotting studies were demonstrated (Yemul et al. 2016).

Unfortunately, due to the lack of cycloaliphatic or aromatic rigid structures, the resulting polymers usually have a low strength; hence they have been considered for non-structural, additive applications, such as plasticizers or stabilizers. The replacement of the diglycidyl ether of bisphenol A, which allows covering all applications by a single biobased epoxy monomer, is still difficult. Biobased rigid aromatic epoxy monomers are needed in order to enable replacement of bisphenol A and obtain a compromise between easy processing and good properties; in this respect, natural sources of aromatic moieties, bearing reactive groups suitable for the introduction of epoxy moieties, include (i) eugenol extractable from natural plant oils, (ii) polyphenolic cross-linked polymers such as tannins and lignin that can be functionalized or depolymerized into smaller fragments prior to epoxidation steps, (iii) terpenes, terpenoids and their synthetic derivatives, such as carvacrol, which can be synthetically obtained from limonene or directly isolated from oregano and thyme essential oils (Ng et al. 2017).

Recently a fully biobased epoxy resin from fatty acids and lignin was designed to counterbalance for that deficit of stiffness: lignin reacts with the epoxidized vegetable oil, and it gets incorporated in the structure; epoxy groups still available after the reaction were cured, as usual, upon addition of a diamine (Ortiz et al. 2020).

Glycidyl esters of epoxidized fatty acids derived from soybean oil and linseed oil proved to yield mechanically stronger polymerized materials than materials obtained using epoxidized oils. They were synthesized using epichlorohydrin to have higher oxirane content and increased reactivity. They have the potential for fabrication of structural epoxy matrixes (Wang and Schuman 2012).

Sugar-based glucopyranoside- and glucofuranoside-based renewable epoxy resin were also investigated: they can be useful if bending stresses higher then the tensile ones are required (Niedermann et al. 2015).

Epoxidized vegetable oils can also undergo oxirane ring-opening with hydrochloric acid, hydrobromic acid, methanol, hydrogen, and water; hence, the resulting polyol structures can be halogenated, methoxylated, and vicinal hydroxy groups are also possible (Guo et al. 2000; Petrović 2010). These variegated biobased polyols can further react with diisocyanates made from petroleum-resources to provide polyurethanes comparable in many aspects with those obtained from petrochemical polyols (Lligadas et al. 2010).

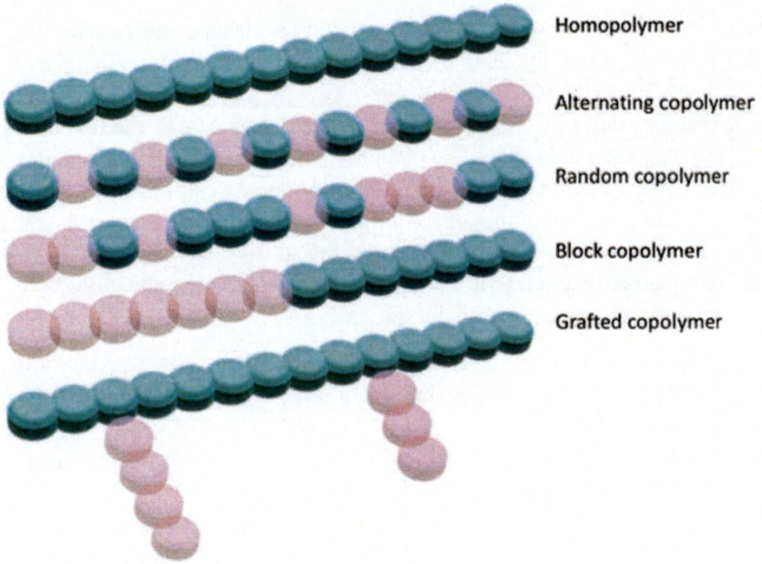

Fig. 6.14 Common types of polymers

6.7 Copolymerizarion and Blending

Processability and tailor-made modifications of the properties of polymers have attracted the attention of both academia and industry to meet specific needs. Copolymerization and blending and are the frequently used methods to alter the properties of polymers.

Copolymerization is the polymerization of more than one monomer species into copolymers; it follows that copolymers consist of at least two types of constituent units. Copolymers architecture can be linear or nonlinear, as illustrated in Fig. 6.14.

Linear copolymers are classified as alternating, periodic, random (statistical), and block copolymers according to the composition and arrangement of the repeating units. In stereoblock copolymers, the units differ only in the tacticity of the monomers.

Nonlinear copolymers are classified as branched copolymers with varying architectures. Graft copolymers are characterized by side chains structurally distinct from the main chain. For example, the main chain repeating unit is the monomer, and branches are formed from another species monomer: the side-chains have constitutional or configurational features that differ from those in the main chain. Moreover, the individual chains of a graft copolymer may be homopolymers or copolymers.

Blending is obtained when at least two thermoplastics macromolecular substances (polymers or copolymers) are mechanically mixed together in the melt state to create a new material with different and hopefully improved physical properties. Polymer blending aims at the suppression of the drawbacks of the parent components. Normally, the physical and mechanical properties of polymer blends can be tailor-made by the judicious selection of the components, optimization of the blend composition and adequate tuning of the preparation conditions. Plastic blends parallel metal alloys and their development represents an easy and cost-effective method of developing versatile polymeric materials for commercial applications.

Polymer blends can be broadly classified as:

- heterogeneous (immiscible polymer blends): in this case, that is the most common, two glass transition temperatures will be observed because the blend is not a single-phase structure
- homogeneous (miscible polymer blends): the single-phase structure results in a single glass transition temperature.
- compatible polymer blends: they are immiscible polymer blend that exhibits macroscopically uniform physical properties because of the strong interactions between the component polymers.

Physical blending is often challenged by the compatibility of most polymer pairs; the performance may be impaired by blending, and quantitative analyses of miscibility-structure-properties correlations (Imre and Pukánszky 2013) are needed to get insights into the chemistry of the process.

According to the rule of mixtures (Eq. 6.1), the properties of homogeneous polymer blends (σb) are obtained from the linear combination of the respective properties of the components (σ1 and σ2) multiplied by their volumetric fraction in the blend (f represents the volumetric fraction of component 1):

$$\sigma b = f\sigma 1 + (1 - f)\sigma 2 \tag{6.1}$$

The mechanical behavior of immiscible polymeric blends generally deviates from the rule of mixtures.

Most polymer pairs are thermodynamically immiscible because of their poor interactions. Compatibility in such polymeric systems is a function of the chemical structure and interfacial interactions. The vast majority of biopolymer-based blends available on the market are compatibilized. Different strategies of compatibilization can be distinguished. In the case of nonreactive compatibilization, pre-made amphiphilics, (Imre et al. 2019) able to interact with the respective polymers in both phases, are often used. The compatibilizer is often a diblock or triblock copolymer with a chemical structure identical to that of the main components, which exerts its activity at the components interface in order to enhance interfacial adhesion and to decrease the interfacial tension. Similarly, the use of low molecular weight compounds, ionomers, or a third polymer at least partially miscible with both blend components has also been explored (Imre and Pukánszky 2013).

Reactive methods are more efficient than nonreactive compatibilization because the blocky structures possess reactive groups that act as compatibilizers during blending (Imre and Pukánszky 2013). Reactive groups can eventually be created by the chemical modification of one of the components. Biopolymers often contain various reactive groups; hence the reactive compatibilization is plausible and convenient. The addition of a compound reactive towards the functional groups of one component and miscible with the other component results in the in situ formation of grafted or block-copolymers acting as versatile compatibilizers. The grafting of an element with anhydrides, resulting in the formation of carboxylic groups prone to react with hydroxyls in the other component, is a typical example. Unsaturated anhydrides, and maleic anhydride, in particular, are very useful in this respect.

Recently, nanofillers proved to be expedient to improve the phase morphologies of immiscible blends and act simultaneously as compatibilizers and nanoreinforcements. They can also provide tailor-made barrier, thermal, and mechanical properties, as well as shape memory (Mochane et al. 2020).

The commercial development of new copolymers and blends of biobased polymers seems endless. In the following, we particularly turn our attention to copolymers and blends of fully biobased components. We limit our analysis to a few examples of recent advances in this hotly debated field of research.

6.7.1 Copolymers

PLA copolymers are intensively studied. They are generally synthesized by ROP of LA and other monomers, initiated with hydroxyl- and amine-functionalized initiators.

Copolymers of glycolic and lactic acids (PGLA) are extensively used in surgical sutures, in tissue engineering, and the development of drug delivery systems.

The ROP of LA and GA requires harsh conditions and tin (II) 2-ethylhexanoate as the catalyst to increase the copolymer randomness. The higher reactivity of glycolic acid would result in longer glycolic acid blocks and a decreased elasticity of the polymer. The replacement of the tin-based catalyst with nontoxic 'biometals' such as magnesium or zinc analogs was successfully studied (Nifant'ev et al. 2018).

Enantiomeric alternating Poly(L-LA-co-GA)] and Poly(D-LA-co-GA) with an average molar mass of 5×10^3 g/mol were reported. Wide-angle X-ray diffractometry indicated that the stereocomplex crystalline modification is different from those reported for enantiomeric homopolymers of PLLA and PDLA. The stereocomplex formation did not largely influence the FTIR peak shape and frequency. The sufficiently high melting temperature values, the high mechanical performance, and high thermal stability suggest that stereocomplex crystallization between enantiomeric alternating lactic acid-based copolymers is a versatile method for the preparation of high-performance biodegradable materials with a wide range of physical properties (Tsuji et al. 2019).

Graft copolymerization of L-lactide onto chitosan using $Ti(OBu)_4$ as a catalyst in DMSO at 90 °C in nitrogen atmosphere gave graft copolymers with increased hydrophilicity and biodegradation (Luckachan and Pillai 2006).

Other studied cyclic monomers for LA copolymerization include caprolactone, a cyclic carbonate monomer, dipropargyl glycolide, and allyl glycidyl (Oh 2011; Gaucher et al. 2010).

For successful biological and biomedical applications of PLA and its copolymers, one serious challenge is associated with their hydrophobicity. The hydrophobicity of PLA may impair its use in sutures, implants for bone fixation, drug delivery vehicles, and tissue engineering scaffolds. This issue was addressed by the synthesis of PLA-based copolymers comprising amphiphilic monomers such as poly(meth) acrylates, poly(ethylene glycol), polypeptides, polysaccharides, and polyurethanes (Oh 2011). These PLA amphiphilic block copolymers are synthesized by radical, ring- opening, or condensation polymerization methods. They are of utmost interest as drug delivery and imaging platforms of self-assembled nanoparticles and tissue engineering of cross-linked hydrogels. In fact, the carboxylic acid (–COOH), amino ($–NH_2$), and hydroxyl (–OH) functional groups are utilized for bioconjugation with targeting biomolecules to achieve active targeting of specific cells.

A new stereo pentablock copolymers (PLLA-PDLA-PBS-PDLA-PLLA) was synthesized by two-step ROP of D- and L-lactides in the presence of bis-hydroxyl-terminated PBS prepolymer that has been prepared by the ordinary polycondensation. The direct connection of the PLLA and PDLA blocks allows easy formation of the stereocomplex crystals. At the same time, the introduction of the semicrystalline PBS block is effective not only for altering the crystallization kinetics but also for conveying an elastomeric property (Lee et al. 2016).

Copolymerization of PHA solve the drawbacks of PHA blending with natural raw materials such as starch, cellulose, and lignin or with other bioplastics such as PLA: blending enables the facile modification of physical and mechanical properties; anyhow these properties would fall within the extremes yielded by the blended polymers, and this represents a limitation; furthermore, the blending of polymers is complicated by poor interfacial interactions, which result in deterioration of properties during material aging. The bacterial development of new PHA copolymers involves a complicated genetic engineering process: for example, poly(3- hydroxybutyrate-co-4-hydroxybutyrate) can be obtained using specific bacteria and nutrient source, with elongated cells expected to provide more space for the biopolymer; anyhow, the most promising route to drastically modify the PHA properties is the PHA chemical functionalization and the synthesis of copolymers in various block/graft architecture (Samui and Kanai 2019); in this way, a hydrophobic polymer can be rendered amphiphilic. The grafting of poly(N-vinylpyrrolidone) groups onto PHBV backbones was used to improve thermal stability and hydrophilicity, restrict the crystallization process, and decrease the flexibility of PHBV (Wang et al. 2008b).

Tailor-made physical properties, and superior mechanical properties of PBS can be achieved by random copolymerization with different types of comonomer units such as adipic acid, terephthalic acid, methyl succinic acid, 2,2-dimethyl succinic acid, benzyl succinic acid, ethylene glycol, and 1,3-propanediol, etc. (Xu and Guo 2010).

Poly(butylene succinate-co-terephthalate (PBST) and Polybutylene Succinate Adipate are the most common PBS copolymers. Copolymerization was also explored to obtain a tunable biodegradation rate of PBS. Random copolymers obtained by copolycondensation with tin-based catalyst were prevalently reported.

The synthesis of a variety of multiblock copolymers such as poly(butylene thiodiglycolate) (Gigli et al. 2012a), poly(thiodiethylene succinate) (Soccio et al. 2012), multiblock copolyesters comprising butylene succinate and triethylene succinate with different block lengths (Gualandi et al. 2012b), a multiblock bioresorbable copolyester, poly(butylene/diethylene glycol succinate (Gualandi et al. 2012a), poly(butylene diglycolate) (Gigli et al. 2013), and poly[(butylene terephthalate)-co-poly(butylene succinate)-block -poly(ethylene glycol)] (Wang et al. 2004) was explored for biomedical application, in most cases. Recently, sugar-based subunits (Zakharova et al. 2015) were introduced along the PBS backbone.

Random copolymerization usually decreases the melting point, degree of crystallinity, heat distortion temperature, and tensile strength of the copolymer. On the converse, the elongation at break and impact strength usually rise with copolymerization.

PGA copolymers, such as poly(lactic-co-glycolic acid) with lactic acid (Tsuji et al. 2019), poly(glycolide-co-caprolactone) with ε-caprolactone and poly (glycolide-co-trimethylene carbonate) with trimethylene carbonate are precious materials in the biomedical field. Biocompatible and biodegradable poly(lactic-co-glycolic acid) (PLGA) nanosystems provide tumor-targeting and diagnostic properties since PLGA can be modified with therapeutic cargos in the internal space or adsorbed onto the surface of PLGA nanoparticles. The introduction of imaging modalities to PLGA-based nanoparticles can enable drug delivery guided by in vivo imaging (Kim et al. 2019).

6.7.2 Blends

PLA blending epitomizes biopolymer modification through the blending technique in order to achieve suitable properties for different applications; in fact, PLA blending has received significant attention over the past decade. PLA-based blends are usually developed for applications demanding better performance as engineering polymers whose high price can be better tolerated. Among the family of biodegradable polyesters, PLA is not toxic, and it has a high mechanical performance. However, the thermal stability of PLA is generally not high enough in many commercial polymers applications. In order to enhance it, an enantiomeric polymer blending of PLLA and PDLA showed higher thermal stability, compared with pure PLLA and PDLA (Tsuji and Fukui 2003). Another goal of PLA blending is the increase of PLA degradation. A variety of PLA mixtures are available, as detailed in the following. Melt-blending of PLA with a small amount of poly(aspartic acid-co-lactide) or poly(sodium aspartate-co-lactide) accelerated the hydrolytic degradation of PLA (Shinoda et al. 2003).

Polysaccharides represent a major part of FW resources and they are a cheap blending feedstock. Blends of starch and PLA in the presence of various water contents and various compatibilizers and plasticizers, including diisocyanate, dioctyl maleate, and poly(vinyl alcohol), have been investigated to improve the interfacial interactions (Ke and Sun 2003; Yu et al. 2006) but the poor miscibility is still an issue (Hamad et al. 2018).

Edible coatings or films made of chitosan have been used in packaging; blending of chitosan with PLA was studied to improve the water vapor barrier of chitosan (Suyatma et al. 2004). The improved strength and ductility of PLA films fabricated with 10 wt.% chitosan compared to those of neat PLA films can make them competitive candidates for food packing applications; nonetheless, the decomposition of chitosan during melt mixing makes the fabrication of this system challenging (Hamad et al. 2018).

PLA/biopolyester blends are very common. Since PCL has a low glass transition temperature, blending PLA with PCL was successfully researched in an attempt to decrease the brittleness of PLA. In particular, the addition of a small amount of low-molecular-weight PCL and long mixing time improved both the tensile strength and ductility of PLA/PCL blends (Cock et al. 2013). PLA/PCL blends compatibilized with lysine triisocyanate showed improved mechanical performance (tensile strength and elongation-at-break) compared with the non-compatibilized blend, while the use of dicumyl peroxide gave a blend with characteristics similar to those of high impact polystyrene (HIPS) and acrylonitrile butadiene styrene (ABS) (Harada et al. 2008; Semba et al. 2006).

Due to the commercial availability of PLA and PBS and the increasing requirements for using biobased chemical building blocks, the application potential of PLA/PBS blends has been intensively researched in the last two decades (Su et al. 2019). PBS usually has a lower melting point and is less brittle than PLA. This combination of properties makes PBS/PLA blends especially interesting for bioplastics producers but also for 3D printing applications. Due to the biodegradability, good thermal resistance and melt processability of PBS, the aliphatic copolyester has been used to improve the melt processability and ductility of PLA (Park et al. 2010). Unfortunately, the improvement of ductility of PLA/PBS blends was at the cost of tensile strength, which markedly decreased (Hassan et al. 2013a). Blends consisting of PLA and PBS are characterized by a thermodynamically favored biphasic composition that often restricts their applications. Better compatibility for tailored and improved material properties was the focus of interest of many academicians. Lysine triisocyanate was proposed for the reactive processing to enhance the compatibility and improve the mechanical performance of PLA/PBS blends (Harada et al. 2007) thanks to the reaction of the isocyanate groups and the terminal hydroxyl or carboxyl groups of PLA and PBS. Maleic anhydride (Persenaire et al. 2014) and twice functionalized organoclay (Chen et al. 2005) were also successfully investigated for the compatibilization of PLA/PBS blends. The biodegradation and disintegration of PLA/PBS blends and their recycling are of priority interest. Other PBS blends include those with poly(L-lysine) (Tan et al. 2014), poly(3-hydroxybutyrate-co-valerate) and PHB (Ma et al. 2012), and PCL (Gumede et al. 2018).

PLA blends with polybutylene succinate-co-adipate, an aliphatic biodegradable polyester prepared from a butylene succinate adipate random copolymer, were investigated. Again, blending with PBAS was explored in order to enhance the ductility and injectability of PLA (Pivsa-Art et al. 2015): higher impact strengths and improved biodegradability were reported, but the poor compatibility of PLA and PBSA resulted in poor tensile strength and ductility. PLA/PBSA blends, compatibilized by reactive extrusion with triphenyl phosphite (Ojijo et al. 2012a, 2013), resulted in both improved tensile properties and its impact strength. The polymer chain extension due to the compatibilizer resulted in ligament-like fibrils located at the interphase, leading to some level of interfacial adhesion.

Polybutylene adipate-co-terephthalate (PBAT) was considered a good candidate for PLA blends because it is a fully biodegradable and tough aliphatic-aromatic copolyester (Jiang et al. 2006). Compared with neat PLA, the non-compatibilized PLA/PBAT blends show improved ductility and toughness but lower tensile strength and elastic modulus. To obtain an improvement in all properties, compatibilizers, and tetrabutyl titanate, in particular, were necessary (Lin et al. 2012). PLA/PBAT blends have an important potential as fully compostable high-performance materials.

PLA/PHA blends (Sudesh et al. 2000; He et al. 2014; Zhou et al. 2015; Zembouai et al. 2014) were investigated with the aim of obtaining a wide range of physical properties and improved processibility (Ohkoshi et al. 2000), but, above all, to increase both the biocompatibility and biodegradability of neat PLA and the thermal stability of neat PHA.

PHA performance can be improved by blending with natural materials or other biopolymers, such as cellulose derivatives, starch, and PCL, or by chemical modification, such as block and graft copolymerizations (Li et al. 2016a, b).

The compatibility of PLA/PHBH blends was enhanced using a reactive epoxy as a bifunctional compatibilizer (Zhou et al. 2015), but the poor toughness was the major drawback.

Poly(hydroxybutyrate-co-hydroxyhexanoate) (PHBHHx) and its dominant blends with PHB are promising from the biomedical point of view: the presence of PHBHHx in PHB reduced crystallization; films with a fairly regular and smooth surface allowed cell attachment and growth, thus strongly improving the biocompatibility of PHB, thereby demonstrating the feasibility of using this polymer blend as scaffold materials for tissue engineering (Kai et al. 2003). The blends of PHB or poly((R)-3-hydroxybutyrate-co-3-hydroxyvalerate) were studied to improve the inherent brittleness, and reduce high production costs associated with microbial polyesters (Yu et al. 2006). The experimental evidence indicates that chitosan is miscible with either PHB or poly((R)-3-hydroxybutyrate-co-3-hydroxyvalerate) at all compositions (Cheung et al. 2002).

PBS characteristichs have been improved through blending with PHB, PLA, PCL, poly(vinylphenol), polyethylene succinate, and fibers (Zeng et al. 2012).

6.8 Improvement of Bioplastic Properties to Bridge the Gap to Conventional Plastics

The biobased polymers market boom is partially hindered by their prohibitive expensiveness and their scarce competitiveness with fossilbased plastics. It follows that only novel technological advances will promote their use. The production of bioplastics is a key strategy that can maximize the use of FW and increase the potential revenue of the entire FW processing, thereby decreasing the environmental burdens arising from waste disposal (e.g., water contamination and GHG emissions).

The improvement of bioplastic properties to bridge the gap to conventional plastics is crucial.

Mechanical reinforcement should be pursued since brittleness and rigidity are the common problems faced by bioplastics. External plasticization, which increases polymer chain mobility, leading to improved flexibility, suffers from the tendency of plasticizers to leach from the plastic or phase separate from the polymer. Internal plasticization via copolymerization or grafting a second monomer to the polymer structure is more efficient than external plasticization, and no plasticizer leaching occurs. The most common, practical, and cost-effective technique to improve polymer mechanical performance is, however, blending with other polymers.

Copolymerization, grafting, blending, modification of the crystallization rate, and the use of fillers have been used to enhance bioplastic flexibility and toughness. These strategies are also used to improve the thermal properties of the biopolymers (Zhao et al. 2020).

Bioplastic barrier properties, discussed in Chap. 9 (Sec. 9.4), are crucial for food packaging; they can be improved by various strategies. Lamination with barrier petroleum-based and non-biodegradable plastics is not a fully biobased option, even if it is effective. Metallization by vacuum coating silicon (or aluminum) dioxide onto PLA film, increased the oxygen and water vapor barrier of up to 8–10 and 6–8 times, compared to those of pristine PLA (Zhao et al. 2018). Nanofillers such as nanoclay and zeolites decrease moisture/gas diffusion (Priolo et al. 2015; Zhao et al. 2020). The introduction of hydrophobic materials, waxes, and hydrolyzed keratin improves the moisture barrier (Morillon et al. 2002). Increasing polymer crystallinity using a nucleating agent is another option to enhance the moisture and gas barriers (Jost and Langowski 2015). Section 10.3.1 deals with biobased barriers useful to prevent migration from conventional packaging materials.

As regards processability, PLA and PHAs cannot be easily processed as normal thermoplastics through conventional plastic processing techniques because their chain linearity results in low resistance to stretching of the melt polymer; this, in turn, causes issues that require film blowing and blow molding or adjustment on processing parameters or the presence of compatibilizers (Wei and McDonald 2015). PLA and PHAs degrade at temperatures close to their melting points, making thermal processing very changeling due to a narrow processing window (Larsson et al. 2016). PLA and PHAs processability can be improved by: (i) long chain branching, (ii) grafting, (iii) peroxide induced crosslinking through reactive melt processing (Wei and McDonald 2015), and (iv) chain extension by using multifunctional chain extenders such as tris(nonylphenyl) phosphite and polycarbodiimide and epoxy-functionalized oligomeric acrylic copolymer (Tiwary and Kontopoulou 2018; Jaszkiewicz et al. 2014). These techniques increase polymer molecular weight and molecular weight distribution and polymer chain entanglement, thereby increasing resistance to untangling during thermal processing. Other techniques to improve processability include blending with elastomers, the use of plasticizers (epoxidized soybean oil, acetyl tributyl citrate, poly(ethylene glycol), limonene (Panaitescu et al. 2017; Arrieta et al. 2014)), nucleation and reinforcement with the addition of nanoclay (Zhao et al. 2020).

6.9 Conclusions

While global bioplastic production keeps sharply increasing because of consumers' environmental awareness, new policies and legislation, as well as technological advancements, the estimated market share of bioplastics still has not exceeded 1% of the global plastics production. Increasing the market share of bioplastics would reduce the dependence on and the environmental impact of fossilbased plastics in a biobased society within the framework of a circular economy approach.

Bibliography

Abdel-Rahman MA, Tashiro Y, Sonomoto K (2013) Recent advances in lactic acid production by microbial fermentation processes. Biotechnol Adv 31:877–902

Abdul Khalil HPS, Saurabh CK, Adnan AS et al (2016) A review on chitosan-cellulose blends and nanocellulose reinforced chitosan biocomposites: properties and their applications. Carbohydr Polym 150:216–226

Ahmed T, Shahid M, Azeem F et al (2018) Biodegradation of plastics: current scenario and future prospects for environmental safety. Environ Sci Pollut Res 25:7287–7298

Albertsson AC, Varma IK (2003) Recent developments in ring opening polymerization of lactones for biomedical applications. Biomacromolecules 4:1466–1486

Albuquerque PBS, Malafaia CB (2018) Perspectives on the production, structural characteristics and potential applications of bioplastics derived from polyhydroxyalkanoates. Int J Biol Macromol 107:615–625. https://doi.org/10.1016/J.IJBIOMAC.2017.09.026

Alonso S, Herrero M, Rendueles M, Díaz M (2010) Residual yoghurt whey for lactic acid production. Biomass Bioenergy 34:931–938. https://doi.org/10.1016/j.biombioe.2010.01.041

Amaro TMMM, Rosa D, Comi G, Iacumin L (2019) Prospects for the use of whey for polyhydroxyalkanoate (PHA) production. Front Microbiol 10:992

Amulya K, Venkateswar Reddy M, Venkata Mohan S (2014) Acidogenic spent wash valorization through polyhydroxyalkanoate (PHA) synthesis coupled with fermentative biohydrogen production. Bioresour Technol 158:336–342. https://doi.org/10.1016/j.biortech.2014.02.026

Amulya K, Jukuri S, Venkata Mohan S (2015) Sustainable multistage process for enhanced productivity of bioplastics from waste remediation through aerobic dynamic feeding strategy: process integration for up-scaling. Bioresour Technol 188:231–239. https://doi.org/10.1016/j.biortech.2015.01.070

Arrieta MP, López J, Hernández A, Rayón E (2014) Ternary PLA-PHB-Limonene blends intended for biodegradable food packaging applications. Eur Polym J 50:255–270. https://doi.org/10.1016/j.eurpolymj.2013.11.009

Athanasiou KA, Niederauer GG, Agrawal CM (1996) Sterilization, toxicity, biocompatibility and clinical applications of polylactic acid/polyglycolic acid copolymers. Biomaterials 17:93–102. https://doi.org/10.1016/0142-9612(96)85754-1

Auras R, Harte B, Selke S (2004) An overview of polylactides as packaging materials. Macromol Biosci 4:835–864. https://doi.org/10.1002/mabi.200400043

Auras R, Lim L-T, Selke SEM, Tsuji H (2010a) Modification of PLA properties by process aids and other additives. In: Poly(lactic acid), synthesis, structures, properties, processing, and applications. Wiley, Hoboken, p 499

Auras R, Lim LT, Selke SEM, Tsuji H (2010b) Poly(lactic acid): synthesis, structures, properties, processing, and applications. Wiley, Hoboken

Avérous L, Pollet E (2012) Environmental silicate nano-biocomposites (Google eBook)

Awasthi D, Wang L, Rhee MS et al (2018) Metabolic engineering of Bacillus subtilis for production of D-lactic acid. Biotechnol Bioeng 115:453–463. https://doi.org/10.1002/bit.26472

Azim H, Dekhterman A, Jiang Z, Gross RA (2006) Candida antarctica lipase B-catalyzed synthesis of poly(butylene succinate): shorter chain building blocks also work. Biomacromolecules 7:3093–3097. https://doi.org/10.1021/bm060574h

Barbosa V, Ramires EC, Razera IAT, Frollini E (2010) Biobased composites from tannin-phenolic polymers reinforced with coir fibers. Ind Crops Prod 32:305–312. https://doi.org/10.1016/j.indcrop.2010.05.007

Bautista M, de Ilarduya A, Alla A, Muñoz-Guerra S (2015) Poly(butylene succinate) Ionomers with Enhanced Hydrodegradability. Polymers (Basel) 7:1232–1247. https://doi.org/10.3390/polym7071232

Bechthold I, Bretz K, Kabasci S et al (2008) Succinic acid: a new platform chemical for biobased polymers from renewable resources. Chem Eng Technol 31:647–654

Benninga H (1990) A history of lactic acid making: a chapter in the history of biotechnology. Springer, Dordrecht

Berwig KH, Baldasso C, Dettmer A (2016) Production and characterization of poly(3-hydroxybutyrate) generated by Alcaligenes latus using lactose and whey after acid protein precipitation process. Bioresour Technol 218:31–37. https://doi.org/10.1016/j.biortech.2016.06.067

Bozell JJ, Petersen GR (2010) Technology development for the production of biobased products from biorefinery carbohydrates – the US Department of Energy's "top 10" revisited. Green Chem 12:539–554. https://doi.org/10.1039/b922014c

Briassoulis D, Giannoulis A (2018) Evaluation of the functionality of bio-based food packaging films. Polym Test 69:39–51. https://doi.org/10.1016/j.polymertesting.2018.05.003

Budak K, Sogut O, Aydemir Sezer U (2020) A review on synthesis and biomedical applications of polyglycolic acid. J Polym Res 27:208

Bugnicourt E, Cinelli P, Lazzeri A, Alvarez V (2014) Polyhydroxyalkanoate (PHA): Review of synthesis, characteristics, processing and potential applications in packaging. Express Polym Lett 8:791–808. https://doi.org/10.3144/expresspolymlett.2014.82

Burgos N, Tolaguera D, Fiori S, Jiménez A (2014) Synthesis and characterization of lactic acid oligomers: evaluation of performance as poly(lactic acid) plasticizers. J Polym Environ 22:227–235. https://doi.org/10.1007/s10924-013-0628-5

Carothers WH, Arvin JA (1929) Studies on polymerization and ring formation. II. Polyesters. J Am Chem Soc 51:2560–2570. https://doi.org/10.1021/ja01383a042

Carothers WH, Dorough GL, van Natta FJ (1932) Studies of polymerization and ring formation. X. the reversible polymerization of six-membered cyclic esters. J Am Chem Soc 54:761–772. https://doi.org/10.1021/ja01341a046

Castro-Aguirre E, Iñiguez-Franco F, Samsudin H et al (2016) Poly(lactic acid) – mass production, processing, industrial applications, and end of life. Adv Drug Deliv Rev 107:333–366

Chaiwutthinan P, Leejarkpai T, Kashima DP, Chuayjuljit S (2013) Poly(lactic acid)/poly(butylene succinate) blends filled with epoxy functionalised polymeric chain extender. Adv Mater Res 664:644–648

Chen GQ, Patel MK (2012) Plastics derived from biological sources: present and future: a technical and environmental review. Chem Rev 112:2082–2099

Chen GX, Kim HS, Kim ES, Yoon JS (2005) Compatibilization-like effect of reactive organoclay on the poly(l-lactide)/poly(butylene succinate) blends. Polymer (Guildf) 46:11829–11,836. https://doi.org/10.1016/j.polymer.2005.10.056

Chen Q, Liang S, Thouas GA (2013) Elastomeric biomaterials for tissue engineering. Prog Polym Sci 38:584–671

Chen Y, Hua WQ, Zhang ZC et al (2019) An efficient, food contact accelerator for stereocomplexation of high-molecular-weight poly(L -lactide)/ poly(D -lactide) blend under nonisothermal crystallization. Polymer (Guildf) 170:54–64. https://doi.org/10.1016/j.polymer.2019.03.003

Chen G-Q, Chen X-Y, Wu F-Q, Chen J-C (2020) Polyhydroxyalkanoates (PHA) toward cost competitiveness and functionality. Adv Ind Eng Polym Res 3:1–7. https://doi.org/10.1016/j.aiepr.2019.11.001

Cheung MK, Wan KPY, Yu PH (2002) Miscibility and morphology of chiral semicrystalline poly-(R)-(3-hydroxybutyrate)/chitosan and poly-(R)-(3-hydroxybutyrate-co-3-hydroxyvalerate)/chitosan blends studied with DSC,1HT1 andT1? CRAMPS. J Appl Polym Sci 86:1253–1258. https://doi.org/10.1002/app.11091

Chiew YL, Shimada S (2013) Current state and environmental impact assessment for utilizing oil palm empty fruit bunches for fuel, fiber and fertilizer – a case study of Malaysia. Biomass Bioenergy 51:109–124. https://doi.org/10.1016/j.biombioe.2013.01.012

Cock F, Cuadri AA, García-Morales M, Partal P (2013) Thermal, rheological and microstructural characterisation of commercial biodegradable polyesters. Polym Test 32:716–723. https://doi.org/10.1016/j.polymertesting.2013.03.015

Cowie JMG (2007) Polymers: chemistry and physics of modern materials.

Cunha AG, Gandini A (2010) Turning polysaccharides into hydrophobic materials: a critical review. Part 1. Cellulose. Cellulose 17:875–889

Ehsani M, Khodabakhshi K, Asgari M (2014) Lactide synthesis optimization: Investigation of the temperature, catalyst and pressure effects. E-Polymers 14:353–361. https://doi.org/10.1515/epoly-2014-0055

EU (2015) From the sugar platform to biofuels and biochemicals. Final Rep Eur Comm Dir Energy 183. doi: contract no. ENER/C2/423-2012/SI2.673791

European Bioplastics (2018) What are bioplastics? https://docs.european-bioplastics.org/publications/fs/EuBP_FS_What_are_bioplastics.pdf

European Bioplastics (2020) Bioplastics market development update 2020. https://www.european-bioplastics.org/market/

Fabbri M, Gigli M, Gamberini R et al (2014) Hydrolysable PBS-based poly(ester urethane) s thermoplastic elastomers. Polym Degrad Stab 108:223–231. https://doi.org/10.1016/j.polymdegradstab.2014.03.033

Farah S, Anderson DG, Langer R (2016) Physical and mechanical properties of PLA, and their functions in widespread applications – a comprehensive review. Adv Drug Deliv Rev 107:367–392

Fiedler S, Steinbüchel A, Rehm BH (2002) The role of the fatty acid β-oxidation multienzyme complex from Pseudomonas oleovorans in polyhydroxyalkanoate biosynthesis: Molecular characterization of the fadBA operon from P. oleovorans and of the enoyl-CoA hydratase genes phaJ from P. oleovorans and Pseudomonas putida. Arch Microbiol 178:149–160. https://doi.org/10.1007/s00203-002-0444-0

Flory PJ, Leutner FS (1952) United States patent office 2,589,687 method of preparing linear polyesters

Folino A, Karageorgiou A, Calabrò PS, Komilis D (2020) Biodegradation of wasted bioplastics in natural and industrial environments: A review. Sustain 12:1–37. https://doi.org/10.3390/su12156030

Follonier S, Goyder MS, Silvestri AC et al (2014) Fruit pomace and waste frying oil as sustainable resources for the bioproduction of medium-chain-length polyhydroxyalkanoates. Int J Biol Macromol 71:42–52. https://doi.org/10.1016/j.ijbiomac.2014.05.061

Forte A, Zucaro A, Basosi R, Fierro A (2016) LCA of 1,4-butanediol produced via direct fermentation of sugars from wheat straw feedstock within a territorial biorefinery. Materials (Basel) 9:1–22. https://doi.org/10.3390/MA9070563

Fujimaki H, Matsumine H, Osaki H et al (2019) Dedifferentiated fat cells in polyglycolic acid-collagen nerve conduits promote rat facial nerve regeneration. Regen Ther 11:240–248. https://doi.org/10.1016/j.reth.2019.08.004

Garlotta D (2001) A literature review of poly(lactic acid) J Polym Environ 9:63–84. https://doi.org/10.1023/A:1020200822435

Gaucher G, Marchessault RH, Leroux JC (2010) Polyester-based micelles and nanoparticles for the parenteral delivery of taxanes. J Control Release 143:2–12

Germec M, Ozcan A, Turhan I (2019) Bioconversion of wheat bran into high value-added products and modelling of fermentations. Ind Crops Prod 139. https://doi.org/10.1016/j.indcrop.2019.111565

Ghorpade VM, Gennadios A, Hanna MA (2001) Laboratory composting of extruded poly(lactic acid) sheets. Bioresour Technol 76:57–61

Gigli M, Lotti N, Gazzano M et al (2012a) Macromolecular design of novel sulfur-containing copolyesters with promising mechanical properties. J Appl Polym Sci 126:686–696. https://doi.org/10.1002/app.36856

Gigli M, Lotti N, Gazzano M et al (2012b) Novel eco-friendly random copolyesters of poly(butylene succinate) containing ether-linkages. React Funct Polym 72:303–310. https://doi.org/10.1016/j.reactfunctpolym.2012.02.013

Gigli M, Lotti N, Gazzano M et al (2013) Synthesis and characterization of novel poly(butylene succinate)-based copolyesters designed as potential candidates for soft tissue engineering. Polym Eng Sci 53:491–501. https://doi.org/10.1002/pen.23289

Gigli M, Fabbri M, Lotti N et al (2016) Poly(butylene succinate)-based polyesters for biomedical applications: a review in memory of our beloved colleague and friend Dr. Lara Finelli. Eur Polym J 75:431–460

Gilding DK, Reed AM (1979) Biodegradable polymers for use in surgery-polyglycolic/poly(actic acid) homo- and copolymers: 1. Polymer (Guildf) 20:1459–1464. https://doi.org/10.1016/0032-3861(79)90009-0

Göktürk E, Pemba AG, Miller SA (2015) Polyglycolic acid from the direct polymerization of renewable C1 feedstocks. Polym Chem 6:3918–3925. https://doi.org/10.1039/c5py00230c

Gottschalk G (1986) Bacterial metabolism. Springer, New York

Gualandi C, Soccio M, Govoni M et al (2012a) Poly(butylene/diethylene glycol succinate) multiblock copolyester as a candidate biomaterial for soft tissue engineering: Solid-state properties, degradability, and biocompatibility. J Bioact Compat Polym 27:244–264. https://doi.org/10.1177/0883911512440536

Gualandi C, Soccio M, Saino E et al (2012b) Easily synthesized novel biodegradable copolyesters with adjustable properties for biomedical applications. Soft Matter 8:5466–5476. https://doi.org/10.1039/c2sm25308a

Gumede TP, Luyt AS, Müller AJ (2018) Review on PCL, PBS, and PCL/PBS blends containing carbon nanotubes. Express Polym Lett 12:505–529

Guo A, Cho Y, Petrović ZS (2000) Structure and properties of halogenated and nonhalogenated soy-based polyols. J Polym Sci Part A Polym Chem 38:3900–3910. https://doi.org/10.1002/1099-0518(20001101)38:21<3900::AID-POLA70>3.0.CO;2-E

Hamad K, Kaseem M, Yang HW et al (2015) Properties and medical applications of polylactic acid: a review. Express Polym Lett 9:435–455. https://doi.org/10.3144/expresspolymlett.2015.42

Hamad K, Kaseem M, Ayyoob M et al (2018) Polylactic acid blends: the future of green, light and tough. Prog Polym Sci 85:83–127. https://doi.org/10.1016/j.progpolymsci.2018.07.001

Harada M, Ohya T, Iida K et al (2007) Increased impact strength of biodegradable poly(lactic acid)/poly(butylene succinate) blend composites by using isocyanate as a reactive processing agent. J Appl Polym Sci 106:1813–1820. https://doi.org/10.1002/app.26717

Harada M, Iida K, Okamoto K et al (2008) Reactive compatibilization of biodegradable poly(lactic acid)/poly(ε-caprolactone) blends with reactive processing agents. Polym Eng Sci 48:1359–1368. https://doi.org/10.1002/pen.21088

Hartmann MH (1998) High molecular weight polylactic acid polymers. In: Biopolymers from renewable resources. Springer, Berlin/Heidelberg, pp 367–411

Hassan E, Wei Y, Jiao H, Muhuo Y (2013a) Dynamic mechanical properties and thermal stability of poly(lactic acid) and poly(butylene succinate) blends composites. J Fiber Bioeng Informatics 6:85–94. https://doi.org/10.3993/jfbi03201308

Hassan MA, Yee LN, Yee PL et al (2013b) Sustainable production of polyhydroxyalkanoates from renewable oil-palm biomass. Biomass Bioenergy 50:1–9

He Y, Hu Z, Ren M et al (2014) Evaluation of PHBHHx and PHBV/PLA fibers used as medical sutures. J Mater Sci Mater Med 25:561–571. https://doi.org/10.1007/s10856-013-5073-4

Hege CS, Schiller SM (2014) Non-toxic catalysts for ring-opening polymerizations of biodegradable polymers at room temperature for biohybrid materials. Green Chem 16:1410–1416. https://doi.org/10.1039/c3gc42044b

Husseinsyah S, Zakaria MM (2011) The effect of filler content on properties of coconut shell filled polyester composites. S. Polymer (Guildf) 6:87–97

Imre B, Pukánszky B (2013) Compatibilization in bio-based and biodegradable polymer blends. Eur Polym J 49:1215–1233

Imre B, García L, Puglia D, Vilaplana F (2019) Reactive compatibilization of plant polysaccharides and biobased polymers: review on current strategies, expectations and reality. Carbohydr Polym 209:20–37. https://doi.org/10.1016/j.carbpol.2018.12.082

Jacobson GB, Shinde R, Contag CH, Zare RN (2008) Sustained release of drugs dispersed in polymer nanoparticles. Angew Chemie Int Ed 47:7880–7882. https://doi.org/10.1002/anie.200802260

Jaszkiewicz A, Bledzki AK, Duda A et al (2014) Investigation of processability of chain-extended polylactides during melt processing – compounding conditions and polymer molecular structure. Macromol Mater Eng 299:307–318. https://doi.org/10.1002/mame.201300115

Jenkins AD, Stepto RFT, Kratochvíl P, Suter UW (1996) Glossary of basic terms in polymer science (IUPAC Recommendations 1996). Pure Appl Chem 68:2287–2311. https://doi.org/10.1351/pac199668122287

Jha A, Kumar A (2019) Biobased technologies for the efficient extraction of biopolymers from waste biomass. Bioprocess Biosyst Eng 42:1893–1901

Jiang L, Wolcott MP, Zhang J (2006) Study of biodegradable polylactide/poly(butylene adipate-*co* -terephthalate) blends. Biomacromolecules 7:199–207. https://doi.org/10.1021/bm050581q

Johansson C, Bras J, Mondragon I et al (2012) Renewable fibers and bio-based materials for packaging applications – a review of recent developments. BioResources 7:2506–2552. https://doi.org/10.15376/biores.7.2.2506-2552

John G, Nagarajan S, Kumar P et al (2019) Natural monomers: a mine for functional and sustainable materials – occurrence, chemical modification and polymerization. Prog Polym Sci 92:158–209

Jost V, Langowski HC (2015) Effect of different plasticisers on the mechanical and barrier properties of extruded cast PHBV films. Eur Polym J 68:302–312. https://doi.org/10.1016/j.eurpolymj.2015.04.012

Kai Z, Ying D, Guo-Qiang C (2003) Effects of surface morphology on the biocompatibility of polyhydroxyalkanoates. Biochem Eng J 16:115–123. https://doi.org/10.1016/S1369-703X(03)00029-9

Karp SG, Igashiyama AH, Siqueira PF et al (2011) Application of the biorefinery concept to produce l-lactic acid from the soybean vinasse at laboratory and pilot scale. Bioresour Technol 102:1765–1772. https://doi.org/10.1016/j.biortech.2010.08.102

Ke T, Sun X (2003) Melting behavior and crystallization kinetics of starch and poly(lactic acid) composites. J Appl Polym Sci 89:1203–1210. https://doi.org/10.1002/app.12162

Khardenavis AA, Vaidya AN, Kumar MS, Chakrabarti T (2009) Utilization of molasses spentwash for production of bioplastics by waste activated sludge. Waste Manag 29:2558–2565. https://doi.org/10.1016/j.wasman.2009.04.008

Kim H-S, Yang H-S, Kim H-J (2005) Biodegradability and mechanical properties of agro-flour-filled polybutylene succinate biocomposites. https://doi.org/10.1002/app.21905

Kim KT, Lee JY, Kim DD et al (2019) Recent progress in the development of poly(lactic-co-glycolic acid)-based nanostructures for cancer imaging and therapy. Pharmaceutics 11:280

Koch D, Mihalyi B (2018) Assessing the change in environmental impact categories when replacing conventional plastic with bioplastic in chosen application fields. Chem Eng Trans 70:853–858. https://doi.org/10.3303/CET1870143

Koller M, Bona R, Braunegg G et al (2005) Production of polyhydroxyalkanoates from agricultural waste and surplus materials. Biomacromolecules 6:561–565. https://doi.org/10.1021/bm049478b

Komesu A, Martinez PFM, Lunelli BH et al (2015) Lactic acid purification by reactive distillation system using design of experiments. Chem Eng Process Process Intensif 95:26–30. https://doi.org/10.1016/j.cep.2015.05.005

Kouisni L, Fang Y, Paleologou M et al (2011) Kraft lignin recovery and its use in the preparation of lignin-based phenol formaldehyde resins for plywood. Cellul Chem Technol 45:515–520

Kowalski A, Libiszowski J, Biela T et al (2005) Kinetics and mechanism of cyclic esters polymerization initiated with tin(II) octoate. Polymerization of ε-caprolactone and L,L-Lactide co-initiated with primary amines. Macromolecules 38:8170–8176. https://doi.org/10.1021/ma050752j

Kwan TH, Hu Y, Lin CSK (2018) Techno-economic analysis of a food waste valorisation process for lactic acid, lactide and poly(lactic acid) production. J Clean Prod 181:72–87. https://doi.org/10.1016/j.jclepro.2018.01.179

Larsson M, Markbo O, Jannasch P (2016) Melt processability and thermomechanical properties of blends based on polyhydroxyalkanoates and poly(butylene adipate-: Co -terephthalate). RSC Adv 6:44354–44363. https://doi.org/10.1039/c6ra06282b

Laurichesse S, Avérous L (2014) Chemical modification of lignins: towards biobased polymers. Prog Polym Sci 39:1266–1290

Lee CW, Na C, Kimura Y, Masutani K (2016) ABCBA pentablock copolymers consisting of Poly(l-lactide) (PLLA: A), Poly(d-lactide) (PDLA: B), and Poly(butylene succinate) (PBS: C): effects of semicrystalline PBS segments on the stereo-crystallinity and properties. Macromol Mater Eng 301:1121–1131. https://doi.org/10.1002/mame.201600158

Lemoigne M (1926) Products of dehydration and of polymerization of β-hydroxybutyric acid. Bull Soc Chim Biol 8:770–782

Li BH, Yang MC (2006) Improvement of thermal and mechanical properties of poly(L-lactic acid) with 4,4-methylene diphenyl diisocyanate. Polym Adv Technol 17:439–443. https://doi.org/10.1002/pat.731

Li X, Chen Y, Zhao S et al (2015) Efficient production of optically pure l-lactic acid from food waste at ambient temperature by regulating key enzyme activity. Water Res 70:148–157. https://doi.org/10.1016/j.watres.2014.11.049

Li Z, Tan BH, Lin T, He C (2016a) Recent advances in stereocomplexation of enantiomeric PLA-based copolymers and applications. Prog Polym Sci 62:22–72

Li Z, Yang J, Loh XJ (2016b) Polyhydroxyalkanoates: Opening doors for a sustainable future. NPG Asia Mater 8:265

Liang Q, Qi Q (2014) From a co-production design to an integrated single-cell biorefinery. Biotechnol Adv 32:1328–1335

Lin S, Guo W, Chen C et al (2012) Mechanical properties and morphology of biodegradable poly(lactic acid)/poly(butylene adipate-co-terephthalate) blends compatibilized by transesterification. Mater Des 36:604–608. https://doi.org/10.1016/j.matdes.2011.11.036

Lligadas G, Ronda JC, Galiá M, Cádiz V (2010) Plant oils as platform chemicals for polyurethane synthesis: current state-of-the-art. Biomacromolecules 11:2825–2835

Lowe CE (1954) Preparation of high molecular weight polyhydroxyacetic ester. US Patent 2668162 (2 February 1954), assigned to E. I. du Pont Co. (1954)

Luckachan GE, Pillai CKS (2006) Chitosan/oligo L-lactide graft copolymers: effect of hydrophobic side chains on the physico-chemical properties and biodegradability. Carbohydr Polym 64:254–266. https://doi.org/10.1016/j.carbpol.2005.11.035

Lunt J (1998) Large-scale production, properties and commercial applications of poly lactic acid polymers. Polym Degrad Stab 59:145–152. https://doi.org/10.1016/S0141-3910(97)00148-1

Lv T, Zhou C, Li J et al (2018) New insight into the mechanism of enhanced crystallization of PLA in PLLA/PDLA mixture. J Appl Polym Sci 135. https://doi.org/10.1002/app.45663

Ma P, Hristova-Bogaerds DG, Lemstra PJ et al (2012) Toughening of PHBV/PBS and PHB/PBS blends via in situ compatibilization using dicumyl peroxide as a free-radical grafting initiator. Macromol Mater Eng 297:402–410. https://doi.org/10.1002/mame.201100224

Madhavan Nampoothiri K, Nair NR, John RP (2010) An overview of the recent developments in polylactide (PLA) research. Bioresour Technol 101:8493–8501. https://doi.org/10.1016/j.biortech.2010.05.092

Mazzuco R, Sadick NS (2016) The use of poly-L-lactic acid in the Gluteal area. Dermatologic Surg. 42:441–443

Mehta R, Kumar V, Bhunia H, Upadhyay SN (2005) Synthesis of poly(lactic acid): a review. J Macromol Sci Polym Rev 45:325–349

Middleton JC, Tipton AJ (2000) Synthetic biodegradable polymers as orthopedic devices. Biomaterials 21:2335–2346. https://doi.org/10.1016/S0142-9612(00)00101-0

Mirabella N, Castellani V, Sala S (2014) Current options for the valorization of food manufacturing waste: a review. J Clean Prod 65:28–41

Mochane MJ, Sefadi JS, Motsoeneng TS et al (2020) The effect of filler localization on the properties of biopolymer blends, recent advances: a review. Polym Compos 41:2958–2979

Moncada J, Matallana LG, Cardona CA (2013) Selection of process pathways for biorefinery design using optimization tools: a colombian case for conversion of sugarcane bagasse to ethanol, Poly-3-hydroxybutyrate (PHB), and energy. Ind Eng Chem Res 52:4132–4145. https://doi.org/10.1021/ie3019214

Morais C, Freitas F, Cruz MV et al (2014) Conversion of fat-containing waste from the margarine manufacturing process into bacterial polyhydroxyalkanoates. Int J Biol Macromol 71:68–73. https://doi.org/10.1016/j.ijbiomac.2014.04.044

Morillon V, Debeaufort F, Blond G et al (2002) Factors affecting the moisture permeability of lipid-based edible films: a review. Crit Rev Food Sci Nutr 42:67–89. https://doi.org/10.1080/10408690290825466

Naranjo JM, Cardona CA, Higuita JC (2014) Use of residual banana for polyhydroxybutyrate (PHB) production: case of study in an integrated biorefinery. Waste Manag 34:2634–2640. https://doi.org/10.1016/j.wasman.2014.09.007

Ng F, Couture G, Philippe C et al (2017) Bio-based aromatic epoxy monomers for thermoset materials. Molecules 22. https://doi.org/10.3390/molecules22010149

Nguyen CM, Kim JS, Nguyen TN et al (2013) Production of l- and d-lactic acid from waste Curcuma longa biomass through simultaneous saccharification and cofermentation. Bioresour Technol 146:35–43. https://doi.org/10.1016/j.biortech.2013.07.035

Niedermann P, Szebényi G, Toldy A (2015) Novel high glass temperature sugar-based epoxy resins: Characterization and comparison to mineral oil-based aliphatic and aromatic resins. Express Polym Lett 9:85–94. https://doi.org/10.3144/expresspolymlett.2015.10

Nifant'ev IE, Shlyakhtin AV, Bagrov VV et al (2018) Efficient synthetic approach to copolymers of glycolic and lactic acids for biomedical applications. Mendeleev Commun 28:412–414. https://doi.org/10.1016/j.mencom.2018.07.024

Notta-Cuvier D, Odent J, Delille R et al (2014) Tailoring polylactide (PLA) properties for automotive applications: effect of addition of designed additives on main mechanical properties. Polym Test 36:1–9. https://doi.org/10.1016/j.polymertesting.2014.03.007

Numata K, Morisaki K, Tomizawa S et al (2013) Synthesis of poly- and oligo(hydroxyalkanoate)s by deep-sea bacteria, Colwellia spp., Moritella spp., and Shewanella spp. Polym J 45:1094–1100. https://doi.org/10.1038/pj.2013.25

Obruca S, Benesova P, Petrik S et al (2014a) Production of polyhydroxyalkanoates using hydrolysate of spent coffee grounds. Process Biochem 49:1409–1414. https://doi.org/10.1016/j.procbio.2014.05.013

Obruca S, Petrik S, Benesova P et al (2014b) Utilization of oil extracted from spent coffee grounds for sustainable production of polyhydroxyalkanoates. Appl Microbiol Biotechnol 98:5883–5890. https://doi.org/10.1007/s00253-014-5653-3

Oh JK (2011) Polylactide (PLA)-based amphiphilic block copolymers: synthesis, self-assembly, and biomedical applications. Soft Matter 7:5096–5108

Ohkoshi I, Abe H, Doi Y (2000) Miscibility and solid-state structures for blends of poly[(S)-lactide] with atactic poly[(R,S)-3-hydroxybutyrate]. Polymer (Guildf) 41:5985–5992. https://doi.org/10.1016/S0032-3861(99)00781-8

Ohkouchi Y, Inoue Y (2006) Direct production of l(+)-lactic acid from starch and food wastes using Lactobacillus manihotivorans LMG18011. Bioresour Technol 97:1554–1562. https://doi.org/10.1016/j.biortech.2005.06.004

Ojijo V, Malwela T, Sinha Ray S, Sadiku R (2012a) Unique isothermal crystallization phenomenon in the ternary blends of biopolymers polylactide and poly[(butylene succinate)-co-adipate] and nano-clay. Polymer (Guildf) 53:505–518. https://doi.org/10.1016/j.polymer.2011.12.007

Ojijo V, Sinha Ray S, Sadiku R (2012b) Role of specific interfacial area in controlling properties of immiscible blends of biodegradable polylactide and poly[(butylene succinate)-co-adipate]. ACS Appl Mater Interfaces 4:6690–6701. https://doi.org/10.1021/am301842e

Ojijo V, Sinha Ray S, Sadiku R (2013) Toughening of biodegradable polylactide/poly(butylene succinate- co -adipate) blends via in situ reactive compatibilization. ACS Appl Mater Interfaces 5:4266–4276. https://doi.org/10.1021/am400482f

Omay D, Guvenilir Y (2013) Synthesis and characterization of poly(d,l-lactic acid) via enzymatic ring opening polymerization by using free and immobilized lipase. Biocatal Biotransformation 31:132–140. https://doi.org/10.3109/10242422.2013.795148

Ortiz P, Vendamme R, Eevers W (2020) Fully biobased epoxy resins from fatty acids and lignin 1–11

Panaitescu DM, Nicolae CA, Frone AN et al (2017) Plasticized poly(3-hydroxybutyrate) with improved melt processing and balanced properties. J Appl Polym Sci 134. https://doi.org/10.1002/app.44810

Park SB, Hwang SY, Moon CW et al (2010) Plasticizer effect of novel PBS ionomer in PLA/PBS ionomer blends. Macromol Res 18:463–471. https://doi.org/10.1007/s13233-010-0512-2

Pastore G, Gabrielli S, Cecchi T et al (2021) A new and efficient lactic acid polymerization by multimetallic cerium complexes: a poly(lactic acid) suitable for biomedical applications. RSC Adv 11:10592–10598. https://doi.org/10.1039/d0ra10637b

Pawar RP, Tekale SU, Shisodia SU et al (2014) Biomedical applications of poly(lactic acid). Rec Pat Regen Med 4:40–51. https://doi.org/10.2174/2210296504666140402235024

Persenaire O, Quintana R, Lemmouchi Y et al (2014) Reactive compatibilization of poly(l-lactide)/poly(butylene succinate) blends through polyester maleation: from materials to properties. Polym Int 63:1724–1731. https://doi.org/10.1002/pi.4700

Petrović ZS (2010) Polymers from biological oils. Contemp Mater 1:39–50. https://doi.org/10.5767/anurs.cmat.100101.en.039p

Pivsa-Art W, Pivsa-Art S, Fujii K et al (2015) Compression molding and melt-spinning of the blends of poly(lactic acid) and poly(butylene succinate-co-adipate). J Appl Polym Sci 132:n/a–n/a. https://doi.org/10.1002/app.41856

Pohlmann A, Fricke WF, Reinecke F et al (2006) Genome sequence of the bioplastic-producing "Knallgas" bacterium Ralstonia eutropha H16. Nat Biotechnol 24:1257–1262. https://doi.org/10.1038/nbt1244

Pooth V, van Gaalen K, Trenkamp S et al (2020) Comprehensive analysis of metabolic sensitivity of 1,4-butanediol producing Escherichia coli toward substrate and oxygen availability. Biotechnol Prog 36. https://doi.org/10.1002/btpr.2917

Priolo MA, Holder KM, Guin T, Grunlan JC (2015) Recent advances in gas barrier thin films via layer-by-layer assembly of polymers and platelets. Macromol Rapid Commun 36:866–879. https://doi.org/10.1002/marc.201500055

Quinaya SHD, Alzate CAC (2014) Plantain and banana fruit as raw material for glucose production. J Biotechnol 185:S34. https://doi.org/10.1016/j.jbiotec.2014.07.115

Quirino RL, Garrison TF, Kessler MR (2014) Matrices from vegetable oils, cashew nut shell liquid, and other relevant systems for biocomposite applications. Green Chem 16:1700–1715

Rajpurohit SR, Dave HK (2019) Analysis of tensile strength of a fused filament fabricated PLA part using an open-source 3D printer. Int J Adv Manuf Technol 101:1525–1536. https://doi.org/10.1007/s00170-018-3047-x

RamKumar Pandian S, Deepak V, Kalishwaralal K et al (2010) Optimization and fed-batch production of PHB utilizing dairy waste and sea water as nutrient sources by Bacillus megaterium SRKP-3. Bioresour Technol 101:705–711. https://doi.org/10.1016/j.biortech.2009.08.040

Reddy CSK, Ghai R, Rashmi, Kalia VC (2003) Polyhydroxyalkanoates: an overview. Bioresour Technol 87:137–146

Reddy G, Altaf M, Naveena BJ et al (2008) Amylolytic bacterial lactic acid fermentation – a review. Biotechnol Adv 26:22–34

Rehm BHA (2003) Polyester synthases: natural catalysts for plastics. Biochem J 376:15–33

Reis M, Albuquerque M, Villano M, Majone M (2011) Mixed culture processes for polyhydroxyalkanoate production from agro-industrial surplus/wastes as feedstocks. In: Comprehensive biotechnology, 2nd edn. Elsevier, pp 669–683

Rivero CP, Hu Y, Kwan TH et al (2016) Bioplastics from solid waste. Curr Dev Biotechnol Bioeng Solid Waste Manag:1–26. https://doi.org/10.1016/B978-0-444-63,664-5.00001-0

Salusjärvi L, Havukainen S, Koivistoinen O, Toivari M (2019) Biotechnological production of glycolic acid and ethylene glycol: current state and perspectives. Appl Microbiol Biotechnol. https://doi.org/10.1007/s00253-019-09640-2

Samui AB, Kanai T (2019) Polyhydroxyalkanoates based copolymers. Int J Biol Macromol 140:522–537. https://doi.org/10.1016/j.ijbiomac.2019.08.147

Sanchez-Vazquez SA, Hailes HC, Evans JRG (2013) Hydrophobic polymers from food waste: resources and synthesis. Polym Rev 53:627–694. https://doi.org/10.1080/15583724.2013.834933

Sandhya M, Aravind J, Kanmani P (2013) Production of polyhydroxyalkanoates from Ralstonia eutropha using paddy straw as cheap substrate. Int J Environ Sci Technol 10:47–54. https://doi.org/10.1007/s13762-012-0070-6

Sanko V, Sahin I, Aydemir Sezer U, Sezer S (2019) A versatile method for the synthesis of poly(glycolic acid): high solubility and tunable molecular weights. Polym J 51:637–647. https://doi.org/10.1038/s41428-019-0182-7

Sasa LA, Yearley EJ, Jablin MS, et al (2011) Shear-induced metastable states of end-grafted polystyrene. https://doi.org/10.1103/PhysRevE.84.021803

Semba T, Kitagawa K, Ishiaku US, Hamada H (2006) The effect of crosslinking on the mechanical properties of polylactic acid/polycaprolactone blends. J Appl Polym Sci 101:1816–1825. https://doi.org/10.1002/app.23589

Serafim LS, Lemos PC, Albuquerque MGE, Reis MAM (2008) Strategies for PHA production by mixed cultures and renewable waste materials. Appl. Microbiol. Biotechnol. 81:615–628

Shinoda H, Asou Y, Kashima T et al (2003) Amphiphilic biodegradable copolymer, poly(aspartic acid-co-lactide): acceleration of degradation rate and improvement of thermal stability for poly(lactic acid), poly(butylene succinate) and poly(−Caprolactone). Polym Degrad Stab 80:241–250. https://doi.org/10.1016/S0141-3910(02)00404-4

Simmons H, Tiwary P, Colwell JE, Kontopoulou M (2019) Improvements in the crystallinity and mechanical properties of PLA by nucleation and annealing. Polym Degrad Stab 166:248–257. https://doi.org/10.1016/j.polymdegradstab.2019.06.001

Singhvi M, Zendo T, Sonomoto K (2018) Free lactic acid production under acidic conditions by lactic acid bacteria strains: challenges and future prospects. Appl Microbiol Biotechnol 102:5911–5924

Skoczinski P, Carus M, de Guzman D, Käb H, Chinthapalli R, Ravenstijn J, Baltus W, Raschka A (2021) Bio-based Building Blocks and Polymers – Global Capacities, Production and Trends 2020–2025

Soccio M, Lotti N, Finelli L et al (2008) Influence of transesterification reactions on the miscibility and thermal properties of poly(butylene/diethylene succinate) copolymers. Eur Polym J 44:1722–1732. https://doi.org/10.1016/j.eurpolymj.2008.03.022

Soccio M, Lotti N, Gazzano M et al (2012) Molecular architecture and solid-state properties of novel biocompatible PBS-based copolyesters containing sulphur atoms. React Funct Polym 72:856–867. https://doi.org/10.1016/j.reactfunctpolym.2012.08.002

Södergård A, Stolt M (2002) Properties of lactic acid based polymers and their correlation with composition. Prog Polym Sci 27:1123–1163

Su S, Kopitzky R, Tolga S, Kabasci S (2019) Polylactide (PLA) and its blends with Poly(butylene succinate) (PBS): a brief review. Polymers (Basel) 11:1193. https://doi.org/10.3390/polym11071193

Sudesh K, Abe H, Doi Y (2000) Synthesis, structure and properties of polyhydroxyalkanoates: biological polyesters. Prog Polym Sci 25:1503–1555

Suyatma NE, Copinet A, Tighzert L, Coma V (2004) Mechanical and barrier properties of biodegradable films made from chitosan and poly (lactic acid) blends. J Polym Environ 12:1–6. https://doi.org/10.1023/B:JOOE.0000003121.12800.4e

Takahashi H, Ishikawa H, Tanaka A (2017) Regenerative medicine for Parkinson's disease using differentiated nerve cells derived from human buccal fat pad stem cells. Hum Cell 30:60–71. https://doi.org/10.1007/s13577-017-0160-3

Tan SG, Chow WS (2010) Biobased epoxidized vegetable oils and its greener epoxy blends: a review. Polym Plast Technol Eng 49:1581–1590

Tan L, Hu J, Ye S et al (2014) Crystallization and degradation behaviors of poly(butylene succinate)/poly(Z-l-lysine) composites. Thermochim Acta 575:279–284. https://doi.org/10.1016/j.tca.2013.11.016

Tang Y, Bu L, He J, Jiang J (2013) l(+)-Lactic acid production from furfural residues and corn kernels with treated yeast as nutrients. Eur Food Res Technol 236:365–371. https://doi.org/10.1007/s00217-012-1865-x

Tanzi MC, Verderio P, Lampugnani MG et al (1994) Cytotoxicity of some catalysts commonly used in the synthesis of copolymers for biomedical use. J Mater Sci Mater Med 5:393–396. https://doi.org/10.1007/BF00058971

Tashiro Y, Matsumoto H, Miyamoto H et al (2013) A novel production process for optically pure l-lactic acid from kitchen refuse using a bacterial consortium at high temperatures. Bioresour Technol 146:672–681. https://doi.org/10.1016/j.biortech.2013.07.102

Tiwary P, Kontopoulou M (2018) Tuning the rheological, thermal, and solid-state properties of branched PLA by free-radical-mediated reactive extrusion. ACS Sustain Chem Eng 6:2197–2206. https://doi.org/10.1021/acssuschemeng.7b03617

Tokiwa Y, Calabia BP, Ugwu CU, Aiba S (2009) Biodegradability of plastics. Int J Mol Sci 10:3722–3742

Totaro G, Sisti L, Celli A et al (2016) Poly(butylene succinate) bionanocomposites: a novel bio-organo-modified layered double hydroxide for superior mechanical properties. RSC Adv 6:4780–4791. https://doi.org/10.1039/c5ra24031j

Tsang YF, Kumar V, Samadar P et al (2019) Production of bioplastic through food waste valorization. Environ Int 127:625–644

Tsuji H, Fukui I (2003) Enhanced thermal stability of poly(lactide)s in the melt by enantiomeric polymer blending. Polymer (Guildf) 44:2891–2896. https://doi.org/10.1016/S0032-3861(03)00175-7

Tsuji H, Yamasaki M, Arakawa Y (2019) Stereocomplex formation between enantiomeric alternating lactic acid-based copolymers as a versatile method for the preparation of high performance biobased biodegradable materials. ACS Appl Polym Mater 1:1476–1484. https://doi.org/10.1021/acsapm.9b00232

Ueno T, Ozawa Y, Ishikawa M et al (2003) Lactic acid production using two food processing wastes, canned pineapple syrup and grape invertase, as substrate and enzyme. Biotechnol Lett 25:573–577. https://doi.org/10.1023/A:1022888832278

van den Oever M, Molenveld K (2017) Replacing fossil based plastic performance products by bio-based plastic products-technical feasibility. N Biotechnol 37:48–59. https://doi.org/10.1016/j.nbt.2016.07.007

Van Wegen RJ, Ling Y, Middelberg APJ (1998) Industrial production of polyhydroxyalkanoates using escherichia coli: an economic analysis. Chem Eng Res Des 76:417–426. https://doi.org/10.1205/026387698524848

Vasala A, Panula J, Neubauer P (2005) Efficient lactic acid production from high salt containing dairy by-products by Lactobacillus salivarius ssp. salicinius with pre-treatment by proteolytic microorganisms. J Biotechnol 117:421–431. https://doi.org/10.1016/j.jbiotec.2005.02.010

Venkateswar Reddy M, Venkata Mohan S (2012) Influence of aerobic and anoxic microenvironments on polyhydroxyalkanoates (PHA) production from food waste and acidogenic effluents using aerobic consortia. Bioresour Technol 103:313–321. https://doi.org/10.1016/j.biortech.2011.09.040

Venkateswar Reddy M, Amulya K, Rohit MV et al (2014) Valorization of fatty acid waste for bioplastics production using Bacillus tequilensis: Integration with dark-fermentative hydrogen production process. Int J Hydrog Energy 39:7616–7626. https://doi.org/10.1016/j.ijhydene.2013.09.157

Venus J (2006) Utilization of renewables for lactic acid fermentation. Biotechnol J 1:1428–1432. https://doi.org/10.1002/biot.200600180

Venus J, Richter K (2006) Production of lactic acid from barley: strain selection, phenotypic and medium optimization. Eng Life Sci 6:492–500. https://doi.org/10.1002/elsc.200520136

Vert M (2009) Degradable and bioresorbable polymers in surgery and in pharmacology: beliefs and facts. J Mater Sci Mater Med 20(2):437–446

Vishnu C, Naveena BJ, Altaf M et al (2006) Amylopullulanase - A novel enzyme of L. amylophilus GV6 in direct fermentation of starch to L(+) lactic acid. Enzyme Microb Technol 38:545–550. https://doi.org/10.1016/j.enzmictec.2005.07.010

Voirin C, Caillol S, Sadavarte NV et al (2014) Functionalization of cardanol: towards biobased polymers and additives. Polym Chem 5:3142–3162

Wada H (2007) Biomechanics at micro- and nanoscale levels. World Scientific

Wang R, Schuman TP (2012) Vegetable oil-derived epoxy monomers and polymer blends: a comparative study with review. Express Polym Lett 7:272–292. https://doi.org/10.3144/expresspolymlett.2013.25

Wang L, Chen J, Liu H et al (2004) Synthesis and evaluation of biodegradable segmented multiblock poly(ether ester) copolymers for biomaterial applications. Polym Int 53:2145–2154. https://doi.org/10.1002/pi.1645

Wang T, Zhang L, Li D et al (2008a) Mechanical properties of polyurethane foams prepared from liquefied corn stover with PAPI. Bioresour Technol 99:2265–2268. https://doi.org/10.1016/j.biortech.2007.05.003

Wang W, Zhang Y, Zhu M, Chen Y (2008b) Effect of graft modification with poly(N-vinylpyrrolidone) on thermal and mechanical properties of poly (3-hydroxybutyrate-co-3-hydroxyvalerate). J Appl Polym Sci 109:1699–1707. https://doi.org/10.1002/app.27983

Wang M, Xu Z, Qiu T et al (2010) Kinetics of lactic acid fermentation on food waste by lactobacillus bulgaricus. Adv Mater Res 113:1235–1238

Water JJ, Bohr A, Boetker J et al (2015) Three-dimensional printing of drug-eluting implants: preparation of an antimicrobial polylactide feedstock material. J Pharm Sci 104:1099–1107. https://doi.org/10.1002/jps.24305

Wei L, McDonald AG (2015) Peroxide induced cross-linking by reactive melt processing of two biopolyesters: poly(3-hydroxybutyrate) and poly(<scp>l</scp> -lactic acid) to improve their melting processability. J Appl Polym Sci 132:n/a–n/a. https://doi.org/10.1002/app.41724

Weise B, Huysman S, Manvi P, Theunissen L (2018) PBS-based fibres for renewable textiles. Bioplastics Magazine. https://www.bioplasticsmagazine.com/en/magazine/online-archive/data/20180524.php. Accessed 14 Dec 2019

Werpy T, Petersen G (2004) Top value added chemicals from biomass. Volume I. US NREL medium: ED; size, 76 pp. https://doi.org/10.2172/15008859

Williams DF (1986) Definitions in Biomaterials. in Proceedings of a Consensus Conference of the European Society for Biomaterials (Chester: Elsevier).

Wu Y, Ma H, Zheng M, Wang K (2015) Lactic acid production from acidogenic fermentation of fruit and vegetable wastes. Bioresour Technol 191:53–58. https://doi.org/10.1016/j.biortech.2015.04.100

Xie Q, Han L, Shan G et al (2016) Polymorphic crystalline structure and crystal morphology of enantiomeric poly(lactic acid) blends tailored by a self-assemblable aryl amide nucleator. ACS Sustain Chem Eng 4:2680–2688. https://doi.org/10.1021/acssuschemeng.6b00191

Xu J, Guo BH (2010) Poly(butylene succinate) and its copolymers: research, development and industrialization. Biotechnol J 5:1149–1163

Yamada T, Jung DY, Sawada R, Tsuchiya T (2008) Intracerebral microinjection of stannous 2-ethylhexanoate affects dopamine turnover in cerebral cortex and locomotor activity in rats. J Biomed Mater Res Part B Appl Biomater 87:381–386. https://doi.org/10.1002/jbm.b.31115

Yamane K, Sato H, Ichikawa Y et al (2014) Invited review development of an industrial production technology for high-molecular-weight polyglycolic acid. Polym J 46:769–775. https://doi.org/10.1038/pj.2014.69

Yemul OS, Kamble RD, Thamke VR et al (2016) Preparation of biodegradable bioepoxy resins from waste vegetable oil for sustainable development. J Surf Sci Technol 32:35–45. https://doi.org/10.18311/jsst/2016/6600

Yu L, Dean K, Li L (2006) Polymer blends and composites from renewable resources. Prog Polym Sci 31:576–602. https://doi.org/10.1016/j.progpolymsci.2006.03.002

Yue H, Ling C, Yang T et al (2014) A seawater-based open and continuous process for polyhydroxyalkanoates production by recombinant Halomonas campaniensis LS21 grown in mixed substrates. Biotechnol Biofuels 7. https://doi.org/10.1186/1754-6834-7-108

Zahari MAKM, Ariffin H, Mokhtar MN et al (2015) Case study for a palm biomass biorefinery utilizing renewable non-food sugars from oil palm frond for the production of poly(3-hydroxybutyrate) bioplastic. J Clean Prod 87:284–290. https://doi.org/10.1016/j.jclepro.2014.10.010

Zakharova E, Alla A, Martínez De Ilarduya A, Muñoz-Guerra S (2015) Bio-based PBS copolyesters derived from a bicyclic d-glucitol. RSC Adv 5:46395–46404. https://doi.org/10.1039/c5ra03844h

Zembouai I, Bruzaud S, Kaci M et al (2014) Poly(3-hydroxybutyrate-co-3-hydroxyvalerate)/polylactide blends: thermal stability, flammability and thermo-mechanical behavior. J Polym Environ 22:131–139. https://doi.org/10.1007/s10924-013-0626-7

Zeng JB, Huang CL, Jiao L et al (2012) Synthesis and properties of biodegradable poly(butylene succinate-co-diethylene glycol succinate) copolymers. Ind Eng Chem Res 51:12258–12265. https://doi.org/10.1021/ie300133a

Zhang C, Su H, Baeyens J, Tan T (2014) Reviewing the anaerobic digestion of food waste for biogas production. Renew Sustain Energy Rev 38:383–392

Zhang H, Bai H, Deng S et al (2019) Achieving all-polylactide fibers with significantly enhanced heat resistance and tensile strength via in situ formation of nanofibrilized stereocomplex polylactide. Polymer (Guildf) 166:13–20. https://doi.org/10.1016/j.polymer.2019.01.040

Zhao Y, Huang C, Bei R et al (2018) Barrier functionality of SiOx layers and their effect on mechanical properties of SiOx/PLA composite films. J Coat Technol Res 15:505–514. https://doi.org/10.1007/s11998-017-0005-8

Zhao X, Cornish K, Vodovotz Y (2020) Narrowing the gap for bioplastic use in food packaging: an update. Environ Sci Technol 54:4732. https://doi.org/10.1021/acs.est.9b03755

Zhou YX, Huang ZG, Diao XQ et al (2015) Characterization of the effect of REC on the compatibility of PHBH and PLA. Polym Test 42:17–25. https://doi.org/10.1016/j.polymertesting.2014.12.014

Chapter 7
Biocomposites from Food Waste

Teresa Cecchi

The development and production of high-performance engineering materials from renewable resources are pursued by industry and academia investigators worldwide.

In general, composite materials consist of two (or more) separate and distinct components: the matrix and the filler. The final multi-phase finished structure differentiates composites from mixtures and solid solutions (Fazeli et al. 2019). The matrix is often important to protect the filler from environmental degradation and mechanical damage. The physical and/or chemical properties of the composite can be very different from those of the individual constituent materials. The composite structure may be preferred for many reasons. For example, it might be stronger, lighter, better from the mechanical point of view, or less expensive when compared to constituent materials. Furthermore, the interfacial bonding between the filler and the polymer matrix is easily predicted to alter the mechanical properties of the composite compared to the unfilled polymer.

The filler, which is often a reinforcement phase of stiff, strong material, regularly fibrous in nature, is embedded in the continuous matrix phase. Conventional reinforcing fillers comprise calcium carbonate, glass fibers, talc, kaolin, mica, silica, and graphite.

Natural, biodegradable, nontoxic plant fibers such as cotton, flax, hemp, sisal, kenaf, and the like, including byproducts from food crops or fibers from recycled wood or waste paper, have been deeply investigated (Vijayan and Krishnamoorthy 2019) since the strength and stiffness of the fibers may be exploited (Bashir and Manusamy 2015). Carbon nanomaterials, cellulose, clays, and nanometals, have also been used to improve bioplastic mechanical properties (Rivera-Briso and Serrano-Aroca 2018).

Improved awareness that nonrenewable resources are becoming limited and our inevitable dependence on renewable resources have arisen (Fazeli et al. 2019). Hence, food solid waste materials from both vegetal and animal sources have turned out to be an attractive biofiller from the economic and environmental points of view, and represent an eco-friendly, circular, and sustainable alternative to conventional

T. Cecchi, C. De Carolis, *Biobased products from food sector waste*,
https://doi.org/10.1007/978-3-030-63436-0_7

reinforcing fillers. The typical chemical complexity of food waste (FW) can be profitably exploited for the production of a vast gamut of biocomposite.

Even if the term biocomposite does mean that all parts of the composites are biobased, the trend is moving in that direction. However, the terms biocomposites, green composites, and ecocomposites cover composite materials made from the combination of:

- biopolymers (thermosettings and thermoplastics) with biobased fillers
- fossilbased plastics with biobased fillers
- biopolymers with inorganic fillers

The biocomposites are prepared from opted precursors to achieved specific properties.

At least one of the constituents is based on biobased renewable resources. Anyhow only biopolymers with natural fillers are fully environmentally friendly (John and Thomas 2008). In terms of life cycle analysis (LCA), if both matrix and filler are obtained from byproducts or even better waste, their use may represent as such a reduction in the environmental impact of the whole process. The hydrophilic character of biopolymers facilitates the successful development of environmentally friendly biocomposites, as most biobased fillers are also hydrophilic in nature.

Currently, green biocomposites are already available in markets for various applications such as automotive, construction, and buildings components. The volume and number of utilizations of biocomposites have infiltrated and vanquished new showcases for regular items or refined specialty applications perseveringly.

Solid, dependable, lightweight, eco-accommodating, and bio-degradable materials are the focus of interest of many researchers. Biofillers prepared from FW, such as natural fibers from crop processing byproducts (e.g., wheat straw, corn stalks, hemp, flax, coconut, oil palm fibers, banana, pineapple, and sugarcane wastes), eggshell, shellfish shell, and mussel shell, among others have received considerable attention in the last few years (Ku et al. 2011).

Upcycling of the FW to produce new biocomposites with new physical/chemical characteristics (i) avoids the cost of FW disposal and potential pollutant emissions; (ii) reduces biobased composites price and this is crucial since expensiveness prevents the take-off of this sector; (iii) avoids taking edible resources as a starting material for biobased composites; (iv) contributes to the biodegradability of the new material and CO_2 neutrality; (v) results in a biobased output different from the usual ones (bio-fuels or bio-energy).

Material science put forth more materials day by day, and the material database grows exponentially. This chapter illustrates an overview of the science, technology, applications, and challenges in relation to biocomposites recently produced from various FW; a case study concerning biocomposites from polylactic acid (PLA), the most important biopolymer, and inorganic fillers is discussed. Chapter 8 details the mechanical testing and characterization of bioplastics and biocomposites.

7.1 Biocomposites from Biobased Thermosetting Polymers Containing FW Biofillers

7.1.1 Phenolic Resins

Various natural fibers were used as biobased phenolic resins reinforcement.

Cashew nut shell liquid-based novalac or resole resins, discussed in Sect. 6.6.1, were used as a matrix to obtain biocomposites with oil palm fibers and clay for the production of roofing material (Tiamiyu and Ibitoye 2012). Rice husks (Ofem et al. 2012), and coffee husks (Bisanda et al. 2003) were also used as reinforcement. Natural fibers from these FWs are usually pretreated with a 5–12% sodium hydroxide solution to increase the number of hydroxyl groups on the surface of the cellulosic fiber.

Lignin modified phenol formaldehyde resins are used as matrices in natural fiber biocomposites based on flax and kenaf fibers (Mahendran et al. 2013).

Barbosa et al. synthesized phenol–formaldehyde–tannin resin and used the tannin–phenolic matrix in the preparation of biocomposites reinforced with coir fibers from coconut waste (Barbosa et al. 2010). It can be expected that lignin fibers derived from other FW can be used for the reinforcement since the lignin content of this fiber results in a high concentration of aromatic rings, which amplifies the compatibility with the matrix.

7.1.2 Epoxy and Polyurethane Resins

In order to improve the thermal and mechanical properties of epoxidized vegetable oil thermosetting polymers discussed in Sect. 6.6.2, vegetable oil-based biocomposites can be obtained by adding fillers.

Flax fiber is commonly used in the field of natural fiber biocomposites based on epoxidized soybean oil thermoset (Adekunle et al. 2010) to increase their strength and stiffness (Liu and Erhan 2008); in this respect, flax fiber could potentially substitute for glass or carbon fibers as reinforcements: the longer the fiber length is, the better the mechanical properties are (Liu et al. 2006).

Flax fiber biocomposites with a methacrylated epoxidized sucrose soyate resin was synthesized, and the nonlinear viscoelastic behavior due to the bond type and structure was studied; it is important for structural applications as the long-term creep behavior of these materials turns into a significant issue for applications in housing and automotive industries (Amiri et al. 2016). Soybean oil-based resin was also composited with both flax mat and recycled paper; the biocomposite was used for a monolithic roof structure, suitable for residential construction; the biobased prototype roof satisfied strength, stiffness, failure, deflection, and safety criteria (Hu et al. 2007).

Oil-based (corn, soybean, fish, and linseed) resins and agricultural fibers (corn stove, wheat straw, and switchgrass) green composites were studied; with agricultural fiber loadings of 75 wt%, they have thermal stabilities up to 275 °C. An increase in the level of unsaturation of the base oil resulted in improvements in the thermal and mechanical properties of the biocomposites (Pfister and Larock 2012).

Banana fibers were used in biobased castor oil polyurethanes biocomposites as reinforcement. The banana fibers were treated with 10 wt.% sodium hydroxide solutions for 1 h to increase the number of hydroxyl groups on the surface of the cellulosic fiber. Mechanical properties of the polyurethane biocomposites increased with increasing fiber volume fraction and length for the untreated and treated banana fiber and were better for the treated banana fiber composites due to the stronger interfacial interactions and adhesion between the treated fibers and the polyurethane matrix (Merlini et al. 2011).

Keratin fibers have a hollow core; they are a light and tough material. They were incorporated into acrylated epoxidized soybean oil. The lightweight biocomposites have a density $\rho \approx 1$ g/cm^3, when the keratin feather fibers volume fraction is 30%. The presence of a significant volume of air in the hollow structure of the fibers is the rationale for a dielectric constant of the biocomposite material as low as 1.7, significantly lower than the conventional silicon dioxide or polymer dielectric insulators. The coefficient of thermal expansion (67.4 ppm/°C) was adequate for electronic application and similar to the value of materials used in printed circuit boards. The incorporation of keratin feather fibers in the soy oil polymer improved the mechanical properties such as storage modulus, fracture toughness, and flexural properties. The biocomposite characteristics, which could be further tuned by fiber selection, make it an eligible material for electronic, automotive, and aeronautical applications (Hong and Wool 2005).

7.2 Biocomposites from Biobased Thermoplastics Containing FW Biofillers

Biobased fillers in combination with biodegradable thermoplastics provide fully biobased and biodegradable solutions. Polybutylene succinate (PBS), Polyhydroxyalkanoates (PHA), and Polylactic acid (PLA) are the most common matrices.

PBS shows a good binding to natural fibers without any additional bonding agent (Nova-Institue 2015). PBS-based biocomposites were prepared by blending the polyester with organic materials, like chitosan (Costa-Pinto et al. 2014) and carboxymethylchitosan (Meesap et al. 2010), collagen (Kosorn et al. 2010), or inorganic ones such as calcium phosphate (Patntirapong et al. 2015), hydroxyapatite (Kosorn et al. 2010), fluoroapatite (Niu et al. 2014). These biocomposites were proposed as scaffolds for tissue engineering and bone regeneration; hence cell proliferation and inflammatory responses were often investigated.

PHA-based biocomposites were deeply studied. Potato peel waste fermented residue was utilized for the preparation of biocomposites of tunable properties with Polyhydroxybutyrate (PHB). The biocomposite had worse mechanical strength but a higher biodegradation rate as compared with pure PHB. The degradation rate increased with increasing fiber content (Wei et al. 2015), and this offers new prospects for fast degrading biomaterials in agricultural and horticultural applications.

Lignocellulosic particles from byproducts of the coffee industry (coffee parchment and husk) were incorporated with PHB as a promising reinforcing filler in the biocomposites. The modification of thermal, mechanical, microstructural, and physical properties of the biocomposite was evaluated. It was found that its thermal stability and mechanical properties were enhanced in comparison to those of the neat polymer, while there was an inverse relationship between the degree of crystallinity and inclusion of coffee parchment. The water absorption of the biocomposite was directly proportional to the coffee waste particle added (Reis et al. 2015).

The suitability of other lignocellulosic wastes such as almond shells, rice husk, and seagrass as fillers in PHB/fiber biocomposites applications were tested. All the fillers showed a reinforcing effect. Fibers did not affect the crystallinity and the disintegration rate of PHB but decreased the thermal stability and the water barrier performance of the biocomposites were. The addition of almond shell resulted in the best balance of properties in terms of permeability and mechanical properties (Sánchez-Safont et al. 2018).

Dried distillers grains with solubles, a major co-product of the bioethanol industry, served as filler for the production of "green" composites with enhanced biodegradability, using biodegradable poly(butylene adipate-co-terephthalate) (PBAT) and PHAs as polymer matrices (Muniyasamy et al. 2013; Madbouly et al. 2014).

Aranberri and co-workers developed biocomposites using PLA, polybutyrateadipate terephthalate (PBAT), and a PLA/thermoplastic copolyester blend as biopolymeric matrixes; chicken feathers were used as fillers with a content of 50 and 60 wt %. The studied biocomposites were manufactured with a torque rheometer. Due to the low tensile strength of chicken feathers, the resulting materials were worse than the neat polymers as regards the mechanical properties. Anyhow, biocomposites resulted in lightweight and thermal-insulating materials when compared with neat bioplastics. The alkali treatment of chicken feathers and the addition of a plasticizer like polyethylene glycol increased the wettability and the adhesion between chicken feathers and the matrix. The developed biocomposite can be utilized for the panel component, flooring, or building material, thereby valorizing chicken feather waste as a renewable material (Aranberri et al. 2017).

PLA is actually one of the most popular matrices for fully biobased composites. PLA/banana fiber biocomposites were prepared by the melt blending method. A silane coupling agent (triethoxy-vinylsilane) and chemical modification enabled the conjugation of banana fiber onto PLA chains. The stability and mechanical properties of the PLA were dramatically increased: the tensile and flexural strengths of the biocomposites were, respectively, 78.6 and 65.4 MPa, 2 and 1.66 times higher compared to pristine PLA. The heat deflection temperature increased from 62 °C to

139 °C; the incorporation of banana fiber would reduce the production cost of materials while meeting the demands of environmental protection agencies (Shih and Huang 2011). Better thermal stability was confirmed in another study concerning PLA/banana fiber biocomposites: the thermograms showed higher glass transition and melting transition temperatures due to the more restricted macromolecular movement (Jandas et al. 2012).

The nanocrystalline cellulose from waste oil palm empty fruit bunch fiber was funtionalized and used in reinforced PLA/PHBV bionanocomposite films, prepared by solution casting techniques; these bionanocomposites have potential application as packaging material due to improved mechanical properties and lower oxygen permeability rate as compared to the polymer blend (Dasan et al. 2017).

Pecan nutshells are a waste obtained from the pecan nut production; they were used as a cheap filler for incorporation by melt blending into PLA up to 7.5% wt. Both untreated and defatted fillers were used. Some mechanical and thermal characteristics of the biocomposites worsened, but biocomposites formulated with defatted pecan nut powder can be used as food containers, packaging trays, or disposable items (Álvarez-Chávez et al. 2017).

As regards PLA/coconut shell powder biocomposites, the increase of the coconut shell powder content decreased the tensile strengths and elongations at break of the biocomposites, but increased their modulus of elasticity. The presence of a coupling agent, 3-aminopropyltriethoxysilane, gave a stronger filler–matrix interaction, which in turn resulted in enhanced tensile strengths, modulus of elasticity, and thermal stability. The coconut shell powder had a nucleating effect (Chun et al. 2012).

Farm dairy effluent represents an untapped side stream that can provide both the PLA monomer (lactic acid from lactose) and a biofiller to produce PLA/farm dairy effluent biocomposite. Dairy effluents contain cellulose fibres, carbohydrates, proteins, urea, and mineral salts (ammonium, phosphates, chlorides, potassium, calcium, and magnesium). PLA was compounded with the dried and grannulated dairy effluent. The Young's modulus of the biocomposite was higher than the neat PLA; however, the tensile strength generally decreased. Improved stiffness and UV accelerated degradation were obtained (Le Guen et al. 2017).

Cow dung of average 4 mm size was blended with PLA at different ratios (0–50 wt%). The results signified an improvement in the flexural properties, whereas the tensile and impact strength as well as the thermal stability decreased with the addition of 50% cow dung due to micro-cracking and voids formation at higher filler content (Ketabchi et al. 2017).

PLA /chicken feather fiber biocomposites were processed using a twin-screw extruder and an injection molder. The tensile moduli of the biocomposites with different filler content (2, 5, 8, and 10 wt%) were higher than that of pure PLA, and a maximum value of 4.2 GPa was attained with 5 wt% of CFF without any substantial weight increment. Uniform dispersion of chicken feather fiber in the PLA matrix was assessed by scanning electron microscopy. The addition of chicken feather fiber enhanced the thermal stability of the biocomposites as compared to pure PLA (Liu et al. 2009).

Maleic-anhydride-grafted polylactide (PLA-g-MA) and treated cross-linked spent coffee grounds were compounded. The dispersion of coffee grounds in the PLA-g-MA matrix was highly homogeneous due to ester formation and resulted in branched and cross-linked macromolecules with good processability and water resistance (Wu 2015).

Cecchi and co-workers (Cecchi et al. 2019) used ceramic FW powder fillers from eggshells and mussel shells to obtain PLA biocomposites that contain an unprecedentedly high amount of filler, equal to 140 over 100 parts of PLA. The volatile organic compounds emitted from PLA and, for the very first time, from its biocomposites were analyzed via headspace-solid phase microextraction-gas chromatography-mass spectrometry (HS-SPME-GC-MS). The molecular fingerprint of the volatiles is very simple and comprises only three aldehydes, a ketone, and two lactides. Scanning electron microscopy (Fig. 7.1) indicated a sufficiently strong biofiller/biopolymer interface; hence both the eggshells and the mussel shells can be considered suitable fillers.

Durian is a tropical fruit well renowned for its nutritional value. Delignification of durian rind gave celluose, proposed as reinforcement element in biocomposite materials (Penjumras et al. 2014).

Collagen from goatskins was used for the preparation of multifunctional biocomposite films. The flexible, semi-transparent, conductive, and magnetic

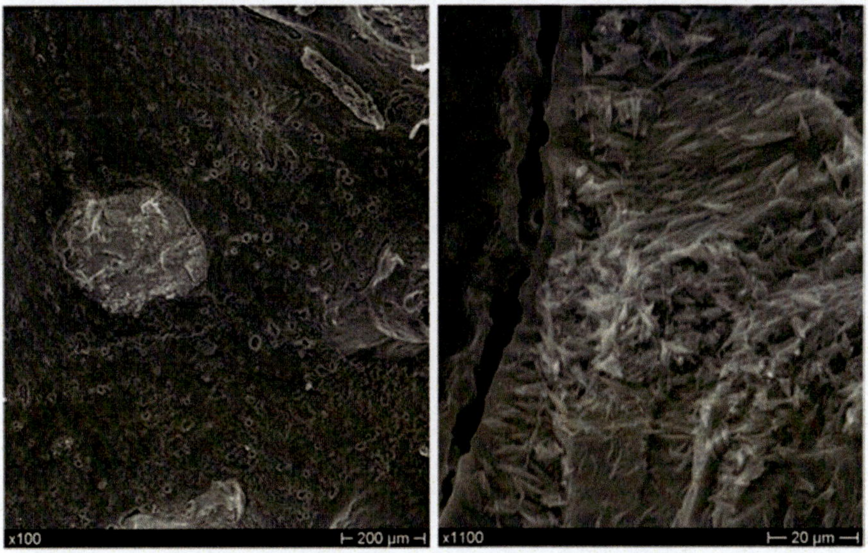

Fig. 7.1 Scanning electron microscopy images of PLA/eggshell powder biocomposite. Reprinted by permission from Springer Nature: Environmental Science and Pollution Research. Unprecedented high percentage of food waste powder filler in poly lactic acid green composites: synthesis, characterization, and volatile profile. Cecchi T., Giuliani A., Iacopini F., Santulli C., Sarasini F., Tirillò J., 26:7263–7271, 2019, https://www.springer.com/journal/11356 (Cecchi et al. 2019)

biocomposite films of micron-thickness were prepared combing collagen extracted from trimmed waste, chitosan, and conductive magnetic graphitic nanocarbon obtained by simple collagen waste carbonization. The electrical conductivity of the biocomposite increased with increasing chitosan and magnetic graphitic nanocarbon concentration. The highest tensile strength of the film was obtained with 10 wt.% addition of magnetic graphitic nanocarbon. This biocomposite can be utilized for magnetic tracking and actuation ferromagnetic like behavior of the graphitic nanocarbon (Ashokkumar et al. 2012).

7.3 Biocomposites from FossilBased Plastics and Elastomers Containing FW Biofillers

The integration of renewable resources driven from FW into fossilbased thermoplastics (such as PP, LDPE, HDPE, etc.), thermosettings (such as epoxy, phenolic, polyurethane, and polyester resins) but also elastomers is interesting from both the environmental and technical points of view (Bashir and Manusamy 2015). Biocomposites obtained from the incorporation of FW biofillers into fossilbased polymers are not fully biodegradable due to the fossilbased component. Their physical, chemical, mechanical, and thermal properties are extremely variable. For example, eggshell-based filler increased the stiffness of PP and performed better than neat calcium carbonate (Toro et al. 2007). Surface modification of clam-shells by furfural enabled a good dispersion of modified powder in PP matrix, which in turn resulted in improved heterogeneous nucleation, thermal stability, and mechanical properties; at the maximum incorporation content (15 wt.%) a good balance between stiffness and toughness of PP biocomposites was achieved (Yao et al. 2013). Improved heterogeneous nucleation for PP crystallization was also observed when mussel shell powder was used as a filler (Li et al. 2012a). The use of shellfish shell as a filler (Li et al. 2012b) is significant not only to obtain new biocomposites but also to solve the serious environmental hazard related to their disposal.

Wastes from plants have been often utilized as biofillers; vegetal fibers are meant to reinforce the polymer. The poor interfacial adhesion between the polar-hydrophilic plant waste and nonpolar-hydrophobic polymers is usually improved using coupling agents such as maleate, silane, a copolymer of ethylene and methyl acrylate, namely ethylene/n-butyl acrylate/glycidyl methacrylate copolymer. The use of vegetable waste can result in higher hygroscopicity of biocomposites, which increases as the filler content increases. Again, myriads of modifications of the polymer matrixes are reported (Bashir and Manusamy 2015).

Some examples of biocomposites from fossilbased plastics and elastomers containing FW biofillers arc briefly detailed Table 7.1 below. Better mechanical properties compared to the neat polymer are usually claimed by the Authors.

Table 7.1 Biocomposites from fossil based plastics and FW biofillers

Biofiller	Matrix	Parcle size	Biofiller content	References
Barley husk	PP	50–300μm	40%	Bledzki et al. (2010)
Cassava pulp	Natural rubber	150–250μm	5 parts per hundred rubber	Ruangudomsakul et al. (2013)
Clam shell waste	PP	–	15%	Yao et al. (2013)
Coconut shell	PP	50–300μm	40%	Bledzki et al. (2010)
Coconut shell	Epoxy resin	<100μm	30%	Sarki et al. (2011)
Coconut shell	Polyester resin	304μm	0, 15, 30, 45, 60 parts per hundreds of total polymer	Husseinsyah and Zakaria (2011)
Coconut shell ash	Epoxy resin	53μm	5, 10, 15, 20, and 25 wt. %	Imoisili et al. (2012)
Eggshell	PP	8.4μm	40%	Toro et al. (2007)
Eggshell	PP	–	–	Kumar et al. (2014)
Eggshell	Styrene Butadiene rubber	Nanoparticles	5, 10, 15 parts per hundred rubber	Saeb et al. (2010)
Mussel shell	PP	40–500	3	Li et al. (2012a)
Popato peel	PP	100–1000μm	10–40%	Sugumaran et al. (2015)
Sago pith	PVA	–	40–50%	Yee et al. (2011)
Post-harvest tomato plant powder	PVA-co-ethylene	63–125μm	2–10%	Nisticò et al. (2017)
Rapeseed stalks	PP	0.420 mm	30, 45, 60%	Zabihzadeh et al. (2011)
Rice husk	Natural rubber latex foam	<300μm	0–10 parts per hundred rubber	Ramasamy et al. (2013)
Scallop shell	PP	–	5%	Lin et al. (2013)
Shellfish shell	PP	40–500 nm	2%	Li et al. (2012b)
Shrimp shells	LDPE	75μm	5, 10, 15, 20, 25, wt. %	Hussein et al. (2011)
Sugarcane bagasse fiber	Recycled PET	<3 mm	5%	Corradini et al. (2009)
Sugarcane bagasse fiber	PS	300 and 425μm	60–80%	Al Bakri et al. (2013)
Tea mill waste fiber	PP and HDPE	0.354 mm	0–50%	Dönmez Çavdar et al. (2011)

7.4 Biocomposites from Biobased Polymers Containing Inorganic Fillers. Case Study: PLA

Due to its intrinsic characteristics, PLA has a key position in the market of biopolymers, being one of the most promising candidates for future developments. PLA can be obtained from fermentable sugars contained in FW and from non-food cellulosic biomass and agricultural wastes, allowing carbon neutral cycles, as detailed in Sect. 6.2. It is used for various applications in the biomedical, packaging, textile sectors, but it still suffers from some shortcomings. In this section, we will deal with PLA biocomposites with inorganic fillers, as a case study. Its modification via the addition of reinforcing fibers, micro- and/or nanofillers can lead to specific end-use characteristics to diversify its uses by moving from the industrial sector of disposables to technical applications of durables (Murariu and Dubois 2016).

The incorporation of hydrophilic mineral fillers may trigger a decrease in mechanical properties due to the forces of attraction among the filler particles and the poor adhesion between the filler and the hydrophobic polymer matrix. Moreover, the hydrophilic nature of the filler may induce deleterious water absorption. Filler characteristics such as particle size, size distribution, specific surface area, and particle shape regulate segregation, aggregation, and the orientation of anisotropic particles. A stiff interphase, under the effect of external load, determines an inhomogeneous stress distribution. Inorganic fillers are sometimes surface-treated by additives able to transform the original hydrophilic surface into a hydrophobic one to mitigate mechanical properties worsening and increase moisture resistance. This way, water absorption, surface energy and melt viscosity decrease, whereas dispersion and processing characteristics improve. Surface modifiers such as stearic acid and other fatty acids or their salts, silanes, titanates, zirconates, etc., can be used (Rothon 1999; Zuiderduin et al. 2002).

For the fabrication of electrical/electronic devices or automotive items, flame retardant properties are of primary interest. Inorganic flame retardants such as aluminum and other metal hydroxides, organo-modified clays, and modified montmorillonite were profitably exploited (Cheng et al. 2012).

Talc is one of the most considered fillers for mixing with PLA in studies and patents (Harris and Lee 2008; Ouchiar et al. 2015; Yu et al. 2012; Kaynak and Erdogan 2016; Zou et al. 2015; Shi et al. 2010, 2015; Xiao et al. 2010). The addition of talc into PLA can speed up PLA crystallization because of a marked nucleating effect of the inorganic filler (Harris and Lee 2008) (Ouchiar et al. 2015): a 65-fold decrease of the isothermal crystallization half-times were reported upon addition of only 2% talc with respect to the neat PLA. This is interesting from the engineering point of view because it can lead to faster cycles of processing/injection molding, and the increased crystallinity is a key-parameter for the ultimate level of resistance at high temperature; the finer talc led to higher degrees of crystallinity even at 1 wt. % (Petchwattana et al. 2014) whereas the thermal stability and heat distortion temperature significantly increased in highly filled biocomposites. By adding 30 wt. % talc, the Young's modulus increased from a value of 4.1 GPa for the neat PLA up to 9.8 GPa for biocomposites (Shakoor and Thomas 2014). It follows that the presence of talc in PLA increases the stiffness, dimensional stability, and barrier properties while reducing the molding time and production costs.

Glass fibers are among the most common and performant reinforcements used in biocomposites. Bioactive glasses in particular are deeply studied for bone regeneration and dental defects because they influence cell behavior and function. When used to reinforce PLA, the resulting biocomposites show improved tensile and flexural strength and better impact properties with respect to neat PLA: the mechanical properties of bioactive glass fiber/PLA biocomposites can be increased to the level of bone tissue with filler percentages up to volume 40% (Pirhonen et al. 2001) and make them interesting for biomedical applications (Varila et al. 2012; Roether et al. 2002), and tissue-engineered chondral implants (Ahtiainen et al. 2015). Superior thermal properties and an impressive increase of the heat deflection temperature contribute to allowing their use as an eco-friendly alternative to fossil plastics.

Calcium phosphate ceramics show potential in the bone repair field, but their brittleness is a major limitation. PLA–hydroxyapatite biocomposites (Li et al. 2008; Shen et al. 2009; Armentano et al. 2010; Rakmae et al. 2012; Talal et al. 2009; Thanh et al. 2015; Zhou et al. 2006; Shikinami et al. 2005; Zhang et al. 2005), are of high interest especially for orthopedic applications since they combine the bone-bonding ability of hydroxyapatite with the absorbability of PLA. The surface modification of hydroxyapatite with silane coupling agents (Zhang et al. 2005) improved the interfacial adhesion and the mechanical characteristics of the bone fixation biocomposites. Carbon fibers characterized by great biocompatibility and high strength were associated with hydroxyapatite to meet the demands of the fixture of weight-bearing bones (Shen et al. 2009). Interestingly, due to its alkalinity, hydroxyapatite can neutralize the lactic acid formed from the degradation of PLA by hydrolysis. This avoids the negative effects of acidity (inflammatory reaction) on the body tissues (Armentano et al. 2010). PLLA/ β-tricalcium phosphate scaffolds were prepared by powder mixing/compression molding/low-temperature treating/particulate leaching technique to avoid harmful residual solvents and surfactants. The highly homogeneous scaffolds showed a good ability to form bone-like apatite in vitro and suitable cell adhesion (Kang et al. 2008).

Micro- and nanoparticles of calcium carbonate were used to produce, respectively, PLA micro- and nano- biocomposites. The treatment of these fillers with calcium stearate (1 wt. %) enhanced the mechanical properties of PLA-based biocomposites, especially for nanosized calcium carbonate up to 30 wt. % but a decreased thermal stability with respect to neat PLA, was observed (Kim et al. 2008).

Biocomposites suitable for the realization of bone implants were obtained from PDLLA and an association of β-tricalcium phosphate and calcium carbonate ($CaCO_3$) (Abert et al. 2016). Biodegradable skull implants comprised, on the inside, a biocomposite of PDLLA with $CaCO_3$ for faster degradation and the ingrowth of bone cells; on the outside, a biocomposite of PLLA with calcium phosphate ensured better mechanical stability (Schiller et al. 2004). Carbonated amorphous calcium phosphates was shown to be able to compensate acidity due to lactic acid during PLA degradation (Schiller and Epple 2003).

A morphological, mechanical, and biological comparison between calcium carbonate and β-tricalcium phosphate as additives of 3D printed scaffolds with polylactic acid matrix for bone tissue regeneration demonstrated that both additives promoted an increase of the porosity, hydrophilicity, and surface roughness of the scaffolds, resulting in an increase of the metabolic activity of human osteoblastic osteosarcoma cells. The synergistic effects of both additives favored cell adhesion (Donate et al. 2020).

Composites of PLA–calcium sulfate ($CaSO_4$) were also considered in the biomedical sector for bone reparation, in situ forming scaffolds, or for implant materials comprising therapeutic agents (Murariu and Dubois 2016). PLA can be effectively melt-blended with previously dried gypsum (Murariu et al. 2007), a by-product of the lactic acid (LA) production process at neutral pH: lactic acid obtained by fermentation is treated with calcium hydroxide, then the broth containing calcium lactate is filtered to remove cells. After that, the filtrate is carbon treated, evaporated, and acidified with sulphuric acid to get LA and calcium sulfate (Singhvi et al. 2018) in a weight ratio 1:1. Hence, it represents a logical filler choice for cost reduction of PLA due to its availability as a waste stream from LA production.

PLA–barium sulfate ($BaSO_4$) biocomposites (Nuutinen et al. 2003; Välimaa 2013; Lämsä et al. 2006) were developed for radiopaque pancreatic stents.

A PLA biocomposite film, embedding nanoparticles of titanium dioxide (TiO_2) with photocatalytic activity, was efficiently used for the volatile organic compounds (VOCs) removal. A maleic anhydride compatibilizer was used to increase the amount of nano-TiO_2 incorporated into the PLA matrix and, in turn, the efficiency of photocatalytic activity (Tharasawatpipat et al. 2015). Poly(hydroxybutyrate-co-valerate)/poly(ε-caprolactone) blends filled with 1 wt % TiO_2 nanoparticles showed better thermal stabilities but delayed evolution of the degradation products(Mofokeng and Luyt 2015).

The use of carbonaceous fillers deserves special attention. It has attracted great scientific and industrial interest due to the high technological potential of such materials in many applications (Kasgoz et al. 2014; Spitalsky et al. 2010; Han and Fina 2011). Carbon black (CB), graphite and graphite/graphene derivatives, single and multiwalled carbon nanotubes (CNT), and fullerenes improve nucleating, mechanical, thermal, and flame retardant properties of PLA; moreover, tailored electrical conductivity (Murariu et al. 2014; Wu et al. 2010a, b; Tsuji et al. 2007; Bourbigot et al. 2011; Xu et al. 2011; Hapuarachchi and Peijs 2010; Frąckowiak and Kozlowski 2014) makes these biocomposites eligible for the fabrication of sensors with sensitivity to strain, temperature or organic solvents or items showing electromagnetic shielding (Frackowiak et al. 2015).

The excellent tensile strength, low density, high thermal and chemical stability, and electrical conductivity of carbon fibers (Huang 2009) make them an eligible reinforcement of PLA in the biomedical sector, as well as for particular engineering uses (Chłopek et al. 2007; Cheng et al. 2015). Thunga et al. (2014) described the possibility to use bio-renewable fibers such lignin as precursors for conversion to carbon fibers.

The huge potential of nanofillers is a hotly debated field and can be illustrated in the case of carbonaceous fillers. In nanocomposites, at least one filler dimension is in the nanometer scale. Physical and thermal properties of nanocomposites are superior to composites filled with micron-sized particles of the same filler (Alamri and Low 2012) due to the considerable surface area developed when nanoparticles are randomly distributed in the matrix. The increased interphase area results in increased interphase forces that promote adhesion, and improve mechanical properties (Ye et al. 2007). Due to their intrinsic characteristics, CNTs (Spitalsky et al. 2010; Han and Fina 2011) are among the most considered nanofillers for mixing with various polymers, including PLA. Both the dispersion state (Mittal et al. 2015) and the surface functionalization of CNTs (Rahmat and Hubert 2011; Wu et al. 2008) influence the electrical percolation threshold and thermal stability of PLA–CNT biocomposites. The sensing performance of an e-nose relying on quantum resistive sensors, based, among others, on carbon nanotube-PLA conductive polymer bionanocomposite, was able to detect several volatile lung cancer biomarkers (Chatterjee et al. 2013) and other organic vapors (Kumar et al. 2012; Castro et al. 2011). The combination of fillers of different geometry can dramatically improve the specific properties of PLA biocomposites: for example, the electrical conductivity of PLA foams can be enhanced using a combination of CNTs and CB in which the CNTs play the role of 'bridge', connecting CB particles and forming better network structure (Wu et al. 2015). The exceptional structural features, charge transport, antistatic, thermal, optical, and mechanical properties (Singh et al. 2011) of graphite make it an exceptional either micro- or nano-filler for various polymers, including PLA (Murariu et al. 2014; Sullivan et al. 2014; Fukushima et al. 2010; Li et al. 2014; Murariu et al. 2010; Wei et al. 2013; Zenkiewicz et al. 2011; Zhu et al. 2011), to produce biocomposites designed with specific end-use properties. Graphite consists of graphene (Singh et al. 2011) nanosheets, which is tantamount to saying that graphene is an isolated single-atom plane of graphite. Graphite effectiveness as a filler is higher where the graphite layers are separated and functionalized (Park and Ruoff 2009); hence many researchers focus on the exfoliation of previously intercalated graphite and to its adequate functionalization (Tong et al. 2013).

Graphite/graphene derivatives (intercalated/expandable graphite (Wei et al. 2013; Zhu et al. 2011), expanded graphite (Murariu et al. 2014; Fukushima et al. 2010; Murariu et al. 2010), exfoliated graphite nanoplatelets (Sullivan et al. 2014)) were associated with PLA matrix to produce nanobiocomposites via melt-compounding (Murariu et al. 2014; Fukushima et al. 2010) or mixing in solution (Xu et al. 2010). Better morphology, thermal, mechanical, and electrical properties were revealed in the case of nanobiocomposites, using graphite nanoplatelets (small stacks of graphene of 10 μm diameter), compared to microbiocomposites using micrometric graphite particles (of 500μm diameter) (Kim and Jeong 2010). Graphene nanoparticles were also tested as filler for many biopolymers different from PLA; for example, the addition of up to 6 wt% of graphene nanoparticles to PHBV resulted in reinforced solution-cast film samples, with increased crystallinity and better thermal degradation behavior (Sridhar et al. 2013).

PLA–expanded graphite biocomposites were characterized by improved rigidity, thermal stability, antistatic and conductive electrical properties, and crystallization kinetics with respect to the neat PLA matrix. The flame retardant properties of PLA–graphite biocomposites are remarkable (Murariu et al. 2014; Fukushima et al. 2010; Wei et al. 2013). The best flame retardant results were reported at 5 wt.% and 10 wt. % graphite loading. Ammonium polyphosphate showed synergistic effects in this respect (Zhu et al. 2011), and flame retardant properties were greatly improved with respect to the biocomposites containing only ammonium polyphosphate or expandable graphite.

It has to be emphasized that the list of inorganic micro- and nanofillers used to produce micro- and nanobiocomposites of PLA is very long and includes layered silicates (Bordes et al. 2009; Ojijo et al. 2014), mica (Gregorova et al. 2012; Souza et al. 2014), kaolin (Ouchiar et al. 2015, 2016; Segura González et al. 2015), zeolites (Gregorova et al. 2012; Hao et al. 2016), wollastonite (Gregorova et al. 2012), boehmite alumina (Agwuncha et al. 2015), organo-modified layered silicates, sepiolite, halloysite, polyhedral oligomeric silsesquioxanes, silica, and zinc oxide, among other nanoparticles (Murariu and Dubois 2016; Urayama et al. 2003).

7.5 Challenging Barriers in the Biocomposite Field

Biocomposite materials have been deeply studied by academicians. Fully degradable biocomposites are optimal for use in agriculture, horticulture. Packaging is the leading application for biocomposites since they offer unique haptic and optic properties. The automotive sector is interested in biocomposites for saving weight in car interiors. The automotive and construction sectors, are looking for lightweight materials with a low carbon footprint.

Biocomposites implementations as valid alternatives for conventional materials are relatively slow because of some intrinsic limitations due to the inherent characteristics of their constituents.

First of all, the inconsistency of the properties of natural fillers, including those obtained from FW, is a major drawback. The properties of agroindustrial waste vary from one harvesting season to another, from the pedoclimatic conditions, or even from one plant to another. Mechanical properties are in turn influenced. Mixing batches of fibers from various harvest and proper selection of the biocomposite constituents were proposed as possible solutions (AL-Oqla et al. 2015a, b). Insufficient availability of data regarding the performance of biocomposites for various industrial applications is due to the big diversity of their constituents, and it represents the most challenging barrier in this field.

Another weakness of using biofillers from FW as reinforcement for the production of biocomposite materials might be the poor compatibility between polymeric matrices (usually non-polar) and natural fibers (polar) (AL-Oqla et al. 2015a), resulting in poor adhesion between the biofiller and the biopolymer matrix. Poor bonding will negatively affect the mechanical properties of the biocomposites. The improper compatibilities between the fillers and matrices might make the performance unmanageable and challenging to be predicted. In order to promote the adhesion between the filler and thermoplastic matrix, chemical "coupling" agents can be used (Aridi et al. 2016). Surface modifications of natural fibers to improve their adhesion with different matrices has also been exploited. For example, an alkaline treatment breaks down the fiber bundles into small and shorter fibers due to the disruption of hydrogen bonding between the fibers (Fuqua et al. 2012; Faruk et al.

2012) and this is expected to enhance compatibility of the filler with the matrix. It is clear that the future production of biocomposite materials from renewable resources relies on the study of coupling agents, compatibilization mechanisms, surface modification, flame retardants, and advanced processing techniques to achieve long-term stability for real practices (Yu et al. 2006; Siročić et al. 2009; Wang et al. 2012).

An additional drawback of natural fibers and FW-based filler is high moisture sensitivity. Moisture absorption may cause deterioration, dimensional changes in the resulting biocomposites, reduced interfacial adhesion, and increased susceptibility to rotting because of microbial or fungal damage. Moisture can also trigger depolymerization via hydrolytic reactions. These issues place a big challenge to use biocomposites for packaging applications and during shipment and long-term storage (Al-Oqla et al. 2015). The moisture content of fibers from agroindustrial waste is usually in the range of 6–12 wt%, but it should be reduced to less than 3 wt% to enable the production of molded parts of good quality. Surface treatment can decrease the water sensitivity of hydrophilic fillers (Jawaid and Abdul Khalil 2011).

The low thermal stability of FW-based fillers (except those of ceramic nature) is another disadvantage. The degradation temperature of most natural fibers lies in the range of 150–220 °C. Thermal degradation and shrinking of the fibers would lead to changes in the physical and/or chemical structures of the biofiller due to depolymerization, color changes, hydrolysis, oxidation, dehydration, decarboxylation, and recrystallization (Murariu and Dubois 2016; Thakur et al. 2012).

As a concluding remark, it has to be emphasized that FW-based biofillers are undoubtedly renewable resources; their supply is potentially unlimited; however, the general energy expenditure throughout collecting, upcycling, and processing has to be taken into account within the life cycle assessment framework (see Chap. 11).

Bibliography

Abert J, Amella A, Weigelt S, Fischer H (2016) Degradation and swelling issues of poly-(d,l-lactide)/β-tricalcium phosphate/calcium carbonate composites for bone replacement. J Mech Behav Biomed Mater 54:82–92. https://doi.org/10.1016/j.jmbbm.2015.09.016

Adekunle K, Åkesson D, Skrifvars M (2010) Biobased composites prepared by compression molding with a novel thermoset resin from soybean oil and a natural-fiber reinforcement. J Appl Polym Sci 116:1759–1765. https://doi.org/10.1002/app.31634

Agwuncha SC, Ray SS, Jayaramudu J et al (2015) Influence of boehmite nanoparticle loading on the mechanical, thermal, and rheological properties of biodegradable polylactide/poly(ϵ-caprolactone) blends. Macromol Mater Eng 300:31–47. https://doi.org/10.1002/mame.201400212

Ahtiainen K, Sippola L, Nurminen M et al (2015) Effects of chitosan and bioactive glass modifications of knitted and rolled polylactide-based 96/4 L/D scaffolds on chondrogenic differentiation of adipose stem cells. J Tissue Eng Regen Med 9:55–65. https://doi.org/10.1002/term.1614

Al Bakri AMM, Liyana J, Norazian MN et al (2013) Mechanical properties of polymer composites with sugarcane bagasse filler. Adv Mater Res:739–744

Alamri H, Low IM (2012) Effect of water absorption on the mechanical properties of nano-filler reinforced epoxy nanocomposites. Mater Des 42:214–222. https://doi.org/10.1016/j.matdes.2012.05.060

Al-Oqla FM, Salit MS, Ishak MR, Aziz NA (2015) A novel evaluation tool for enhancing the selection of natural fibers for polymeric composites based on fiber moisture content criterion. BioResources 10:299–312. https://doi.org/10.15376/biores.10.1.299-312

AL-Oqla FM, Sapuan SM, Ishak MR, Nuraini AA (2015a) Decision making model for optimal reinforcement condition of natural fiber composites. Fibers Polym 16:153–163. https://doi.org/10.1007/s12221-015-0153-3

AL-Oqla FM, Sapuan SM, Ishak MR, Nuraini AA (2015b) Predicting the potential of agro waste fibers for sustainable automotive industry using a decision making model. Comput Electron Agric 113:116–127. https://doi.org/10.1016/j.compag.2015.01.011

Álvarez-Chávez CR, Sánchez-Acosta DL, Encinas-Encinas JC et al (2017) Characterization of extruded poly(lactic acid)/pecan nutshell biocomposites. Int J Polym Sci 2017:13–17. https://doi.org/10.1155/2017/3264098

Amiri A, Yu A, Webster D, Ulven C (2016) Bio-based resin reinforced with flax fiber as thermorheologically complex materials. Polymers (Basel) 8. https://doi.org/10.3390/polym8040153

Aranberri I, Montes S, Azcune I et al (2017) Fully biodegradable biocomposites with high chicken feather content. Polymers (Basel) 9:593. https://doi.org/10.3390/polym9110593

Aridi NAM, Sapuan SM, Zainudin ES, AL-Oqla FM (2016) Mechanical and morphological properties of injection-molded rice husk polypropylene composites. Int J Polym Anal Charact 21:305–313. https://doi.org/10.1080/1023666X.2016.1148316

Armentano I, Dottori M, Fortunati E et al (2010) Biodegradable polymer matrix nanocomposites for tissue engineering: a review. Polym Degrad Stab:2126–2146

Ashokkumar M, Sumukh KM, Murali R et al (2012) Collagen-chitosan biocomposites produced using nanocarbons derived from goatskin waste. Carbon NY 50:5574–5582. https://doi.org/10.1016/j.carbon.2012.08.006

Barbosa V, Ramires EC, Razera IAT, Frollini E (2010) Biobased composites from tannin-phenolic polymers reinforced with coir fibers. Ind Crops Prod 32:305–312. https://doi.org/10.1016/j.indcrop.2010.05.007

Bashir ASM, Manusamy Y (2015) Recent developments in biocomposites reinforced with natural biofillers from food waste. Polym Plast Technol Eng 54:87–99. https://doi.org/10.1080/03602559.2014.935419

Bisanda ETN, Ogola WO, Tesha JV (2003) Characterisation of tannin resin blends for particle board applications. Cem Concr Compos 25:593–598

Bledzki AK, Mamun AA, Volk J (2010) Barley husk and coconut shell reinforced polypropylene composites: the effect of fibre physical, chemical and surface properties. Compos Sci Technol 70:840–846. https://doi.org/10.1016/j.compscitech.2010.01.022

Bordes P, Pollet E, Avérous L (2009) Nano-biocomposites: biodegradable polyester/nanoclay systems. Prog Polym Sci 34:125–155

Bourbigot S, Fontaine G, Gallos A, Bellayer S (2011) Reactive extrusion of PLA and of PLA/carbon nanotubes nanocomposite: processing, characterization and flame retardancy. Polym Adv Technol 22:30–37. https://doi.org/10.1002/pat.1715

Castro M, Kumar B, Feller JF et al (2011) Novel e-nose for the discrimination of volatile organic biomarkers with an array of carbon nanotubes (CNT) conductive polymer nanocomposites (CPC) sensors. Sensors Actuators B Chem 159:213–219. https://doi.org/10.1016/j.snb.2011.06.073

Cecchi T, Giuliani A, Iacopini F et al (2019) Unprecedented high percentage of food waste powder filler in poly lactic acid green composites: synthesis, characterization, and volatile profile. Environ Sci Pollut Res 26:7263–7271. https://doi.org/10.1007/s11356-019-04187-1

Chatterjee S, Castro M, Feller J-F (2013) An E-nose made of carbon nanotube-based quantum resistive sensors for the detection of eighteen polar/nonpolar VOC biomarkers of lung cancer textile-based transdermal therapy using hydrogel of non-toxic biopolymer view project cellulose nanofibers reinforced recyclable thermoset composites development view project. Artic J Mater Chem B. https://doi.org/10.1039/C3TB20819B

Cheng KC, Bin YC, Guo W et al (2012) Thermal properties and flammability of polylactide nanocomposites with aluminum trihydrate and organoclay. Carbohydr Polym 87:1119–1123. https://doi.org/10.1016/j.carbpol.2011.08.065

Cheng KC, Lin YH, Guo W et al (2015) Flammability and tensile properties of polylactide nanocomposites with short carbon fibers. J Mater Sci 50:1605–1612. https://doi.org/10.1007/s10853-014-8721-2

Chłopek J, Morawska-Chochół A, Bajor G et al (2007) The influence of carbon fibres on the resorption time and mechanical properties of the lactide-glycolide co-polymer. J Biomater Sci Polym Ed 18:1355–1368. https://doi.org/10.1163/156856207782246858

Chun KS, Husseinsyah S, Osman H (2012) Mechanical and thermal properties of coconut shell powder filled polylactic acid biocomposites: Effects of the filler content and silane coupling agent. J Polym Res 19. https://doi.org/10.1007/s10965-012-9859-8

Corradini E, Ito EN, Marconcini JM et al (2009) Interfacial behavior of composites of recycled poly(ethyelene terephthalate) and sugarcane bagasse fiber. Polym Test 28:183–187. https://doi.org/10.1016/j.polymertesting.2008.11.014

Costa-Pinto AR, Martins AM, Castelhano-Carlos MJ et al (2014) In vitro degradation and in vivo biocompatibility of chitosan–poly(butylene succinate) fiber mesh scaffolds. J Bioact Compat Polym 29:137–151. https://doi.org/10.1177/0883911514521919

Dasan YK, Bhat AH, Ahmad F (2017) Polymer blend of PLA/PHBV based bionanocomposites reinforced with nanocrystalline cellulose for potential application as packaging material. Carbohydr Polym 157:1323–1332. https://doi.org/10.1016/j.carbpol.2016.11.012

Donate R, Monzón M, Ortega Z et al (2020) Comparison between calcium carbonate and β-tricalcium phosphate as additives of 3D printed scaffolds with polylactic acid matrix. J Tissue Eng Regen Med 14:272–283. https://doi.org/10.1002/term.2990

Dönmez Çavdar A, Kalaycioğlu H, Mengeloğlu F (2011) Tea mill waste fibers filled thermoplastic composites: the effects of plastic type and fiber loading. J Reinf Plast Compos 30:833–844. https://doi.org/10.1177/0731684411408752

Faruk O, Bledzki AK, Fink HP, Sain M (2012) Biocomposites reinforced with natural fibers: 2000–2010. Prog Polym Sci 37:1552–1596

Fazeli M, Florez JP, Simão RA (2019) Improvement in adhesion of cellulose fibers to the thermoplastic starch matrix by plasma treatment modification. Compos Part B Eng 163:207–216. https://doi.org/10.1016/j.compositesb.2018.11.048

Frąckowiak S, Kozlowski M (2014) Shape specific electrically conductive fillers in Poly(lactid acid) composites for strain sensing applications. Sens Lett 12:1299–1302. https://doi.org/10.1166/sl.2014.3312

Frackowiak S, Ludwiczak J, Leluk K et al (2015) Foamed poly(lactic acid) composites with carbonaceous fillers for electromagnetic shielding. Mater Des 65:749–756. https://doi.org/10.1016/j.matdes.2014.10.009

Fukushima K, Murariu M, Camino G, Dubois P (2010) Effect of expanded graphite/layered-silicate clay on thermal, mechanical and fire retardant properties of poly(lactic acid). Polym Degrad Stab 95:1063–1076. https://doi.org/10.1016/j.polymdegradstab.2010.02.029

Fuqua MA, Huo S, Ulven CA (2012) Natural fiber reinforced composites. Polym Rev 52:259–320

Gregorova A, Machovsky M, Wimmer R (2012) Viscoelastic Properties of mineral-filled poly(lactic acid) composites. Int J Polym Sci 2012. https://doi.org/10.1155/2012/252981

Han Z, Fina A (2011) Thermal conductivity of carbon nanotubes and their polymer nanocomposites: a review. Prog Polym Sci 36:914–944

Hao YH, Huang Z, Wang JW et al (2016) Improved thermal stability of poly (l-lactide) with the incorporation of zeolite ZSM-5. Polym Test 49:46–56. https://doi.org/10.1016/j.polymertesting.2015.11.010

Hapuarachchi TD, Peijs T (2010) Multiwalled carbon nanotubes and sepiolite nanoclays as flame retardants for polylactide and its natural fibre reinforced composites. Compos Part A Appl Sci Manuf 41:954–963. https://doi.org/10.1016/j.compositesa.2010.03.004
Harris AM, Lee EC (2008) Improving mechanical performance of injection molded PLA by controlling crystallinity. J Appl Polym Sci 107:2246–2255. https://doi.org/10.1002/app.27261
Hong CK, Wool RF (2005) Development of a bio-based composite material from soybean oil and keratin fibers. J Appl Polym Sci 95:1524–1538. https://doi.org/10.1002/app.21044
Hu B, Dweib M, Wool RP, Shenton HW (2007) Bio-based composite roof for residential construction. J Archit Eng 13:136–143. https://doi.org/10.1061/(ASCE)1076-0431(2007)13:3(136)
Huang X (2009) Fabrication and Properties of carbon fibers. Materials (Basel) 2:2369–2403. https://doi.org/10.3390/ma2042369
Hussein AA, Sultan AA, Matoq QA (2011) Mechanical behaviour of low density polyethylene/shrimp shells composite. J Basrah Res 37:5–11
Husseinsyah S, Zakaria MM (2011) The effect of filler content on properties of coconut shell filled polyester composites. S Polymer (Guildf) 6:87–97
Imoisili PE, Ibegbulam CM, Adejugbe TI (2012) Effect of concentration of coconut shell ash on the tensile properties of epoxy composites. Pacific J Sci Technol 13:463–468
Jandas PJ, Mohanty S, Nayak SK (2012) Renewable resource-based biocomposites of various surface treated banana fiber and poly lactic acid: characterization and biodegradability. J Polym Environ 20:583–595. https://doi.org/10.1007/s10924-012-0415-8
Jawaid M, Abdul Khalil HPS (2011) Cellulosic/synthetic fibre reinforced polymer hybrid composites: a review. Carbohydr Polym 86:1–18
John MJ, Thomas S (2008) Biofibres and biocomposites. Carbohydr Polym 71:343–364
Kang Y, Yin G, Yuan Q et al (2008) Preparation of poly(l-lactic acid)/β-tricalcium phosphate scaffold for bone tissue engineering without organic solvent. Mater Lett 62:2029–2032. https://doi.org/10.1016/j.matlet.2007.11.014
Kasgoz A, Akin D, Ayten AI, Durmus A (2014) Effect of different types of carbon fillers on mechanical and rheological properties of cyclic olefin copolymer (COC) composites. Compos Part B Eng 66:126–135. https://doi.org/10.1016/j.compositesb.2014.05.010
Kaynak C, Erdogan AR (2016) Mechanical and thermal properties of polylactide/talc microcomposites: before and after accelerated weathering. Polym Adv Technol 27:812–822. https://doi.org/10.1002/pat.3721
Ketabchi MR, Yasin FM, Abdullah LC et al (2017) Performance of cow dung reinforced biodegradable poly(lactic acid) biocomposites for structural applications. J Polym Environ 26:474–486. https://doi.org/10.1007/s10924-017-0963-z
Kim I-H, Jeong YG (2010) Polylactide/exfoliated graphite nanocomposites with enhanced thermal stability, mechanical modulus, and electrical conductivity. J Polym Sci Part B Polym Phys 48:850–858. https://doi.org/10.1002/polb.21956
Kim HS, Park BH, Choi JH, Yoon JS (2008) Mechanical properties and thermal stability of poly(L-lactide)/calcium carbonate composites. J Appl Polym Sci 109:3087–3092. https://doi.org/10.1002/app.28229
Kosorn W, Thavornyutikarn B, Phumsiri B et al (2010) Hydrolyzed poly(butylene succinate) scaffolds coated with bioactive agent. J Met Mater Miner 20:95–99
Ku H, Wang H, Pattarachaiyakoop N, Trada M (2011) A review on the tensile properties of natural fiber reinforced polymer composites. Compos Part B Eng 42:856–873. https://doi.org/10.1016/j.compositesb.2011.01.010
Kumar B, Castro M, Feller JF (2012) Poly(lactic acid)-multi-wall carbon nanotube conductive biopolymer nanocomposite vapour sensors. Sensors Actuators B Chem 161:621–628. https://doi.org/10.1016/j.snb.2011.10.077
Kumar R, Dhaliwal JS, Kapur GS, Shashikant (2014) Mechanical properties of modified biofiller-polypropylene composites. Polym Compos 35:708–714. https://doi.org/10.1002/pc.22714

Lämsä T, Jin H, Mikkonen J et al (2006) Biocompatibility of a new bioabsorbable radiopaque stent material (BaSO4 containing poly-L,D-lactide) in the rat pancreas. Pancreatology 6:301–305. https://doi.org/10.1159/000092772

Le Guen MJ, Thoury-Monbrun V, Castellano Roldán JM, Hill SJ (2017) Assessing the Potential of farm dairy effluent as a filler in novel PLA biocomposites. J Polym Environ 25:419–426. https://doi.org/10.1007/s10924-016-0824-1

Li J, Lu XL, Zheng YF (2008) Effect of surface modified hydroxyapatite on the tensile property improvement of HA/PLA composite. Appl Surf Sci 255:494–497. https://doi.org/10.1016/j.apsusc.2008.06.067

Li HY, Tan YQ, Zhang L et al (2012a) Bio-filler from mussel shell: preparation and its effects on polypropylene composites properties. Wuji Cailiao Xuebao/J Inorg Mater 27:977–983. https://doi.org/10.3724/SP.J.1077.2012.11653

Li HY, Tan YQ, Zhang L et al (2012b) Bio-filler from waste shellfish shell: preparation, characterization, and its effect on the mechanical properties on polypropylene composites. J Hazard Mater 217–218:256–262. https://doi.org/10.1016/j.jhazmat.2012.03.028

Li X, Xiao Y, Bergeret A et al (2014) Preparation of polylactide/graphene composites from liquid-phase exfoliated graphite sheets. Polym Compos 35:396–403. https://doi.org/10.1002/pc.22673

Lin Z, Guan Z, Chen C et al (2013) Preparation, structures and properties of shell/polypropylene biocomposites. Thermochim Acta 551:149–154. https://doi.org/10.1016/j.tca.2012.10.009

Liu Z, Erhan SZ (2008) "Green" composites and nanocomposites from soybean oil. Mater Sci Eng A 483–484:708–711. https://doi.org/10.1016/j.msea.2006.12.186

Liu Z, Erhan SZ, Akin DE, Barton FE (2006) "Green" composites from renewable resources: Preparation of epoxidized soybean oil and flax fiber composites. J Agric Food Chem 54:2134–2137. https://doi.org/10.1021/jf0526745

Liu T, Lam P-M, Zhao Y et al (2009) Mechanical and thermal properties of chicken feather fiber/PLA green composites. Compos Part B Eng 40:650–654. https://doi.org/10.1016/j.compositesb.2009.04.011

Madbouly SA, Schrader JA, Srinivasan G et al (2014) Biodegradation behavior of bacterial-based polyhydroxyalkanoate (PHA) and DDGS composites. Green Chem 16:1911–1920. https://doi.org/10.1039/c3gc41503a

Mahendran AR, Wuzella G, Aust N et al (2013) Processing and characterization of natural fibre reinforced composites using lignin phenolic binder. Polym Polym Compos 21:199–206. https://doi.org/10.1177/096739111302100401

Meesap P, Uppanan P, Thavornyutikarn B et al (2010) Surface hydrolyzed poly(butylene succinate) microsphere incorporated carboxymethylchitosan scaffolds for cartilage tissue engineering

Merlini C, Soldi V, Barra GMO (2011) Influence of fiber surface treatment and length on physico-chemical properties of short random banana fiber-reinforced castor oil polyurethane composites. Polym Test 30:833–840. https://doi.org/10.1016/j.polymertesting.2011.08.008

Mittal G, Dhand V, Rhee KY et al (2015) A review on carbon nanotubes and graphene as fillers in reinforced polymer nanocomposites. J Ind Eng Chem 21:11–25

Mofokeng JP, Luyt AS (2015) Morphology and thermal degradation studies of melt-mixed poly(hydroxybutyrate-co-valerate) (PHBV)/poly(ε-caprolactone) (PCL) biodegradable polymer blend nanocomposites with TiO2 as filler. J Mater Sci 50:3812–3824. https://doi.org/10.1007/s10853-015-8950-z

Muniyasamy S, Reddy MM, Misra M, Mohanty A (2013) Biodegradable green composites from bioethanol co-product and poly(butylene adipate-co-terephthalate). Ind Crops Prod 43:812–819. https://doi.org/10.1016/j.indcrop.2012.08.031

Murariu M, Dubois P (2016) PLA composites: from production to properties. Adv Drug Deliv Rev 107:17–46. https://doi.org/10.1016/j.addr.2016.04.003

Murariu M, Da Silva Ferreira A, Degée P et al (2007) Polylactide compositions. Part 1: effect of filler content and size on mechanical properties of PLA/calcium sulfate composites. Polymer (Guildf) 48:2613–2618. https://doi.org/10.1016/j.polymer.2007.02.067

Murariu M, Dechief AL, Bonnaud L et al (2010) The production and properties of polylactide composites filled with expanded graphite. Polym Degrad Stab 95:889–900. https://doi.org/10.1016/j.polymdegradstab.2009.12.019

Murariu M, Laoutid F, Dubois P et al (2014) Pathways to biodegradable flame retardant polymer (nano)composites. In: Polymer green flame retardants. Elsevier, pp 709–773

Nisticò R, Evon P, Labonne L et al (2017) Post-harvest tomato plants and urban food wastes for manufacturing plastic films. J Clean Prod 167:68–74. https://doi.org/10.1016/j.jclepro.2017.08.160

Niu Y, Cao L, Wei J et al (2014) Development of a bioactive composite of nano fluorapatite and poly(butylene succinate) for bone tissue regeneration. J Mater Chem B 2:1174–1181. https://doi.org/10.1039/c3tb21371d

Nova-Institue (2015) Biobased Polybutylene Succinate (PBS) – an attractive polymer for biopolymer compounds. Succinity:1–11. https://doi.org/10.1016/j.jallcom.2012.03.012

Nuutinen JP, Clerc C, Törmälä P (2003) Mechanical properties and in vitro degradation of self-reinforced radiopaque bioresorbable polylactide fibres. J Biomater Sci Polym Ed 14:665–676. https://doi.org/10.1163/156856203322274923

Ofem MI, Umar M, Ovat FA (2012) Mechanical properties of rice husk fiiled cashew nut shell liquid resin composites. J Mater Sci Res 1. https://doi.org/10.5539/jmsr.v1n4p89

Ojijo V, Sinha Ray S, Sadiku R (2014) Concurrent enhancement of multiple properties in reactively processed nanocomposites of polylactide/poly[(butylene succinate)- *co* -adipate] Blend and Organoclay. Macromol Mater Eng 299:596–608. https://doi.org/10.1002/mame.201300306

Ouchiar S, Stoclet G, Cabaret C et al (2015) Comparison of the influence of talc and kaolinite as inorganic fillers on morphology, structure and thermomechanical properties of polylactide based composites. Appl Clay Sci 116–117:231–240. https://doi.org/10.1016/j.clay.2015.03.020

Ouchiar S, Stoclet G, Cabaret C, Gloaguen V (2016) Influence of the filler nature on the crystalline structure of polylactide-based nanocomposites: new insights into the nucleating effect. Macromolecules 49:2782–2790. https://doi.org/10.1021/acs.macromol.5b02746

Park S, Ruoff RS (2009) Chemical methods for the production of graphenes. Nat Nanotechnol 4:217–224. https://doi.org/10.1038/nnano.2009.58

Patntirapong S, Singhatanadgit W, Meesap P et al (2015) Stem cell adhesion and proliferation on hydrolyzed poly(butylene succinate)/β-tricalcium phosphate composites. J Biomed Mater Res Part A 103:658–670. https://doi.org/10.1002/jbm.a.35214

Penjumras P, Rahman RBA, Talib RA, Abdan K (2014) Extraction and characterization of cellulose from Durian Rind. Agric Agric Sci Procedia 2:237–243. https://doi.org/10.1016/j.aaspro.2014.11.034

Petchwattana N, Covavisaruch S, Petthai S (2014) Influence of talc particle size and content on crystallization behavior, mechanical properties and morphology of poly(lactic acid). Polym Bull 71:1947–1959. https://doi.org/10.1007/s00289-014-1165-7

Pfister DP, Larock RC (2012) Cationically cured natural oil-based green composites: effect of the natural oil and the agricultural fiber. J Appl Polym Sci 123:1392–1400. https://doi.org/10.1002/app.33636

Pirhonen E, Grandi G, Törmälä P (2001) Bioactive glass fiber/polylactide composite. Key Eng Mater 192–195:725–728. https://doi.org/10.4028/www.scientific.net/kem.192-195.725

Rahmat M, Hubert P (2011) Carbon nanotube-polymer interactions in nanocomposites: a review. Compos Sci Technol 72:72–84. https://doi.org/10.1016/j.compscitech.2011.10.002

Rakmae S, Ruksakulpiwat Y, Sutapun W, Suppakarn N (2012) Effect of silane coupling agent treated bovine bone based carbonated hydroxyapatite on in vitro degradation behavior and bioactivity of PLA composites. Mater Sci Eng C 32:1428–1436. https://doi.org/10.1016/j.msec.2012.04.022

Ramasamy S, Ismail H, Munusamy Y (2013) Preparation and characterization of rice husk powder incorporated natural rubber latex foam. Adv Mater Res 626:523–529. https://doi.org/10.4028/www.scientific.net/AMR.626.523

Reis KC, Pereira L, Melo ICNA et al (2015) Particles of coffee wastes as reinforcement in polyhydroxybutyrate (PHB) based composites. Mater Res 18:546–552. https://doi.org/10.1590/1516-1439.318114

Rivera-Briso A, Serrano-Aroca Á (2018) Poly(3-Hydroxybutyrate-co-3-Hydroxyvalerate): Enhancement Strategies for Advanced Applications. Polymers (Basel) 10:732. https://doi.org/10.3390/polym10070732

Roether JA, Gough JE, Boccaccini AR et al (2002) Novel bioresorbable and bioactive composites based on bioactive glass and polylactide foams for bone tissue engineering. J Mater Sci Mater Med 13:1207–1214. https://doi.org/10.1023/a:1021166726914

Rothon RN (1999) Mineral fillers in thermoplastics: Filler manufacture and characterisation. Adv Polym Sci 139:68–107. https://doi.org/10.1007/3-540-69220-7_2

Ruangudomsakul W, Ruksakulpiwat C, Ruksakulpiwat Y (2013) The study of using bio-filler from cassava pulp in natural rubber composites. Adv Mater Res 747:371–374. https://doi.org/10.4028/www.scientific.net/AMR.747.371

Saeb MR, Ramezani-Dakhel H, Sarami R (2010) Improving thermo-mechanical properties of styrene butadiene rubber nanocomposites using eggshell bio-filler. In: Fourth International Conference on Quantum, Nano and Micro Technologies, ICQNM 2010, pp 10–13

Sánchez-Safont EL, Aldureid A, Lagarón JM et al (2018) Biocomposites of different lignocellulosic wastes for sustainable food packaging applications. Compos Part B Eng 145:215–225. https://doi.org/10.1016/j.compositesb.2018.03.037

Sarki J, Hassan SB, Aigbodion VS, Oghenevweta JE (2011) Potential of using coconut shell particle fillers in eco-composite materials. J Alloys Compd 509:2381–2385. https://doi.org/10.1016/j.jallcom.2010.11.025

Schiller C, Epple M (2003) Carbonated calcium phosphates are suitable pH-stabilising fillers for biodegradable polyesters. Biomaterials 24:2037–2043. https://doi.org/10.1016/S0142-9612(02)00634-8

Schiller C, Rasche C, Wehmöller M et al (2004) Geometrically structured implants for cranial reconstruction made of biodegradable polyesters and calcium phosphate/calcium carbonate. Biomaterials 25:1239–1247. https://doi.org/10.1016/j.biomaterials.2003.08.047

Segura González EA, Olmos D, González-Gaitano G et al (2015) Effect of kaolin nanofiller and processing conditions on the structure, morphology, and biofilm development of polylactic acid. J Appl Polym Sci 132:n/a–n/a. https://doi.org/10.1002/app.42676

Shakoor A, Thomas NL (2014) Talc as a nucleating agent and reinforcing filler in poly(lactic acid) composites. Polym Eng Sci 54:64–70. https://doi.org/10.1002/pen.23543

Shen L, Yang H, Ying J et al (2009) Preparation and mechanical properties of carbon fiber reinforced hydroxyapatite/polylactide biocomposites. J Mater Sci Mater Med 20:2259–2265. https://doi.org/10.1007/s10856-009-3785-2

Shi QF, Mou HY, Gao L et al (2010) Double-melting behavior of bamboo fiber/talc/poly (lactic acid) composites. J Polym Environ 18:567–575. https://doi.org/10.1007/s10924-010-0252-6

Shi X, Zhang G, Phuong TV, Lazzeri A (2015) Synergistic effects of nucleating agents and plasticizers on the crystallization behavior of Poly(lactic acid). Molecules 20:1579–1593. https://doi.org/10.3390/molecules20011579

Shih YF, Huang CC (2011) Polylactic acid (PLA)/banana fiber (BF) biodegradable green composites. J Polym Res 18:2335–2340. https://doi.org/10.1007/s10965-011-9646-y

Shikinami Y, Matsusue Y, Nakamura T (2005) The complete process of bioresorption and bone replacement using devices made of forged composites of raw hydroxyapatite particles/poly l-lactide (F-u-HA/PLLA). Biomaterials 26:5542–5551. https://doi.org/10.1016/j.biomaterials.2005.02.016

Singh V, Joung D, Zhai L et al (2011) Graphene based materials: past, present and future. Prog Mater Sci 56:1178–1271

Singhvi M, Zendo T, Sonomoto K (2018) Free lactic acid production under acidic conditions by lactic acid bacteria strains: challenges and future prospects. Appl Microbiol Biotechnol 102:5911–5924

Siročić AP, Hrnjak-Murgić Z, Jelenčić J (2009) Effect of the filler and the compatibilizer on the properties of filled blends. Compos Interfaces 16:85–95. https://doi.org/10.1163/156855409X402876

Souza DHS, Andrade CT, Dias ML (2014) Isothermal crystallization kinetics of poly(lactic acid)/synthetic mica nanocomposites. J Appl Polym Sci 131. https://doi.org/10.1002/app.40322

Spitalsky Z, Tasis D, Papagelis K, Galiotis C (2010) Carbon nanotube-polymer composites: chemistry, processing, mechanical and electrical properties. Prog. Polym. Sci. 35:357–401

Sridhar V, Lee I, Chun HH, Park H (2013) Graphene reinforced biodegradable poly(3-hydroxybutyrate-co-4-hydroxybutyrate) nano-composites. Express Polym Lett 7:320–328. https://doi.org/10.3144/expresspolymlett.2013.29

Sugumaran V, Vimal KK, Kapur GS, Narula AK (2015) Preparation and morphological, thermal, and physicomechanical properties of polypropylene-potato peel biocomposites. J Appl Polym Sci 132:n/a–n/a. doi: https://doi.org/10.1002/app.42445

Sullivan EM, Oh YJ, Gerhardt RA et al (2014) Understanding the effect of polymer crystallinity on the electrical conductivity of exfoliated graphite nanoplatelet/polylactic acid composite films. J Polym Res 21. https://doi.org/10.1007/s10965-014-0563-8

Talal A, Waheed N, Al-Masri M et al (2009) Absorption and release of protein from hydroxyapatite-polylactic acid (HA-PLA) membranes. J Dent 37:820–826. https://doi.org/10.1016/j.jdent.2009.06.014

Thakur VK, Singha AS, Thakur MK (2012) Biopolymers based green composites: mechanical, thermal and physico-chemical characterization. J Polym Environ 20:412–421. https://doi.org/10.1007/s10924-011-0389-y

Thanh DTM, Trang PTT, Huong HT et al (2015) Fabrication of poly (lactic acid)/hydroxyapatite (PLA/HAp) porous nanocomposite for bone regeneration. Int J Nanotechnol 12:391–404. https://doi.org/10.1504/IJNT.2015.067897

Tharasawatpipat C, Suwannahong K, Kruenate J, Kreetachat T (2015) Removal of VOCs by photocatalytic oxidation using nano-TiO2/PLA biocomposite. J Environ Biol 36:617–621

Thunga M, Chen K, Grewell D, Kessler MR (2014) Bio-renewable precursor fibers from lignin/polylactide blends for conversion to carbon fibers. Carbon NY 68:159–166. https://doi.org/10.1016/j.carbon.2013.10.075

Tiamiyu AO, Ibitoye SA (2012) Effect of clay addition on service properties of a developed OPF-CNSL-formaldehyde roofing material. Constr Build Mater 36:358–364. https://doi.org/10.1016/j.conbuildmat.2012.04.130

Tong XZ, Song F, Li MQ et al (2013) Fabrication of graphene/polylactide nanocomposites with improved properties. Compos Sci Technol 88:33–38. https://doi.org/10.1016/j.compscitech.2013.08.028

Toro P, Quijada R, Yazdani-Pedram M, Arias JL (2007) Eggshell, a new bio-filler for polypropylene composites. Mater Lett 61:4347–4350. https://doi.org/10.1016/j.matlet.2007.01.102

Tsuji H, Kawashima Y, Takikawa H, Tanaka S (2007) Poly(l-lactide)/nano-structured carbon composites: conductivity, thermal properties, crystallization, and biodegradation. Polymer (Guildf) 48:4213–4225. https://doi.org/10.1016/j.polymer.2007.05.040

Urayama H, Ma C, Kimura Y (2003) Mechanical and thermal properties of poly(L-lactide) incorporating various inorganic fillers with particle and whisker shapes. Macromol Mater Eng 288:562–568. https://doi.org/10.1002/mame.200350004

Välimaa (2013) US6524345B1 – surgical implant

Varila L, Lehtonen T, Tuominen J et al (2012) In vitro behaviour of three biocompatible glasses in composite implants. J Mater Sci Mater Med 23:2425–2435

Vijayan R, Krishnamoorthy A (2019) Review on natural fiber reinforced composites. In: Materials today: proceedings. Elsevier, pp 897–906

Wang B, Wang X, Tang G et al (2012) Preparation of silane precursor microencapsulated intumescent flame retardant and its enhancement on the properties of ethylene-vinyl acetate copolymer cable. Compos Sci Technol 72:1042–1048. https://doi.org/10.1016/j.compscitech.2012.03.022

Wei P, Bocchini S, Camino G (2013) Flame retardant and thermal behavior of polylactide/expandable graphite composites. Polimery 58:361–364. https://doi.org/10.14314/polimery.2013.361

Wei L, Liang S, McDonald AG (2015) Thermophysical properties and biodegradation behavior of green composites made from polyhydroxybutyrate and potato peel waste fermentation residue. Ind Crops Prod 69:91–103. https://doi.org/10.1016/j.indcrop.2015.02.011

Wu CS (2015) Renewable resource-based green composites of surface-treated spent coffee grounds and polylactide: characterisation and biodegradability. Polym Degrad Stab 121:51–59. https://doi.org/10.1016/j.polymdegradstab.2015.08.011

Wu D, Wu L, Zhang M, Zhao Y (2008) Viscoelasticity and thermal stability of polylactide composites with various functionalized carbon nanotubes. Polym Degrad Stab 93:1577–1584. https://doi.org/10.1016/j.polymdegradstab.2008.05.001

Wu D, Wu L, Zhou W et al (2010a) Crystallization and biodegradation of polylactide/carbon nanotube composites. Polym Eng Sci 50:1721–1733. https://doi.org/10.1002/pen.21695

Wu D, Wu L, Zhou W et al (2010b) Relations between the aspect ratio of carbon nanotubes and the formation of percolation networks in biodegradable polylactide/carbon nanotube composites. J Polym Sci Part B Polym Phys 48:479–489. https://doi.org/10.1002/polb.21909

Wu D, Lv Q, Feng S et al (2015) Polylactide composite foams containing carbon nanotubes and carbon black: synergistic effect of filler on electrical conductivity. Carbon NY 95:380–387. https://doi.org/10.1016/j.carbon.2015.08.062

Xiao H, Yang L, Ren X et al (2010) Kinetics and crystal structure of poly(lactic acid) crystallized nonisothermally: effect of plasticizer and nucleating agent. Polym Compos 31:2057–2068. https://doi.org/10.1002/pc.21004

Xu J-Z, Chen T, Yang C-L et al (2010) Isothermal crystallization of poly(<scp>l</scp> -lactide) induced by graphene nanosheets and carbon nanotubes: a comparative study. Macromolecules 43:5000–5008. https://doi.org/10.1021/ma100304n

Xu Z, Niu Y, Wang Z et al (2011) Enhanced nucleation rate of polylactide in composites assisted by surface acid oxidized carbon nanotubes of different aspect ratios. ACS Appl Mater Interfaces 3:3744–3753. https://doi.org/10.1021/am200932q

Yao ZT, Chen T, Li HY et al (2013) Mechanical and thermal properties of polypropylene (PP) composites filled with modified shell waste. J Hazard Mater 262:212–217. https://doi.org/10.1016/j.jhazmat.2013.08.062

Ye Y, Chen H, Wu J, Ye L (2007) High impact strength epoxy nanocomposites with natural nanotubes. Polymer (Guildf) 48:6426–6433. https://doi.org/10.1016/j.polymer.2007.08.035

Yee TW, Rahman WAWA, Sin LT (2011) Properties and morphology of poly(vinyl alcohol) blends with sago pith bio-filler as biodegradable composites. J Vinyl Addit Technol 17:184–189. https://doi.org/10.1002/vnl.20265

Yu L, Dean K, Li L (2006) Polymer blends and composites from renewable resources. Prog Polym Sci 31:576–602. https://doi.org/10.1016/j.progpolymsci.2006.03.002

Yu F, Liu T, Zhao X et al (2012) Effects of talc on the mechanical and thermal properties of polylactide. J Appl Polym Sci 125:E99–E109. https://doi.org/10.1002/app.36260

Zabihzadeh SM, Omidvar A, Marandi MAB et al (2011) Effect of filler loading on physical and flexural properties of rapeseed STEM/PP composites. BioResources 6:1475–1483. https://doi.org/10.15376/biores.6.2.1475-1483

Zenkiewicz M, Richert J, Rytlewski P, Richert A (2011) Comparative analysis of shungite and graphite effects on some properties of polylactide composites. Polym Test 30:429–435. https://doi.org/10.1016/j.polymertesting.2011.03.004

Zhang SM, Liu J, Zhou W et al (2005) Interfacial fabrication and property of hydroxyapatite/polylactide resorbable bone fixation composites. Curr Appl Phys 5:516–518. https://doi.org/10.1016/j.cap.2005.01.023

Zhou DS, Zhao KB, Li Y et al (2006) Repair of segmental defects with nano-hydroxyapatite/collagen/PLA composite combined with mesenchymal stem cells. J Bioact Compat Polym 21:373–384. https://doi.org/10.1177/0883911506068554

Zhu H, Zhu Q, Li J et al (2011) Synergistic effect between expandable graphite and ammonium polyphosphate on flame retarded polylactide. Polym Degrad Stab 96:183–189. https://doi.org/10.1016/j.polymdegradstab.2010.11.017

Zou G-X, Jiao Q-W, Zhang X et al (2015) Crystallization behavior and morphology of poly(lactic acid) with a novel nucleating agent. J Appl Polym Sci 132:n/a-n/a. https://doi.org/10.1002/app.41367

Zuiderduin WCJ, Westzaan C, Huétink J, Gaymans RJ (2002) Toughening of polypropylene with calcium carbonate particles. Polymer (Guildf) 44:261–275. https://doi.org/10.1016/S0032-3861(02)00769-3

Part III
Characterization of Biobased Products

Chapter 8
Mechanical Characterization of Biobased Products from Food Waste

Teresa Cecchi

Biobased materials are increasingly being developed and used in a wide assortment of applications such as packaging, building products, medical devices, and electronics. For each function, the biopolymer must be investigated for processability and tested to guarantee suitability and conformity to regulations and legislation.

Product developers must characterize their biobased materials, which can be used as bulk polymers, films, fibers, or nanomaterials. The mechanical characterization is of utmost importance.

Test descriptions detailed in this chapter aim at providing a descriptive summary to enhance test understanding.

8.1 Tensile Testing

The ability of bioplastics to withstand tensile forces is important, as they are increasingly used in many different sectors. The tensile testing of plastics is standardized in ASTM D 638 (ASTM International 2014a), "Standard Test Method for Tensile Properties of Plastics". ASTM D 638 applies to rigid plastic samples with thickness ranging between 1.00 mm and 14 mm.

Mechanical loading of any material results in deformation, according to its specific mechanical behavior, which, in turn, depends on physical and chemical properties.

Tensile tests are performed by applying a tensile force to a sample specimen and measuring various properties of the specimen under stress. These tests (tensile strength, elongation at break, and modulus of elasticity) are conducted on a tensile testing machine on dumbbell-shaped (or "dog bone" shaped) test specimens. Specimens are placed in the (manual or pneumatic) grips of the testing machine; the crosshead speed is set at a specific value. The test object is drawn from two opposite directions: this way, the length increases, and the diameter shrinks. The amount of load as a function of the length increase is recorded during the test. Conditioning

T. Cecchi, C. De Carolis, *Biobased products from food sector waste*,
https://doi.org/10.1007/978-3-030-63436-0_8

of the test specimens at 23 ± 2 °C with a relative humidity of 50 ± 5% for at least 40 h before testing is required. Testing is performed in the same conditions. At least five specimens for individual determinations of tensile properties must be made on each sample to be tested.

Through tensile tests, many different tensile properties can be calculated.

Stress and strain are the parameters that describe material behavior under mechanical loading.

If a tensile force is used to stretch the material, the Tensile Stress (nominal) is the force (F) per unit area acting on a plane in the material.

$$Tensile\,Stress(MPa) = \frac{Force(N)}{Thickness(mm) \cdot Width(mm)}$$

The tensile strength (nominal) represents the ability of a material to withstand a lengthwise stress; it is calculated by dividing the maximum load sustained by the specimen during the test by the average original cross-sectional area in the gage length segment of the specimens; the tensile strength can be calculated at yield or at break; more accurately, for plastics that exhibit yield points, it is defined as the stress at the first local maximum in the stress–strain curve:

$$Tensile\,strength(MPa) = \frac{Force_{localmax}(N)}{Thickness(mm) \cdot Width(mm)}$$

The deformation due to the effect of stresses is called elongation or strain, according to the apparatus used to measure it. The deformation will occur in the gage section whose crosssectional area is less than that of the shoulders and grip section.

Percent Elongation is the change in gage length relative to the original specimen gage length, expressed as a percent using an extension indicator (extensometer) that is able to measure the distance between two designated points within the gage length of the test specimen as the specimen is stretched.

Percent Elongation at Yield is calculated by reading the extension (change in gage length) at the yield point expressed as a percent.

Percent Elongation at Break is computed by reading the extension at the point of specimen rupture expressed as a percent.

Nominal Strain is the percent change in grip separation relative to the original grip separation. It is calculated using a crosshead extension indicator able to assess the change in the separation of the grips, that is, the crosshead movement.

Nominal Strain at break is of particular interest; it is the strain at the moment of rupture relative to the original grip separation.

Elongation values are needed for engineering design, but they are meaningful only if deformation within the specimen gage length is uniform. In the case of necking (the localized reduction in cross-section, which may occur in a material under tensile stress) or non-uniform deformation within the specimen gage length, nominal strain values are reported even if they are of qualitative utility only.

The modulus of elasticity (also known as elastic modulus or Young's modulus) is the ratio of tensile stress (nominal) to corresponding strain below the proportional limit of a material. Since strain is dimensionless, the modulus of elasticity is expressed in force per unit area. It is obtained by dividing the difference in stress, corresponding to any segment of the section on the straight line (obtained by the least square fit) in the stress-strain plot, by the corresponding difference in strain. Elastic modulus values shall be calculated whenever possible but, for materials where no proportionality is evident, the secant value shall be calculated by dividing the corresponding stress (nominal) by the designated strain. The tensile modulus of elasticity is proportional to the stiffness of the material.

It has to be emphasized that only plastics 1.0 mm or greater in thickness are eligible for this test. Tensile properties determined according to Test Method D638 need dumbbell specimens of appropriate dimensions for sheet, plate, and molded plastics. Dimensions of specimens are also detailed for rigid tubes and rods. The physical dimensions of specimens must be measured according to the ASTM D5947 protocol (ASTM International 2018a) since the size of the samples will affect the results.

ASTM D 638 (ASTM International 2014a) does not cover the determination of tensile properties of rigid plastic samples less than 1.0 mm in thickness. Since tensile properties can vary with specimen thickness, tensile properties of thin plastic sheeting can be obtained using ASTM D882 (ASTM International 2018b). Materials that fail by tearing are not eligible for this test since they give anomalous and unreliable data.

8.2 Hardness Testing

The hardness testing of plastics is standardized in ASTM D2240 (ASTM International 2017a).

The term hardness can be defined as the resistance that a material offers against penetration of a harder body, which wears or scratches the softer material that is tested. It's a crucial parameter for engineers if one takes into account the adverse cosmetic effects of worn or scratched surfaces and moving parts.

There are five methods of hardness measurement: Rockwell, Brinell, Vickers, Knoop, and Shore. The hardness testing of plastics is usually measured by the Shore and Rockwell scales for softer and harder plastic materials, respectively. The hardness value is obtained from the penetration of the durometer indenter foot into the sample specimen, which increases with decreasing resistance of the plastic toward indentation.

The Rockwell scale is commonly used for hard thermosets, as indicated by ASTM D785 (ASTM International 2015a).

There are several Shore hardness scales for measuring the hardness of different materials. Shore A scale is the preferred method for soft vulcanized rubber, natural rubber, nitriles, thermoplastic elastomers, wax, felt, and leathers. Instead, Shore B, Shore C, and Shore O scales are not commonly used. Shore D scale is used for hard rubber and plastic.

Both Shore A and D scales are from 0 to 100, and when Shore D is 58, the corresponding Shore A is 100. A correspondence table between them is available.

8.3 Tear Strength

Testing is performed by pulling specimens apart using a tensile test machine. The maximum force required to tear the tested sample specimen is used to calculate the tear strength. Tear strength is defined as force per unit thickness necessary to start a tear through the specimen tested. This test provides a measure of resistance to tearing and reveals the possible anisotropy of a material.

ASTM D624 (ASTM International 2020a) is the testing standard for measuring the tear strength of thermoset rubbers, thermoplastic elastomers, and silicones.

ASTM D1938 (ASTM International 2019) is the standard test method for tear-propagation resistance (Trouser Tear) of plastic film and thin sheeting by a single-tear method. The method is not applicable for brittle film or sheeting material.

Conditioning of the test specimens at 23 ± 2 °C with a relative humidity of 50 ± 5% for at least 40 h before testing is required. Testing is performed in the same conditions. At least five specimens for individual determinations of tear strength must be made on each sample to be tested.

ASTM D1922 (ASTM International 2020b) is the standard test method for propagation tear resistance of plastic film and thin sheeting by pendulum method.

8.4 Flexural Testing

The flexural testing of plastics is standardized in ASTM D790 (ASTM International 2017b).

A bar of rectangular cross-section, held by two supports, is loaded by a loading nose midway between the supports. The specimen is deflected until rupture occurs or until a maximum strain of 5.0% is reached, whichever occurs first. Conditioning of the test specimens at 23 ± 2 °C with a relative humidity of 50 ± 5% for at least 40 h before testing is required. Testing is performed in the same conditions. At least five specimens for individual determinations of flexural properties must be made on each sample to be tested.

The flexural properties obtained are different. When a homogeneous elastic material is tested in flexure, the maximum stress in the outer surface of the sample specimen is at the midpoint. This is the Flexural Stress, and it may be calculated for any point on the load-deflection curve. The Flexural Strength is the maximum flexural stress sustained by the test specimen during a bending test.

The Tangent Modulus of Elasticity, or "modulus of elasticity," is the stress to the flexural strain ratio within the elastic limit. It is calculated from the tangent to the initial straight-line portion of the load-deflection curve.

A Secant Modulus is calculated as the slope of the straight line that joins the origin and a selected point (chosen at prespecified stress or strain in accordance with the appropriate material specification or by customer contract) on the actual stress-flexural strain. Similarly, a chord modulus may be calculated from two discrete points (in accordance with the appropriate material specification or by customer contract) on the load-deflection curve.

8.5 Impact Resistance Testing

The impact resisting testing of plastic is standardized in ASTM D256 (ASTM International 2018c).

The Izod pendulum impact test indicates the energy to break standard specimens under specific parameters of specimen mounting, notching, size, and pendulum velocity-at-impact.

The presence of a notch generates a stress concentration that increases the probability of a brittle, rather than a ductile, fracture. Four similar methods are available. They all use the same testing machine and specimen dimensions, but correspondence tables, useful for correlating the results from the different test methods, are not available. The testing apparatus comprises a massive base with a vise for holding the specimen, and a pendulum-type hammer, connected through a rigid frame and bearings to the base. It also has a device to measure the energy loss and indicate the breaking energy of the specimen. Conditioning of the test specimens at 23 ± 2 °C with a relative humidity of 50 ± 5% for at least 40 h before testing is required. Testing is performed in the same conditions. At least five specimens for individual determinations of impact resistance must be made on each sample to be tested. The Notched Izod impact strength is a relative measure of the materials resistance to impact.

8.6 Density Testing

Density measurements are performed according to ASTM D792 (ASTM International 2020c) that applies to plastic sheets, rods, tubes, or molded items.

Solid plastics can be tested in water (method A) or, in specific cases, in other liquids (method B).

Specific gravity or Relative Density is "the ratio of the mass in air of a unit volume of the impermeable portion of the material at 23°C to the mass in air of equal density of an equal volume of gas-free distilled water at the same temperature". Density is the mass in air in kilograms per cubic meter of an impermeable portion of the material at 23 °C, that is the mass per unit volume of a material.

Density testing is important because density can identify a material; density changes can be caused by crystallinity variations, absorption of solvent, thermal history, or a low degree of uniformity among different specimens.

Conditioning of the test specimens at 23 ± 2 °C with a relative humidity of 50 ± 5% for at least 40 h before testing is required. Testing is performed in the same conditions.

The method involves weighing a one-piece specimen of 1–50 g in air and in air-free distilled water, using a sinker if the density of the specimen is lower than that of water and taking into account the weigh the sample holder (and sinker, if used) immersed under the same conditions. Specimens can be wet by, but otherwise not affected by water.

8.7 Compression Testing

Compression measurements are performed according to ASTM D695 (ASTM International 2015b) that represents the standard test method for compressive properties of rigid plastics.

Loading a specimen at a relatively low and uniform rate enables the study of the behavior of materials under crushing pressure. The specimen is compressed by compressive plates, and deformation is recorded as a function of the applied load. Compressive stress and strain are measured. The resulting stress-strain diagram provides information on the elastic limit, proportional limit, yield point, yield strength, compressive strength. An extensometer is used to determine the modulus of elasticity, following the usual procedure for other stress-strain curves.

Specimens are typically blocks (12.7 × 2.7 × 25.4 mm) but cylinders are also used (12.7 mm in diameter and 25.4 mm long).

8.8 Creep

Creep, also called "cold flow", is a slow but permanent deformation phenomenon over time due to stress; ASTMD 2990 (ASTM International 2017c) provides standard test methods for tensile, compressive, and flexural creep and creep rupture of plastics under specified environmental conditions. For measurements of creep-rupture, tension is the selected stress mode.

These test methods measure the extension or compression as a function of time and time-to-rupture or failure of a specimen subject to constant tensile or compressive load under specified environmental conditions. The creep modulus and strength of materials are calculated from creep and creep-rupture tests data. They are relevant to engineers because they allow them to predict the behaviour of materials under long-term loads and to predict dimensional changes that may occur as a result of such loads.

Selection of temperatures for creep and creep-rupture testing depends on the intended use of the test results.

8.9 Fatigue

Fatigue is related to the weakening of materials from repeated use, leading to eventual failure. ASTM D7791 applies to rigid and semi-rigid plastic materials. It is used to determine the fatigue limits under repeated uniaxial stress for a large number of cycles (ASTM International 2017d). It is no surprise that plastics will fail at stress levels well below their tensile or compressive strengths when subjected to cyclic loading. Accelerated tests aim at getting end results sooner, but the heat has to be promptly removed to avoid results that might not be equitable to normal service conditions since temperature has a strong influence on the mechanical properties.

Once the specimens are prepared in accordance with the standard, they are loaded into the test machine. The test is performed with a cycling of a proper frequency because properties can vary with specimen depth and test frequency. Test frequency can be up to 25 Hz, but a frequency of 5 Hz or less is recommended.

The tests should be conducted at the same temperature and humidity each time, as prescribed by the test standard; hence a temperature chamber and a humidity chamber for non-ambient temperature and humidity levels are recommended.

If the specimen does not exhibit an elastic behaviour, plastic deformation will occur during fatigue testing and caution shall be taken when interpreting the results.

This test method can be used with a procedure for fatigue testing in tension and another procedure for fatigue testing in compression (only for rigid plastics). Uniaxial loading systems with tension and compression capabilities are used to determine these properties.

8.10 Friction

The "Standard Test Method for Static and Kinetic Coefficients of Friction of Plastic Film and Sheeting" is ASTM D1894 (ASTM International 2014b).

The static and kinetic coefficients respectively indicate the static (starting) and kinetic (moving) resistance of one surface being dragged across another. They can be obtained from the force necessary to get the sled started, and to maintain motion, respectively. In order to calculate these coefficients, a 64 mm square specimen is attached to a sled of specified weight that is pulled across a second surface (254 mm × 127 mm) at a specific speed.

The static coefficient of friction is the initial force to the sled weight ratio. The kinetic coefficient of friction corresponds to the average force needed to achieve a uniform sliding of the surfaces the sled weight ratio.

Atomic Force Microscopy in friction mode proved to be a powerful tool for studying nano-friction. Surface properties depend on the nature of the polymer (Maurice et al. 2014).

8.11 Wear

The laboratory testing of the wear of materials during sliding is described in the standard ASTM G99 "Standard Test Method for Wear Testing with a Pin-on-Disk Apparatus" (ASTM International 2017c). Materials are tested in pairs: one, a pin, is positioned perpendicular to the other, usually a flat circular disk, and the test machine causes a circular sliding path on the disk surface when the pin specimen is pressed against the disk at a specified load. The amount of wear is reported as volume loss in cubic millimeters for the pin and the disk separately.

Wear results are determined by measuring linear dimensions and converting linear measures of wear to wear volume. If loss of mass is measured, the density of the material is used to convert the mass loss value to volume loss.

8.12 Mechanical Performance of Bioplastics

Among bioplastics, only a few are commercialized, and they only have a < 1% plastic market share because they are seldomly able to address industrial applications. Apart from drop-in bioplastics (biobased plastics chemically identical to the fossil-based analog), PHAs and PLA are the most used bioplastics because they are reasonably able to match the functionalities of common fossilbased plastics, namely LDPE, HDPE, PP, PS, PVC, PU, and PET. The mechanical properties of bioplastics are probably the most important of all the physical properties of bioplastic for most real-field applications.

The mechanical performance of PBS makes it a good competitor of PP and PE; it is highly crystalline and has good processability and thermal stability even if ductility decreases in time due to the cold crystallization phenomenon. PLA has high strength and high modulus, similar to PS. However, its brittleness and low heat deformation temperature limit its applicability. The thermal degradation under melt processing and the slow biodegradation rate are additional issues. Hydrolytic sensitivity prevents PLA packaging of moisture-sensitive long shelf-life products frozen baked; PLA films laminated with barrier materials (PP, PE) or coated with inorganic silicon and aluminum oxides can be used in this respect.

PHAs have good heat resistance and gas-barrier properties similar to polyvinyl chloride (PVC) and PET thermoplastics. PHAs are quite rigid, and the use of PHB is limited by its brittleness and poor processability. PHAs are less thermally stable than conventional plastics. Copolymerization is useful in this respect. Application in the packaging sector requires laminating with paper and other polymers and plasticization, copolymerization, blending, and reinforcement with inorganic or organic fillers (Zhao et al. 2020).

Disclaimer Readers are referred to the most recent and up-to-date ASTM standards available on https://www.astm.org, the only authentic reference. The information provided in this chapter does not constitute legal advice. The Author does not accept any liability as regards the contents of this Chapter.

Bibliography

ASTM International (2014a) ASTM D638 – 14 Standard Test Method for Tensile Properties of Plastics. https://www.astm.org/Standards/D638.htm. Accessed 13 Mar 2021

ASTM International (2014b) ASTM D1894 – 14 Standard Test Method for Static and Kinetic Coefficients of Friction of Plastic Film and Sheeting. https://www.astm.org/Standards/D1894.htm. Accessed 13 Mar 2021

ASTM International (2015a) ASTM D785 – 08(2015) Standard Test Method for Rockwell Hardness of Plastics and Electrical Insulating Materials. https://www.astm.org/Standards/D785.htm. Accessed 13 Mar 2021

ASTM International (2015b) ASTM D695 - 15 Standard Test Method for Compressive Properties of Rigid Plastics. https://www.astm.org/Standards/D695.htm. Accessed 28 Feb 2021

ASTM International (2017a) D2240 – 15 Standard Test Method for Rubber Property — Durometer Hardness. ASTM Int 05:1–13. https://doi.org/10.1520/D2240-15E01. Accessed 13 Mar 2021

ASTM International (2017b) ASTM D790 – 17 Standard Test Methods for Flexural Properties of Unreinforced and Reinforced Plastics and Electrical Insulating Materials. https://www.astm.org/Standards/D790.htm. Accessed 13 Mar 2021

ASTM International (2017c) ASTM D2990 - 17 Standard Test Methods for Tensile, Compressive, and Flexural Creep and Creep-Rupture of Plastics. https://www.astm.org/Standards/D2990.htm. Accessed 28 Feb 2021

ASTM International (2017c) G99-17: Standard Test Method for Wear Testing with a Pin-on-disk Apparatus. https://doi.org/10.1520/G0099-17. Accessed 13 mar 2021

ASTM International (2017d) ASTM D7791 - 17 Standard Test Method for Uniaxial Fatigue Properties of Plastics. https://www.astm.org/Standards/D7791.htm. Accessed 28 Feb 2021

ASTM International (2018a) ASTM D5947 - 18 Standard Test Methods for Physical Dimensions of Solid Plastics Specimens. https://www.astm.org/Standards/D5947.htm. Accessed 28 Feb 2021

ASTM International (2018b) ASTM D882 – 18 standard test method for tensile properties of thin plastic sheeting. https://www.astm.org/Standards/D882.htm. Accessed 31 Dec 2019

ASTM International (2018c) ASTM D256 – 10(2018) standard test methods for determining the izod pendulum impact resistance of plastics. https://www.astm.org/Standards/D256.htm. Accessed 26 Dec 2019

ASTM International (2019) ASTM D1938 - 19 Standard Test Method for Tear-Propagation Resistance (Trouser Tear) of Plastic Film and Thin Sheeting by a Single-Tear Method. https://www.astm.org/Standards/D1938.htm. Accessed 28 Feb 2021

ASTM International (2020a) ASTM D624 - 00(2020) Standard Test Method for Tear Strength of Conventional Vulcanized Rubber and Thermoplastic Elastomers. https://www.astm.org/Standards/D624.htm. Accessed 28 Feb 2021

ASTM International (2020b) ASTM D1922 - 15(2020) Standard Test Method for Propagation Tear Resistance of Plastic Film and Thin Sheeting by Pendulum Method. https://www.astm.org/Standards/D1922.htm. Accessed 28 Feb 2021

ASTM International (2020c) ASTM D792 - 20 Standard Test Methods for Density and Specific Gravity (Relative Density) of Plastics by Displacement. https://www.astm.org/Standards/D792.htm. Accessed 28 Feb 2021

Maurice B, Ahmad F, Sophie B (2014) Formulation and surface properties of bio-based polymer films for pharmaceutical applications. In: Key engineering materials. Trans Tech Publications Ltd, Zurich, pp 829–835

Zhao X, Cornish K, Vodovotz Y (2020) Narrowing the Gap for Bioplastic Use in Food Packaging: An Update. Environ. Sci. Technol. https://doi.org/10.1021/acs.est.9b03755

Chapter 9
Physico-chemical Characterization of Bioplastics and Biocomposites

Teresa Cecchi

The identification and comparison of different physico-chemical characteristics of materials are particularly important from the commercial point of view to make comparisons between different lots, to assess compliance with the specifications or intended use.

A number of different analytical techniques and approaches are used in this respect. In the following, we will detail the most useful ones.

9.1 Morphology. Scanning Electron Microscopy (SEM)

Qualitative information about the physical state, surface and internal structures of bioplastics, and biocomposites can be obtained using microscopic methods. Among them, Scanning Electron Microscopy (SEM) is the most common.

A scanning electron microscope makes use of a focused beam of electrons to produce images of a sample. The surface is scanned by electrons which interact with atoms in the sample specimen. The output signals contain information about the surface morphology and composition. Usually, secondary electrons emitted by atoms excited by the scanning electron beam and back scattered electrons are detected using the Everhart-Thornley detector. The number of secondary electrons depends on specimen topography. This way, a resolution better than 1 nm can be achieved. Anyhow, specimens must be electrically conductive, hence an ultrathin coating of electrically conducting metal (gold and gold/palladium alloy being the most common materials) is usually deposited on non-conducting samples by low-vacuum sputter coating or by high-vacuum evaporation (Álvarez-Chávez et al. 2017; Tiimob et al. 2016; Madera-Santana et al. 2015), prior to scanning.

T. Cecchi, C. De Carolis, *Biobased products from food sector waste*,
https://doi.org/10.1007/978-3-030-63436-0_9

SEM is usually done in order to analyze the microstructure and morphology of specimens through the analysis of micrographs in different magnifications.

SEM analysis of bioplastics and biocomposites was used for diverse reasons.

Firstly, SEM is a crucial tool to investigate filler adhesion and dispersion in biocomposites (Cecchi et al. 2017; Thunga et al. 2014; Sarki et al. 2011; Koutsomitopoulou et al. 2014; Liu et al. 2009; Quiles-Carrillo et al. 2018; Wu 2015) as illustrated in Chap. 7 (Fig. 7.1) and copolymers (Chłopek et al. 2007; Raza et al. 2019; Quiles-Carrillo et al. 2018). In this respect, the study of the morphology of impact fractured specimens of bioplastics and biocomposites is noteworthy because the microscopic analysis on the fracture surfaces of the samples reveals the failure surface structure and failure behavior induced by tensile test (Johari et al. 2016; Jandas et al. 2012; Liu et al. 2009; Bashir and Manusamy 2015; Tiimob et al. 2016). The interfacial adhesion between the phase of the filler and that of the matrix can be inferred from SEM micrographs of the fractured surface of the biocomposites (Scaffaro et al. 2018). It is usually found that particle size and surface roughness are important for the adhesion of a filler in a polymer matrix, and fatty components of the filler may form a thin film on the filler surface that prevents good adhesion. Actually, samples can also be fractured cryogenically (Álvarez-Chávez et al. 2017; Madera-Santana et al. 2015).

Secondly, SEM analysis can reveal the morphology of fillers used for biocomposites. For example, the morphological analysis of fillers such as cellulose fibers was useful to determine the surface roughness and texture needed to adhere to the surface of the polylactic acid (PLA) matrix. It was concluded that the surface smoothness has a negative effect on adhesion since it precludes physical bonding (Johari et al. 2016). The morphology of various fillers, such as *Posidonia oceanica* dead leaves (Scaffaro et al. 2018), hemp fibers (Song et al. 2013), and olive pit powder in PLA-matrix biocomposites (Koutsomitopoulou et al. 2014) were also studied by SEM imaging.

Thirdly, SEM analysis enabled the investigation of the change in biocomposite surface morphology in studies concerning biodegradability in soil (Wu 2015) or after invasion by bacteria *Burkholderia cepacia* (Jandas et al. 2012). Composting was also studied by SEM: the rate of bioplastics biodegradation at the micro-scale was studied via the comparison of SEM micrographs acquired at different times. SEM analysis highlighted the microbial activity of degradation on the bioplastic: the surface lost its smoothness, and cracks were evident (Adamcová et al. 2018). The presence of the filler increased biodegradation (Wu 2015).

SEM analysis can also be useful for assessing the eventual biphasic morphology of bioplastic blends (Su et al. 2019) that would be witnessed by different discrete domains in the micrographs. SEM imaging was also used to assess the improved filler–matrix interaction when a silane coupling agent (3-aminopropyltriethoxysilane) was used for the preparation of PLA/coconut shell powder biocomposites (Chun et al. 2012).

9.2 Crystallinity and Thermal Stability. Differential Scanning Calorimetry (DSC) and Thermogravimetric Analysis (TGA)

Important characteristics of bioplastics are mechanical, optical, and barrier properties which, in turn, depend on the physical properties (crystallinity, density, phase change temperatures, polarity, etc.).

Polymers may be either semicrystalline or amorphous. Amorphous material has no organization in the solid state. No polymer is organized into a fully crystalline material because imperfections and amorphous areas in semicrystalline polymers may be momentous. The degree of crystallinity usually ranges between 10% and 80%.

Crystallinity is the state which diffracts X-ray and exhibits a first-order transition known as melting that has a discontinuity in the volume-temperature dependence. The heat of transition is called the enthalpy of fusion or melting. Percent crystallinity has a big influence on optical, mechanical, thermal, and chemical properties, such as transparency, hardness, density, and diffusion.

Crystallinity is associated with the partial alignment of molecular chains, which fold together and form ordered regions called lamellae, which, in turn, compose larger spheroidal structures named spherulites or dense, flat regions called crystallites. The amorphous regions give the material certain elasticity and impact resistance, while higher crystallinity increases hardness but also brittleness.

Polymer crystallization usually results from cooling from the melt, mechanical stretching, or solvent evaporation.

The glass transition temperature (T_g) is the macro-manifestation of the polymer chain's flexibility and crystallinity. It increases with increasing rigidity of the polymer chain and represents the temperature at which the polymer changes from a hard, glassy, and relatively brittle state to a soft, flexible, rubbery state. This boundary is not only important for the product performance but also for its processing, because below the glass transition temperature, the available energy is insufficient to consent to the molecules' harmonized mobility. Actually, a wide temperature window for thermoplastic processing makes the polymeric material suitable for extrusion, injection molding, thermoforming, and film blowing (Xu and Guo 2010). During heating, the glass transition occurs before melting. Amorphous polymers do not exhibit a melt temperature (T_m), they only have a T_g. On the converse, semicrystalline polymers exhibit both T_m and T_g since there are both the amorphous and crystalline regions. The flexibility of amorphous polymers is drastically reduced when they are cooled below the T_g because a dimensional change in the polymer causes temporary distortions of the strong valence bonds.

Higher T_g underscores higher degrees of polymerization and better thermal stability of the polymer.

The Flory–Fox equation (Eq. 9.1) relates the number-average molecular weight of the polymer, M_n, to its glass transition temperature, T_g, according to the following relationship:

$$T_g = T_{g,\infty} - K / M_n \quad (9.1)$$

where $T_{g,\infty}$ is the maximum theoretical T_g at an infinitely high molecular weight and K is an empirical parameter related to the free volume present in the polymer sample. During heating, cold crystallization at a temperature (T_c) above T_g may occur; it is an intriguing exothermic process in which the polymer chains gain sufficient mobility to fold and arrange themselves into a crystalline structure. Cold crystallization shows if the complete crystallization did not occur during a rapid cooling cycle from the melt and stable nuclei had no time to grow.

Polymers experience major changes in properties during heating or cooling: they either absorbs or releases heat energy. Heat-induced phase transition behavior of bioplastics and biocomposites can be easily studied by Differential scanning calorimetry (DSC) (Shih and Huang 2011).

DSC is a thermoanalytical technique in which a sample and reference are kept at the same temperature during their heating; the heat required, which depends on the sample and reference heat capacities, is measured as a function of temperature.

In DSC measurements, a certain amount of sample is inserted inside hermetic sample pans and put in the DSC micro-furnace. The reference heat capacity is known over the explored temperature range. From the heating cycle, T_g, T_m, and T_c can be determined. The T_g value is obtained from the midpoint of the DSC heating curve deflection from baseline. Enthalpy values can be determined by integrating the area of the cold crystallization and melting peaks (Nisticò et al. 2018). Results may be influenced by the heating rate and sample heat history.

The crystallinity of the sample can be obtained from the comparison of the melting enthalpy obtained for the tested sample with that of a reference sample of the same kind but with 100% crystallinity (Fukushima et al. 2005). The T_c can be a useful parameter to highlight the internal stress introduced during the bioplastic fabrication and released at that temperature.

For example, DSC results suggested that poly(e-caprolactone) (PCL) was able to crystallize to a certain extent within PCL/PDLA blends and that PLA was partially miscible with poly(butylene adipate) (Kfoury et al. 2013). Miscibility is often inferred from the fact that a single T_g is detected in the blend, and its value is between the T_g of the components (Imre and Pukánszky 2013).

In biocomposites, T_c and T_m are not strongly affected by the presence of the filler (Chun et al. 2012); anyhow, when a silane coupling agent (3-aminopropyltriethoxysilane) was used to improve the filler–matrix interaction, the thermal stabilities of the biocomposites increased. For PLA/coconut shell powder biocomposites, the presence of the filler and the use of the silane coupling agent increased the T_g and crystallinities due to enhanced interfacial bonding. The presence of a peak T_c indicated that the filler has a nucleating effect (Chun et al. 2012).

Thermogravimetric analysis (TGA) is the technique of choice to study the thermal stability of of bioplastics and biocomposites. Both the TGA thermogram and its derivative differential thermo gravimetric (DTG) curves can help in determining the nature and the degree of the sample thermal degradation. The thermogravimeter is an instrument that measures the change in weight of a specimen upon heating. The mass change indicates thermal degradation. Weight loss during heating results from decomposition with the emission of gaseous products and can be recorded as a function of the temperature in thermogravimetric curves. In this respect, the composition of the materials in terms of organic and inorganic components is important.

The thermogravimeter is an extremely accurate balance in which the sample is progressively heated from room temperature up to over 1000 °C with programmable temperature ramps. The results of the thermogravimetry of polymeric materials are usually expressed in terms of percentages of the residual weight of the sample specimen as a function of the temperature.

The TGA analysis method is used to obtain the decomposition onset temperature (T_{onset}), that is, the temperature of initial degradation of the polymeric matrix.

The maximum decomposition temperature (T_{max}) characterized by the maximum weight loss and the decomposition temperature at 50% weight loss (T_{50}) are also important parameters to assess the thermal stability of a biomaterial.

The thermal stabilities of coffee ground and bamboo flour PLA biocomposites (Baek et al. 2013), collagen-chitosan biocomposites (Ashokkumar et al. 2012), and dried farm dairy effluent PLA biocomposites were investigated by TGA (Le Guen et al. 2017).

TGA analysis of the coconut shell powder filled PLA biocomposite evidenced the thermal decomposition of hemicellulose at 300 °C, but and increased thermal stability of the biocomposite was inferred from the fact that the total weight loss at 600 °C decreased with filler content, probably because the char may act as a protective barrier (Chun et al. 2012).

In TGA, waste plant fibers, which are potential fillers, usually release absorbed moisture at about 40–140 °C (Shih and Huang 2011), degradation of cellulose and hemicellulose occurs between 250 and 400 °C, while lignin degrades over 400 °C (Baek et al. 2013).

DSC and TGA analyses demonstrated that coconut oil could serve as a nontoxic plasticizer that improves the flexibility of PLA films without affecting their thermal stability or impairing the polymer processing (Bhasney et al. 2017).

TGA and DSC also proved to be useful in biodegradability tests (Alshehrei 2017) and in the study of copolymerization of bioplastics (Xu and Guo 2010).

9.3 Nuclear Magnetic Resonance (NMR)

All nuclei that contain an odd number of protons and an odd number of neutrons and nuclei for which the sum of proton and neutron is odd have an intrinsic nuclear magnetic moment and angular momentum (analogous to the classical angular momentum of a spinning sphere); in other words, they have a nonzero nuclear spin. Common examples are ^{1}H and ^{13}C nuclei. They represent nuclear magnets, and there is no energetic difference for any particular orientation in the space. On the converse, in the presence of an external magnetic field, there is a high-energy state (opposite to the field) and a low-energy state (aligned with the field); in thermal equilibrium, the low-energy orientation is preferred.

If the spin is equal to 1/2 (like ^{1}H), there are two possible orientations of the nucleus in a magnetic field.

The energy gap between the levels increases with increasing magnetic field. If the sample is irradiated with electromagnetic radiation of equal energy to the energy difference between the two levels (ΔE), there is a shift in the orientation of the nuclear spin that will rotate from aligned with the field to the opposite direction.

When the nucleus relaxes to its original state, it emits an electromagnetic signal whose energy corresponds to the energy gap between the low and high energy states. The NMR signal of a certain nucleus is displaced into the spectrum at higher or lower frequencies depending on its chemical environment: this happens because the electrons are charged particles. Hence, when they are immersed in a magnetic field, they move and generate a weaker field, which is opposed to the applied one. The greater the electron density, the greater the shielding will be. In this way, the nuclei will feel a lower magnetic field, and therefore the signal will be shifted towards lower frequencies than those of the nuclei with lower electron density.

NMR analysis enables the investigation of the chemical nature of the sample.

^{1}H-NMR allows to determine the extent of polymerization because signals from the repetitive units of the polymer are different from those of the corresponding chemical moieties in end groups of the polymeric chain and from their ratio the degree of polymerization (DP), that is, the number of monomeric units in a macromolecule can be calculated.

The number average molecular weight (Mn) is the average molecular weight of all the polymer chains in the sample and is defined by:

$$\mathrm{Mn} = \Sigma N_i \mathrm{M}_i \,/\, \Sigma N_i \qquad (9.2)$$

where M_i represents the molecular weight of a specific chain and N_i is the number of chains of that molecular weight. It follows that Mn is the the ratio of the total weight of polymer to the total number of molecules.

Mn is obtained by multiplying the mass of each repeating unit by the DP.

Actually, ^{1}H-NMR allows determining Mn until 5000 Da. Above this value, the signals of the end-groups are difficult to measure. ^{1}H-NMR enabled the determination of DP and Mn of the prepolymer used to synthesize PLLA for biomedical applications via safe multimetallic cerium complexes and ^{13}C-NMR confirmed the PLLA configuration; results were compared with the absolute molecular mass obtained from gel permeation chromatography, another important analytical strategy that is also able to shed light on polymers chain linearity (Pastore et al. 2021).

NMR analyses are also convenient for following the polymerization reaction (Xie et al. 1999), for the evaluation of the bioplastic purity or the eventual presence of side products (Li et al. 2005), for the determination of a specific copolymer content (Xu and Guo 2010), for the characterization of a potential natural plasticizer for bioplastics (coconut oil) (Bhasney et al. 2017), for the assessment of the influence of mechanical processing of PLA (injection and extrusion/injection) on PLA chains (Kfoury et al. 2013; Farah et al. 2016), to evaluate the content of cellulose in different fillers (Le Guen et al. 2017), in biodegradability tests (Alshehrei 2017), for the study of blends (Yu et al. 2006) and compatibilization in bioplastic blends (Imre and Pukánszky 2013).

Both proton and carbon-13 NMR are expedient for the determination of tacticity (Kapelski 2013), (Pastore et al. 2021), which represents the relative stereochemistry of adjacent chiral centers within a macromolecule.

9.4 Barrier and Permeation Properties, Transmission Rate Measurement

Polymers are permeable to gases, and their barrier or permeation properties may impair their suitability for a specific application. These properties are very important factors in determining the suitability of polymer for a wide range of applications in the packaging, medical, cosmetics, or food sectors.

In packaging, for example, ineffective barrier properties may reduce the shelf life of the enclosed product that becomes vulnerable to surrounding environmental factors.

Barrier and permeation properties depend on material characteristics, morphology, orientation, crystallinity, thickness, temperature, and additives. They are crucial for product development, manufacture, and marketing.

Barrier and permeation properties are evaluated by measurement of transmission rate for gases (quantity of gas that is able to pass through a specific area of the plastic material in the unit time). The main gases that affect the stability of most products are water vapor and oxygen.

The water vapor transmission rate is calculated according to ASTM D1653, for organic coating films (ASTM International 2013), ASTM E96 for materials (ASTM International 2016), and ASTM F1249 for plastic film and sheeting using a modulated infrared sensor (ASTM International 2020a).

The water vapor transmission rate is particularly important in packaging because some products need protection from outside air moisture, others require that the moisture contained should not be allowed to evaporate through the package.

The oxygen transmission Rate is measured via ASTM D3985, for flexible barrier materials using a coulometric sensor (ASTM International 2017a): at selected temperature and humidity conditions that particularly influence the measure, a barrier film (typical test area: 50 cm^2) is sealed between a chamber containing oxygen and a chamber void of oxygen and a coulometric sensor quantifies the oxygen that is transmitted through the material in mol/(m^2s).

Customized tests study how other gases and vapors – such as hydrogen (H_2), methane (CH_4), oxygen (O_2), nitrogen (N_2), Argon (Ar), carbon dioxide (CO_2), and water (H_2O) – move across a material.

In Sect. 6.8 the strategies used to improve barrier properties, crucial for food packaging, and to bridge the gap to conventional plastics are discussed.

9.5 Biodegradation and Composting

The end of life scenarios of bioplastics is an emerging field of research.

In our "disposable" culture, biodegradable and compostable packaging materials are two of the new green living trends.

Anyhow only a few people are aware of the difference between these two terms, which are not equivalent and should not be used interchangeably to avoid confusion in the marketplace.

Unsubstantiated claims to biodegradability and compostability were common in the past as a consequence of the lack of well-identified testing methods and environmental requirements.

Three international standards (the European Standard EN 13432:2000, ISO 17088:2012, and the US Standard ASTM D6400, see references in Table 9.1) outline the same test schemes and lay down criteria for what can or cannot be described as compostable as regards plastics and products made from plastics under an aerobic and thermophilic environment, in municipal and industrial aerobic composting facilities. Logos and certificates issued by certification bodies allow demonstrating the conformity of a product with the standard compostability norms. False and misleading environmental claims are being pursued.

These standards are intended to ensure that the materials will break down in industrial composting conditions.

A compostable material must

1. be biodegradable, and this means that at least 90% of the materials have to be broken down biochemically within 6 months, into CO_2, water, and minerals with the eventual production of biomass with the help of microorganisms. Biodegradability is determined by measuring the CO_2 produced by the sample under controlled composting conditions according to ISO 14855-1:2012 and ISO 14855-2:2018;
2. be disintegrable, and this involves the physical decomposition of a product into tiny pieces. After 12 weeks, at least 90% of the product should undergo a physical decomposition into tiny pieces and should be able to pass through 2 × 2 mm mesh; disintegration is evaluated at pilot-scale by simulating an authentic composting environment according to ISO 16929:2019;
3. have a chemical composition characterized by low levels of heavy metals (less than specified values for certain elements such as Zn, Cu, Ni, Cd, Pb, Hg, Cr, Mo, Se, As, Co) and of hazardous substances to the environment (such as F)
4. not be ecotoxic, which involves the absence of negative effects of the final compost in the environment.

Table 9.1 Extract of latest standards in respect of biodegradability, disintegration, compostability

Standard	Title
ISO 14855-1:2012 (ISO 2012a)	Determination of the ultimate aerobic biodegradability of plastic materials under controlled composting conditions—Method by analysis of evolved carbon dioxide—Part 1: General method
ISO 14855-2:2018 (ISO 2018)	Determination of the ultimate aerobic biodegradability of plastic materials under controlled composting conditions—Method by analysis of evolved carbon dioxide—Part 2: Gravimetric measurement of carbon dioxide evolved in a laboratory-scale test
ISO 16929:2019 (ISO 2019a)	Plastics—Determination of the degree of disintegration of plastic materials under defined composting conditions in a pilot-scale test
ISO 20200:2015 (ISO 2015)	Plastics—Determination of the degree of disintegration of plastic materials under simulated composting conditions in a laboratory-scale test
ISO 14851:2019 (ISO 2019)	Determination of the ultimate aerobic biodegradability of plastic materials in an aqueous medium—Method by measuring the oxygen demand in a closed respirometer
ISO 14852:2018 (ISO 2018)	Determination of the ultimate aerobic biodegradability of plastic materials in an aqueous medium—Method by analysis of evolved carbon dioxide
ISO 17556:2019 (ISO 2019b)	Plastics—Determination of the ultimate aerobic biodegradability of plastic materials in soil by measuring the oxygen demand in a respirometer or the amount of carbon dioxide evolved
ISO 16221:2001 (ISO 2001)	Water quality—Guidance for determination of biodegradability in the marine environment
ISO 17088:2012 (ISO 2012b)	Specifications for compostable plastics
ISO 18606:2013 (ISO 2013)	Packaging and the environment—Organic recycling
EN 17033:2018 (CEN 2018)	Plastics—Biodegradable mulch films for use in agriculture and horticulture—Requirements and test methods
EN 14995:2006 (CEN 2006)	Plastics—Evaluation of compostability—Test scheme and specifications
EN 13432:2000 (CEN 2000)	Packaging—Requirements for packaging recoverable through composting and biodegradation—Test scheme and evaluation criteria for the final acceptance of packaging
ASTM D6400 – 19 (ASTM International 2019a)	Standard specification for labeling of plastics designed to be aerobically composted in municipal or industrial facilities
ASTM D5988-18 (ASTM International 2018a)	Standard test method for determining aerobic biodegradation of plastic materials in soil
ASTM D5929-18 (ASTM International 2018b)	Standard test method for determining biodegradability of materials exposed to source-separated organic municipal solid waste mesophilic composting conditions by respirometry
ASTM D6954-18 (ASTM International 2018c)	Standard guide for exposing and testing plastics that degrade in the environment by a combination of oxidation and biodegradation

(continued)

Table 9.1 (continued)

Standard	Title
ASTM D6691-17 (ASTM International 2017b)	Standard test method for determining aerobic biodegradation of plastic materials in the marine environment by a defined microbial consortium or natural sea water inoculum
ASTM D7991-15 (ASTM International 2015a)	Standard test method for determining aerobic biodegradation of plastics buried in sandy marine sediment under controlled laboratory conditions
ASTM D7475-20 (ASTM International 2020b)	Standard test method for determining the aerobic degradation and anaerobic biodegradation of plastic materials under accelerated bioreactor landfill conditions
ASTM D5526-18 (ASTM International 2018d)	Standard test method for determining anaerobic biodegradation of plastic materials under accelerated landfill conditions
ASTM D5511-18 (ASTM International 2018e)	Standard test method for determining anaerobic biodegradation of plastic materials under high-solids anaerobic-digestion conditions
ASTM D7473-12 (ASTM International 2012)	Standard test method for weight attrition of plastic materials in the marine environment by open system aquarium incubations
ASTM D5338-15 (ASTM International 2015b)	Standard test method for determining aerobic biodegradation of plastic materials under controlled composting conditions, incorporating thermophilic temperatures

It is clear that for compostable materials, a high degree of biodegradation and disintegration is needed on specified limited time-scales, without any harmful effect of the compost quality on the environment (Imre and Pukánszky 2013). Actually, during biodegradation, it is important to focus on the fate of additives, including catalysts, fillers, and colorants, among others, used to optimize the technical properties of biodegradable plastics. They can be released via migration, leaching, emission, degradation, or fragmentation. Nevertheless, a comprehensive chemical analysis of heavy metals and any toxic chemicals embedded in such a complex matrix is difficult and expensive to achieve. Thus, the final compost might contain a mixture of plastic-derived chemicals.

It is clear that a biodegradable material may not be compostable, while every compostable material is at least biodegradable.

Biodegradation is possible only if monomers or specific additives are susceptible to a microbial attack via specific enzymatic processes. Biodegradation is linked to a significant change in the chemical structure of a material caused by biochemical (enzymatic and non-enzymatic) activity, while disintegration is the material breakdown into very small fragments (Su et al. 2019). The biodegradability is, in the first place, influenced by the chemical structure of the polymers; in particular, microorganisms cannot easily attack polymers with a C–C backbone (i.e., polyolefins). Polyesters are biodegradable, but the material structure (crystallinity, orientation, morphology, and shape) influences biodegradation and disintegration (Kim and Park 1999): for example, it is clear that amorphous or less-ordered regions degrade more easily than crystalline regions. Under controlled conditions, the biodegradability of polymers can be tailored by the optimization of parameters, such as temperature, oxygen, and water content and the presence of specific microorganisms (EU 2018). The process is favored by darkness, high humidity, adequate minerals, and other nutrients. Biodegragation does not necessarily occur under all conditions. For example, anaerobic conditions in landfills usually prevent many biological processes necessary for mineralization. Enzymatic degradation starts from the polymers' surface because enzymes cannot penetrate the polymer, and degradation products with low molecular weights easily dissolve in the surrounding aqueous medium (Mochizuki and Hirami 1997). Actually, there are diverse environments for biodegradation, such as soil, water environments, digester plants, household composting units, and industrial composting facilities.

Polymers can also be degraded by photodegradation and thermal degradation. These different processes, along with biodegradation, can act either independently or in combination. Photodegradation is initiated by the absorption of light energy by reactive groups of the polymer, leading to smaller fragments. Similarly, heating of the polymer in the presence of oxygen results in thermo-oxidative degradation: bonds are broken down with the formation of radicals and volatiles.

While industrial composting conditions are characterized by elevated temperatures (55–60 °C) combined with high relative humidity, the presence of oxygen, and periodical mixing under standardized conditions, home-composting is an uncontrolled process strongly dependent on the geographical and climatological parameters as well as on individual actions taken by households and plastics claimed to be compostable may not be suitable for home composting (EU 2018). For this reason, there is no general standard for the biodegradability of plastics in these uncontrolled environments. For example, neat PLA needs industrial composting conditions at high temperatures for quick biodegradation and is not eligible for home composting or biodegradation in soil in a reasonable time frame. On the converse, PBS has the ability to biodegrade at ambient temperatures (<35 °C) (Nova-Institute 2019).

Worldwide, standards and draft standards for biodegradability, disintegration, and compostability of plastics under controlled parameters have been recently released and are illustrated in Table 9.1.

Biodradable bioplastics were indicated in Fig. 6.1 of Chap. 6. It is clear that not all biobased products are biodegradable and compostable, and not all fossilbased plastics will not biodegrade or compost.

As regards polylactic acid, it can be recycled to its monomer by thermal depolymerization or hydrolysis; since it is a thermoplastic aliphatic polyester, it undergoes simple hydrolytic cleavage of the ester moieties of the polymer backbone with only a little or even no assistance from enzymes resulting in the formation of non-harmful and not toxic monomers (Pawar et al. 2014) but it can also be attacked by microorganisms. Heterogeneous PLA composting with small amounts (<30% by weight) of pre-composted yard waste was effective (Ghorpade et al. 2001). However, it has been emphasized that PLA does not biodegrade as fast as other organic wastes during composting, worsening its acceptance in industrial composting facilities; to enable its biodegradation in comparable time frames with other organic materials, bioaugmentation, that is, the addition of specific microbial strains such as *Geobacillus thermoleovorans*, proved to be a promising technique to accelerate the biodegradation of compostable plastics (Castro-Aguirre et al. 2018). PLA composite and blends were also explored to increase PLA biodegradability. PLA/PHA blends (Zembouai et al. 2014) were investigated with the aim of obtaining a wide range of physical properties and improved processibility (Ohkoshi et al. 2000), but, above all, to increase both the biocompatibility and biodegradability of neat PLA and the thermal stability of neat PHA. The compatibility of PLA/PHBH blends was enhanced using a reactive epoxy as a bifunctional compatibilizer (Zhou et al. 2015), but the poor toughness is still the major drawback. The hydrophobicity of PLA partly impairs its biodegradability. PLA was blended with more hydrophilic polymers, such as starch, chitosan, and cellulose to enhance biodegradability. Contrasting results as regards the PHA/PLA blends were reported, while faster degradation in soil was noticed for PLA/PBS blends (Hubbe et al. 2021).

Polybutylene succinate (PBS) is a bioplastic that biodegrades into water and carbon dioxide with the microorganism under the soil. It was found that *Amycolatopsis* sp. HT-6 and *Penicillium* sp. strain 14-3 are able to degrade PBS. Several thermophilic actinomycetes such as *Microbispora rosea*, *Excellospora japonica* and *E. viridilutea* can degrade samples of emulsified PBS (Tokiwa et al. 2009). The biodegradability of PHA and of various blends of different types of PHA makes them good candidates in the packaging and biomedical sectors (Koller 2018).

Polyglycolic acid (PGA) biodegradability into nontoxic compounds is particularly appreciated in biomedical applications even if removal of the tin (II) based catalyst is needed since it is known to be toxic (Budak et al. 2020).

Actually, in various circumstances, composite production aimed at obtaining a material with improved biodegradability; for example, dried distillers grains with solubles, a major co-product of the biothanol industry, have been used for the production of "green" composites with enhanced biodegradability, using biodegradable poly(butylene adipate-co-terephthalate) (PBAT) and PHAs as polymer matrices (Muniyasamy et al. 2013; Madbouly et al. 2014). A fully biodegradable PLA/PBS composite reinforced with wood flour (Ludwiczak et al. 2019) is another example of this largely used strategy.

9.6 Determination of the Biobased Carbon Content

The market chain is strongly interested in showing that the material is indeed based on recently living organic matter. A renewable resource is something that is replenish to replace the portion depleted at a rate that exceeds depletion according to ASTM E2114 (ASTM International 2019b). A biobased material is an organic substance in which carbon derives from carbon dioxide recently fixed via photosynthesis in a renewable resource. Biobased materials can be both plant-based and animal-based. The quantification of the biobased carbon content of a product is crucial in assessing the renewability of the naturally replenished supply used to produce it. This is essential because a specific building block may be fossilbased or biobased, as in the case of drop-in platform molecules.

The determination of the biobased carbon content can be accomplished, taking into account that ^{14}C is only found in biobased materials: renewable carbon was directly in equilibrium with CO_2 in the atmosphere at the time of harvesting of a crop. On the converse, ^{14}C can not be found in fossil fuels because it has a half-life of about 5700 years. It follows that the relative ^{14}C levels are proportional to the biomass content of the material. Standard Test Methods for discriminating between carbon resulting from contemporary input and that derived from fossilbased input are described in ASTM D6866 (ASTM International 2021). These test methods apply to any product that can be combusted in the presence of oxygen to produce CO_2 gas. Radioactive-isotope analysis can determine the proportion of carbon in a material's organic compounds that comes from recently living, non-petroleum sources, even if the original biobased material has been chemically, thermally, or physically modified.

9.7 Chemical Compatibility

Both packaging and recent high-profile uses of bioplastics formulated into durable automotive, electronics, and consumer goods necessitate chemical compatibility of the material with the intended environment. Testing related to chemical compatibility is crucial. A current test used for plastics, such as ASTM D543 (ASTM International 2020c), is usually adopted for evaluating the resistance to chemical reagents. Prediction of the actual performance on the basis of test results depends upon the similarity between the testing and the end-use conditions. The effect of chemical reagents on sample properties shall be established by making measurements on standard specimens suitable for the described tests before and after immersion or stress, or both. Fluids can affect plastics by attacking chemically or absorbing into critical zones. Chemically reactive fluids can cause chain scission, chemical modification, and cross-linking by hydrolysis and oxidation. Adsorption of chemicals into micro-yielded or stress-dilated zones weakens mechanical strength (Wright 1996). Crystallinity obviously improves chemical resistance. The selection of test conditions (manner and duration of contact with reagents, temperature of the system, applied stress, and other factors) should mimic those encountered in the particular application. Unfortunately, while chemical compatibility charts are available for fossilbased plastics, literature data are very scarce as regards bioplastics (Heikkinen et al. 2018); the chemical resistance of PBS, biobased polyamides, and biobased polyethylene terephthalate was graded good as that of petroleum-based plastic, while that of PLA, PHA, and thermoplastic starch was rated poor (Zhao et al. 2020).

Bibliography

Adamcová D, Radziemska M, Zloch J et al (2018) Sem analysis and degradation behavior of conventional and bio-based plastics during composting. Acta Univ Agric Silvic Mendelianae Brun 66:349–356. https://doi.org/10.11118/actaun201866020349

Alshehrei F (2017) Biodegradation of synthetic and natural plastic by microorganisms. J Appl Environ Microbiol 5:8–19. https://doi.org/10.12691/JAEM-5-1-2

Álvarez-Chávez CR, Sánchez-Acosta DL, Encinas-Encinas JC et al (2017) Characterization of extruded poly(lactic acid)/pecan nutshell biocomposites. Int J Polym Sci 2017:13–17. https://doi.org/10.1155/2017/3264098

Ashokkumar M, Sumukh KM, Murali R et al (2012) Collagen-chitosan biocomposites produced using nanocarbons derived from goatskin waste. Carbon N Y 50:5574–5582. https://doi.org/10.1016/j.carbon.2012.08.006

ASTM International (2012) ASTM D7473-12 Standard Test Method for Weight Attrition of Plastic Materials in the Marine Environment by Open System Aquarium incubations. https://www.astm.org/Standards/D7473.htm. Accessed 5 Aug 2020

ASTM International (2013) ASTM D1653-13 Standard Test Methods for Water Vapor Transmission of Organic Coating Films. https://www.astm.org/Standards/D1653.htm. Accessed 3 Mar 2021

ASTM International (2015a) ASTM D7991-15 Standard Test Method for Determining Aerobic Biodegradation of Plastics Buried in Sandy Marine Sediment under Controlled Laboratory Conditions. https://www.astm.org/Standards/D7991.htm. Accessed 5 Aug 2020

ASTM International (2015b) ASTM D5338-15(2021) Standard Test Method for Determining Aerobic Biodegradation of Plastic Materials Under Controlled Composting Conditions, Incorporating Thermophilic Temperatures. https://www.astm.org/Standards/D5338.htm. Accessed 3 Mar 2021

ASTM International (2016) ASTM E96 / E96M-16 Standard Test Methods for Water Vapor Transmission of Materials. https://www.astm.org/Standards/E96.htm. Accessed 3 Mar 2021

ASTM International (2017a) ASTM D3985-17 Standard Test Method for Oxygen Gas Transmission Rate Through Plastic Film and Sheeting Using a Coulometric Sensor. https://www.astm.org/Standards/D3985.htm. Accessed 3 Mar 2021

ASTM International (2017b) ASTM D6691-17 Standard Test Method for Determining Aerobic Biodegradation of Plastic Materials in the Marine Environment by a Defined Microbial Consortium or Natural Sea Water Inoculum. https://www.astm.org/Standards/D6691.htm. Accessed 3 Mar 2021

ASTM International (2018a) ASTM D5988-18 Standard Test Method for Determining Aerobic Biodegradation of Plastic Materials in Soil. https://www.astm.org/Standards/D5988.htm. Accessed 3 Mar 2021

ASTM International (2018b) ASTM D5929-18 Standard Test Method for Determining Biodegradability of Materials Exposed to Source-separated Organic Municipal Solid Waste Mesophilic Composting Conditions by Respirometry. https://doi.org/10.1520/D5929-18. Accessed 3 Mar 2021

ASTM International (2018c) ASTM D6954-18 Standard Guide for Exposing and Testing Plastics that Degrade in the Environment by a Combination of Oxidation and Biodegradation. https://www.astm.org/Standards/D6954.htm. Accessed 3 Mar 2021

ASTM International (2018d) ASTM D5526-18 Standard Test Method for Determining Anaerobic Biodegradation of Plastic Materials under Accelerated Landfill Conditions. https://www.astm.org/Standards/D5526.htm. Accessed 3 Mar 2021

ASTM International (2018e) ASTM D5511-18 Standard Test Method for Determining Anaerobic Biodegradation of Plastic Materials under High-Solids Anaerobic-Digestion Conditions. https://www.astm.org/Standards/D5511.htm. Accessed 3 Mar 2021

ASTM International (2019a) ASTM D6400-19 Standard Specification for Labeling of Plastics Designed to be Aerobically Composted in Municipal or Industrial Facilities. https://www.astm.org/Standards/D6400.htm. Accessed 3 Mar 2021

ASTM International (2019b) ASTM E2114-19 Standard Terminology for Sustainability Relative to the Performance of Buildings. https://www.astm.org/Standards/E2114.htm. Accessed 3 Mar 2021

ASTM International (2020a) ASTM F1249-20 Standard Test Method for Water Vapor Transmission Rate Through Plastic Film and Sheeting Using a Modulated Infrared Sensor. https://www.astm.org/Standards/F1249. Accessed 3 Mar 2021

ASTM International (2020b) ASTM D7475-20 standard test method for determining the aerobic degradation and anaerobic biodegradation of plastic materials under accelerated bioreactor landfill conditions. https://www.astm.org/Standards/D7475.htm. Accessed 5 Aug 2020

ASTM International (2020c) ASTM D543-20 Standard Practices for Evaluating the Resistance of Plastics to Chemical Reagents. https://www.astm.org/Standards/D543.htm. Accessed 3 Mar 2021

ASTM International (2021) ASTM D6866-21 Standard Test Methods for Determining the Biobased Content of Solid, Liquid, and Gaseous Samples Using Radiocarbon Analysis. https://www.astm.org/Standards/D6866.htm. Accessed 3 Mar 2021

Baek BS, Park JW, Lee BH, Kim HJ (2013) Development and application of green composites: using coffee ground and bamboo flour. J Polym Environ 21:702–709. https://doi.org/10.1007/s10924-013-0581-3

Bashir ASM, Manusamy Y (2015) Recent developments in biocomposites reinforced with natural biofillers from food waste. Polym Plast Technol Eng 54:87–99. https://doi.org/10.1080/03602559.2014.935419

Bhasney SM, Patwa R, Kumar A, Katiyar V (2017) Plasticizing effect of coconut oil on morphological, mechanical, thermal, rheological, barrier, and optical properties of poly(lactic acid): a promising candidate for food packaging. J Appl Polym Sci 134:1–12. https://doi.org/10.1002/app.45390

Budak K, Sogut O, Aydemir Sezer U (2020) A review on synthesis and biomedical applications of polyglycolic acid. J Polym Res 27

Castro-Aguirre E, Auras R, Selke S et al (2018) Enhancing the biodegradation rate of poly(lactic acid) films and PLA bio-nanocomposites in simulated composting through bioaugmentation. Polym Degrad Stab 154:46–54. https://doi.org/10.1016/j.polymdegradstab.2018.05.017

Cecchi T, Montani IIT, Montani VIAG (2017) Synthesis characterization, and volatile profile of polylactic acid (pla) mussel shell powder composites authors

CEN (2000) EN 13432:2000 Packaging – requirements for packaging recoverable through composting and biodegradation – test scheme and evaluation criteria for the final acceptance of packaging

CEN (2006) EN 14995:2006 Plastics – evaluation of compostability – test scheme and specifications. http://store.uni.com/catalogo/en-14995-2006. Accessed 5 Aug 2020

CEN (2018) EN 17033:2018 Plastics – biodegradable mulch films for use in agriculture and horticulture – requirements and test methods. https://standards.cen.eu/dyn/www/f?p=204:110:0::::FSP_PROJECT,FSP_ORG_ID:41401,6230&cs=19E53F436D5E8A6FF49358DA8C195A6E2. Accessed 5 Aug 2020

Chłopek J, Morawska-Chochół A, Bajor G et al (2007) The influence of carbon fibres on the resorption time and mechanical properties of the lactide-glycolide co-polymer. J Biomater Sci Polym Ed 18:1355–1368. https://doi.org/10.1163/156856207782246858

Chun KS, Husseinsyah S, Osman H (2012) Mechanical and thermal properties of coconut shell powder filled polylactic acid biocomposites: effects of the filler content and silane coupling agent. J Polym Res 19. https://doi.org/10.1007/s10965-012-9859-8

EU (2018) Communication from the COMMISSION to the EUROPEAN Parliament, The Council, The European Economic and Social Committee and The Committee of the regions a European strategy for plastics in a circular economy

Farah S, Anderson DG, Langer R (2016) Physical and mechanical properties of PLA, and their functions in widespread applications – a comprehensive review. Adv Drug Deliv Rev 107:367–392

Fukushima K, Furuhashi Y, Sogo K et al (2005) Stereoblock poly(lactic acid): synthesis via solid-state polycondensation of a stereocomplexed mixture of poly(L-lactic acid) and poly(D-lactic acid). Macromol Biosci 5:21–29. https://doi.org/10.1002/mabi.200400121

Ghorpade VM, Gennadios A, Hanna MA (2001) Laboratory composting of extruded poly(lactic acid) sheets. Bioresour Technol 76:57–61

Heikkinen ITS, Kauppinen C, Liu Z et al (2018) Chemical compatibility of fused filament fabrication-based 3-D printed components with solutions commonly used in semiconductor wet processing. Addit Manuf 23:99–107. https://doi.org/10.1016/j.addma.2018.07.015

Hubbe MA, Lavoine N, Lucia L, Dou C (2021) Formulating bioplastic composites for biodegradability, recycling, and performance: a review. BioResources 16(1):2021–2083

Imre B, Pukánszky B (2013) Compatibilization in bio-based and biodegradable polymer blends. Eur Polym J:1215–1233

ISO (2019) ISO 14851:2019. Determination of the ultimate aerobic biodegradability of plastic materials in an aqueous medium – Method by measuring the oxygen demand in a closed respirometer.

ISO (2001) ISO 16221:2001 – Water quality – guidance for determination of biodegradability in the marine environment. https://www.iso.org/standard/30205.html. Accessed 5 Aug 2020

ISO (2012a) ISO 14855-1:2012 – determination of the ultimate aerobic biodegradability of plastic materials under controlled composting conditions – method by analysis of evolved carbon dioxide – Part 1: general method. https://www.iso.org/standard/57902.html. Accessed 5 Aug 2020

ISO (2012b) ISO 17088:2012. Specfications for compostable plastics

ISO (2013) ISO 18606:2013 Packaging and the environment – Organic recycling

ISO (2015) ISO 20200:2015 – Plastics – determination of the degree of disintegration of plastic materials under simulated composting conditions in a laboratory-scale test. https://www.iso.org/standard/63367.html. Accessed 5 Aug 2020

ISO (2018) ISO 14852:2018 Determination of the ultimate aerobic biodegradability of plastic materials in an aqueous medium – Method by analysis of evolved carbon dioxide

ISO (2019a) ISO 16929:2019 – Plastics – determination of the degree of disintegration of plastic materials under defined composting conditions in a pilot-scale test. https://www.iso.org/standard/72473.html. Accessed 5 Aug 2020

ISO (2019b) ISO 17556:2019 – Plastics – determination of the ultimate aerobic biodegradability of plastic materials in soil by measuring the oxygen demand in a respirometer or the amount of carbon dioxide evolved. https://www.iso.org/standard/74993.html. Accessed 5 Aug 2020

Jandas PJ, Mohanty S, Nayak SK (2012) Renewable resource-based biocomposites of various surface treated banana fiber and poly lactic acid: characterization and biodegradability. J Polym Environ 20:583–595. https://doi.org/10.1007/s10924-012-0415-8

Johari AP, Mohanty S, Kurmvanshi SK, Nayak SK (2016) Influence of different treated cellulose fibers on the mechanical and thermal properties of poly(lactic acid). ACS Sustain Chem Eng 4:1619–1629. https://doi.org/10.1021/acssuschemeng.5b01563

Kapelski A (2013) Stereocontrolled ring-opening polymerization of lactide monomers by lewis-acidic metal complexes. Shaker Verlag Gmbh, Germany

Kfoury G, Raquez J-M, Hassouna F et al (2013) Recent advances in high performance poly(lactide): from "green" plasticization to super-tough materials via (reactive) compounding. Front Chem 1:1–46. https://doi.org/10.3389/fchem.2013.00032

Kim YJ, Park OO (1999) Miscibility and biodegradability of poly(butylene succinate)/poly(butylene terephthalate) blends. J Environ Polym Degrad 7:53–66. https://doi.org/10.1023/A:1021846219775

Koller M (2018) Biodegradable and biocompatible polyhydroxy-alkanoates (PHA): auspicious microbial macromolecules for pharmaceutical and therapeutic applications. Molecules 23

Koutsomitopoulou AF, Bénézet JC, Bergeret A, Papanicolaou GC (2014) Preparation and characterization of olive pit powder as a filler to PLA-matrix bio-composites. Powder Technol 255:10–16. https://doi.org/10.1016/j.powtec.2013.10.047

Le Guen MJ, Thoury-Monbrun V, Castellano Roldán JM, Hill SJ (2017) Assessing the potential of farm dairy effluent as a filler in novel PLA biocomposites. J Polym Environ 25:419–426. https://doi.org/10.1007/s10924-016-0824-1

Li H, Wu J, Brunel S et al (2005) Polymerization of lactides and lactones by metal-free initiators. Ind Eng Chem Res 44:8641–8643. https://doi.org/10.1021/ie058022n

Liu T, Lam P-M, Zhao Y et al (2009) Mechanical and thermal properties of chicken feather fiber/PLA green composites. Compos Part B Eng 40:650–654. https://doi.org/10.1016/j.compositesb.2009.04.011

Ludwiczak J, Frackowiak S, Leluk K, Hanus-Lorenz B (2019) A fully biodegradable pla/pbs composite reinforced with wood flour. Cellul Chem Technol 53:955–962. https://doi.org/10.35812/CelluloseChemTechnol.2019.53.93

Madbouly SA, Schrader JA, Srinivasan G et al (2014) Biodegradation behavior of bacterial-based polyhydroxyalkanoate (PHA) and DDGS composites. Green Chem 16:1911–1920. https://doi.org/10.1039/c3gc41503a

Madera-Santana TJ, Freile-Pelegrín Y, Encinas JC et al (2015) Biocomposites based on poly(lactic acid) and seaweed wastes from agar extraction: evaluation of physicochemical properties. J Appl Polym Sci 132:1–8. https://doi.org/10.1002/app.42320

Mochizuki M, Hirami M (1997) Structural effects on the biodegradation of aliphatic polyesters. Polym Adv Technol 8:203–209. https://doi.org/10.1002/(SICI)1099-1581(199704)8:4<203::AID-PAT627>3.0.CO;2-3

Muniyasamy S, Reddy MM, Misra M, Mohanty A (2013) Biodegradable green composites from bioethanol co-product and poly(butylene adipate-co-terephthalate). Ind Crop Prod 43:812–819. https://doi.org/10.1016/j.indcrop.2012.08.031

Nisticò R, Evon P, Labonne L et al (2018) Post-harvest tomato plants and urban food wastes for manufacturing plastic films. J Clean Prod 167:68–74. https://doi.org/10.1016/j.jclepro.2017.08.160

Nova-Institute (2019) Biodegradable, bio-based polymers in various environments – renewable Carbon. http://bio-based.eu/downloads/biodegradable-bio-based-polymers-in-various-environments/. Accessed 14 Dec 2019

Ohkoshi I, Abe H, Doi Y (2000) Miscibility and solid-state structures for blends of poly[(S)-lactide] with atactic poly[(R,S)-3-hydroxybutyrate]. Polymer (Guildf) 41:5985–5992. https://doi.org/10.1016/S0032-3861(99)00781-8

Pastore G, Gabrielli S, Cecchi T et al (2021) A new and efficient lactic acid polymerization by multimetallic cerium complexes: a Poly(lactic acid) suitable for biomedical applications. RSC Adv 11:10592–10598. https://doi.org/10.1039/d0ra10637b

Pawar RP, Tekale SU, Shisodia SU et al (2014) Biomedical applications of poly(lactic acid). Rec Pat Regen Med 4:40–51. https://doi.org/10.2174/2210296504666140402235024

Quiles-Carrillo L, Montanes N, Sammon C et al (2018) Compatibilization of highly sustainable polylactide/almond shell flour composites by reactive extrusion with maleinized linseed oil. Ind Crop Prod 111:878–888. https://doi.org/10.1016/j.indcrop.2017.10.062

Raza ZA, Noor S, Khalil S (2019) Recent developments in the synthesis of poly(hydroxybutyrate) based biocomposites. Biotechnol Prog 35:1–15. https://doi.org/10.1002/btpr.2855

Sarki J, Hassan SB, Aigbodion VS, Oghenevweta JE (2011) Potential of using coconut shell particle fillers in eco-composite materials. J Alloys Compd 509:2381–2385. https://doi.org/10.1016/j.jallcom.2010.11.025

Scaffaro R, Lopresti F, Botta L (2018) PLA based biocomposites reinforced with Posidonia oceanica leaves. Compos Part B Eng 139:1–11. https://doi.org/10.1016/j.compositesb.2017.11.048

Shih YF, Huang CC (2011) Polylactic acid (PLA)/banana fiber (BF) biodegradable green composites. J Polym Res 18:2335–2340. https://doi.org/10.1007/s10965-011-9646-y

Song Y, Liu J, Chen S et al (2013) Mechanical properties of poly (lactic acid)/hemp fiber composites prepared with a novel method. J Polym Environ 21:1117–1127. https://doi.org/10.1007/s10924-013-0569-z

Su S, Kopitzky R, Tolga S, Kabasci S (2019) Polylactide (PLA) and its blends with poly(butylene succinate) (PBS): a brief review. Polymers (Basel) 11:1193. https://doi.org/10.3390/polym11071193

Thunga M, Chen K, Grewell D, Kessler MR (2014) Bio-renewable precursor fibers from lignin/polylactide blends for conversion to carbon fibers. Carbon N Y 68:159–166. https://doi.org/10.1016/j.carbon.2013.10.075

Tiimob BJ, Jeelani S, Rangari VK (2016) Eggshell reinforced biocomposite – an advanced "green" alternative structural material. J Appl Polym Sci 133:1–10. https://doi.org/10.1002/app.43124

Tokiwa Y, Calabia BP, Ugwu CU, Aiba S (2009) Biodegradability of plastics. Int J Mol Sci 10:3722–3742

Wright DC (1996) Environmental stress cracking of plastics. Smithers Rapra Technology

Wu CS (2015) Renewable resource-based green composites of surface-treated spent coffee grounds and polylactide: characterisation and biodegradability. Polym Degrad Stab 121:51–59. https://doi.org/10.1016/j.polymdegradstab.2015.08.011

Xie W, Chen D, Fan X et al (1999) Lithium chloride as catalyst for the ring-opening polymerization of lactide in the presence of hydroxyl-containing compounds. J Polym Sci Part A Polym Chem 37:3486–3491. https://doi.org/10.1002/(SICI)1099-0518(19990901)37:17<3486::AID-POLA6>3.0.CO;2-2

Xu J, Guo BH (2010) Poly(butylene succinate) and its copolymers: research, development and industrialization. Biotechnol J 5:1149–1163

Yu L, Dean K, Li L (2006) Polymer blends and composites from renewable resources. Prog Polym Sci 31:576–602. https://doi.org/10.1016/j.progpolymsci.2006.03.002

Zembouai I, Bruzaud S, Kaci M et al (2014) Mechanical recycling of Poly(3-Hydroxybutyrate-co-3-Hydroxyvalerate)/Polylactide based blends. J Polym Environ 22:449–459. https://doi.org/10.1007/s10924-014-0684-5

Zhao X, Cornish K, Vodovotz Y (2020) Narrowing the gap for bioplastic use in food packaging: an update. Environ Sci Technol 54:4732. https://doi.org/10.1021/acs.est.9b03755

Part IV
Safety and Sustainability of Biobased Products from Food Waste

Chapter 10
Assessment of the Safety of BioBased Products

Teresa Cecchi

Before the fossil era, at the beginning of the twentieth century, a large part of the resources used by society in the production of goods was renewable. The availability and versatility of chemical building blocks from crude oil obscured the interest in wood and other natural materials. The impetus for the resurgence in the use of renewable resources has come not only from the sustainability concept but also from the potential for improved safety and environmental quality of biobased outputs due to their allegedly less hazardous nature (Chambers and Muecke 2010).

The European Green Deal is the most recent response to climate and environmental-related challenges (EU 2019). Resource-intensive sectors such as plastics are the focus of interest of policymakers, but a regulatory framework for biodegradable and biobased plastics is not ready yet. Academic research is pioneering in this respect. In the following, we will detail research evidence concerning the safety of biobased products.

Figure 10.1 outlines the twelve principles of green chemistry (Kharisov et al. 2019). Many of them are inspiring for the multiproduct biorefinery fed by food waste (FW), anyhow, in this chapter, we will focus on the possibility to avoid harmful synthetic methods and catalysts in order to reduce the product toxicity due to the presence of unsafe additives or volatile organic compounds.

10.1 Volatile Organic Compounds (VOCs) from Bioplastics

Volatile organic compounds (VOCs) are of concern as air pollutants. A clear definition of VOCs can be found in the Directive 2004/42/CE of the European Parliament and of the Council. "'Volatile organic compound (VOC)' means any organic compound having an initial boiling point less than or equal to 250°C measured at a standard pressure of 101,3 kPa" (The European Parliament and the Council of the European Union 2004). Their indoor presence was correlated to negative effects on human health and wellbeing and to the "sick building syndrome" (Cecchi 2014a;

T. Cecchi, C. De Carolis, *Biobased products from food sector waste*,
https://doi.org/10.1007/978-3-030-63436-0_10

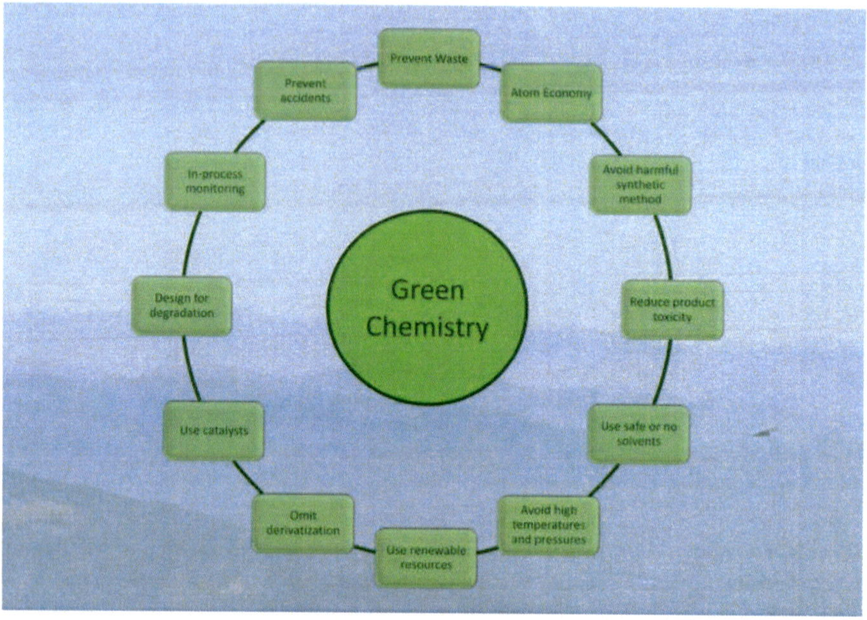

Fig. 10.1 The 12 principles of green chemistry

Adamová et al. 2019) characterized by allergies, eye irritation, nose and throat malfunction, and tiredness; hence the analysis of VOCs is crucial for assessing the safety of both common materials and innovative bioplastics to be used in the building sector.

As regards bioplastics the VOCs fingerprint is essential for (i) understanding the chemical processes involved in polymer synthesis, stability, and degradation, (ii) detecting toxic compounds, including volatile catalyst, antioxidants, plasticizers, and other additives (iii) detecting substances with distinct odor and taste whose potential migration from bioplastics may lead to the sensorial deterioration of packaged foods and beverages. The emissions depend on the chemical composition of the material and climate conditions. Primary emissions, greatest after the synthesis of the material, are due to nonbound VOCs that can originate from solvents, technological agents, and unreacted raw materials such as residual monomers, low molecular weight oligomers, and polymerization byproducts. Secondary emissions are due to chemically or physically bound VOCs formed during polymer decomposition, degradation, and oxidation, or reactions of the primary VOCs (Wolkoff 2003; Cecchi 2014b).

Protocols for VOC determination should detect compounds in the part per trillion to part per billion range. On-site, direct-reading methods are less precise and accurate than collection methods, which can typically achieve a more sensitive quantification limit since they involve the collection of the compound into a container or its concentration onto a sorbent for later analysis.

As regards VOC emissions from bioplastics, there is a dearth in the literature since it is allegedly felt that their biobased source guarantees their safety and the absence of any health concerns. Since the use of biobased materials for building construction and for the automotive sector has become increasingly prevalent in recent years, indoor air quality concerns related to VOC release have to be properly addressed. In fact, indoor air quality issues and the reduction of the level of indoor air contaminants within the interior of the building are the focus of interest of many building rating systems and areas of legislation (Chambers and Muecke 2010; Wei et al. 2015). European Collaborative Action ECA "Indoor Air Quality & its Impact on Man" stimulated a multidisciplinary collaboration of European scientists.

Both EPA Compendium of Methods for the Determination of Toxic Organic Compounds in Ambient Air (TO methods) and ISO standards for indoor products emission testing are available; many countries have established mandatory or voluntary assessment programs regarding the release of VOCs from construction materials and furnishing, but reference legislation that sets limits to the concentrations of pollutants in indoor air has not been issued yet. Most of the available analytical methods rely on either gas chromatography (GC) or high-performance liquid chromatography (HPLC). The ISO 16000 series of standards are of particular interest. ISO 16000-6:2011 (ISO 2011), ISO 16000-9:2006 (ISO 2006a), ISO 16000-10:2006 (ISO 2006b), ISO 16000-11:2006 (ISO 2006c) are used for VOCs determination indoors. Most ISO standards have the corresponding ASTM standard.

The emission test chamber is widely used to test VOC emissions. It is a proper chamber working at constant temperature and humidity, with a specific rate of air exchange inside, where it is possible to place the sample whose emissions are studied. The airflow containing the emissions passes through a glass or stainless steel sorbent tube, usually packed with a porous polymer stable up to 400°C, properly designed for volatiles trapping. The thermal desorption unit enables the desorption of VOCs from the sorbent cartridges; VOCs are quantitatively transferred to the gas chromatograph, where they are separated during the chromatographic run.

The test chamber equipment is very expensive, and the procedure is very time-consuming; for these reasons, academicians studied the possibility to use the Solid Phase Micro Extraction (SPME) technique for the sampling step.

Compared to several sampling techniques, SPME shows the advantages of being solvent-free, cheap, and easy to use. The low cost of the very compact and lightweight device, the possibility of automation, and the simplicity of coupling with chromatographic techniques are further advantages of the SPME sampling strategy. This environmentally and user-friendly technique allows for integrating sampling, enrichment, and transfer to the chromatographic equipment. VOCs are usually released from a sample sealed in a vial. The SPME manual device is illustrated in Fig. 10.2. The holder has a needle containing a retractable fiber that pierces the vial septum; the fiber, coated with an extracting phase, is then exposed to the sample. Volatiles are extracted at a specific temperature for a specific time because the extracted amount of each analyte is a function of temperature and time. The curve, at constant temperature, flattens with increasing adsorption time; higher reproducibility is obviously expected in this flat equilibrium region; subsequently, the fiber is retracted and transferred to the GC injection port where thermal desorption of the adsorbates occurs. The extracting phase can extract different kinds of analytes from different kinds of media. Absorbing fibres should be conditioned before use. The

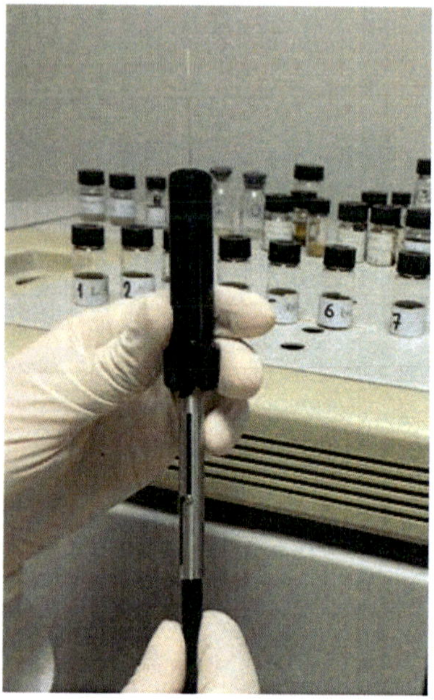

Fig. 10.2 SPME device

divinylbenzene/carboxen/polydimethylsiloxane (DVB/CAR/PDMS) fiber is almost universal because it is able to adsorb analytes in a wide range of polarity. The optimization of adsorption time and temperature provides high concentration factors. If the detection relies on a mass spectrometer, VOCs can be easily determined both qualitatively and quantitatively. The identification of the constituents is usually based on the comparison of the retention times with those of authentic samples, on computer matching against commercial spectral libraries (http://webbook.nist.gov/chemistry/) (Cecchi and Alfei 2013), and by comparison of their Kovats retention indices relative to n-alkanes, calculated using straight-chain alkanes mixture, with the averaged values reported in the bibliography for chromatographic columns similar to that used (van Den Dool and Dec. Kratz 1963).

Noteworthy, sampling via SPME is independent of humidity if it is not excessively high (Harper 2000). Anyhow, partition equilibrium composition depends on analyte volatility, sampling temperature, and time because competitive adsorption may occur; hence it is essential to adjust the design of experiment in order to pinpoint the best extraction conditions. The main drawback of SPME is the fact that absolute quantity determination is not straightforward (Pawliszyn 2011): the quantity of analyte extracted by the fiber is proportional to its concentration in the sample as long as equilibrium is reached; anyhow, determination of amounts of target compounds relative to total peak area easily allows the comparison of quantities of the same compound in different samples under constant conditions.

Emissions from PLA and PLA composites were sampled via SPME and measured through gas chromatography coupled to mass spectrometry (SPME-GC-MS) (Cecchi et al. 2019; Salazar et al. 2017).

The identification and quantification of VOCs formed after extrusion and after thermoforming of some commercial types of PLA via the SPME-GC-MS apparatus was studied. The extruded film released more analytes (namely acetaldehyde, acetone,2-methyl-2-propanol, acetic acid,2,3-pentanedione,2,4-dimethyl-2-pentanol,2,3,4-trimethyl-hexane, and2,2,4,6,6-pentamethyl-heptane) and higher amounts of VOCs than thermoformed samples. After 6 months of PLA storage at ambient conditions, the presence of VOCs decreased (Salazar et al. 2017). From a commercial point of view, these results are remarkable: the detected VOCs have distinct odor and taste and their migration may lead to the sensorial deterioration of packaged foods and beverages; interestingly, VOCs in PLA samples are not produced during storage at room temperature, but only during its processing at high temperatures.

A SPME-GC-MS apparatus was used to study the volatile fingerprint of both PLA obtained from (S)-lactic acid and the same PLA compounded with ceramic food waste powder fillers, such as eggshells and mussel shells. Volatiles from biocomposites can originate from the matrix polymer or from the natural-filler or from a combination of the two (Cecchi et al. 2019).

The molecular fingerprint of VOCs from PLA and from its composites is outlined in Table 10.1. It details the odor type (Dunkel et al. 2009) and the chemical

Table 10.1 Odor type, experimental and literature retention indices (RI), average GC peak areas and standard deviation (triplicate experimental design), and chemical class of VOCs analyzed by HS-SPME-GC-MS

Analyte	Heptanal	6-methyl-5-hepten-2-one	Nonanal	Trans (R,S) lactide (meso)	Cis (S,S) lactide	Decanal
PLA area count	1.98E+05	3.86E+05	2.18E+06	9.64E+04	5.26E+05	7.30E+05
PLA area count ST DEV	2.96E+04	6.38E+04	1.94E+05	2.68E+04	9.29E+04	2.60E+05
PLA/mussel shell area count	nd	nd	2.15E+05	7.42E+04	3.39E+05	2.24E+05
PLA/mussel shell area count ST DEV	nd	nd	1.12E+05	4.46E+04	1.21E+05	3.16E+04
PLA/egglshell area count	4.14E+05	3.96E+05	3.69E+06	7.22E+04	6.39E+05	2.28E+06
PLA/egglshell area count ST DEV	1.04E+05	1.02E+05	1.82E+06	1.98E+04	2.13E+05	1.10E+06
RI lit	902	999	1115	na	na	1214
RT exp	906	1009	1126	1168	1199	1225
Chemical class	Aldehyde	Ketone	Aldehyde	Ester	Ester	Aldehyde
Odor type	Green	Citrus	Aldehydic	na	na	Aldehydic

Adapted by permission from Springer Nature: Environmental Science and Pollution Research. Unprecedented high percentage of food waste powder filler in poly lactic acid green composites: synthesis, characterization, and volatile profile. Cecchi T., Giuliani A., Iacopini F., Santulli C., Sarasini F., Tirillò J., 26:7263–7271, 2019. https://www.springer.com/journal/11356

class of each VOC, its experimental and literature retention indices (van Den Dool and Dec. Kratz 1963), as well as its abundance.

Interestingly, priority air pollutants released by fossil plastics that pose a significant public health threat (such as phenylalkanes, phthalates, and bisphenol A) are missing from the volatile profile of PLA and its composites. It is rewarding to observe that VOCs known to be causative factors of hormone disruption or reproductive dysfunction are similarly missing.

The volatile profile is very simple. Fig. 10.3 shows the chromatographic profiles for raw PLA from (S)-lactic acid, PLA compounded with eggshell powder, and PLA compounded with mussel shell powder. Cis-(S,S)-lactide and trans-(R,S)- lactide are easily predicted to be released from the PLA matrix. The presence of cis-(S,S)-lactide can be taken for granted since it is the dimer of the (S)-lactic acid used to obtain PLA; anyhow, at high temperature, an inversion of configuration may occur. This explains the presence of smaller amounts of trans-(R,S)-lactide. Heptanal, nonanal, and decanal are typical food odorants that do not pose a concern to public health: they were actually found also in the volatile fraction of both fillers. Compounding PLA with mussel shells seems to reduce the VOC emission since the chromatographic volatile profile of the filler is almost flat, and the presence of the filler reduces the surface of PLA available for the emission of its volatiles. In this respect, the filler behaves as a kind of PLA matrix diluent, decreasing its emissions.

It cannot be excluded, anyhow, that at a higher temperature, PLA and its composites are not exempt from other emissions due to thermal degradation that would not occur at ambient temperature (Wojtyła et al. 2017); it was demonstrated that PLA is one of the lowest emitters if used as a 3D printer filament (Azimi et al. 2016);

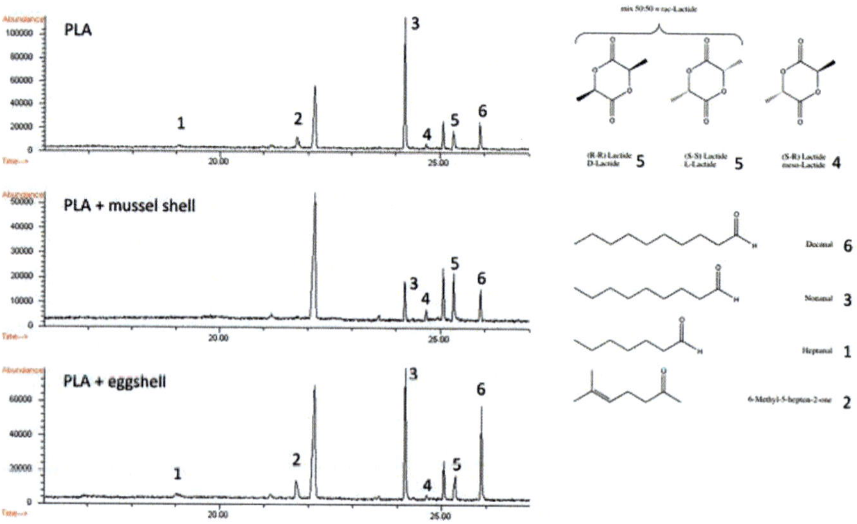

Fig. 10.3 Chromatograms of pure PLA, PLA/eggshell powder composite, and PLA/mussel shell powder composite.

anyhow, the high temperature of the 3D printer nozzle may be responsible for the emission of some the toxic or irritant VOCs from the PLA filaments, even if VOCs apportionment, between PLA matrix and additives in the filament, can be difficult; the specific source of the volatiles was unsure and, actually, out of hundreds of VOCs identified, only 5 ((S,S)-Lactide, acetaldehyde, 1-butanol, formaldehyde, and decanal) were consistently detected for all PLA filaments and thereby linked to the PLA matrix (Davis et al. 2019).

The detrimental influence of increasing temperature on the VOCs emission was demonstrated for neat PLA, PBS, and their bio-composites with pineapple flour and destarched cassava flour. VOC and formaldehyde were measured via thermal extractor analysis, field and laboratory emission cell techniques, and 20 L small chamber test with sorbent tubes and GC-MS analyses (Kim et al. 2011).

The detection of VOC (volatile oligomers) via SPME-GC-MS was also used to monitor the changes in PLA matrix exposed to outdoor soil environment. Lactic acid and lactoyl lactic acid appeared in the volatile profile of PLA films not before 12 months, thereby indicating that degradation in soil is a slow process due to hydrolysis and microbial metabolism (Gallet et al. 2000).

Main volatiles from a lignin/PLA/PHB composite, identified from pyrolysis GC-MS, were sulfur compounds (Ház et al. 2018).

Naturally occurring di- and multifunctional carboxylic acids were used to crosslink epoxidized sucrose soyate; the resulting highly functional biobased epoxy resin was a VOC-free thermoset with excellent thermal and mechanical properties. Fossilbased thermosets usually release volatile solvents or monomers, used to achieve favorable properties and low viscosity during the curing process. For this reason, these fully green VOC-free thermosets can replace fossilbased ones; noteworthy, they could be degraded either thermally or in alkaline media (Ma and Webster 2015).

However, it is felt that the VOCs release from bioplastics is far from being suitably studied; hence it is a cogent research need.

10.2 Catalyst Residues and Green Catalysts

Intensive research efforts about catalysts used for the synthesis of bioplastics focus on a number of different criteria for green industrial applications. First of all, the catalyst should be effective with small loadings, and it should not be harmful to the environment (during biodegradation) and the health since many materials have potential application in the biomedical sector.

As regards renewable and biodegradable PLA, nontoxic, highly active green catalysts for its polymerization would be highly desirable. Actually, tin(II)octanoate, $Sn(Oct)_2$, is the commonly used catalyst for the industrial ring-opening polymerization of lactide described in Chap. 6; its homogeneous nature impairs the complete removal of the catalyst from the final product in which it is incorporated. Its growing use is leading to an increased intake of residual tin into the environment.

Due to its known toxicity for several aquatic organisms and many adverse effects on enzymes, receptors, and whole organisms, replacement strategies in agricultural applications are of great interest. Even if $Sn(Oct)_2$ has been approved by the U.S. FDA (Howard et al. 2019), its genotoxicity and cytotoxicity are major drawbacks in biomedical applications (Parwe et al. 2017; Tanzi et al. 1994). The remnants of tin-based catalysts hinder safe PLA application for medical devices and food handling packages (Howard et al. 2019). The development of safe catalysts for biopolymers synthesis is of great interest. Cerium trichloride heptahydrate is a non-toxic and easy to handle Lewis acid catalyst in organic reactivity; a new and efficient lactic acid polymerization by multimetallic cerium complexes yielded a poly(lactic acid) suitable for biomedical applications, in high yields up to 95% and a molecular weight ranging from 9000 to 145000 g/mol (Pastore et al. 2021).

Iron and zinc are prominent metal candidates as tin substitutes. Three biocompatible iron(II) hybrid guanidine complexes rely on a metal ion with a clear biological role to polymerize lactide to long-chain polylactide with monomer/initiator ratios of more than 5000:1 via the coordination–insertion mechanism and with an increased polymerization activity compared to $Sn(Oct)_2$ (Rittinghaus et al. 2019). Achiral iron complexes proved to be stereoselective under mild reaction conditions for the ring-opening polymerization of lactide for the production of thermally stable and industrially relevant stereocomplexes (Marin et al. 2019).

Four new zinc guanidine complexes have been tested for the polymerization of non-purified lactide under industrially relevant conditions: PLA with molecular weights over 71,000 g/mol were achieved with a polymerization rate that compares to that of $Sn(Oct)_2$ (Schäfer et al. 2019). Similarly, zinc L-Proline complex was used as a safe catalyst in bulk polymerization of L-lactide and gave high molecular weight PLLA (Parwe et al. 2017); zinc amido-oxazolinate complexes proved to be active and isoselective initiators for the ring-opening polymerization of rac-lactide, producing isotactic stereoblock polylactides (Abbina and Du 2014). The potential zinc complexes prompted the investigation of analogous magnesium and calcium catalysts due to their low toxicity and cost.

Divalent metals (i.e., Mg, Ca, and Zn) with monoanionic ancillary scaffolds or neutral ligands were also successfully tested (Wheaton et al. 2009; Chen et al. 2007).

Germanium is an unharmful element that is able to mimic the tin catalytic activity to yield high molar mass PLA due to their similar properties. Germanium complexes, tested in combination with zinc and copper, showed high bulk polymerization activities for lactide at 150 °C. Germanium provides the active site for the ring opening polymerization of lactide probably through the coordination–insertion mechanism (Rittinghaus et al. 2020).

The fully biobased poly(propylene succinate), with thermal stability similar to that of its fossilbased analog, was obtained starting from the biobased 1,4-butanediol and tetraisopropyl orthotitanate as a catalyst (Parcheta and Datta 2018).

Biocatalysis is regarded as both a green and sustainable technology. The spectacular advances in molecular biology and biotechnology resulted in protein engineering that has enabled the optimization of existing enzymes and the invention of entirely new biocatalytic reactions (Sheldon and Woodley 2017).

For example, bacteria have been engineered in a microbial plastic factory to produce lactate-based polyesters in a single-step metal-free system (Cheng et al. 2014). In this research field, the literature is still scarce, but it is felt that the quantum leap from conceptualization of this idea to showcase the potential of this technology is not far.

10.3 Migration: Risks and Opportunities from Biobased Food Contact Materials

Food contact materials (FCMs) are all items envisioned to come into contact with food (or water for human consumption) during its production, processing, transportation, storage, preparation, and serving. FCMs include packaging and containers, kitchenware, cutlery, and dishes.

The safety of FCMs must be assessed as chemicals can migrate from the materials into food or water: any potential transfer should not (i) have adverse effects on human health and on the taste and/or odor of foods, (ii) change the composition of the food in an unacceptable way. Factors that influence migration include the kind of food and its fat content, the nature of FCM, the temperature, and the duration of the contact. The migration of chemicals from the packaging into the food deserves academic and industrial attention. It is the result of diffusion, dissolution, and processes involving the mass transport phenomena, often described by Fick's second law, of low molecular mass compounds initially present in the package into a food sample or food simulant. Migration studies are usually performed using standardized food simulants and conditions specified in European food packaging regulations (Valdés et al. 2014).

In Europe, the safety of FCM is evaluated by the European Food Safety Authority (EFSA), which provides FCMs manufacturers with opinions on substances to be used in food contact materials. The U.S. Food and Drug Administration (FDA) laws and regulations similarly aim at protecting the consumers' health throughout the supply chain.

Current fossilbased polymers used as packaging materials face environmental and disposal issues. Actually, food package stability provided by fossilbased plastics during the shelf-life of the product is an advantage, but it becomes a disadvantage when the packages enter the post-use phase. Moreover, problems of food contamination due to substances migration from packaging are always the focus of attention. The world of packaging has great opportunities to develop a circular economy and industrial symbiosis. Bioplastics have been seen as an opportunity in this respect, notwithstanding their usually higher cost. Biodegradable biobased food packaging deserves attention because it could reduce material waste, elongate shelf life, and enhance food quality.

The gas and water vapor permeability are crucial properties of bioplastics in the packaging sector (they were discussed in Sect. 6.8); longshelf-life foods such as snacks and many other dry foods require high-barrier packaging, whereas fresh fruits, vegetables, and ready-to-eat meals need permeable packaging to stay fresh. Most bioplastics have lower water barrier properties due to their hydrophilicity but similar oxygen barrier characteristics as conventional plastics; hence their principal applications are breathable packagings for short shelf-life products. Still, PLA has an oxygen transmission rate at least one order of magnitude below those of the synthetic polymers used in the same field as PET and PS (Siracusa 2012; Zhao et al. 2020)

Biobased plastics have to comply with the same regulations with respect to food safety as fossilbased plastics; many biobased plastics such as biodegradable polyesters, starch, cellulose-based bioplastics, and drop-in bioplastics have been proven safe for food contact use (van den Oever et al. 2017).

Anyhow, bioplastics are not additive-free. For example, plasticizers are added to provide the necessary workability to bioplastics featuring fragility and brittleness during thermoformation and end-use application. Since bioplastics are usually less stable and have a lower diffusion barrier than conventional plastics, additional stabilizers and additives are often used to make their performance adequate and to overcome these problems. Furthermore, bioplastics are seldomly VOC-free (Ma and Webster 2015).

There is a dearth in the scientific literature as regards a comprehensive comparative evaluation of the migration behavior of chemicals from biobased and fossil-fuel based plastics. The differences between conventional and biobased plastics are not clear since the physico-chemical properties of biobased and fossil based materials might result in higher or lower migration rates of additives. Comparative studies describing the relative or absolute migration rates of the additives are scarce. The vast majority of studies describe migration from PLA and starch-based polymers (Avella et al. 2005; Conn et al. 1995). The low migration of lactic acid and its dimer, the lactide, from PLA and the fact that they do not pose serious health concerns, according to their toxicological data, made PLA one of the Generally Recognized As Safe (GRAS) substances (Conn et al. 1995) even if the development of safe catalyst candidates as tin substitutes and their use in green industrial applications is mandatory, as discussed above.

Zhu et al. (2014) described experimental results and a suitable model for the migration of the plasticizer triacetin from starch acetate into food simulants, but the authors concluded that they are still far from understanding migration on a molecular level and theoretical models covering these aspects would be desirable. The monomers of cellulose-based, starch-based, PHB, and PLA polymers are judged to be of no health concern (Clarke et al. 2012; Conn et al. 1995) at variance with the toxicological profile of many other monomers and additives used in the production of packaging materials, e.g., bisphenols (used to harden polycarbonate plastics and epoxy coatings), phthalates (used as plasticizers), vinyl chloride and acrylamide that pose serious environmental concerns and health risks to population (Plank and Trela 2018).

Since plasticizers are also needed in bioplastics, researchers tested the ability of some biobased products to mimic the role of conventional plasticizers in fully green bioplastics. For example, in an attempt to avoid the use of phthalates to mitigate the brittleness of biopolymers, oligomers of lactic acid (with a molar mass of around 1000 Da) are blended with commercial PLA to improve ductility, obtaining transparent and ductile materials, able for films manufacturing. These oligomers are proposed as innovative, and fully compatible and biodegradable plasticizers for PLA to avoid the migration of conventional plasticizers (Burgos et al. 2014). Among other possible PLA biobased plasticizers, cardanol acetate, obtained as a by-product of cashew nut extraction by the use of nontoxic and low environmental impact reagents, proved to be as effective as conventional plasticizers. In addition, plasticizer migration tests showed a lower weight loss of PLA plasticized by cardanol acetate compared to diethylhexyl phthalate plasticized PLA (Greco and Maffezzoli 2016).

Lower migration compared to fossil plastics was assessed for PLA nanocomposites with organoclays in packaging for the cosmetics industry. Total overall migration levels were well below the total legislative migration limit (10 mg/dm^2), established by EU-Plastics Regulation 10/2011 (European Commission 2020), and the nanocomposite migration extracts stimulated minimal toxicity to the skin (Connolly et al. 2019).

Compliance with legislation for food packaging materials was also demonstrated for the overall migration levels in both non-polar and polar simulants of transparent nanocomposite-based packaging film comprised of poly(3-hydroxybutyrate-co-3-hydroxyvalerate) and cellulose nanocrystal methyl ester (Yu et al. 2014).

Nano-packaged foods are momentous. Mixing nanoparticles into packaging materials was used to improve their temperature, humidity, and gas barrier resistance or to introduce in active packaging antimicrobial, oxygen, or ultra-violet scavengers. Possible interaction of nanomaterials with food products and potential risks and health hazards are possible as a result of migration of nanoparticles into food materials (Adeyeye 2019). Even if the migration of nanoclay from starch-nanocomposites was reported to be of no concern (Avella et al. 2005), Houtman and colleagues demonstrated the in vitro toxicity of nanoclays with cytotoxic effects on two different cell lines (Houtman et al. 2014); cytotoxic and genotoxic effects were independently confirmed (Maisanaba et al. 2013; Sharma et al. 2010); there is a need for further testing under more standardized conditions to perform an adequate risk assessment of nanomaterials used in bionanocomposites.

10.3.1 Biobased Barriers to Prevent Migration from Conventional Packaging Materials

Biobased barriers to mitigate migration from common packaging materials were cleverly envisioned, devised, and tested. Cellulose-based packaging materials are very common due to their lightweight, stability, and affordable price. However, the use of recycled paper and board may result in the migration of undesirable substances into the packed goods. To reduce this kind of migration, the application of functional barriers in the packaging is crucial: biobased or biodegradable polymers proved to be effective (Walzl et al. 2019). In order to enhance the safety of fiber-based packaging, a biobased composite barrier layer was used to protect against mineral oil migration through paperboard. Starch-clay composite coatings were effective barriers against the migration of mineral oil and could hence improve the barrier properties and safety of fiber-based primary food packaging (Koivula et al. 2016). Biobased coatings, such as modified starches (cationic starch and cationic waxy starch), plant and animal proteins (gluten and gelatine), were evaluated as potential barriers to chemical contaminants from recycled paper and board for food packaging; diffusion studies into the solid food simulant poly 2,6-diphenyl-p-phenylene oxide confirmed that all the tested biopolymers slowed down migration (Guazzotti et al. 2014). Two different types of paperboard produced from primary and secondary cellulosic fibers were upgraded by coating them with alginate and chitosan; these two biodegradable biomaterials decreased (i) air permeability, (ii) water vapor transmission, (iii) permeation of volatile compounds, (iv) migration of mineral oil saturated hydrocarbons and mineral oil aromatic hydrocarbons while improving grease resistance (Kopacic et al. 2018).

As regards metal cans, tomato processing wastes were transformed into a novel bio-lacquer coating intended for metal food packaging: this way, consumers protection is achieved via the upcycling of food waste from the food chain itself; this strategy meets current demands for sustainable production (Montanari et al. 2017).

10.3.2 Biobased Products for Active and Smart Food Packaging

Food package is increasingly conceived not only as a simple container: actually, it is progressively perceived as adding value to the product. Smart packaging is an active research field and a key area of development in new multifunctional materials because it could be used to highlight biochemical or microbial changes in foods, to prevent food counterfeit and adulteration, to provide better food protection and preservation, marketing, and smart communication to consumers. In active packaging the sustained release to foodstuff of active agents is an intended, intentional, and favorable case of migration to obtain a deliberate interaction of the packaging with the food. Intelligent packaging systems should not release substances into the food since they are used to inform consumers about the conditions of the packaged food.

Food industry is subject to pressure from different stakeholders: producers, retailers, and customers have different priorities. By this scenario, the use of agricultural and food processing wastes in the packaging industry is getting increasing interest because it provides both reductions in environmental pollution by the packaging disposal and waste recovery. Food technology researchers are engaged with developing intelligent packing for better preservation and increased shelf life of food products. In this context, nontoxic pH indicator films produce a visual response to pH changes and represent a sensor that provides visual information to the customers concerning food deterioration. Anthocyanins from the blueberry by-product in starch-based films (Luchese et al. 2017), blueberry and blackberry pomace extracts in chitosan-based antioxidant smart films (Kurek et al. 2018) anthocyanins from red cabbage in a PVA/chitosan film are examples of intelligent food packaging (Pereira et al. 2015).

Biobased products find application as active agents for incorporation into biobased packaging, especially as antimicrobials because microbial spoilage is the major cause of senescence, gas build-up, and off-odors; rancidity, color loss/change, nutrient losses, dehydration can also be addressed (López-Gómez et al. 2009).

Gelatin and methyl cellulose were used as matrixes in antimicrobial active and edible films with citrus extract, which shows antibacterial activity (Iturriaga et al. 2012). The addition of grape seed extracts, nisin, and EDTA to soy protein isolate edible films reduced the populations of major foodborne pathogens and improved the tensile properties of the soy protein isolate edible films (Sivarooban et al. 2008). The antimicrobial activity of active films based on barley bran protein and gelatin was due to the addition of grapefruit seed extracts; the active films were used in packed salmon fish (Song et al. 2012).

Chitosan can be obtained from seafood waste utilization as detailed in Sect. 5.9. Chitosan and nano-chitosan edible coatings are food preservative with antimicrobial activity; chitosan-based films are widely studied for fishery products applications. (Yu et al. 2019); extracts of grape pomace were incorporated into chitosan edible films providing antioxidant properties and promising shelf-life extension (Ferreira et al. 2014). Gelatin and whey protein are additional FW (discussed in Chap 2) that have the potential to be used as carriers for antioxidants or antibacterial agents in coatings and films; noteworthy, the presence of lactoperoxidase in whey results in a broad spectrum antibacterial activity (Yu et al. 2019).

The use of mint extracts or pomegranate peel extracts in chitosan and polyvinyl alcohol active films increased protection against UV light and their tensile strength without significantly affecting their puncture strength and barrier properties; the extracts conferred antioxidant properties and effectiveness against Gram- positive food bacteria to the active films (Kanatt et al. 2012).

Three different concentrations of lemon extract, thymol, and lysozyme were incorporated in PLA, PCL, and LDPE; the antimicrobial activity of the resulting nine active films was tested. Antimicrobials incorporated into PLA and low-density polyethylene (LDPE) lost some antimicrobial activity due to higher processing temperature. On the other hand, the lower processing temperature of PCL prevented the degradation of antimicrobial activity. In particular, lysozyme showed higher thermal stability (Del Nobile et al. 2009).

Olive leaves extract rich in polyphenols with high antimicrobial efficiency on many food pathogens was successfully incorporated into PLA films. The antimicrobial activities against *Staphylococcus aureus* increased with increasing extract concentration in the film. Moreover, the amount of extract in the film positively correlated to the water vapor permeability and the water solubility; the degradation rates of films also increased, thereby conjugating antimicrobial effects and improved properties (Erdohan et al. 2013).

Antioxidant piperidine conjugated star-shaped PLLA (star-PLLA-PPD) proved to be a novel polyfunctional free radical stabilizing additive for bioplastics. A model study on the free radical scavenging ability of this active ingredient in PLA blends was carried out. The star-PLLA-PPD efficiently scavenges free radicals in the PLA matrix upon irradiation with doses generally used for food and medical device sterilization, thereby protecting molecular weight reduction when compared with the neat PLA. The migration of star-PLLA-PPD is significantly lower than piperidine, thereby confirming the importance of the new radical scavenger for bioplastics in the food packaging and biomedical sectors (Haema et al. 2017).

Biobased and biodegradable active films based on a PLA/PHB blend with lactic acid oligomers as plasticizers used carvacrol as an active agent for antimicrobial active packaging. In this case, migration of the active additive resulted in the microbiological protection of foodstuff with high-fat content. These results underlined the specific antimicrobial properties of these bio-films, suggesting their applicability in active food packaging (Burgos et al. 2017).

The potential of biocomposites for active food packaging is one of the most-studied approaches in the last years. Improved mechanical, barrier, antioxidant, and antimicrobial properties are sought with a focus on safety and risk associated with fillers and additives, migration properties, and possible human ingestion. For example, *Thymus Vulgaris* leaf extract was used as a stabilizer and reducing agent in the green synthesis of zinc oxide silver nanoparticles, used as an active agent in poly(3-hydroxybutyrate-co-3-hydroxyvalerate)-chitosan to fabricate a novel degradable biopolymer nanocomposite characterized by good mechanical properties and antimicrobial activity and with the lowest migration rate to improve the shelf life of poultry items (Zare et al. 2019). Biocomposites with rice straw paper incorporating 30% activated carbon enabled ethylene scavenging; ethylene is a plant hormone responsible for softening, ripening, color change, and senescence of many kinds of fruits; its environmentally-friendly scavenging is extremely important in the fresh fruit and vegetable industry since common ethylene scavengers are based on potassium permanganate that is not allowed to come into contact with foodstuff because of its toxicity and color (Sothornvit and Sampoompuang 2012).

To sum up, the production of active packaging systems releasing natural antimicrobials allowing to extend the shelf life of food products, and the development of new high-barrier packaging materials (often using cellulose nanocrystals) are among the latest multifaceted innovations. Aesthetic aspects, communication, and information will be essential issues since innovations are considered to be outstanding when it comes to lower resource use and to improvements in sustainability and safety. In this perspective, developing biobased materials and bioplastics is of great relevance.

Regulations 1935/2004/EC and 450/2009/EC pioneered the development of general requirements and specific safety and marketing issues related to active packaging materials, the intentional migration of active elements to food, criteria for the safety evaluation procedure, admissable active elements identification, and maximum release of the active component; the latter could exceed the overall migration requirements indicated in the legislation, and it should not be considered in the calculation of the overall migration limit (Dainelli et al. 2008); European Commission 2004; European Commission 2009; Llorens et al. 2012).

Consolidated texts of Regulations enable tracking of the evolution of legal acts via incorporation of successive amendments and corrigenda. Normal migration is covered by the specific legislation applicable to those materials, such as the EU Regulation 10/2011 (European Commission 2020).

Guidelines concerning the Commission Regulation (EC) No 450/2009 were provided to deal with the interpretation and application of certain features in the referred legislation, such as toxicological properties and the extent to which the original substance or breakdown products could transfer into food.

Similarly, new active packaging materials used in food-contact applications are subject to pre-market regulatory clearance by the US Food and Drug Administration if they are deemed "food additives."

Even if smart and active packaging still remains more expensive than "traditional" packaging, new technologies and industrial development are expected to quicken cost reduction.

10.4 Registration, Evaluation, Authorisation and Restriction of Chemicals (REACH) Regulation for Biobased Products

Companies are gradually interested in the registration of biobased substances under European Union's Registration, Evaluation, Authorisation and Restriction of Chemicals (REACH) Regulation (The European Parliament and the Council of the European Union 2006).

The principle is "one substance, one registration"; hence data sharing among all registrants by the joint submission of registration data to the European Chemical Agency (ECHA) is mandatory.

Persistence in the environment, bioaccumulation, and toxicity are the key targets of the hazard assessment required by the REACH regulation consolidated version for all chemicals that are produced or imported in Europe in amounts exceeding 10 tonnes per year. The regulation heavily impacts most companies (manufacturers of chemicals, importers of individual chemicals, mixtures for onwards sale or finished products from outside the EU, and downstream users) across the EU; they must identify risks and inform users about the risk management measures. ECHA's website provides useful information about already registered substances: if the substance has already been registered for the same use by any company, it should not be registered again. REACH establishes procedures for collecting and assessing information on the properties and hazards of substances, and companies have the burden of proof. If the risks cannot be managed, according to the ECHA's scientific committee judgment, authorities can restrict the use of substances or make it subject to prior authorization or definitively ban them. In the long run, less dangerous alternatives should replace the most hazardous substances. In principle, REACH applies to all chemicals but, by this scenario, REACH may favors manufacturers of safe biobased substitutes for substances of very high concern: in this respect, REACH may actually be an opportunity rather than the administrative obstacle it is often perceived to be.

Substances that are discarded as waste as defined in the consolidated text of Directive 2008/98/EC ("Waste Framework Directive") (EU 2008) are exempted from REACH, but REACH applies for any substance, mixture, or article obtained from the recovery of waste and that meets the end of waste criteria: they generally need to be registered, except recovered substances already registered. However, for a recovered substance, it is sometimes possible to get an exemption from REACH registration.

Among substances exempted from registration we may find many biobased products. In fact, exemptions from registration apply to substances that present minimum risk (listed in Annex IV of the REACH Regulation) or substances for which registration is deemed inappropriate or unnecessary, often occurring in nature and not chemically modified (listed in Annex V of the REACH Regulation), to polymers, and to substances that are adequately regulated under other legislation, like substances used in food or feeding stuff or in medicinal products, where the relevant criteria are met. Even if a substance is exempted from the obligation to register it may still be subject to authorization or restriction provisions under REACH Regulation (ECHA 2016).

In particular, compost and biogas, as well as byproducts, unless they are imported or placed on the market themselves, are clearly mentioned among the exemptions from the obligation to register in Annex V of the REACH Regulation.

It is clear that biobased companies might qualify for exemptions that may lessen the registration burden. Readers are referred to the text of the REACH regulation, the only authentic legal reference.The information provided in this chapter does not constitute legal advice. The Author does not accept any liability as regards the contents of this document.

Bibliography

Abbina S, Du G (2014) Zinc-catalyzed highly isoselective ring opening polymerization of rac -lactide. ACS Macro Lett 3:689–692. https://doi.org/10.1021/mz5002959

Adamová T, Hradecký J, Prajer M (2019) VOC emissions from spruce strands and hemp shive: in search for a low emission raw material for bio-based construction materials. Materials (Basel) 12:2026. https://doi.org/10.3390/ma12122026

Adeyeye SAO (2019) Food packaging and nanotechnology: safeguarding consumer health and safety. Nutr Food Sci 49:1164–1179

Avella M, De Vlieger JJ, Errico ME et al (2005) Biodegradable starch/clay nanocomposite films for food packaging applications. Food Chem 93:467–474. https://doi.org/10.1016/j.foodchem.2004.10.024

Azimi P, Zhao D, Pouzet C et al (2016) Emissions of ultrafine particles and volatile organic compounds from commercially available desktop three-dimensional printers with multiple filaments. Environ Sci Technol 50:1260. https://doi.org/10.1021/acs.est.5b04983

Burgos N, Tolaguera D, Fiori S, Jiménez A (2014) Synthesis and characterization of lactic acid oligomers: evaluation of performance as poly(lactic acid) plasticizers. J Polym Environ 22:227–235. https://doi.org/10.1007/s10924-013-0628-5

Burgos N, Armentano I, Fortunati E et al (2017) Functional properties of plasticized bio-based poly(lactic acid)_poly(Hydroxybutyrate) (PLA_PHB) films for active food packaging. Food Bioprocess Technol 10:770–780. https://doi.org/10.1007/s11947-016-1846-3

Cecchi T (2014a) Identification of representative pollutants in multiple locations of an Italian school using solid phase micro extraction technique. Build Environ 82:655–665. https://doi.org/10.1016/j.buildenv.2014.10.009

Cecchi T (2014b) Head space – solid phase micro extraction profile of volatile organic compounds emitted from parquet samples. J Wood Chem Technol 34:211–224. https://doi.org/10.1080/02773813.2013.861849

Cecchi T, Alfei B (2013) Volatile profiles of Italian monovarietal extra virgin olive oils via HS-SPME-GC-MS: newly identified compounds, flavors molecular markers, and terpenic profile. Food Chem 141:2025. https://doi.org/10.1016/j.foodchem.2013.05.090

Cecchi T, Giuliani A, Iacopini F et al (2019) Unprecedented high percentage of food waste powder filler in poly lactic acid green composites: synthesis, characterization, and volatile profile. Environ Sci Pollut Res 26:7263–7271. https://doi.org/10.1007/s11356-019-04187-1

Chambers M, Muecke M (2010) Biobased products and the LEED® rating system. J Green Build 5:91–107. https://doi.org/10.3992/jgb.5.4.91

Chen HY, Tang HY, Lin CC (2007) Ring-opening polymerization of l-lactide catalyzed by a biocompatible calcium complex. Polymer (Guildf) 48:2257–2262. https://doi.org/10.1016/j.polymer.2007.02.030

Cheng HN, Gross RD, Smith PB (2014) Green polymer chemistry: biocatalysis and materials II. American Chemical Society, Washington, DC

Clarke K, Tchabanenko K, Pawlosky R et al (2012) Oral 28-day and developmental toxicity studies of (R)-3-hydroxybutyl (R)-3-hydroxybutyrate. Regul Toxicol Pharmacol 63:196–208. https://doi.org/10.1016/j.yrtph.2012.04.001

Conn RE, Kolstad JJ, Borzelleca JF et al (1995) Safety assessment of polylactide (PLA) for use as a food-contact polymer. Food Chem Toxicol 33:273–283. https://doi.org/10.1016/0278-6915(94)00145-E

Connolly M, Zhang Y, Brown DM et al (2019) Novel polylactic acid (PLA)-organoclay nanocomposite bio-packaging for the cosmetic industry; migration studies and in vitro assessment of the dermal toxicity of migration extracts. Polym Degrad Stab 168:108938. https://doi.org/10.1016/j.polymdegradstab.2019.108938

Dainelli D, Gontard N, Spyropoulos D et al (2008) Active and intelligent food packaging: legal aspects and safety concerns. Trends Food Sci Technol 19:S103–S112. https://doi.org/10.1016/j.tifs.2008.09.011

Davis AY, Zhang Q, Wong JPS et al (2019) Characterization of volatile organic compound emissions from consumer level material extrusion 3D printers. Build Environ 160:106209. https://doi.org/10.1016/j.buildenv.2019.106209

Del Nobile MA, Conte A, Buonocore GG et al (2009) Active packaging by extrusion processing of recyclable and biodegradable polymers. J Food Eng 93:1–6. https://doi.org/10.1016/j.jfoodeng.2008.12.022

Dunkel M, Schmidt U, Struck S et al (2009) SuperScent--a database of flavors and scents. Nucleic Acids Res 37:D291–D294. https://doi.org/10.1093/nar/gkn695

ECHA (2016) Guidance on registration. ECHA, Helsinki. https://doi.org/10.2823/969

Erdohan ZÖ, Çam B, Turhan KN (2013) Characterization of antimicrobial polylactic acid based films. J Food Eng 119:308–315. https://doi.org/10.1016/j.jfoodeng.2013.05.043

EU (2008) Directive 2008/98/EC of the European Parliament and of the Council of 19 November 2008 on waste and repealing certain Directives. EU, Brussels Consolidated text available at https://eur-lex.europa.eu/legal-content/EN/TXT/?uri=CELEX%3A02008L0098-20180705. Accessed 7th Mar 2021

EU (2019) Communication from the commission to the European parliament, the European Council, The Council, The European Economic and Social Committee and The Committee of The Regions, The European Green Deal. EU, Brussels, pp 1–24. https://eur-lex.europa.eu/legal-content/EN/TXT/?uri=CELEX%3A52019DC0640. Accessed 7th Mar 2021.

European Commission (2004) Regulation (EC) No 1935/2004 on materials and articles intended to come into contact with food and repealing Directives 80/590/EEC and 89/109/EEC. Off J Eur Union L338(4):4–17 Consolidated text available at https://eur-lex.europa.eu/legal-content/EN/TXT/?uri=CELEX%3A02004R1935-20090807. Accessed 7th Mar 2021

European Commission (2009) Commission Regulation (EC) No. 450/2009 of 29 May 2009 on active and intelligent materials and articles intended to come into contact with food. Off J Eur Union L 135:3–11

European Commission (2020) COMMISSION REGULATION (EU) 2020/1245 of 2 September 2020 amending and correcting Regulation (EU) No 10/2011 on plastic materials and articles intended to come into contact with food. In: Off. J. Eur. Union. https://eur-lex.europa.eu/legal-content/EN/TXT/PDF/?uri=CELEX:32020R1245&from=EN. Accessed 6 Mar 2021

Ferreira AS, Nunes C, Castro A, et al (2014) Influence of grape pomace extract incorporation on chitosan films properties. Carbohydr Polym 113:490–499. https://doi.org/10.1016/j.carbpol.2014.07.032

Gallet G, Lempiäinen R, Karlsson S (2000) Characterization by solid phase microextraction-gas chromatography-mass spectrometry of matrix changes of poly(L-lactide) exposed to outdoor soil environment. Polym Degrad Stab 71:147–151. https://doi.org/10.1016/S0141-3910(00)00165-8

Greco A, Maffezzoli A (2016) Cardanol derivatives as innovative bio-plasticizers for poly-(lactic acid). Polym Degrad Stab 132:213–219. https://doi.org/10.1016/j.polymdegradstab.2016.02.020

Guazzotti V, Marti A, Piergiovanni L, Limbo S (2014) Bio-based coatings as potential barriers to chemical contaminants from recycled paper and board for food packaging. Food Addit Contam Part A Chem Anal Control Expo Risk Assess 31:402–413. https://doi.org/10.1080/19440049.2013.869360

Haema K, Piroonpan T, Taechutrakul S et al (2017) Piperidine-conjugated polyfunctional star-shaped PLLA as a novel bio-based antioxidant additive for bioplastics. Polym Degrad Stab 143:145–154. https://doi.org/10.1016/j.polymdegradstab.2017.07.001

Harper M (2000) Sorbent trapping of volatile organic compounds from air. J Chromatogr A 885:129–151

Ház A, Michal J, Lisý A (2018) Compositions of volatile organic compounds emitted from natural and synthetic polymers. Earth in a trap? 2018 Anal Methods Fire Environ Sci 2018:1–6

Houtman J, Maisanaba S, Puerto M et al (2014) Toxicity assessment of organomodified clays used in food contact materials on human target cell lines. Appl Clay Sci 90:150–158. https://doi.org/10.1016/j.clay.2014.01.009

Howard IC, Hammond C, Buchard A (2019) Polymer-supported metal catalysts for the heterogeneous polymerisation of lactones. Polym Chem 10:5894–5904. https://doi.org/10.1039/c9py01472a

ISO (2006a) ISO 16000-9:2006 – indoor air — part 9: determination of the emission of volatile organic compounds from building products and furnishing — Emission test chamber method. https://www.iso.org/standard/38203.html. Accessed 6 Aug 2020

ISO (2006b) ISO 16000-10:2006 – indoor air — part 10: determination of the emission of volatile organic compounds from building products and furnishing — Emission test cell method. https://www.iso.org/standard/38204.html. Accessed 6 Aug 2020

ISO (2006c) ISO 16000-11:2006 – indoor air — part 11: determination of the emission of volatile organic compounds from building products and furnishing — Sampling, storage of samples and preparation of test specimens. https://www.iso.org/standard/38205.html. Accessed 6 Aug 2020

ISO (2011) ISO 16000-6:2011 – indoor air – part 6: determination of volatile organic compounds in indoor and test chamber air by active sampling on Tenax TA sorbent, thermal desorption and gas chromatography using MS or MS-FID. https://www.iso.org/standard/52213.html. Accessed 6 Aug 2020

Iturriaga L, Olabarrieta I, de Marañón IM (2012) Antimicrobial assays of natural extracts and their inhibitory effect against Listeria innocua and fish spoilage bacteria, after incorporation into biopolymer edible films. Int J Food Microbiol 158:58–64. https://doi.org/10.1016/j.ijfoodmicro.2012.07.001

Kanatt SR, Rao MS, Chawla SP, Sharma A (2012) Active chitosan-polyvinyl alcohol films with natural extracts. Food Hydrocoll 29:290–297. https://doi.org/10.1016/j.foodhyd.2012.03.005

Kharisov BI, Kharissova OV, Máximo C et al (2019) Greener synthesis of chemical compounds and materials. https://doi.org/10.1098/rsos.191378

Kim KW, Lee BH, Kim S et al (2011) Reduction of VOC emission from natural flours filled biodegradable bio-composites for automobile interior. J Hazard Mater 187:37–43. https://doi.org/10.1016/j.jhazmat.2010.07.075

Koivula HM, Jalkanen L, Saukkonen E et al (2016) Machine-coated starch-based dispersion coatings prevent mineral oil migration from paperboard. Prog Org Coatings 99:173–181. https://doi.org/10.1016/j.porgcoat.2016.05.017

Kopacic S, Walzl A, Zankel A et al (2018) Alginate and chitosan as a functional barrier for paper-based packaging materials. Coatings 8:235. https://doi.org/10.3390/coatings8070235

Kurek M, Garofulić IE, Bakić MT, et al. (2018) Development and evaluation of a novel antioxidant and pH indicator film based on chitosan and food waste sources of antioxidants. Food Hydrocoll. 84, 238–246. https://doi.org/10.1016/j.foodhyd.2018.05.050

Llorens A, Lloret E, Picouet PA et al (2012) Metallic-based micro and nanocomposites in food contact materials and active food packaging. Trends Food Sci Technol 24:19–29

López-Gómez A, Fernández PS, Palop A et al (2009) Food safety engineering: an emergent perspective. Food Eng Rev 1:84–104. https://doi.org/10.1007/s12393-009-9005-5

Luchese CL, Sperotto N, Spada JC, et al. (2017) Effect of blueberry agro-industrial waste addition to corn starch-based films for the production of a pH-indicator film. Int. J. Biol. Macromol. 104:11–18. https://doi.org/10.1016/j.ijbiomac.2017.05.149

Ma S, Webster DC (2015) Naturally occurring acids as cross-linkers to yield VOC-free, high-performance, fully bio-based, degradable thermosets. Macromolecules 48:7127–7137. https://doi.org/10.1021/acs.macromol.5b01923

Maisanaba S, Puerto M, Pichardo S et al (2013) In vitro toxicological assessment of clays for their use in food packaging applications. Food Chem Toxicol 57:266–275. https://doi.org/10.1016/j.fct.2013.03.043

Marin P, Tschan MJL, Isnard F et al (2019) Polymerization of rac-Lactide using achiral Iron complexes: access to thermally stable Stereocomplexes. Angew Chemie – Int Ed 58:12585–12589. https://doi.org/10.1002/anie.201903224

Montanari A, Bolzoni L, Cigognini IM et al (2017) Tomato bio-based lacquer for sustainable metal packaging. In: Acta horticulturae. International Society for Horticultural Science, Leuven, pp 159–165

Pandey A, Kumar P, Singh V (2010) Application of bioplastics in bulk packaging : a revolutionary. University of Science & Technology, Hisar

Parcheta P, Datta J (2018) Kinetics study of the fully bio-based poly(propylene succinate) synthesis. Functional group approach. Polym Degrad Stab 155:238–249. https://doi.org/10.1016/j.polymdegradstab.2018.07.025

Pastore G, Gabrielli S, Cecchi T. et al (2021) A new and efficient lactic acid polymerization by multimetallic cerium complexes: a poly(lactic acid) suitable for biomedical applications. RSC Adv 11:10592–10598. https://doi.org/10.1039/d0ra10637b

Parwe SP, Warkad SD, Mane MV et al (2017) Investigation of the biocompatibility and cytotoxicity associated with ROP initiator and its role in bulk polymerization of L-lactide. Polymer (Guildf) 111:244–251. https://doi.org/10.1016/j.polymer.2017.01.058

Pawliszyn J (2011) Handbook of solid phase microextraction. Elsevier, Ontario

Pereira VA, de Arruda INQ, Stefani R (2015) Active chitosan/PVA films with anthocyanins from Brassica oleraceae (Red Cabbage) as Time-Temperature Indicators for application in intelligent food packaging. Food Hydrocoll. 43, 180–188. https://doi.org/10.1016/j.foodhyd.2014.05.014

Plank CM, Trela BC (2018) A review of plastics use in winemaking: Haccp considerations. Am J Enol Vitic 69:307–320. https://doi.org/10.5344/ajev.2018.17041

Rittinghaus RD, Schäfer PM, Albrecht P et al (2019) New kids in lactide polymerization: highly active and robust iron guanidine complexes as superior catalysts. ChemSusChem 12:2161–2165. https://doi.org/10.1002/cssc.201900481

Rittinghaus RD, Tremmel J, Růžička A et al (2020) Undiscovered potential: Ge catalysts for lactide polymerization. Chem – Eur J 26:212–221. https://doi.org/10.1002/chem.201903949

Salazar Gonzalez R, Domenek S, Plessis C, Ducruet V (2016) Quantitative determination of volatile organic compounds formed during polylactide processing by MHS-SPME. Polym Degrad Stab 136:80–88. https://doi.org/10.1016/j.polymdegradstab.2016.12.010

Salazar R, Domenek S, Plessis C, Ducruet V (2017) Quantitative determination of volatile organic compounds formed during Polylactide processing by MHS-SPME. Polym Degrad Stab 136:80–88. https://doi.org/10.1016/j.polymdegradstab.2016.12.010

Schäfer PM, McKeown P, Fuchs M et al (2019) Tuning a robust system: N,O zinc guanidine catalysts for the ROP of lactide. Dalt Trans 48:6071–6082. https://doi.org/10.1039/c8dt04938f

Sharma AK, Schmidt B, Frandsen H et al (2010) Genotoxicity of unmodified and organo-modified montmorillonite. Mutat Res – Genet Toxicol Environ Mutagen 700:18–25. https://doi.org/10.1016/j.mrgentox.2010.04.021

Sheldon RA, Woodley JM (2017) Role of biocatalysis in sustainable chemistry. Chem Rev 118:801. https://doi.org/10.1021/acs.chemrev.7b00203

Siracusa V (2012) Food packaging permeability behaviour: a report. Int J Polym Sci 2012:11

Sivarooban T, Hettiarachchy NS, Johnson MG (2008) Physical and antimicrobial properties of grape seed extract, nisin, and EDTA incorporated soy protein edible films. Food Res Int 41:781–785. https://doi.org/10.1016/j.foodres.2008.04.007

Song HY, Shin YJ, Bin SK (2012) Preparation of a barley bran protein-gelatin composite film containing grapefruit seed extract and its application in salmon packaging. J Food Eng 113:541–547. https://doi.org/10.1016/j.jfoodeng.2012.07.010

Sothornvit R, Sampoompuang C (2012) Rice straw paper incorporated with activated carbon as an ethylene scavenger in a paper-making process. Int J Food Sci Technol 47:511–517. https://doi.org/10.1111/j.1365-2621.2011.02871.x

Tanzi MC, Verderio P, Lampugnani MG et al (1994) Cytotoxicity of some catalysts commonly used in the synthesis of copolymers for biomedical use. J Mater Sci Mater Med 5:393–396. https://doi.org/10.1007/BF00058971

The European Parliament and the Council of the European Union (2004) Directive 2004/42/CE of the European Parliament and of the Council of 21 April 2004 on the limitation of emissions of volatile organic compounds due to the use of organic solvents in certain paints and varnishes and vehicle refinishing products and amendi. Off J Eur Unio 87–96. http://eur-lex.europa.eu/LexUriServ/LexUriServ.do?uri=OJ:L:2004:143:0087:0096:EN:PDF

The European Parliament and the Council of the European Union (2006) REACH Regulation. EC (2006) Regulation (EC) No 1907/2006 of the European Parliament and of the Council of 18 December 2006 concerning the Registration, Evaluation, Authorisation, and Restriction of Chemicals (REACH), establishing a European Chemicals Agency, amending Directive 1999/45/ EC and repealing Council Regulation (EEC) No 793/93 and Commission Regulation (EC) No 1488/94 as well as Council Directive 76/769/EEC and Commission Directives 91/155/ EC, 93/67/EEC, 93/105/EC and 2001/21/ EC. J Eur Union 30.12.2006. http://ec.europa.eu/environment/chemicals/reach/reach_intro.htm. Accessed 7th Mar 2021.

Valdés A, Mellinas AC, Ramos M et al (2014) Natural additives and agricultural wastes in biopolymer formulations for food packaging. Front Chem 2:1–10. https://doi.org/10.3389/fchem.2014.00006

van Den Dool H, Dec. Kratz P (1963) A generalization of the retention index system including linear temperature programmed gas-liquid partition chromatography. J Chromatogr A 11:463–471. https://doi.org/10.1016/s0021-9673(01)80947-x

van den Oever M, Molenveld K, van der Zee M, Bos H (2017) Bio-based and biodegradable plastics - Facts and figures. Focus on food packaging in the Netherlands. Rapport nr. 1722. https://doi.org/10.18174/408350

Walzl A, Kopacic S, Bauer W, Leitner E (2019) Characterization of natural polymers as functional barriers for cellulose-based packaging materials. Food Addit Contam Part A 36:976–988. https://doi.org/10.1080/19440049.2019.1600747

Wei W, Ramalho O, Mandin C (2015) Indoor air quality requirements in green building certifications. Build. Environ. 92:10–19

Wheaton CA, Hayes PG, Ireland BJ (2009) Complexes of Mg, Ca and Zn as homogeneous catalysts for lactide polymerization. J Chem Soc Dalt Trans 25:4832–4846. https://doi.org/10.1039/b819107g

Wojtyła S, Klama P, Baran T (2017) Is 3D printing safe? analysis of the thermal treatment of thermoplastics: ABS, PLA, PET, and nylon. J Occup Environ Hyg 14:D80–D85. https://doi.org/10.1080/15459624.2017.1285489

Wolkoff P (2003) Trends in Europe to reduce the indoor air pollution of VOCs. Indoor Air 13:5–11

Yu H, Yan C, Yao J (2014) Fully biodegradable food packaging materials based on functionalized cellulose nanocrystals/poly(3-hydroxybutyrate-co-3-hydroxyvalerate) nanocomposites. RSC Adv 4:59792–59802. https://doi.org/10.1039/c4ra12691b

Yu D, Regenstein JM, Xia W (2019) Bio-based edible coatings for the preservation of fishery products: a review. Crit Rev Food Sci Nutr 59:2481–2493

Zare M, Namratha K, Ilyas S et al (2019) Smart fortified PHBV-CS biopolymer with ZnO–Ag nanocomposites for enhanced shelf life of food packaging. ACS Appl Mater Interfaces 11:48309–48320. https://doi.org/10.1021/acsami.9b15724

Zhao X, Cornish K, Vodovotz Y (2020) Narrowing the gap for bioplastic use in food packaging: an update. Environ Sci Technol 54:4732–4732. https://doi.org/10.1021/acs.est.9b03755

Zhu J, Li X, Huang C et al (2014) Structural changes and triacetin migration of starch acetate film contacting with distilled water as food simulant. Carbohydr Polym 104:1–7. https://doi.org/10.1016/j.carbpol.2013.12.087

Chapter 11
Life Cycle Assessment

Carla De Carolis

11.1 Life Cycle Assessment Introduction

The trend worldwide is geared toward promoting the use of biobased products as part of the circular economy for more sustainable development. Policymakers in many countries are developing goals and strategies for the development of bio-economies based on sustainability goals, ensuring energy supplies supported by Renewable energy sources (RES), and develop competitive, innovative products by using the new approach of biorefineries (EuropaBio, Brussels, Belgium 2011). Nonetheless, there are concerns that the biobased economy may undermine the sustainability of the transition, e.g., from indirect impacts of land use to over-rapid depletion of biomass resources, therefore not all biobased products or exploitation of biomass residual could e considered sustainable (Annick Hedlund-de Witt 2014).

There is a real risk that bioeconomy strategies might be "utopia" because far-reaching strategies are difficult to achieve if not supported by solid investigations of sustainability analyses, at least for most common biobased pathways. The persistent conflict of different uses of biomass among food, biobased products, and energy production could lead to unstable policy with short-term shifts of investments. Therefore it is urgent to identify the correct ratio of biomass fluxes destined to the different mentioned sectors (food, energy, and biobased products). Furthermore, a broader success of new biobased value chains could trigger new societal conflicts over bioeconomy if efficiency gains, cascading use, residue use, and sustainability certification are not sufficient to ensure a sustainable supply of biomass, especially if aimed at large scale projects (Rolf Meyer 2017).

In the interest of ensuring a transition toward a sustainable biobased economy, adequate life cycle assessment investigations with a broad systems perspective are thus required (Staffas L. et al. 2013).

Bit of authors contributed to the life cycle studies in order to include important sustainability indicators (McCormick and Kautto 2013; Baptiste et al. 2018; Martin-Gamboa et al. 2017).

T. Cecchi, C. De Carolis, *Biobased products from food sector waste*,
https://doi.org/10.1007/978-3-030-63436-0_11

It's therefore needed the development and dissemination of efficient methods that including important environmental impact indicators, such as water depletion, indirect land-use change, and impacts on ecosystem quality and biodiversity. The limited publications about social life cycle assessments (SLCAs) and life cycle sustainability assessments (LCSAs) on the Biobased Products indicate that how these are still immature tools; therefore, there is necessary to improve the methods as well as their application in more case studies (Martin et al. 2018).

11.2 Biobased Products and Life Cycle Assessment

Life Cycle Management (LCM) is considered as an integrated concept of methodologies and operational approaches aimed at improving the environmental, social, and economic sustainability of products, technologies, services, policies, and organizations from a life cycle perspective. Life Cycle Management is based on the LCA – Life Cycle Assessment of well know as LCA – Life Cycle Analysis which should be applied to different biobased products. LCA can be defined as the international standardized methodology used to evaluate a product's impact on the environment over the entire period of its life – from cradle (that is, from biomass growth/cultivation or raw material extraction) to the grave (that is, disposal after end uses or recycling whenever possible, etc.) (Heijungs et al. 2010).

During the last decades, the LCM frameworks seem being become common tools strongly used by businesses and policy-making organizations in order to clearly identify their green-strategy and to map a pathway more sustainable to take.

As a new pattern for market growth, a more Circular Economy awareness is arising, several methods and certification schemes to quantify and communicate the impacts and benefits of circular systems have emerged. LCM toolbox can start to provide a scientifically founded package to monitor and heading toward a more sustainable society, even if LCM the community must face important challenges to reach this objective, in particular, in terms of positioning LCM on the political and business agendas, with a further simplifying and standardizing LCM methods without over-streamlining and disregarding the complexity and promoting innovation based on LCM concepts and tools (Benedetto et al. 2018).

Sometimes, life cycle sustainability assessments (LCSAs) applied to the Bioeconomy could seem still immature tools. For instance, applied LCA on food products and respective byproducts circularity, doesn't consider the environmental impact indicator like biodiversity among several cultivated crops. The endangered cultivars and their respective genic pool represent an important input against the genetic erosion toward the safeguard of most species.

Another example, wood, and cellulose are major inputs for many circular products, including part of packaging material often used in the agri-food sector (pallets and paper). Within the assessment methods to quantify the circularity potential, such as Life Cycle Assessment, there are limitations regarding the evaluation of impacts on biodiversity, due to the complexity of the characterization of land use and land-use change, and to the lack of consideration of forest management on landscape-level and labeling, as done through certified forestry management.

Therefore, practitioners and decision-makers prefer to look at other impact categories, such as climate change and water use. Starting from the existing work of UNEP SETAC and other significant existing methods, new methodology relying on a more complex management system was proposed in order to establish characterization factors aiming at integrating biodiversity and forest management practices when required (Vionnet et al. 2017).

At the same time, exploitation of food byproducts lignin-based products (fruit pits, fruit woody-shells, lignin residual) are offering interesting opportunities and challenges for environmentally friendly biobased resins" that's worth considering. The use of lignin, (firstly underutilized by-product from the pulp and rural residues), has the potential to be used as a valuable source for sustainable biobased products. The market assessment of emerging lignin products aims to support R&D process by providing information on challenges and opportunities for such products at the early design stage.

If a side it represents an innovative sector able to have a promising segment of biobased market, the other part it might cause a further competition with the same biomass used for bioenergy sector (agro-pellets) with a further arising of both market prices.

Furthermore, interesting results performing from international projects as "Smartli Project" (https://clicinnovation.fi/project/smartli/) outcomes (Lettner et al. 2017), highlight that the life cycle thinking integrated with most innovative technical R&D processes based on the involvement of stakeholders and on a multi-perspective assessment (economic, environmental and technological) of newly developed products are highly valuable to face the barriers and incentives for the biobased products market diffusion.

Focusing the attention on food waste for new byproducts, Sala points out how is possible to apply supporting tools like Life Cycle Thinking on the food waste by also showing the potential added value and shortcoming in developed countries.

Based on the quantification of food losses and waste, LCT and LCA approaches are potentially applicable in order to support the identification of the best options. The main challenges in order to improve LCT methods are related to the characterization of biotic resources in LCIA and to the integration of all environmental factors since strictly themselves interconnected: eco-system, water, food, land, and energy into LCA, and considering all these issues as correlated (Sala 2017).

Furthermore, an increasing number of industrics are including in their commercial strategies the assessing of the availability of biobased materials in product design.

The use of biobased materials can be considered as a key strategy in the eco-design approach, also contributing to a circular economy at some points. However, several socio-economic factors induced by market conditions or disrupting political structures within society can constrain their availability.

Some authors proposed a method for assessing biobased materials availability constraints, aiming at supporting eco-design strategies. Their method considers five main sources of constraints (environmental, socio-economic, and physical constraints to plants growth), over the materials supply chain, and results in a set of availability indicators (Bach et al. 2017).

Sperling gives further concrete contributions on the sustainability of biobased plastics, trying to quantify the environmental and socio-economic aspects for a concrete circular economy. From his investigation is possible to deduce how is important to base new products sourcing local biobased plastics as well as to always include the associated social aspects by different countries, despite the fact that these socio-economic aspects are often neglected and the availability of such information is very often limited.

During the latest decade, the continuous monitoring of several studies conducted on LCA analysis applied to the emerging biobased products highlighted the need to use a jointed guideline for each biobased product by biomass residuals and fossil-based material (i.e. fossilbased plastic). A common methodology is needed to reduce the methodological gaps existing for the assessment of such materials in continuous evolution for a real and effective circular economy (Spierling et al. 2017).

11.3 Carbon Assessment, Carbon Footprint and Product Environmental Footprint

Biobased products seem an opportunity for sustainable production, despite is necessary to demonstrate their impacts along the whole life chain. Each production and use of products and services can cause different gases to be emitted into the atmosphere. The Kyoto Protocol points out six relevant gases like carbon dioxide (CO_2), methane (CH_4), nitrous oxide (N_2O), hydrofluorocarbons (HFCs), perfluorocarbons (PFCs), and sulfur hexafluoride (SF_6) are responsible for absorbing and emitting radiation at specific wavelengths emitted by the Earth's surface, the atmosphere, and clouds. These phenomena can increase the global temperature causing the greenhouse effect, due to GHGs increasing – Green Houses Gas emissions (Wiedmann and Minx 2007).

Six typologies of GHGs are actually considered, and plenty of methodologies to calculate the impacts have been raised during the last decades. After 2010, a consistent partnership between big companies from different sectors has been coordinated by EU Commission in order to solve the problems caused by the proliferation of environmental labels and certification schemes based on the different calculations of Carbon Assessment.

The Carbon Assessment is used to determine the total Carbon Footprint[1] (expressed as $CO_{2eq.}$) of several products and/or services. Once the boundaries system of the chain is fixed, the Total Carbon Footprint derives by Operational Carbon (Carbon emitted during the lifetime of product/service operation, included

[1] Different definitions of the Carbon Footprint are reported by Wiedmann and Minx (2007)

the fluxes of energy, water, wastes, raw materials in the whole process after fixed the boundary system), plus Embodied Carbon (Carbon emitted during the manufacture, transport, and construction of product or the service created), and fewer the Carbon Offset (Carbon sequestration, or Carbon reduced if Renewable Energy is being used or produced during the process) (GreenRe 2017).

$$\text{Total Carbon Footprint} = (\text{Operational Carbon}) + (\text{Embodied Carbon}) - (\text{Carbon Offset}) \quad (11.1)$$

Despite LCA is being considered a useful tool able to quantify the environmental impacts also on biobased materials, its application and data interpretation demonstrated several limits (Troy et al. 2013).

To biopolymers are often associated with a quick degradation and/or recycling, but their environmental impacts associated with the creation, use, disposal, and biodegradability remain unclear due to the huge variety of biomass residues exploitable as raw materials used for BBPs, different uses, and recycling of BBPs at the end of life, and their disposal procedures. These important factors can influence the real environmental impacts of BBPs beyond a single LCA study. The end-of-life options for biopolymers may prove to alter the product's real impact on the environment. In fact, the recommendations concerning the waste management of PLA in some countries point out that the material should be composted at industrial facilities since the appropriate degradation conditions cannot be met in 'backyard' compost pile (Kale et al. 2007; Vink et al. 2004) but in LCA analysis, the Industrial composting can be comparable to landfilling impacts which may further decrease any GWP benefits that were achieved. Biopolymers are biodegradable and compostable, but not always cities have suitable composting infrastructures, thus most of the biodegradable materials are landfilled (Gruber et al. 2005).

LCA on biobased products carried out by James and Grant (2005) represents one of the studies where composting is an EOL option. This highlights an important gap between the impacts caused by the compostability factor and LCAs simply based on the current polymer disposal practices.

Troy et al. (2013) confirm that has been done little work to assess both BBPs composting and recycling as options in LCA methodologies. BBPs recycling may contribute to further life-cycle benefits by reducing energy and material input into the productive cycle (Piemonte 2011). However, the current infrastructure and logistics may require retro-fitting not always feasible for efficient recycling of certain biopolymers based on PHA and PLA.

Many of the "End of Life" (EOL) Scenarios reported in the literature are based on typical food waste ratios used for bioplastic production advertised as compostable, but it is unclear how consumers are disposing of these biopolymers. In addition, it is not clear what mix of EOL scenarios would provide the greatest benefits to biopolymers' life-cycle environmental impacts. In addition to this unclear BBP's impacts, other studies clearly stating that biopolymers will not break down under landfill conditions, like PLA (Kijchavengkul and Auras 2008; Madival et al. 2009; Kolstad et al. 2012) while other analyses highlight the methane emissions from landfills as a negative impact in the BBP's life cycle (Shen and Patel 2008; Bohlmann

2004; Groot and Borén 2015) with further increasing the GWP associated with biopolymers.

Another important factor to be constantly assured is represented by LCA Databases used during the analyses. The quality of LCA is affected both the quality of the inventory data as well as by the function of determining appropriate system boundaries. The databases are to be scrutinized for accuracy and completeness. The BBP's production methods can change for several production processes, thus continuous updates to inventories must be made.

Vink has continued to provide updated profiles for the creation of NatureWorks' PLA which should be incorporated into any future LCAs of that product, while ecoinvent database with TRACI produced results no comparable to the results from NatureWorks (Vink et al. 2010).

The gap among several LCA studies on BBP's depends on a wide range of variables like factors and assumptions that can influence GPW impacts (system boundaries, GHG emission avoided, and resulting from the end of life treatments). The comparison of results other LCA studies confirms low coherence throughout the literature along latest decades, due to high variability of different technology, energy, and material fluxes used for several BBP's productions, of which the most know are polyhydroxyalkanoate (PHA)/PHB, polylactic acid (PLA) and thermoplastic starch (TPS) (Troy et al. 2013).

Table 11.1 illustrates the comparison of overall ranges of CO_2 equivalents reported among different LCAs of biopolymers that included GHG in their life cycle inventory. The wide variability of LCA studies demonstrates that parameters cannot be compared among all studies. Fifteen LCAs analyses reported CO_2

Table 11.1 Variability of range concerning the $CO_{2eq.}$ emissions data of LCAs analyses on biopolymers

Authors	Range $CO_{2\,eq.}$ (Kg $CO_{2\,eq.}$/Kg BBPs)	Type of BBPs
Kurdikar et al. (2008)	−4; 5,5	PHA/PHB – Polyhydroxyalkanoate
Akiyamaa et al. (2003)	0; 0,5	
Kim and Dale (2005)	0; 4,3	
Pietrini et al. (2007)	0; 4,0	
Yu and Chen (2008)	0,5	
Kendall (2012)	3,1; 5,1	
Vink et al. (2003)	−1,5; 3,5	PLA – Polylactic acid
Bohlmann (2004)	0,8; 2,3	
Vink et al. (2007)	0,3; 2	
James and Grant (2005)	4	
Madival et al. (2009)	5,7; 6	
Vink et al. (2010)	1,3	
Groot and Borén (2015)	0,5; 0,7	
Shen and Patel (2008)	1,1; 1,2	TPS – Thermoplastic starch
Piemonte and Gironi (2011)	1	

equivalent data that could be compared, while others did not report the values converted to CO_2 equivalents. Furthermore, other analyses (James and Grant 2005; Vink et al. 2010; Piemonte and Gironi 2011), provided single data points while others provided ranges values due to scenario changes and uncertainty.

The mitigation of the environmental impacts is strongly influenced by mentioned fluxes used in the production processes, like RES uses (Slater and Gerngross 2000), water consumption, and uses of virtuous agricultural practices which may further reduce the environmental impact for producing biomass feedstock.

The higher variability of LCA analyses on BBPs is also due to recent entrance of Biopolymers into the market if compared to the fossil based polymers that follow a more standardized and proven production processes (Vink et al. 2007). Despite that, the Biobased Industries has obtained consistent gains over a short period (Vink et al. 2007, 2010), to demonstration how biobased Market is arising fastly, thus any comparison between biopolymers and the respective fossilbased polymers must demonstrate improvement in terms of environmental benefits (technology improvements, decreasing of energy and material used) (Landis et al. 2007; Mohanty et al. 2002).

Proliferating of several Carbon footprint methods and certification scheme (i.e.: ISO 14040-44, ISO 14025, WRI GHG protocol, PAS 2050, BP X30, and many more) constrained most of the industries to produce different LCA studies and calculations reporting the environmental performance for the same product in case that product was going to be sold in different countries, with huge efforts both in economic terms that of time burden by most of industries themselves. This situation has been verified inside Europe (i.e. France, Italy, and the United Kingdom), despite they are EU countries that follow the same European regulations (Mugnier et al. 2010; Benedetto et al. 2018).

The harmonized method to calculate the environmental performance of products along their entire supply chain has been performed, giving further development to the Product Environmental Footprint (PEF) and Organization Environmental Footprint (OEF) which represent the official methods adopted by EU Commission since April 2013, (Eu Commission 2013).

According to the official method, the tool able to calculate impacts is the Product Carbon Footprint (PCF). The carbon footprint measures the total GHG emissions caused directly and indirectly by a person, organization, or product. A carbon footprint is measured in tonnes of carbon dioxide equivalent (tCO_{2eq}) and considers the mentioned six Kyoto Protocol GHGs. The PCF is a tool for measuring, managing, and communicating GHGs related to goods and services, therefore it can be applied to any biomass, food, or biobased materials. The PCF applied to BBPs should be based on the LCA calculation tool that would lead to results more replicable, reliable, and verifiable.

PCF also considers the emissions over the whole life of a product or service, from the extraction of raw materials and manufacturing process right through to its use and recycling or disposal.

The aim of that harmonization is continuously implemented and provides industry, member states and all key operators to use a science-based calculation tool able to lead credible and replicable results relating to the impacts generated products or services. The achievement of these ambitious objectives requires, however, the development of product category rules (organization sectoral rules for the implementation of OEF). Further implementations are needed and a number of additional features are to be included, like clear rules on verification and data quality, checking their value and origin, for each product chain, included the BBPs – BioBased Products.

The PEF method explores a new way of communicating the environmental impacts of products as part of the efforts towards a more concrete circular Economy. This new approach not only expresses the actual and potential environmental impact but also includes the potential benefits of the recycling/reuses process, as well as energy recovery valorization of food byproducts and wastes toward to increasing both bioenergy that innovative biobased products.

While LCA approaches understandable for professionals operating in these sectors, this new approach obliges to consider multiple aspects, with particular attention to how this complex message has to beings get across to consumers and how it will contribute to fulfilling the awareness of the complex environmental issues, through a more considerably streamlining communication (EU Commission 2019).

Due to always more emerging biobased sector, it is important that LCA users present transparent data and to provide the real context of their results in order to quantify the impacts of BBPs by using several biomass and different end-uses options, as well as to aid biobased Industries in making effective improvements in their production processes.

11.4 Boosting Sustainable BBPs' Future

Among the biobased processes, there are both the well-proven BBPs processes by using on traditional biomass (i.e. wheat-based, vegetable sugar-based) and also alternative processes based on still unexploited biomass, (i.e. waste-based, lignin-based) (Tsang et al. 2019; Belboom and Anelique 2016; Collins et al. 2019).

Further strategic LCA's contributions for innovative BBPs from lignin-cellulosic byproducts are arriving from several researchers in this three-year term (Banu et al. 2019).

These processes span a large variety of technological opportunities for the future with the challenge to match increasingly with the objectives of the circular economy.

However, to ensure the sustainability of these processes and materials, it is required to develop fully the characterization of the related environmental impacts and to integrate the systematic assessment of several aspects along with the biobased products value chain (i.e. economic and social) (Pratima Bajpai 2019), because most of the biobased processes are not fully still analyzed for complete sustainability, and socio-economic and environmental factors need to be better investigated, like biodiversity, resource availability, impacts on the biobased market, biodegradation time, and regeneration time from waste to resource. All these factors are essential in order to generate added value by the organic waste-based processes, and so would reverse the value chain from end-of-life to new resource, leading to adapt consideration in Life Cycle modeling approaches.

Indeed, if these impacts will be demonstrated, the waste-based processes will turn the wastes into new resources with economic value and so would reverse the value chain from end-of-life to new resources, leading to adapt consideration in Life Cycle modeling approaches.

Taking the example of wood-based lignin biomass (the biggest bio-polymer not fully exploited yet) or rural/agroindustrial residues (fiber) as waste-based or byproducts, they could become more valuable than the main product (fruit, vegetables, cereals), which is leading to rethinking the allocation system from mass to economic allocation with further impacts on the society.

Within this framework of a circular economy, biobased materials represent one of the emerging areas where life cycle management can connect biobased resources supply chain management with waste, waste-water management systems, and energy recovery.

In this context, the UE biobased Industry consortium aims to increase competitiveness and contributing to establishing European countries as a key player in research, demonstration, and deployment of advanced biobased products and biofuels, doing more clearness on the sustainability pathways to be undertaken in the coming years (Biobased Industry Consortium 2017).

Several research projects funded by BBJ JU Horizon 2020 have the plan of including accurate LCAs for a huge pan of biobased products by using food and agroindustrial wastes (Newpack-h2020.eu 2020; Urbiofin.eu 2020; Pérez et al. 2020).

The recent research activities in industrial settings by BBJ consortium are validating the production of at least two new bio-plastics based on PHB- PLA blends by exploitation of potato peel, with greater improved sustainability performance (Chen and Patel 2012), by the addition of natural extracts of antioxidant (Celli 2020), antibacterial and nano-additives (Khaneghah et al. 2018).

The ability to extend the functionalities will be validated in real industrial environments in the coming years, including the co-blending of PHB with PLA, the nanocellulose extraction from wheat straw, the incorporation into PHB-PLA blends, and encapsulation of natural antioxidants/antimicrobials (Jack Miller 2017; Mascheroni et al. 2016; Li et al. 2015; Rampazzo et al. 2017; Piergiovanni et al. 2020).

Great emphasis is focusing on LCA Assessment and the industrial processes in developing; with aims to demonstrate a concrete biodegradability of Biopolymers by the end of 2021, ensuring compliance to the market and regulatory requirements, as well as preparing for future scale-up of the processes to achieve a pre-industrial production and identification of stakeholders perceptions (Newpack-h2020.eu 2020).

One of the next strategic challenges where researchers and industrial consortium are working is expected the management of End of Life (EoL) of biopolymers. Indeed, the biopolymers, as well as rural biomass residues and food wastes, are partially subject to Waste legislation if they are not recycled or used in other production processes. To deeply know their technical specifications will let to know the potential end-of life-options, as is already the case for many products used internationally.

Indeed, shelf life, biodegradability factor, and several quality parameters of most of the biobased products are strongly required by the food industries, as well as the health aspects are important, especially when BBPs are designed for food-packaging (Petersen et al. 1999; Peelman et al. 2013; Weber et al. 2010).

Based on the recent investigations carried out by several European consortiums (Enabling Consortium 2020) there already are several best practices concerning the best uses of biomass residues to innovative BBPs including sectors from Biofertilizers to bio-building construction, from agroforestry based plastics to the nutraceutical sector. The mentioned initiatives are promising for the exploitation of residues and waste organic-based, despite most of their LCAs analyses have to be still well fine-tuned.

The plenty of initiatives obtained by several international projects represent promising results for reaching strategic objectives but should be reached as soon as possible. It probably will be achievable in a medium and long time due to unexpected economic breakdown by COVID-19 pandemic which is strongly affecting the technological investments, as well as all processes and regulations that need to be standardized for most of the biobased products.

The opportunity of biobased materials by the valorization of food byproducts has as soon as started to contribute to the circular economy and is going hand in hand with challenges of the green marketplace, the new resource management with value creation from no-cost waste to high-profit products, and finally, business opportunities are creating new services will place, at last, the consumer at the center of the value chain.

It is expected that all these R&D Initiatives will be able to involve several sectors by agricultural, agro-food, forestry, pulp, and paper, aquatic and bio-waste sectors, and of world-leading companies in the plant breeding, biotechnology, chemistry, energy, and bioprocess engineering sectors in the coming years (Biobased Industry Consortium 2017).

Despite the pandemic conditions by COVID-19 are changing part of the bioeconomy priorities, to capitalize on the vast amount of R&D investments and results on agro-waste to BBPs will make the existing pilot plants/biorefineries look like demonstration facilities used and useful as possible to the whole community, and thus giving concrete opportunities to kick-off the green economy.

Bibliography

Akiyamaa M, Tsugea T, Doia Y (2003) Environmental life cycle comparison of polyhydroxyalkanoates produced from renewable carbon resources by bacterial fermentation. Polym Degrad Stability J 80:2003

Bach V, Berger M, Finogenova N, Finkbeiner M (2017) Assessing the availability of bio-based materials in product design. LCM – life cycle management international conference 2017. Springer Open, Luxembourg

Bajpai P (2019) Environmental impact of biobased polymers. In: Biobased polymers – properties and applications in packaging. Elsevier, Cambridge, MA. isbn:9780128184059

Banu JR, Kavitha S, Kannah RY, Devi TP, Gunasekaran M, Kim S-H, Kumar G (2019) A review on biopolymer production via lignin valorization. Bioresour Technol J 290. https://doi.org/10.1016/j.biortech.2019.121790

Baptiste P, do Carmo BBT, Margni M (2018) Propagating uncertainty in life cycle sustainability assessment into decision-making problems: a multiple criteria decision aid approach. Design sustainable technologies, Products Policies J, Open Access Springer.

Belboom S, Anelique L (2016) Does biobased polymer achieve better environmental impacts than fossil polymer? Comparison of fossil HDPE and biobased HDPE produced from sugar beet and wheat. Biomass Bioenergy J 85:159–167

Benedetto E, Gericke K, Guiton M (2018) Designing sustainable technologies, product and policies. From science to innovation. Springer International Publishing, Cham. https://doi.org/10.1007/978-3-319-66981-6. isbn:978-3-319-66980-9

Bio-based Industry Consortium (2017) SIRA – Strategic Innovation & Research Agenda. Brussels: http://biconsortium.eu

Bohlmann MG (2004) Biodegradable packaging life-cycle assessment. Environ Prog. https://doi.org/10.1002/ep.10053

Celli A (2020) Full valorization of winery waste towards new materials inside the NoAW Project. https://www.newpack-h2020.eu/

Chen G-Q, Patel MK (2012) Plastics derived from biological sources: present and future: a technical and environmental review. Chemical Reviews, pubs.acs.org/CR, American Chemical Society, vol 112

Collins MN, Mărioara Nechifor M, Tanasă F, Zănoagă M, McLoughlin A, Stróżyk MA, Culebras M, Teacă CA (2019) Valorization of lignin in polymer and composite systems for advanced engineering applications – a review. Int J Biol Macromol 131:828–849. https://doi.org/10.1016/j.ijbiomac.2019.03.069

Enabling Project (2020) Best practices atlas for BBPs, EU Project Horizon 2020. www.enabling--project.com/platforms#best-practices-atlas

EU Commission, Environment. Policy background https://ec.europa.eu/environment/eussd/smgp/policy_footprint.htm, 07-08-2019

Eu Commission Recommendation of 9 April 2013 on the use of common methods to measure and communicate the life cycle environmental performance of products and organisations, 2013/179/EU, 2013

EuropaBio (2011) Building a bio-based economy for Europe in 2020, EuropaBio policy guide. European Association for Bioindustries, Brussels

GreenRe (2017). https://greenre.org/carbon-assessment.html. Carbon assessment

Groot WJ, Borén T (2015) Life cycle assessment of the manufacture of lactide and PLA biopolymers from sugarcane in Thailand. Int J Life Cycle Assess 15. https://doi.org/10.1007/s11367-010-0225-y

Gruber P, Henton DE, Lunt J, Randall J (2005) Polylactic acid technology. Nat Fibers Biopolym Biocompos 16:2005

Heijungs R, Huppes G, Guinée J (2010) Life cycle assessment and sustainability analysis of products, materials and technologies. Toward a scientific framework for sustainability life cycle analysis. Polym Degrad Stab 95(3). https://doi.org/10.1016/j.polymdegradstab.2009.11.010

James K, Grant T (2005) LCA of degradable plastic bags. Proceedings, of the 4th Australian LCA conference RMIT University, Melbourne, Australia. Centre of Design at RMIT (Royal Melbourn Institute of Technology)

Kale G, Kijchavengkul T, Auras R, Rubino M, Selke SE, Singh SP (2007) Compostability of bioplastic packaging materials: an overview. Macromol Biosci J 7:255–277

Kendall A (2012) A life cycle assessment of biopolymer production from material recovery facility residuals. J Resour Conserv Recycl 61:69–74

Khaneghah AM, Hashemi SMB, Eş I, Fracassetti D, Limbo S (2018) Efficacy of antimicrobial agents for food contact applications: biological activity, incorporation into packaging, and assessment methods: a review. J Food Protect 81:1142–1156

Kijchavengkul T, Auras R (2008) Compostability of polymers. Polym Int J 57:793–804
Kim S, Dale B (2005) Life cycle assessment study of biopolymers (polyhydroxyalkanoates) – derived from no-tilled corn. Int J Life Cycle Assess 10:200–210
Kolstad JJ, Vink ETH, De Wilde B, Debeer L (2012) Assessment of anaerobic degradation of Ingeo™ polylactides under accelerated landfill conditions. Polym Degrad Stab 97:1131–1141
Kurdikar D, Fournet L, Slater SC, Paster M, Gruys KJ, Gerngross TU, Coulon R (2008) Greenhouse gas profile of a plastic material derived from a genetically modified plant. J Indus Ecol
Landis AE, Miller SA, Theis TL (2007) Life cycle of the corn–soybean agroecosystem for bio-based production. Environ Sci Technol J 41:1457–1464. https://doi.org/10.1021/es0606125
Lettner M, Greiner D, Stern T, Hesser F (2017) Integrated market orientation in technical R&D processes-opportunities and challenges and environmentally friendly bio-based resins. LCM – life cycle international conference 2017. Springer Open, Luxembourg
Li F, Mascheroni E, Piergiovanni L (2015) The potential of nanocellulose in the packaging field: a review. Packag Technol Sci Int J. https://doi.org/10.1002/pts.2121
Madival S, Auras R, Singh SP, Narayan R (2009) Assessment of the environmental profile of PLA, PET and PS clamshell containers using LCA methodology. J Clean Prod 17:1183–1194
Martin M, Røyne F, Ekvall T, Moberg Å (2018) Life cycle sustainability evaluations of bio-based value chains: reviewing the indicators from a Swedish perspective. Sustain J
Martin-Gamboa M, Iribarren D, García-Gusano D, Dufoura J (2017) A review of life-cycle approaches coupled with data envelopment analysis within multi-criteria decision analysis for sustainability assessment of energy system. J Clean Prod 150:164–174
Mascheroni E, Rampazzo R, Ortenzi MA, Piva G, Bonetti S, Piergiovanni L (2016) Comparison of cellulose nanocrystals obtained by sulfuric acid hydrolysis and ammonium persulfate, to be used as coating on flexible food-packaging materials. Cell J 23:779–793
McCormick K, Kautto N (2013) The bioeconomy in europe: an overview. Sustain J 5(6):2589–2608. https://doi.org/10.3390/su5062589
Meyer R (2017) Bioeconomy strategies: contexts, visions, guiding implementation principles and resulting debates. Sustainability. https://doi.org/10.3390/su9061031
Miller J (2017) Nanocellulose: producers, products, and applications: a guide for end users. TAPPI Press, Peachtree Corners. isbn:1595102647
Mohanty AK, Misra M, Drzal LT (2002) Sustainable bio-composites from renewable resources: opportunities and challenges in the green materials world. J Polym Environ 10:19–26
Mugnier E, Mairet A, Boucher J (2010) Product carbon footprinting – a study on methodologies and initiatives. EU Commission
Newpack-h2020.eu (2020). www.newpack-h2020.eu
Peelman N et al (2013) Application of bioplastics for food packaging. Trends Food Sci Technol J 32(2):128–141
Pérez V, Pascual A, Rodrigo A, Torreiro MG, Latorre-Sánchez M, Coll Lozano C, David-Moreno A, Miguel J, Dominguez O, Serna-Maza A, Herrero García N, González Granados I, Roldan-Aguayo R, Ovejero-Roncero D, Molto Marin JL, Smith M, Musinovic H, Raingué A, Belard L, Pascual C, Lebrero R, Muñoz R (2020) Chapter 2 – Integrated innovative biorefinery for the transformation of municipal solid waste into biobased products. In: Waste biorefinery – integrating biorefineries for waste valorisation. https://doi.org/10.1016/B978-0-12-818228-4.00002-2
Petersen K, Nielsen PV, Bertelsen G, Lawther M, Olsen MB, Nilsson NH, Mortensen G (1999) Potential of Biobased materials for food packaging. Trends Food Sci Technol J 10:52–68
Piemonte V (2011) Bioplastic wastes: the best final disposition for energy saving. J Polym Environ. Springer Link 19:988–994
Piemonte V, Gironi F (2011) Land-use change emissions: how green are the bioplastics? Environ Prog Sustain Energy AIChE – The Global Home of Chemical Engineers 30(4):685–691
Piergiovanni L, Fotie G, Amoroso L, Akgun B, Limbo S (2020). From cellulose to nanocellulose: new opportunities for food packaging applications from the most abundant natural biopolymer, 2020. http://www.newpackh2020.eu/docs/NewPack_Workshop%20Presentations/8_PIERGIOVANNI%20et%20al.pdf

Pietrini M, Roes L, Patel MK, Chiellini E (2007) Comparative life cycle studies on poly(3-hydroxybutyrate)-based composites as potential replacement for conventional petrochemical plastics. Biomacromolecules. ACS Publications 8:2210–2218. https://doi.org/10.1021/bm0700892

Rampazzo R, Alkan D, Gazzotti S, Ortenzi MA, Piva G, Piergiovanni L (2017) Cellulose nanocrystals from lignocellulosic raw materials, for oxygen barrier coatings on food packaging films: cellulose nanocrystals from lignocellulosic raw materials. J Packag Technol Sci 30:645–661. https://doi.org/10.1002/pts.2308

Sala S (2017) Bioeconomy contribution to circular economy. International conference on life cycle management. Springer, Luxemburg

Shen L, Patel MP (2008) Life cycle assessment of polysaccharide materials: a review. J Polym Environ 16:154

Slater SC, Gerngross TU (2000) How green are green plastics? Sci Am 283

Spierling S, Knüpffer E, Venkatachalam MM, Behnsen H, Krieg H, Albrecht S, Endres HJ (2017) The sustainability of bio-based plastics—quantifying environmental and socio-economic aspects of a computer mouse for a circular economy. In: LCM – life cycle management international conference. Spring Open, Luxembourg

Staffas L, Gustavsson M, McCormick K (2013) Strategies and policies for the bioeconomy and bio-based. Sustain J 5(6):2751–2769. https://doi.org/10.3390/su5062751.2013

Troy AH, Bilec MM, Landis AE (2013) Sustainability assessments of biobased polymers. J Polym Degrad Stability 98(9):1898–1907

Tsang YF, Kumar V, Samadar P, Yang Y, Lee J, Ok YS, Song H, Kim K-H, Kwon EE, Young YJ (2019) Production of bioplastic through food waste valorization. Environ Int J 127:625–644

Urbiofin.eu (2020). https://www.urbiofin.eu/

Vink ETH, Rabago KR, Glassner DA, Gruberb PR (2003) Applications of life cycle assessment to NatureWorks™ polylactide (PLA) production. Polym Degrad Stability J 80:403–419

Vink ETH, Rabago KR, Glassner DA, Springs B, O'Connor RP, Kolstad J, Gruberb PR (2004) The sustainability of nature works™ polylactide polymers and ingeo™ polylactide fibers: an update of the future, 1st international conference on bio-based polymers (ICBP 2003), November 12–14 2003, Saitama, Japan, vol 4. ICBP

Vink ETH, Glassner DA, Kolstad JJ, Wooley RJ, O'Connor RP (2007, 2007) The eco-profiles for current and near-future NatureWorks® polylactide (PLA) production. Indus Biotechnol 3. https://doi.org/10.1089/ind.2007.3.058

Vink ETH, Davies S, Kolstad JJ (2010) The eco-profile for current Ingeo® polylactide production. Indus Biotechnol J 6:212–224. https://doi.org/10.1089/ind.2010.6.212

Vionnet S, Bauer C, Cockburn D, Brunberg B, Stevenson M (2017) Wood products circularity and the biodiversity, challange. In: LCM – life cycle management international conference. Spring Open, Luxembourg

Weber CJ, Haugaard V, Festersen R, Bertelsen G (2010) Production and applications of biobased packaging materials for the food industry. J Food Addit Contam. https://doi.org/10.1080/02652030110087483

Wiedmann T, Minx J (2007) A definition of 'carbon footprint'. In: Pertsova CC (ed) Ecological economics research trends. Nova Science Publishers, New York. isbn:1-60021-941-1

Witt A H-d (2014) Rethinking sustainable development: considering how different worldviews envision "development" and "quality of life". Sustain J 6(11):2014. https://doi.org/10.3390/su6118310

Yu J, Chen LXL (2008) The greenhouse gas emissions and fossil energy requirement of bioplastics from cradle to gate of a biomass refinery. Environ Sci Technol 42:6961–6966. https://doi.org/10.1021/es7032235

Chapter 12
Digital Revolution Advantages: Efficient Processes and Sustainable Feedstock

Carla De Carolis

12.1 Emergence the Food Value Chains: A Circular and Sustainable Economy Approach

The global economy thus thought so far must change from linear concept "*take-make-dispose*" to the circular concept of bio-economy "*take-make-recycle-reuse*". The linear concept of economy has proved to be unsustainable as well as was has been found to be too resource-intensive – incompatible with our climate change mitigation goals or the earth's finite resources – with consumption nearly twice what the planet can regenerate each year.

There is a huge literature focusing the attention to quantify the total food losses along all the supply chain with providing concrete evidence on the negative impacts (Parfitt et al. 2010; Buzby and Hyman 2012). The UN Food and Agriculture Organization investigated that 1.3 billion food wastes are generated each year along all food chains, amounting to one-third of all food produced globally for human consumption (FAO 2011.1). Notably, the economic impacts of food wastes generated from manufactures industries to catering, from retails to households are roughly around 680$ billion in industrialized countries and 310$ billion in developing countries (FAO 2019). Furthermore, most of food waste and biomass residues generated along the entire food chain are different both as to quality than quantity. From primary production (agricultural residues not exploited) to secondary production or (industrial processing residues or byproducts), and at last, to food wastes generated by household and catering services. In adding to food wastes, their packaging and not consumable material used to support their logistic and shelf-life along all food chain, add pressure on the environment and represent unsustainable costs for the whole community to the international level (Ferreira de Cruz et al. 2012, 2014; Williams et al. 2012). Global manufacturing consumes about 54% of the world's energy and a fifth of its greenhouse gas (GHG) emissions, and Industrial waste makes up to half of the world's total waste generated each year and social

T. Cecchi, C. De Carolis, *Biobased products from food sector waste*,
https://doi.org/10.1007/978-3-030-63436-0_12

externalities were also observed, jobs were lost or displaced, and new ones took too long to emerge in multiple markets (EIA 2016).

The food waste valorization and re-use could seem hard especially in a short-time. Despite this negative impact, the food and beverage industry is characterized by a relatively limited number of multinational companies and plenty of small producers from around the world with a different food value chain from country to country (WEF 2018). This represents a facility for improving the efficiency of the food chains internationally, therefore an opportunity for policymakers and industries to collaborate with aim to implement prevention and mitigation measures to reduce by side the food waste and as well as to re-think and replicate good practices for their valorizing into new biobased products (Morone et al. 2019).

Sustainable production is the challenge for the coming years. The aim of the advanced Governments is to accelerate sustainable production projects as well as to leverage renewable resources as a tool for increasing the bioeconomy development based on green-business. Fortunately, several countries are actively implementing prevention and mitigation measures to reduce food waste (FAO 2014). The opportunities for reaching so have expanded with the advent of Fourth Industrial Revolution technology – an ecosystem that holds a dizzying amount of innovation across digital, physical, and biological fields. Examples of the latter technologies include sustainable and integrated technologies that add value of byproducts, considered industrial side streams, by producing marketable biobased products, from chemicals polymers to animal feed, from biofuels to bioenergy by using a new concept of valorization bio-cascading resources, at last (Buzby and Hyman 2012).

The Fourth Industrial Revolution for sustainable growth is already started for several manufacturing industries across major global regions with the aim to give a solid example as well as exploring the innovative pathways from the perspective of productive systems. As the volume of innovation explodes and interconnection among them, the cost of advanced technologies is plummeting (WEF 2018).

The combined effects of technology and its free-falling cost are accelerating progress in several sectors, exponentially. The technological advancements especially applied to the green sectors are being considered as leverage points for achieving sustainable production from cradle to grave. The results emerging for bio-economy is something entirely new – a framework, for the bio-industry, established on interconnected innovation able to grow the Fourth Industrial Revolution.

One of the most important challenges in the coming years will be to fully leverage digital, physical, and biological tech advancements to be able to predict and capture consumer demand, and connect it seamlessly through production operations and materials sourcing in a blueprint that minimizes the environmental footprint, (Mohamed Samir 2018).

Such innovation framework helps quantify the potential for sustainable value creation so that governments, bio-industries, and businesses can "re-design" and "re-thinking" their growth strategies accordingly. It also enables them to scale their bigger contribution towards the Green Agenda for Sustainable Development, thus enhancing economic and industry competitiveness.

Morone *et al*, explains how the development of sustainable industrial processes based on the valorization of non-preventable food waste is dependent on the refinement of the original resource, which produces value-added products. Refinement commences with the pre-treatment and extraction of components (e.g. proteins, lipids, and bio-active compounds) that can be used as value-added co-products in arising sector of Nutraceutical (Morone et al. 2019).

Other valuable components can be extracted, chemical conversion and bioprocessing can be subsequently applied to produce biobased products such as chemicals, polymers, materials, fertilizers, before being used for biofuels and or bioenergetic purposes. Thinking to use Biomass residues exclusively as an energy vector for RES Policies is being considered reductive by now. This new approach of thinking the waste lets drive the industrial community and focus the strategic investments from the Bioenergy Industry toward a more complex model such as Biorefinery.

That model should represent an example of a sustainable industrial implementation for coming years, therefore should be subsequently evaluated via process design and sustainability analysis.

Optimal Biorefinery models should achieve the minimum energy and water consumption, no generation of waste, high biomass to biobased product conversion yield, low production cost with a lowest environmental impact, or even negative cost-effective of the whole process, and at last, high societal acceptance.

Biorefinery concept grafted into the food processes production could bring an innovating production for green growth which can go a long way towards mitigating negative environmental impacts and decoupling the creation of the gross domestic product (GDP), and conserving the natural resources. The OECD Environmental Outlook to 2050 suggests that technological progress can indeed improve the intensity of economies in the coming decades (OECD 2012).

Indeed, developing and emerging economies can be key players, often heightening the importance of sustainability issues, especially for those industries classified as "Energy-intensive manufacturing" (food and beverage, pulp and paper, basic chemicals, refining, iron, and steel, etc.) (EIA 2016).

Furthermore, the food industry is classified as "energy-intensive" to "low-tech" Investment, (according to the OECD's technology classification based on R&D intensity relative to value-added), and could absorb innovations inside of own processes by including new Biorefinery concept with significant upsides. In fact, analysis by the United Nations Industrial Development Organization (UNIDO) claims that the food chain industries can sustain value-added growth across various stages of economic development.

Which then developments of the Fourth Industrial Revolution hold the most promise for accelerating sustainable production by using "*Biorefinery Approach*" especially in food value chains? Not seems to be there a specific technology able to constitute a strategy accelerated for addressing challenges of sustainability production, but the success key seems to be in the connection of the innovations in several sectors with the aim to improve the efficiency of the processes, and at the same time valorizing the food waste along the chain. Five cross-food industry key trends are emerging with promising impacts among the whole food chain:

New materials from wastes and byproducts (BBPs – Biobased Products): A quarter of all food, measured by energy content, is wasted from "farm to fork", and 8% of the loss occurs both upstream that downstream value chain (FAO 2011; Lipinski et al. 2013). If at this value is also included the potentiality given by using chemical conversion and bioprocessing to produce a platform chemical able to supply new chemical building blocks for new materials (I.e. Biocolourants, Biofertilizers, Bioethanol, composite plastic film, biochar and bioenergy, biofungicides, enzymes) (Morone et al. 2019), the residues value increase fastly due to biomass still not exploited or few valorized to new resources. The biobased materials are poised to play a greater role across industries in the near term. It's expected they can become better than traditional materials, thanks to advances in nano-biotech, green chemistry, and smart lab technology, despite their costs not seems to be cheaper than substituted materials so far. Advances, however, will also depend on the speed with which new materials production and their processing technology can be scaled from research as far as the commercial production by using suitable investments.

Advanced rural systems: The sustainability challenges for food and agriculture in the short and medium-term are particularly acute. Agriculture accounts for 80–90% of freshwater consumption (Ranganathan 2013), and 24% of global GHG emissions (FAO 2013).

Agri-food systems contribute significantly to soil erosion and pollution because of fertilizers, pesticides, deforestation, and over irrigation.

The International Food Policy Research Institute indicates that 5–10 million hectares of cropland are lost annually to severe degradation and that declining yields can be expected over a much larger area (Scherr and Yadav 1996). Notably, Increased demand on land and water for organic feedstock for manufacturing makes agri-food systems a cross-industry issue. Thus, the greatest source of innovation for rural sector is in "precision farming, automated agriculture" and biotech, where the internet of things, data and analytics are coupled with crop science to optimize farming processes, and any decisions on every aspect of the farm, from fertilizer and irrigation to harvesting time, crop selecting and seed spacing. Advances will drive substantial yield gains. Improved planning, could help to address food scarcity, safeguarding human health, the ecosystem rescources, as well as reducing the risk of any investment to be implemented.

Agri-food Factory Efficiency: Agri-food Factory Efficiency: another important issue is factories' automation and their processes, from smart warehousing to advanced biobased products manufacturing by food wastes, to increase resource productivity, shorten supply chains, improving the logistic management, and reduce the energy consumption from non-renewable resources. Improving the efficiency of the factory allows to move closer to demand markets environmentally, but it's necessary don't leave out managing labor market changes.

Traceability: From tracing the origin of byproducts in the food industry, at to trace the food commodities, the technologies such as block chain coupled with sensors and data tags are enabling companies to provide verified information about the materials, processes, and people behind to each single food products and potential biobased products. Building concrete sustainability requires data flows interlinked among food supply chains, eliminating low-value-added processes, ensuring fair earnings for smaller suppliers, and enabling remanufacturing, waste recycling, with exploitation byproducts into new added value biobased products, (WEF 2018).

Digitalization will be always more important moving forward, smart technologies for tracking food quality while also providing healthy choices for consumers are becoming a priority in the near future. But at the single food, several direct and indirect wastes are linked. As household food waste is heterogeneous and thus very complex, artificial intelligence and neural networks are needed to quickly decipher the nature of the waste (or feedstock) so that it can be best used within a Biorefinery. Understanding the nature of incoming feedstock so that follow-on processes can be optimized to deliver the best available output represents one of the maximum priorities for a successful Biorefinery. This could represent a success key for future biorefineries operating by using food waste, working round the clock, envisaging and providing chemicals, materials, and bio-energy, as well as localized employment (Morone et al. 2019).

Forecasts for relating to food consumption to worldwide in 2050 will require a 70% increase in overall food production because of population growth and changes in consumption driven by an expanding middle class (FAO 2012) with demand for red meat and dairy products increasing by up to 80%, therefore of respective food wastes linked to the their production (Searchinger et al. 2013). It's desirable that every opportunity presented by the Fourth Industrial Revolution in a new circular economy context must be used to realize a global food production system that can address challenges with limited environmental impacts, negative cost of processes by valorizing biomass residuals and food waste, with the highest sustainability while harnessing growth, innovation and development opportunities. It estimated that opportunities to create value in agri-food value chains could potentially reach $2.3 trillion annually by 2030 (AlphaBeta 2016).

It will be also necessary to strengthen a key challenge among industry and government so that they working together to promote a strong, risk-based regulatory environment and funding program to encourage the research and commercialization of new technologies and investments among the entire agriculture value chain – from seeds to food packaging and waste" with particular attention to the byproducts valorizing in the view of circular economy (WEF 2018).

Lastly, a holistic systems approach considering prevention and valorization is required, therefore will be essential to stimulate the better connectivity between social scientists, economist, engineers and designers, physical, agronomists, and biology scientists in order to develop integrated strategies for minimizing and valorizing food waste. Such connectivity could minimize the impact of global drivers and mega-trends. At last, education and awareness of citizens with regard to food waste e potential recycling and valorizing needs to be adopted as 'morally right', rather than 'morally wrong', philosophy with respect to food waste. It is also essential to promote the mentioned concepts since primary education up to tap high-level education with appropriate training programs (Morone et al. 2019).

Recently, data, investigations, and all many evidences collected so far, claim that food waste is a real problem, and visibly actual. Many solutions are ready to be used and promising technologies are commercially available, in both social and scientific realms. The key challenge will be their implementation and connectivity in a new concept of the circular economy.

12.2 Support of the Rural Economy in the Horizon 2020 Framework

In the last decade, the search for innovative pathways towards sustainability has been brought to the forefront of international agenda settings. While international organizations and institutions, such as the United Nations, ADB (Asian Developed Bank), Work Bank Group and the European Union (EU), mobilized around the grand challenge of sustainability to counter the climate change and poverty, on both a local and a global scale. During the latest decades, several policy documents and funding schemes started to consider and consolidate the "eco-innovation" as a key concept (or buzzword).

By focusing the attention at European level, eco-innovation is well-applied to food waste reduction as well as to the valorization of biomass residues, so much that it has been framed by the EU research funding programs, including Horizon 2020.

In particular, the EU program HORIZON 2020 has mostly constructed around the notion of "eco-efficiency" transversely applied to several fields including rural and food sector.

Interesting, in the context of Eco-effectiveness, eco-innovation always appeared in close relation to the terms "circular economy". Eco-innovation, that as a buzzword, has evolved through the years in favour of the new rising concept of the 'circular economy' became imperative in the new framework of Horizon 2020 (Colombo et al. 2019), so much that "systemic eco-innovation" and "circular economy appeared to be considered as synonymous (EU Commission 2018).

In the context of the circular economy, HORIZON 2020 responds to the key challenges our planet is facing for the years to come:

- adapting to and mitigating climate change;
- ensuring food security;
- safeguarding the natural resource base,
- promoting alternatives to fossilbased economies
- sustainably using resources while protecting ecosystems.

In addition, agriculture and food systems, forestry, the marine, and the biobased sectors will be at the very heart of the challenges to be addressed in the coming years.

Regarding the contents of the Work Programme, many of the challenges are addressed to the global nature, requiring the development of global solutions and opening up the innovation process to all active players in international cooperation with third countries and relevant international organizations or initiatives.

In this international perspective, the Work Programme of Horizon 2020 identifies 2 priorities for international cooperation that are focused on:

- *Blue Growth*: which intend to launch a flagship initiative for the South Atlantic Ocean, paving the way towards an 'All Atlantic Ocean Research Alliance', as well as a flagship on the "Future of Seas and Oceans" in line with the G7 Initiative, and at the same time to reinforce cooperation with partners in other regions such as the Baltic Sea and the North Sea, the Mediterranean and the Black Sea.
- *Sustainable Food Security*: with the aim to continue supporting flagships initiatives with China and Africa on Food and Nutrition Security and Sustainable Agriculture (EU Commission (a) 2019b).

Concerning the mentioned challenges applied to the EU Context, the Work program of HORIZON 2020 focuses five priorities:

1. *Addressing climate change and resilience on land and sea*: Climate change is one of the biggest challenges Europe and the world are facing. The primary sectors covered by SC2 (Table 12.1) are among the most important sources of CO_2 emissions and are highly vulnerable to climate change. At the same time ecosystems like forests soils, and oceans are major carbon sequestration reservoirs (i.e. Organic Carbon, and salts of Carbonate). The R&I actions in this work program support meeting the ambitious climate targets calibrated on needs for food, feed, biobased products and energy of 10 billion of the global population by 2030.
2. *Making the transition towards a circular bioeconomy*: R&I actions across all SC2 (Table 12.1) sectors support resource-efficient production and distribution systems, value-chains based on new and more efficient use of wastes, residues, and agroindustrial byproducts, as well as new business models able to enhance the EU natural capital. It is essential, therefore, that European Commission's Circular Economy Package is implemented.
3. *Fostering functional ecosystems, sustainable food systems, healthy lifestyles*: Ensuring long-term food and nutrition security and sustainability of food systems requires sustainable management of Environmental resources (i.e., land, soil, water, biodiversity) as providers of terrestrial and aquatic ecosystem services. These services are key to the basis of the whole food chain, nutrition, lifestyle, and health. The investments in this section of the Work Program focus on future-proof EU food systems to make them more sustainable, resilient, responsible, diverse, competitive, and inclusive.

Table 12.1 Horizon 2020 work programmes, eco-innovation fields applied to the food and rural sectors

Section	Work programme[a]	Number of calls for proposal with keyworkd Eco-innovation in Horizon 2020 (2014–2020) (Colombo et al. 2019)	Number of call for proposal with keyword Eco-innovation applied to food and rural sector and in Horizon 2020 (2014–2020) (2014–2020) (our elaboration)
ES – excellent science	1. General introduction	0	
	2. Future and emerging technologies	0	
	3. Marie Skłodowska-curie actions	0	
	4. European research infrastructures	0	
IL – industrial leadership	1. Leadership in enabling and industrial technologies introduction	1	
	2. Information and communication technologies	0	
	3. Nanotechnologies, advanced materials, biotechnology and advanced manufacturing and processing	4	
	4. Space	0	
	5. Access to risk finance	0	
	6. Innovation in small and medium enterprises: SMEs instrument (phase 1 + 2)	8	4
	7. Biobased industries join undertaking BBJ	0	4
SC – societal challenges	1. Health, demographic change and wellbeing	0	
	2. Food security, sustainable agriculture and forestry, marine and maritime and inland water research and the bioeconomy	**6**	**15**
	3. Secure, clean and efficient energy	1	
	4. Smart, green and integrated transport	0	
	5. Climate action, environment, resource efficiency and raw materials	29	4
	6. Europe in a changing world e inclusive, innovative and reflective societies	0	
	7. Secure societies e protecting freedom and security of Europe and its citizens	0	
	8. Spreading excellence and widening participation	0	
	9. Science with and for society	0	
	10. Communication, dissemination and exploitation	0	

EU Commission (c) (2019d)

4. *Boosting major innovations (inland and sea) new products, value chains, and markets*: Innovation is at the core of EU policies. Demonstrating and scaling up new technologies and business models that create breakthrough innovations are crucial for ensuring the long-term competitiveness in several sectors. But scaling up of innovative new technologies represents a high commercial risk for private investors, especially during development stages. Thus public investment is essential during the early stages, to ensure the creation of sustainable value chains, resulting in new jobs, products, and services in the years to come.
5. *Developing smart, connected territories and value chains in rural and coastal areas*: Helping the rural and coastal areas to meet the wide range of economic, environmental, and social challenges of the twenty-first century is one of the key challenges for Europe. This priority addresses the territorial dimension of Research and Investments actions in primary production, the food and biobased industries, most of which are located in rural and coastal areas. R&I activities aim at the better capitalization of territorial assets, taking account of long-term drivers to open new sustainable avenues for business, services, and value chains in support of rural and coastal communities, promoting new partnerships between producers, processors, retailers, and society.

The Horizon 2020 funding program, therefore, embraces several targets in according with the European policy established. The Rural policy focused on food supply takes into account the typical EU rural structure and its potential network of supply. In fact, rural regions make up roughly half of the territory of the EU-28, with just over one-quarter of the population living in these areas. This background lets to understand how is needed to invest in eco-innovative systems of food chains beginning from the rural sector.

By building innovation capacity and implementing innovative solutions, rural and sparsely populated areas could overcome their inherent challenges and remain viable business locations as well as sustainability centres. There are many aspects of EU rural innovation characteristics of innovation ecosystems, the conditions for the local business sectors, industrial diversity, access to physical and digital infrastructure, relations with urban counterparts for example.

The HORIZON 2020 Platforms and HORIZON EUROPE offer a range of calls, tenders, and prizes aimed to promote inter-projects, international cooperation for several fields, and capitalization services to reinforce the different approaches and identifying synergies and good practices.

Notably, the whole funding program HORIZON 2020 has launched by the EU Commission during the last decade, dedicating part of own Investment Agenda €1.3 billion, and organizing specific opportunities for rural, Agroindustrial, Bioenergy, and Biobased sector, through invitations to submit the innovative proposals in several sections (EU Commission (a) 2019b).

In particular, the Work Programme is structured around four thematic sections:

- "Sustainable Food Security",
- "Blue Growth",
- "Rural Renaissance",

- "Food and Natural Resources"
- "Thematic Investment Platform on Circular Bioeconomy".

Within this thematic-section, a number of topics contribute to the Focus Areas:

- "Low Carbon",
- "Circular Economy"
- "Digitisation".

In particular, the *SC2 section of Work Programme*[1] focuses on the sustainable management of commodities land uses, and water managment, with the aim to secure healthy food as well as to preserve public assets such as biodiversity and clean water.

A specific investigation has been made among several sections and work programs of HORIZON 2020, in order to highlight the funding opportunities relating to eco-innovation, with particular attention to the food and rural sectors, Table 12.1.

EU Commission focused more efforts to counter the food waste along all the chain. Annually in the EU Countries roughly 88 million tonnes of food is being wasted along the whole agri-food value chain, from primary production to final consumer, with consequent high environmental, social and economic impacts (EU Commission 2016). The problem is particularly amped up, and therefore worrying for perishable foods. Reducing food waste, primarily through prevention, efficiency along the chain, recycling, and waste valorization into the new products has enormous potential for ensuring sustainable food and nutrition security, reducing environmental impacts by improved resource use efficiency. Nonetheless, reducing food losses along the agri-food value chain represents a target not straightforward to reach, because it's resulting of many and interlinked factors. Many investigations have been carried out to know the causes, and many innovative solutions are already available. It remains, therefore, urgent need for their demonstration and market replication to different levels, with aims to avoid the burden-shifting of food waste from one stage of the agri-food value chain to another.

Part of the calls focused on food waste in Horizon 2020 aim to identify, validate and demonstrate innovative, effective ways to reduce food losses and waste, all along the agri-food value chain from primary production down to final household consumption and disposal. They should consider diverse forms of innovation, e.g., technological, social, organizational, managerial and institutional, etc. that allow key actors to better organize and coordinate their activities, to monitor conditions, to eliminate causes of inefficiency along the whole chain, and hence, discard as little food as possible without compromising on their quality, including safety, and sustainability.

The call for proposals for food waste reduction and valorization, have to take into account "Sustainable Food System Approaches" to tackle the inherent links between ecosystems, food production, the food chain, and consumer health and wellbeing.

---

[1] https://ec.europa.eu/programmes/horizon2020/find-your-area

More in detail, EU Commission expects that SFS Projects aim to cover the following thematic actions:

- to deliver diverse and healthy food from land and sea
- to increase resource efficiency and environmental performance of food systems from primary production to consumers
- to understand the impact of climate change on agriculture, resources, food quality and identify options to manage its effects
- to reduce greenhouse gas emissions and emissions of air pollutants from land use and food production taking into account main drivers such as inputs and consumption patterns.

The solutions arising from SC.2 (Table 12.1., Societal Challenges, point n°2) Proposals expected to deliver significant economic, environmental and social benefits. Investments, for example, are focused on healthy and safe food, promotion of biodiversity for agriculture safeguard, on sustainable management of soils, on increasing animal welfare, as well as on the implementation of Food and Blue Clouds. That investments aim to support food production by 2030 with pioneering a digital revolution and new value chains in rural economies.

New ideas, products, technologies, policy recommendations and social innovations are working in tandem to supply a supportive framework for genuine improvements within the way we produce and consume. This is in line with societal expectations for impact-driven research (EU Commission (a) 2019b).

To test and demonstrating efficacy of introduced innovative approaches and to further improve understanding of the root causes behind the current situation, the EU Commission also expects to fund the projects that include tasks to monitor, and valorize food losses and wastes (and associated economic and environmental costs) along the agri-food value chains into new bio-products, (EU Commission (b) 2019c). Any methods used for this purpose should be compatible with several international legislations on measurement and reporting data on food losses and wastes. In addition, the factors enabling and hindering innovative approaches should be taken into consideration. Further targets will be based on setting-up recommendations, best practice guidelines and toolkits for promising innovative approaches to the reduction of food losses and waste shall be developed, taking into consideration the underlying socio-cultural factors and gender aspects.

The coming proposals fall under the concept of the ‘multi-actor approach’ (EU Commission (a) 2019b), ensuring solid collaboration between relevant actors, from farmers or farmers associations, across agri-food industry (including small businesses), wholesalers and retailers, food-related services, until consumers and policymakers. At last, to maximize the concrete impacts, dedicated work packages for international cooperation with no-EU Countries are also encouraged.

In this respect, the EU funding programs like HORIZON 2020 Green Deal and HORIZON EUROPE are demonstrating of being driving-force able to realize the circular economy with involving several key actors, as public institutions, policymakers, SMEs, industries, and researchers, as well as citizenship and their associations. It has developed in order to provide concrete opportunities to embrace more eco-centric and inclusive approaches to economics, towards stronger sustainability, and the more systematic inclusion opened to plenty of public and private entities and organizations.

The first results of such project activities are actually demonstrating a strong contributor to creating the conditions for dynamic and innovative farming and food sectors managing to turn high-quality products and high environmental standards into a competitive advantage.

Bibliography

AlphaBeta (2016). aluing the SDG prize in food and agriculture: unloking business opportunities to accellerate sustainable and inclusive growth. AlphaBeta, Singapore. www.alphabeta.com

Buzby JC, Hyman J (2012) Total and per capita value of food loss in the United States. Food Policy J 37(5):561–570

Colombo LA, Pansera M, Owen R (2019) The discourse of eco-innovation in the European Union: an analysis of the eco-innovation action plan and horizon 2020. J Clean Prod 214:653–665

EIA (2016) International energy outlook: industrial sector energy consumption with projection to 2040. U.S. Energy Information Administration. Indipendent Statistic and Analysis. DOE/IEA-0484(2016). ISBN 978-0-16-093333-2, www.iea.gov/forecast/ieo

EU Commission (2016) Food waste. Retrieved from EU Commission. https://ec.europa.eu/food/safety/food_waste_en

EU Commission (2018) Horizon 2020, work programme 2016–2017, societal challenge n. 12, climate action, environment, resource efficiency and raw materials, Retrived from Eu Commission. http://ec.europa.eu/research/participants/data/ref/h2020/wp/2016_2017/main/h2020-wp1617-climate_en.pdf

EU Commission (2019a). https://ec.europa.eu/programmes/horizon2020/h2020-sections. Last access in 2020

EU Commission (a) (2019b) Work Programme Horizon 2020. Brussells, European Commission Decision C(2019)4575 of 2 July 2019

EU Commission (b) (2019c) Food waste measurement. Retrieved from Policies and Information and Services of EU Commission. https://ec.europa.eu/food/safety/food_waste/eu_actions/food-waste-measurement_en

EU Commission (c) (2019d) Horizon 2020. Retrieved 09 02 2019, from Electronic Data Interchange AREA (SEDIA) of EU Commission. https://ec.europa.eu/programmes/horizon2020/en

FAO (2011.1) Global food losses and food waste – extent, causes and prevention. ISBN 978-92-5-107205-9. www.fao.org/3/a-i2697e.pdf

FAO (2012) Global agriculture towards 2030/2050. Rome, (Italy), Working paper No. 12-03. www.fao.org/3/a-ap106e.pdf, 2020

FAO (2013) FAO statistical yearbook. Food and Agriculture Organization of the United Nations, Rome (Italy), ISBN 978-92-5-107396-4. Retrieved from www.fao.org/publications, 2020

FAO (2014) Mitigation of food westage, societal costs and benefits. ISBN 978-92-5-108510-3. www.fao.org/3/a-i3989e.pdf, 2020

FAO (2017) Rezaei M, LiuB, Food loss and waste in the food supply chain. NUTFRUIT Mag 71. Retrieved from www.fao.org/documents/card/en/c/30245942-5cdb-42b6-bb1a-98243f108446/

FAO (2019) The state of Food and Agriculture 2019. Moving forward on food loss and waste reduction. Rome. Licence: CC BY-NC-SA 3.0 IGO. www.fao.org/3/ca6030en/ca6030en.pdf, 2019

Ferreira da Cruz N, Simões P, Marques RC (2012) Economic cost recovery in the recycling of packaging waste: the case of Portugal. J Clean Prod 37:8–18

Ferreira da Cruz N, Ferreira S, Cabral M, Simões P, Marques RC (2014) Packaging waste recycling in Europe: is the industry paying for it? Waste Manage J 34(2):298–308

Lipinski B, Hanson C, Waite R, Searchinger T, Lomax J, Kitinoja L (2013) Reducing food loss and waste. Working paper, installment 2 of creating a sustainable food future. World Resources Institute, UNEP, Washington, DC. Available online at www.worldresourcesreport.org, 2020

Morone P, Koutinas A, Gathergood N, Arshadid M, Matharue A (2019) Food waste: challenges and opportunities for enhancing the emerging bio-economy. J Clean Prod 221(1):10–16

OECD (2012) OECD environmental outlook to 2050: the consequences of inaction. ISBN: 9789264122246. Doi.org/10.1787/9789264122246-en, 2012

Parfitt J, Macnaughton S, Barthel M (2010) Food waste within food supply chains: quantification and potential for charge to 2050. Philos Trans R Soc B Biol Sci J 365(1554):3065–3081. https://doi.org/10.1098/rstb.2010.0126

Ranganathan J (2013) The global food challenge explained in 18 graphics. Retrieved from WRI – World Resources Institute. Retrievied from www.wri.org/blog/2013/12/global-food-challenge-explained-18-graphics, 2020

Samir M (2018) Driving the sustainability of production systems with fourth industrial revolution innovation. World Economic Forum White Paper. World Economic Forum. REF 190118 – case 00039558. Retrieved from http://www3.weforum.org/, 2020

Scherr SJ, Yadav S (1996) Land degradation in the developing world: implication for food Agriculture and the environment to 2020. International Food Policy Research Institute. Whashington DC 20036–3006 USA, 1996. www.ifpri.org/sites/default/files/publications/vp14.pdf

Searchinger T, Hanson C, Ranganathan J, Lipinski B, Waite R, Winterbottom R, Dinshaw A, Heimlich R (2013) The great balancing act. Working Paper. Washington, DC 2002, USA. World Resources Institute, http://pdf.wri.org/great_balancing_act.pdf, 2013

WEF (2018) Driving the sustainability of production systems with fourth industrial revolution innovation. White paper, ref 190118 – case 00039558, Geneva. World Economic Forum

Williams H, Wikström F, Otterbring T, Löfgren M, Gustafsson A (2012) Reasons for household food waste with special attention to packaging. J Clean Prod 24:141–148

Chapter 13
Research Gap and Needs

Carla De Carolis

13.1 Introduction

The economic progress of the latest century is proved unsustainable. Since resources are limited as well as the ability to absorb pollution is curbed, our current linear economy has to be eventually replaced with the Circular economy as it majorly demands a sustainable consumption and production of resources through re-using and re-cycling. To accomplish the Circular Economy, systemic drivers such as "bio-economy" needs to be incorporated. Thus wastes generated can be integrated into the Circular Economy through sustainable bio-economy processes.

13.2 From Waste to Wealth Using Green Chemistry: The Way to Long Term Sustainability

The linear economy currently configured is not only unsustainable but also unstable. It is expected the global population will increase to 9.2 billion by 2050, further exacerbating increasing energy and resource demands.

The issue is focused on economic growth which is strongly dependent on raw materials, but we're living on a planet with finite resources. Therefore it's mandatory to strike a balance between growth and resource management.

Economic growth will occur at 1.5–3% per annum, which will have compounding effects on both the depleting 'proven fossil fuel reserves by 2069 (i.e. marketable reserves of oil, gas, coal and for nuclear energy), and ultimately recoverable reserves by 2088 (Stephens et al. 2010).

Not much has been done about it until the governments started giving, at environmental protection, its due importance while drafting the national policies. Sustainable development has become a priority for the world's policymakers too since mankind has been able to destabilize all the major ecosystems of the planet in

T. Cecchi, C. De Carolis, *Biobased products from food sector waste*,
https://doi.org/10.1007/978-3-030-63436-0_13

a few decades. Therefore, the challenge we face is to make a fast transition towards society and an economic system that should be based on recycling and valorizing of wastes to new biomaterials easily degraded able to substitute the conventional materials.

Radical changes are required and it's necessary to rethink the links between re-use, re-cycling, and valorization of food wastes, biomass residuals, in a new perspective of economic prosperity, bolstered by huge investments in new financial tools, technologies, and innovations toward a circular bioeconomy (Venkata Mohan et al. 2019).

Green chemistry is based on the substitution of fossil resource consumption by using green alternatives processes (Biorefineries) for a new supply of biobased products and Bioenergy. In this context, green chemistry contributes to improving plenty of key pathways concerning the valorization by biomass wastes, with a further reduction of fossil carbon intensity.

Recently it has seen significant growth in the demand for chemical products and new materials. However, this demand has not been met with similar growth in green chemistry solutions and practices. While green chemistry and adoption of best practices have certainly occurred, this is demonstrated by an increasing number of success stories, it is still considered for a niche growth. Green Chemistry needs to be still integrated into the Industry and SME sectors, educational systems, as well as in government programs.

Through extensive analysis several barriers have been identified, (Fennelly and Associates 2015) including:

1. the complexity of global supply chains and exiting and consolidated infrastructures;
2. costs and time to scale to new technologies,
3. the incumbency of existing technologies that are cost-effective, high performing but maybe problematic environmentally,
4. concerns about the risks involved in moving to green chemistry solutions (performance, process changes, material incompatibility or costs of recertification and potential for substitutes to be later designated chemicals of concern);
5. limited investment, incentives, education, and metrics for green chemistry.

Despite that, concrete strategies, and bottom-up approaches can be adopted to accelerate green chemistry implementation, development, and adoption, with the view to reaching a point where all production sectors can become sustainable, (GC3. 2015).

Tickner J. and Becker M. identified a wide range of strategies to accelerate green chemistry innovation, which is being identified (Tickner and Becker 2016) in:

- to enhance Market Dynamics: building comprehensive, green chemistry enablers, effective interventions that create market shifts to support green chemistry research, development, and adoption.
- to support Smart Policies: designing for innovative state policies that can effectively support the supply of and demand for green chemistry solutions.

- to foster collaboration: facilitating the flow of information about green chemistry solutions among biomass suppliers (i.e. farmers) and product makers (i.e. chemical industry and SMEs) as well as assembling partnerships to tackle priority challenges can support the collaborations necessary to grow the marketplace for green chemistry solutions (ENABLING Project 2020).
- to inform Marketplace: strengthening the dissemination activities and potential Best Practices about the opportunity given by valorization of biomass residues and their conversion into the new BBPs, the potentiality of green chemistry business benefits from the socio-economic and environmental point of view (ENABLING Project 2020).

As more governments, industries, and companies are adopting green chemistry as base of their production, there is also a critical need to understand the toxicity of chemicals in terms of ecological and health impact. Thus, the biorefineries should furthermore evaluate toxicity and relative impacts throughout the BBP's development lifecycle, minimize the toxic waste generation, and use of toxic raw materials or produced.

The European Chemicals Agency (Echa) is asking EU member states to evaluate 74 substances under the Community Rolling Action Plan (CORAP) for 2020–2022. This contains substances suspected of posing a risk to human health or the environment. Fourteen of these are slated for evaluation in 2020, while 60 substances are planned for evaluation in 2021 and 2022 (ECHA/NR/19/38, 2019).

ViridisChem has built a comprehensive toxicity database on 90 million chemicals and 2.4 billion properties and toxicity data by the implementation of the chemical database available at the global (Vaidya 2019). The ViridisChem toxicity database evaluates toxicity products (covering most organic and inorganic chemicals, bio-polymers, petrochemical-related mixtures, including nutraceuticals, cosmetics, and drug interaction products) by using 44 different endpoints defined by NSF/ANSI 355 standard per United Nations, EPA, and EU-REACH guidelines that recommend comprehensive breakdown of ecological, health and safety hazards. Utilizing this information, it is possible to know the toxicity of any chemicals, even the new and postulated molecules, mixtures/formulations, and processes with further benefits to pharmaceutical, biochemical, agrochemical industries as well as biomass and chemical suppliers. That tool is useful:

- to predict the toxicity of new molecules, and their derivatives in real-time.
- to avoid the use of toxic raw materials, and find lesser toxic chemicals that satisfy the reaction-specific requirements.
- to fully understand the health, safety, and ecological risks of toxic formulations, and find better formulations that offer the same benefits.
- to avoid the use of a cocktail of formulations that may be nontoxic individually, but become very potent when combined.
- to define sustainable processes for products development by avoiding toxic reagents and wastes during every step of the chain.

Considering toxicity aspects, especially in the emerging sector as green chemistry, is extremely important. These solutions will have a huge global addressable market and can offer a tremendous revenue-generating opportunity for most SMEs and industries operating in the (BBP) BioBased Products sector (Vaidya 2019).

Green chemistry, innovative technologies, and processes applied are issues essential to reduce the gap among industrial needs and research applied on the new valorization of organic wastes to new resources.

The research interventions allowing reducing the fossil carbon intensity of our economy need to be better investigated deeply, such as the substitution of fossil resources with alternatives for material and energy supply.

The innovative materials sector and potential substitution of fossil carbon-based products will be mainly based on several biomass and wastes recycled, therefore applied conversion technologies such as fermentation, gasification, pyrolysis will be taken into consideration, as well as further researches are in progress on the lignocellulosic processing and other biobased products (i.e.chemical building blocks) (Gerssen-Gondelach et al. 2014). A further research field that should be deeply investigated is "recycling of carbon" contained in existing and future materials by using photocatalysis, without consuming further resources. As described in e.g. Tahir and Amin, recycling technologies are in the early research phase (Tahir and Amin 2013).

Despite there is plenty of international pro-active policy about the environment, the literature on medium and long-term potential pathways for circular bio-economy is incomplete, as confirmed by Fabian Shipfer (Schipfer et al. 2017). This is due to uncertainly still existing technologies to be applied to biomass waste towards new biobased products in a new concept of biorefinery. Therefore it is also difficult to plan substantial investments for innovative biorefineries without having long-term potential scenarios of development.

A wide literature (Matzenberger et al. 2015; Patel et al. 2007) confirms several options about possible developments concerning bioenergy (conventional biofuels), food and feed sectors (Chap. 3). On the contrary, the literature on possible development scenarios and planning relating to in advanced biomaterial production and consumption is still in the research stage, and stands out as relatively scarce (Patel et al. 2007; Daioglou et al. 2014). Only some researchers and experts tried to outline short-term expectations (Scarlat et al. 2015; Dammer et al. 2013).

For example, the fermentation of lignocellulose to ethanol is expected to be in an R&D and demonstration phase (Gerssen-Gondelach et al. 2014) to reach a ripe technology and stabilization of materials. This makes it useful to discuss the commercialization of lignocellulosic-based materials and biofuels and their planning for 2025–2030, despite electric sector seems to be the best solution for transport sector in the coming years. Meanwhile, the installed production capacities for Bioethanol and BBPs from sugar and starch-based plant utilization are held constant until 2050 and only additional advanced biomaterials production are assumed to be based on wood.

In addition to the above-mentioned literature, a huge of EU Projects, for example "BIOYAWS" Project (Bioways Project 2020), "ENABLING" Project (Enabling

Project 2020) are monitoring the progress of several advanced biobased products by Biomass residues, biomass estimation and example cases of best practices as well as future trends. However, no detailed medium and long-term biobased material scenarios are still well delined so far (Schipfer et al. 2017).

De facto, part of this lack determines the key gap between the research not still ready to face the market and industry sector still waiting for the validated results as solutions of unresolved issues, as:

- Potential capacity of advanced biobased materials to substitute substantial amounts of fossilbased materials in competitive pathways.
- Traceability and monitoring production of more promising advanced biobased materials, their certification, and testing with a guaranty for health and environment in all their life cycles.
- Selection of biomass residues and organic wastes for ensuring a supply of biobased products, and production rate of such biobased products to allow the phasing out of fossilbased counterparts.
- Investigations relating to the possible cascading biomass along BBPs chains, and comparison with existing biomass use for the bioenergy sector.
- Investigations on the potential distribution of additional biomass supply and demand that can be globally distributed among industrialized and developing countries, and analyses on the possible implications of international biomass trade, (Schipfer et al. 2017).

13.3 Exploitation of Non-Food Feedstock as Smart Alternative to Crops Usage for a Sustainable Bioeconomy

During this transition phase, bioeconomy has been considered a requirement for sustainable growth, and it has been defined by various organizations around the globe. The Organization for Economic Cooperation and Development defines it "as a world where biotechnology contributes to a significant share of economic output" (OECD 2009). While FAO – Food and Agricultural Organisation defines bioeconomy as "the production, utilization, and conservation of biological resources, including related knowledge, science, technology, and innovation, to provide information, products, processes, and services across all economic sectors aiming toward a sustainable economy" (FAO 2019).

The German Bioeconomy Council defines bioeconomy as the "knowledge-based production and use of biological resources to provide products, processes, and services in all economic sectors within the frame of a sustainable economic system". Bioeconomy is focusing the attention on implementing innovative biological approaches through collaboration with researchers, stakeholders and policymakers (European Commission 2012a; Council G.B. 2018). Based on the latest definition given by European Commission, bioeconomy can be considered as "the production of renewable biological resources and the conversion of these resources and waste streams into value-added products, such as food, feed, biobased products, and bioenergy" (European Commission 2012b). Also, Obama Administration mentioned bioeconomy strategy as "one based on the use of research and innovation in the biological sciences to create economic activity and public benefit" (White House 2012).

But how many pathways of bioeconomy are real sustainable? It borns the gap among different ways to conceive the economic growth and a more sustainable evolution approach, from an older linear economy concept based on the fossil resources to a circular bio-economy principally based on the new concept to conceive the "wastes as resources". But this needs high expertise of "bio-technological" cascades sectors and processes from organic wastes to new added high-value products (BBPs and Biofuels) (Davis et al. 2016).

The importance of circular bioeconomy is now realized globally and in many nations, including developing countries. The international community and Governments are working on strategies to achieve a greater sustainability level. Many technological sectors applied to eco-biological issues have a strong potential to innovate enabling to reach greater technology readiness levels and ready to support a more sustainable market. India is gearing up to meet the challenge of achieving US$ 100 billion worth circular bioeconomy by 2025 as it already has necessary sources such as booming biotechnology industries and an enthusiastic scientific workforce, (Venkata Mohan et al. 2019). In this context, petroleum-based refinery (mirror of the linear economy) are shifting to new emerging biorefinery (Clark and Deswarte 2008), where non-edible based feedstock/biogenic wastes are being used as raw materials for producing a range of products as: biofuels, industrial biochemicals, and biomaterials, including biopolymers (Clark and Deswarte 2015; De Jong et al. 2013). The conventional biorefineries have still dedicated feedstocks like fuel crops (Corn, Sugarbeet, Sugarcane, Vegetable Oils) but the focus is now shifting towards utilizing the process residuals or the waste generated from daily life. Biomass and its residues, municipal solid waste (MSW), food waste, aquatic waste and industrial gases are among a few potential feedstocks for waste biorefineries (Bioways Project 2020; Venkata Mohan et al. 2019). The greatest challenge for a sustainable biorefinery is also to integrate technologies toward biological systems and bioprocesses especially in those operating across the food waste management systems, (Schieb et al. 2015).

Notably, municipal solid waste (MSW) is one of the major wastes generated around the globe, and approximately 50–60% is represented by biogenic composition. The proper valorization of MSW can help in the simultaneous reduction of the environmental impacts associated with waste management and respective disposal into landfills. Around 1.3 billion tonnes of food is wasted annually accounting for one-third of its total global production that incurs a cost of USD 900 billion to the global economy (Dahuya et al. 2018). Re-directing the biogenic waste, from landfills to waste biorefineries, would lead to incredible employment opportunities in industries and academia especially in the sectors of Agri-food, chemical and healthcare, pharma, and logistics (Clark and Deswarte 2015; Dahuya et al. 2018; Cristóbal et al. 2018).

The agri-food waste biorefinery (González-García et al. 2019; Beltrán-Ramírez et al. 2019) as well as sugar-based and starch-based pulp wastes (Adiletta et al. 2019) and lignocellulosic residues (Hassan et al. 2019), are the focus of extensively discussed by several researchers because represent promising agro-food wastes to industrial scale.

MSW also has enormous potential to be used as feedstock for the production of several commodity chemicals, biofuels, bioenergy and compost through the deployment of appropriate methods and systemic approaches (Fig. 13.1). However, while designing such biorefineries, economic feasibility in line with environmental sustainability, a lower carbon footprint should be taken into consideration (Venkata Mohan et al. 2019). More than 100 good model examples have actually implemented in real Best Practices around 16 European Countries and collected by part-

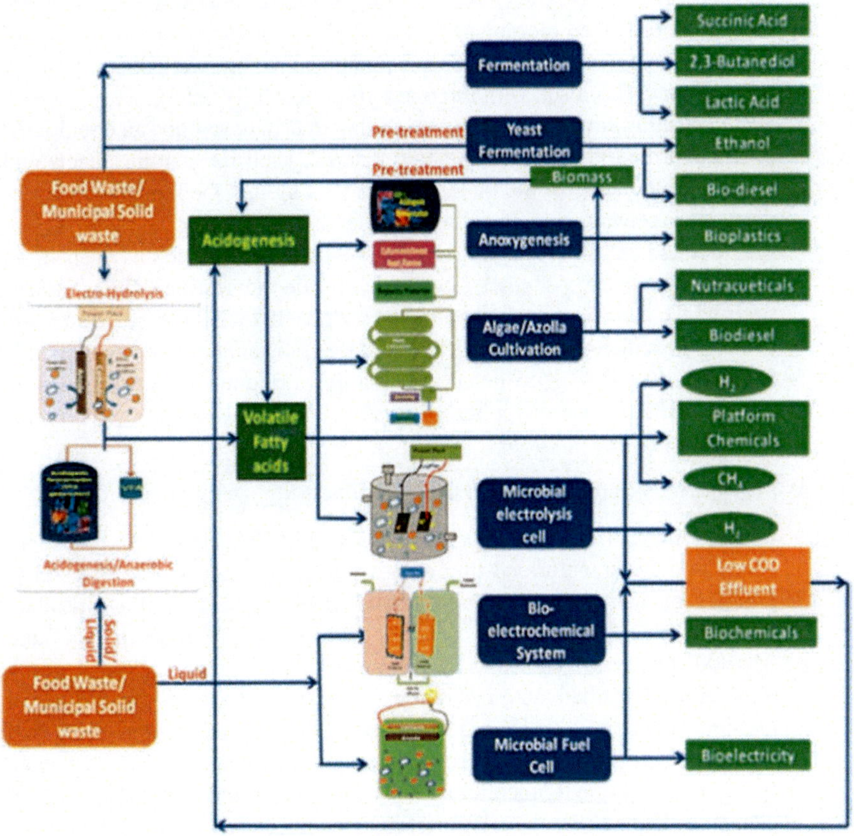

Fig. 13.1 A scheme of food waste biorefinery: spectrum of products that can be generated in a closed loop by the valorization of organic food wastes (Venkata Mohan et al. 2019). (Reprinted from Bioresource Technology Reports, Vol. 7, Venkata Mohan S., Dahiya S., Amulya K., Katakojwala R., Vanithaa T.K.,, "Can circular bioeconomy be fueled by waste biorefineries — A closer look", copyright 2019, with permission from Elsevier)

ners of Enabling Project (2019) in the specific platform (https://www.enablingproject.com/platforms#best-practices-atlas). Best Practices confirm the concrete implementation of the biorefinery concept by using a sustainable bottom-up approach that aims to exploit local biomass wastes. It is the case of agro-industrial by-products listed below:

(a) Vegetable Oil-based Residues:

- A Bulgarian company evolved in a small biorefinery able to produce food vegetable oils and biodiesel from oil crops cultivated in marginal lands (rape, soya, and sunflower). The by-products of processes (glycerin and rare fatty acids) are valorized in-house as new bio-based products for pharmaceutical and cosmetic production (including as stabilizers and foam suppressors). Other process residues are used as biological forage. The company has the possibility to construct stock-breeding farms for pigs and cattle and feed these with produced biological forage. It places the accent on the development of biological stock-breeding. This allows the circular economy of the whole firm to be closed with internal byproducts fully valorized.
- By the refining soybean oil process, an Italian firm developed an epoxidized oil transformed in bio-polyurethane material, used for coating panels and walls. The lightness of the materials, only 1.2 kg/m2, facilitates their transport, management, and installation. The digital printing decorations are incorporated within the resin itself. The biomaterial can become even smart when the low energy sensors are inserted inside them, which allow navigation and the use of contents, making the wall itself a wifi hotspot. In addition to allowing a dialogue with smartphones, these installations can contain sensors of all types (e.g. to detect air quality or acquire images), and with selected APPs, several options to work with virtual reality or with augmented reality are possible.

(b) Winery-based By-products and Grapes Residues:

- Italian wineries (30 wine cooperatives cover 35,000 hectares), and distillery company have transformed themselves into a biorefinery able to use by-products of their processes for BBPs and chemical building blocks production. The biorefinery uses grape wastes from internal processing (around 540,000 tons/year), with a fully recycling of their organic wastes, (only 0.1% of the discarded materials is considered real waste). The byproducts of the agro-industrial process are valorized for polyphenols, enocyanin, and tartaric acid production, destined for pharmaceutical, nutraceutical, and cosmetic sectors, while lignocellulosic wastes are valorized by using bioenergy pathways. Biogas plant (1 MW) is fed by cellulosic wastes for biomethane (advanced Biofuel), internal energy, and digestate production. This last one is transformed into compost and delivered as biofertilizer to the wineries associated with biorefinery. While lignocellulosic residues are valorized into the biomass plant as solid biofuels for heat production.

(c) Apple Peel-based By-products:

- An Italian company patented a sustainable apple-peel-based biomaterial, with reducing 50% of solvents and synthetic polyurethane. Patented apple-based material born from the exploitation of apple peel, an abundant by-product deriving from apple juice industry. More than 3 million tons/year are produced close to the Alpine Regions causing problems for correct disposal of fruit residues deriving from agro-industrial transformation. The company is able to collect apple peel residuals from local farmers associated with the apple juice industry, and valorize into innovative engineering bio-materials applied to the textile sector.

(d) Rural and Agro-industrial Residues for Bioplastics:

- A French Company recently started to working with fibers and by-products, and bio-composites with the creation of a real "green business model". They aim to valorize cereal, seashells, fruit kernels, vegetable fibers, and textile waste to produce original and new compounds that may later be used by other companies to produce original plastic products with a unique design. The company started new production lines working local bio-residues. The aim was to create high-performance and bio-based compounds sold to the packaging industry to develop products for several sectors. The innovation lies in the mixed uses of different biomass by-products as fillers in polymers in order to produce Bio-packaging for several sectors. Biomass from cereal waste (wheat, barley, and corn), Biomass from aquatic co-products (scallops, oysters, algae), Biomass from agro-industry: nutshells kernels (hazelnut, almond, rice, olives), processing of fruits (coffee ground, cacao shells, apple pomace, grape seed) Biomass from vegetal crop-fibers (miscanthus, flax, hemp, wood, cork) results as a suitable input for primary injection molding processes in the production of rigid secondary packaging, regular consumer goods, technical parts, agriculture and horticulture products, cosmetic. The feedstock needs to be ground, dried, and sieved before being used by the plastic industry. The company delivered added value to biomass residues delivered by local farmers, giving benefits to their revenues.

(e) Phyto-stimulants by Animal Feathers:

- A cluster of zootechnical companies and the Faculty of Food and Biochemical Technology of Prague (Republic of Czech) have implemented a promising model for the exploitation of selected animal residues. Feathers account for 5–7% of the total weight of the chickens. Currently, this waste, mixed with other slaughter waste, is usually converted to biogas by anaerobic digestion. Feathers consist of 91% protein (keratin), 1% lipids, and 8% water, and can be better valorized to produce added-valuable products. The business model foresees to process by hydrolysis in the pressure tank. The result is protein hydro-isolated solutions containing amino acids and partially soluble proteins. The resulting hydrolysis products have significant effects on plant growth. When used as toppings, biomass growth, increased nutrient uptake, photosynthesis, increased stress resistance of plants and other positive factors have been demonstrated. The hydrolysis based-technology is currently being tested in a pilot plant. Significant factors for widespread use include the form of waste materials (ratio of quill vs. plumage in the feathers used). One of the targets of the project is also to extract and isolating aspartic acid from hydrolysate. The business model could be also extended to the treatment of other types of animal waste (i.e., animal hair), or to a possible mixture of animal and vegetable waste. Furthermore, an important factor is the ability to offer primary livestock producers a service that will help them to better locate and recover hard-to-treat waste materials for innovative uses.

(f) Potato and Corn Residues for Starch and Cellulose-based Diapers:

- A Germany small enterprise produces sustainable diapers, which consist almost completely of renewable materials and are 100% compostable. The company has developed an innovative method able to replace the SAP – Super Absorbent Polymer with a biodegradable biomaterial by using potato and corn residues. The supply of biomass residues is agreed from local farmers and suppliers. The whole concept of the company can be labelled as "environmentally friendly" and responds to the concept of circular economy. Despite the approach fully responds to the circular economy model, the solution has costs twice the price of normal diapers. Yet, the German company is growing fast, and their demand for bio-based products is expected to increase in the coming years. The idea is capturing the public interest and made its appearance on the media several times while being nominated for the GreenTec-Awards 2017.

(g) Corn Starch and Seed Hulls for Coffe Caps:

- The innovative Coffe Cap product consists of the coffe capsules and an environmentally friendly cellulose-based lid that needs no additional glue for sealing the capsule. The coffee capsule consists of the innovative material made of sunflower seed hulls, mineral filler, and by patented Biopolymer. The hulls of sunflower seeds are a waste product from the extraction of sunflower seeds, while innovative biopolymer is a starch-based and biodegradable plastic. Extra barrier packaging is unnecessary due to excellent oxygen transmission rates of the overall material composition, thus saving further waste. The whole product completely degrades on the home compost within 12 months. Given the ready availability of corn residues in terms of production and processing in Europe, its sugar molecules are used as raw material. The valuable proteins of the plant are not used and can be further processed, for example as animal feed. Hence, there is no conflict with the food industry and a double income for corn producers is generated. Other farmers can specialise in or complement their production (processes) with sunflower seeds and providing opportunities for business diversification.

(h) Biorefinery Model for Zootechnical Sector:

- In specialized rural areas with a higher density of animal farms, valuing animal manure as well as reducing the animal wastes are a major issue, since intensive livestock are causing a significant phosphate surplus in rural areas. A Dutch Company (part of International Group) is creating a sustainable model in the context of the circular economy where food, feed, bio-based products, and fuels are ingredients and solutions of the same model by processing all animal by-products into different valuable products. The International Group and companies affiliated have a strong believe that both edible and in-edible residuals can and should be transformed into valuable products in order to realize economic and ecological sustainability. Each company of the Internatioanl group focuses on processing one or more different residual waste flows and are mostly complementary on a regional scale: biodiesel process from fat waste. Biogas, biomethane, and phosphate-rich biofertilizers by anaerobic digestion process, bio-based products by animal by-products (i.e. keratin, added value proteins). These three processes take place at the same biorefinery place in Netherland. The process is an example of closing the loop on economic and ecological sustainability for urgent challenges in the agricultural and zootechnical sectors. Through chain collaboration and high dedicated technology, this business model enables to create of a stable economic opportunity out of animal and food waste streams like pig manure, animal by-products, and food residues

(i) Potato Residues for Biopolymers and Biomaterials:

- A Dutch Company expert in Biopolymers, started to valorize agricultural residue streams right after World War II, by processing waste streams from the potato industry into cattle feed products. In the late 90s, the business in cattle feed started to decline and from that moment the company started to look for alternative businesses. The company could rely on a strong and established network, availability of biomass/residual streams, and therefore took the opportunity to experiment with bioplastics based on potato waste streams. It kept on investing in innovative production processes and extensive R&D and is now one of the experts in bio-based (biodegradable) compound development and production. Intensive R&D work has been realized in order to process potato wastes streams into valuable granulate. Besides the processing technology, also knowledge and R&D work is done to create a niche market in bioplastic products (same strength, lifespan, biodegradable characteristics). Furthermore, by-products are valorized in other markets using the extensive networks and expertise of the company. For the bio-based company, an increasing profit and turnover is realized since more and more consumers and plastic product companies are willing to use a nonfossil-based granulate. A few years ago, it was responsible for 2 tonnes of bio-based plastic products per month. In 2019 this has been increased to 30 tonnes of bio-based plastic products per month. With increasing scale, the costs of producing the bio-based granulate lower, but it has still a higher cost price than fossil-fuel-based granulate. The business model of company processes based on potato residues of agro-industry into several semi-finished products. The main goal is the production of bio-granulate (bioplastic), processed from potato starch. Bioplastic can be used in industrial processes (extrusion/compounder) to produce all-kind of daily-life plastic products. Based on the characteristics of the end product, the granulate can be made for different product lifespans and/or in a biodegradable variant. The business model sees the collaboration of two main actors: the potato industry, supplying potato residual streams, and Rodenburg, processing waste streams into bio-(degradable) granulate; valorizing by-products into PET-food, wood- and paper binder, lubricate for drilling applications. There is an increasing demand and supply of bioplastic products based on bio-granulate (made out of potato starch). Any plastic product can be produced out of the different bio-granulate semi-finished products. The potato industry, which is the main actor involved in the management of potato processing, has now a financial benefit in valorizing a former waste stream (cost). The biomass comes from Belgium, Netherlands, and Germany, which supply potato residue streams from potato industry. A major strength behind this business model is the fact that Dutch Company has a reliable and well-established network in the agricultural waste stream industry and can therefore benefit from the continuous supply stream.

Further innovative applications are listed in the Enabling Atlas (Enabling Project 2019), like pineapple fibers used for textile materials, wool for bio-packaging solutions, spent brewery grains and dried distilled grains for bioplastic ingredients, and high-demand biochemicals like L+ D-lactic acid, and ethyl lactate, and many other best practices.

Accomplish this vision requires involving insitutions and entities able to deliver key data, to re-define rules, and local waste licensing regulations for transportation, processing, and disposal of food waste quikly. Furthermore, COVID-19 demonstrated how important is to conside the prevention and risks pertaining to health due to transmission of diseases, and potential contaminations. The implication of food and organic waste recycling, disposal or composting, and further valorization through biorefineries, and the public perceptions, and their impacts on local economies must be checked thoroughly. World Health Organization (WHO) confirms Coronaviruses need a live animal or human host to multiply and survive, therefore COVID-19 cannot multiply on the surface of materials, including food packages or biomaterials. Nevertheless, Kampf (Kampf et al. 2020) after having analyzed more than 22 studies related to the several coronaviruses persistence (both deriving from human that veterinary) on inanimate surfaces like metals, paper, ceramics, glass, including plastics, and others, found that human coronaviruses can remain temporarily infectious on these surfaces until to 9 days, depending on the material type. The same authors point out that the coronaviruses may be inactivated by disinfecting the potentially contaminated surfaces. Data about the lifespan of SARS-CoV-2 in different surfaces have been also summarized by Nghiem (Nghiem et al. 2020), and show that the virus can remain viable for 3 h in aerosols, 4 h in copper, 24 h in cardboard, 2–3 days in stainless steel, 3–4 days in solid faces, and 3 days in plastics and sewage. Despite the several CoronaVirus cannot multiply in different materials, first evidence confirm their survival on organic waste and several materials for a certain timeframe. During the pandemic period, several measures and recommendations for solid food waste handling and management have been developed by international organizations (Santos et al. 2021). The World Health Organization (WHO) has provided guidance on how to safely manage fecal waste and wastewater, and to manage both healthcare and household waste generated by people in quarantine (WHO 2020). Likewise, other organizations have conducted debates on virtual platforms (ISWA 2020a, 2020b), and developed guidelines aiming at raising awareness and encouraging local actions related to safe solid waste management (SWM) to protect the environment and public health, including the solid waste workers (CDC 2020; European Commission 2020; SWANA 2020). Such initiatives are of utmost importance, since waste traceability and management infected by SARS CoV-2, along with water and sewage treatment, are still unsolved issues, especially in developing countries where most of the organic food wastes are manually collected and recycled. In a future outlook of the circular economy, the current circumstances force certainly, to reassess habits and approaches to manage organic and food waste in an approach called "reboot" able to ensure that all procedures used for food waste valorizing to new bio-based products follow regulations to "contamination-proof", with particular attention to the biomass residues originated from the urban and zootechnical sector.

13.4 BioBased Product Recycling

Circular Economy represents an economic system based on the recycling products and reuse of raw materials while maintaining the restorative capacity of the natural resources (OECD 2018). The circular Economy replaces 'the end of life concept' with "restoration" by minimization of material use, shifting towards the use of renewable energy, eliminate the use of lesser toxic chemicals and supports zero waste discharge through enhanced modifications in the design of products, systems, materials and business models. The economy's linear model of "take, make and dispose of" is being modified as the "take, make, reuse, recycle and remanufacture" which facilitates a significant reduction of the waste generated (OECD 2018; Carus 2017).

Valorizing the wastes as new resources should be considered by the industries at the design level itself and the most suitable "end of life" option for their product(s) should be implemented also including the toxicity approach. But it's not sufficient. Which will end of life of BBPs be once created? Which will degradation time BBPs be? How will BBPs be recycled? Can new BBPs be re-used? Can BBPs change intended use after the End of their Life Cycle?

Part of the solutions to those questions is possible to find if the circular economy (CE) would be also applied to the BBPs once their uses will be terminated. The concept of CE basically works on the two-cycle principle, represented in the famous "butterfly" (Venkata Mohan et al. 2019). According to this model, the biological resources cycle separately from the abiotic materials, taking into account the regenerative capacity of the biological resources. The focus of the biological cycle is to cascade the extraction of bio-materials/chemicals and returning nutrients to the biosphere wherever possible for example by composting, anaerobic digestion, etc. It means that part of biomaterials is created in order to close the loop themselves, (i.e. Biofertilizers from food wastes) (Rorat and Vandenbulcke 2019).

For several BBP's life cycles, closing their natural cycle analyses could be conflict or in most cases not complete (Belboom and Léonard 2016), despite the perceptions regarding BBPs as bioplastic uses are positive. This is the case of polymers, which represent one of the most used materials with a production of 107 Mty^{-1} globally, and principally derived by fossil fuels (Cherubini and Strømman 2011). Ethylene is the main polymer precursor of several chemicals, e.g. vinyl chloride, ethylbenzene, ethylene oxide, or ethanol, among them several types of its polymer form. The production of polymers consumes fossil fuels but also induces other environmental impacts as their accumulation in the environment, in landfills, or even in the ocean, if they are non-recycled due to their chemical, physical and biological inertness. In this context new Biopolymers replacing those traditional, have to also be thought for reducing, reusing, and recycling the biogenic material. The replacement of traditional polymers with biodegradable biopolymers could be to solve part of the solution, despite biodegradability is not the only criteria to take into account in the environmental aspects of polymers. All the value chains should be analysed (LCA – Life Cycle Assessment) from the biomass supply until the end-of-life, (Ojeda 2013), including the potential re-using and re-cycling and biodegradation.

As an example, the use of sugarcane feedstock for Bioethanol production is used, in turn, as the chemical building block for biobased polymer production in Brazil, (Caldeoron and Arantes 2019; IRENA 2013). The environmental impact of biobased ethylene produced from sugar cane has already been assessed in the literature using LCA and shows a reduction of GHG emissions and fossil fuel consumption when replacing fossil fuel with biobased ethanol, (Liptow and Tillman 2012; Alvarenga et al. 2013a, b; Van Uyvanck et al. 2014).

Other investigations demonstrate LCA's improvement from traditional polymers and respective bio-polymers. Van der Harst and Potting studied the results of ten LCA based on disposal cups produced from fossilbased plastics to Biobased plastics, (Van der Harst and Potting 2013) and showed conflict in results in terms of Global Warming Potential (GWP). The comparison in terms of energy and GHG emissions and energy consumption has been assessed for fossilbased PET and the biobased PEF (polyethylene furandicarboxylate). Results desmontrated environmental benefit of the PEF based on cornstarch, (Eerhart 2012). Furthermore, the sustainability (in terms of GWP and energy consumption) of biopolymers (polylactic acid, poly hydroxyl-alkanoate, and thermoplastic starch) has been further compared to other fossil polymers, and confirming the influence of the end-of-life in the global assessment, (Hottle et al. 2013). In some cases, BBP's LCA could result not optimal due to not eco-friendly technological processes and/or biomass used. The BBPs deriving by dedicated crops can mitigate the environmental impacts o due to direct and/or indirect land-use change (ILUC) as mentioned by Liptow and Tillman (2012).

The main important step for bioproducts is the supply of the selected biomass feedstock, like dedicated cultivations or food wastes or biomass residuals. This initial step is very sensitive Grabowski et al. (2015) explain the importance of specific data relative to crop production to ensure the quality of the environmental impacts of biobased polymers. In fact Belboom and Léonard (2016), shows a reduction of impact of around 60% for both climate change and fossil fuel depletion categories when using biobased HDPE – High Density PolyEthylene (by biobased ethanol deriving from sugar beet and wheat residues) instead of its fossil counterpart HDPE by fossil fuels. But for all other impact categories, fossil HDPE achieves better results than the biobased products.

Further considerations have to be treated for specific BBPs like bioplastics. Approximately 99% of plastics are produced through petrochemicals (CIEL 2017). China's announcement that it would no longer accept international plastic waste for recycling from December 31, 2017, has exacerbated this problem and increased the need for sustainable bioplastic-based solutions by 2030, as 111 Mt of plastic waste will be displaced through that policy changes (Brooks et al. 2018). Therefore need to replace plastic material with respective bioplastic is becoming a crucial point of most international policies.

Currently, there are a wealth of natural biobased polymers as well as monomeric feedstocks for bioplastic production. Karan et al. (2019) summarizes the main classes of currently developed biobased plastics. These include plastics based on starch, polyhydroxyalkanoates (PHAs), polylactic acid (PLA), cellulose, renewable polyethylene, and polyvinyl chloride (PVC), as well as protein-based polymers. Each Bioplastic has got a specific degradability and strength characteristics, therefore specific uses. Because of high amount of blends in which the monomers can be mixed or cross-linked, their properties can be modified through chemical derivatization, as well as by introduction of additives as plasticizers, stabilizers, fillers, processing aids, and colorants. By their combinations can be produced several varieties of plastics with different physical characteristics (e.g., melting point, density, shelf life, biodegradability, UV resistance, transparency, thermoplastic versus thermosetting materials) (Fig. 13.2).

There are several options for bioplastics production. The attractive option is represented by microalgaland cyanobacterial plants for bioplastic production, as these offer a series of advantages that contribute to sustainability goals (i.e., Good Health, Renewable Energy, Economic Growth, Industry, Innovation and Green Infrastructure, Sustainable Cities and Communities, Responsible Consumption and Production).

Biobased plastics are becoming part of an expanding circular bioeconomy. Their production includes both non-degradable and biodegradable plastics. Both are important for sustainable solutions on the basis of different needs (Karan et al. 2019).

Non-biodegradable Bioplastic can also be considered a sustainable solution if it has seen as Carbon sinks stock, and to contribute to carbon capture and storage

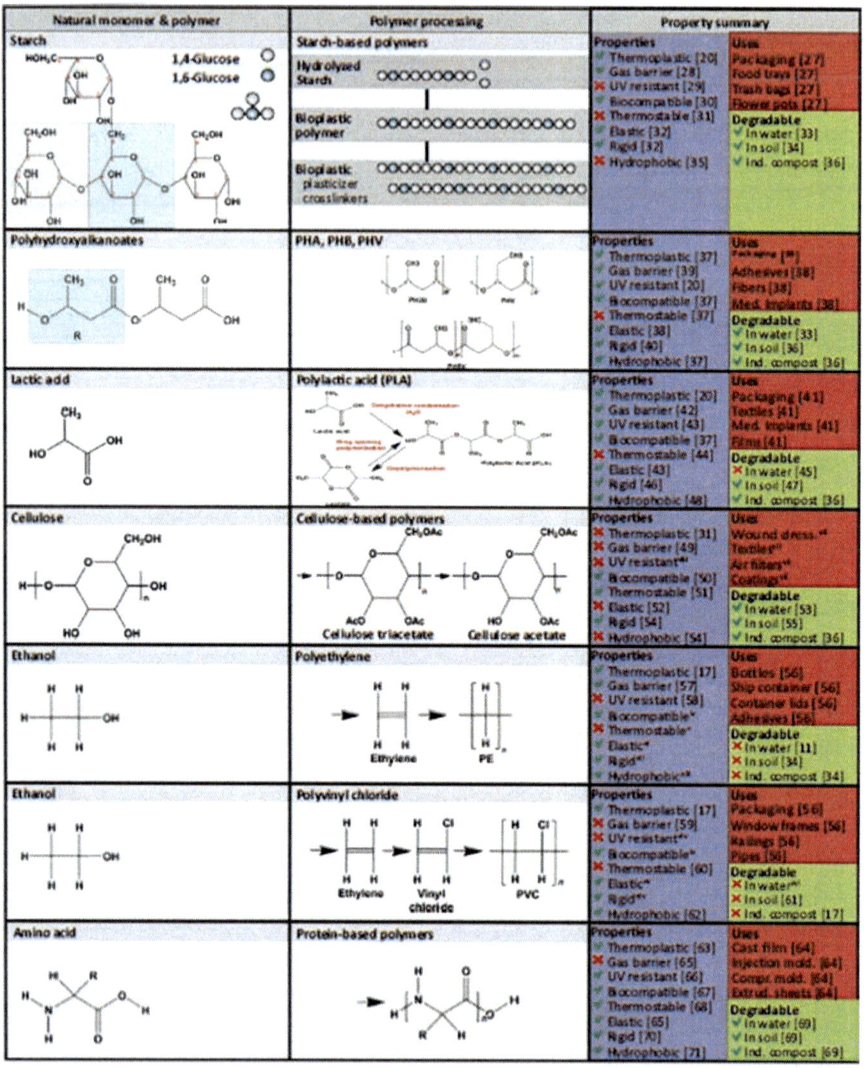

Fig. 13.2 Bioplastic categories, their properties, and uses (Karan et al. 2019). (Reprinted from Trends in Planc Science, Vol. 34, Karan H., Funk C., Grabert M., Oey M., Hankamer B, "Green Bioplastics as Part of a Circular Bioeconomy", Copyright 2019, with permission from Elsevier)

through integration into nondegradable long-term infrastructures including road surfaces, PVC Tube sewer piping, and building materials. Therefore, a replacement of petrochemical-based feedstock with available bio-derived material like biobased polyethylene (bio-PE) could enable this transition. This process should be supported by a legislated accreditation making such infrastructure eligible for carbon credits (Carus and Dammer 2013).

Degradable bioplastics can be instead designed to be either totally biodegradable in a matter of months or years (Rutkowska et al. 2002). They can be used to produce

CH_4 (by using AD – Anaerobic Digestion process) (Bátori et al. 2018), short-to-medium shelflife products that degrade fully to minimize their environmental impact. The timescale over which plastics deteriorate should theoretically be tailored to the product's design. In that context, technical standards are critical for leading environmental-friendly bioplastic design process, and ensurig a sustainable Industrial development for emerging bioplastics.

From the legislative point of view, there are already developed standards regulations for tightly controlled industrial composting systems of Bioplastics. Key Technical Standards are actually in force nowadays: ISO 17088 at the international level (ISO 17088 2012), EN13432 (UNI EN 13432 2002) and EN 14995 (UNI EN 14995 2007) at EU level, and ASTM 6400 (ASTM 6400 2019), ASTM 6868 (ASTM 6868 2019) in the USA. Despite that, further efforts should be carried out in order to establish the degradation time of bioplastics in home composting as well as in terrestrial and aquatic ecosystems without releasing toxic byproducts.

An example of replacement with degradable bio-plastic is represented by food packaging industry PET-based. Petrochemically food packaging PET- based, such as soft-drink bottles, food containers, has been shown to persist in the environment for over 90 years (Edge et al. 1991).

Furthermore, its specific recycling rate is still only around 20% (Geyer et al. 2017) and is not sufficient to avoid environmental impacts like the great seas garbage patch or plastic soup (Lebreton et al. 2018; Van Sebille 2015). Furthermore, degradation is often only partial and in many cases leads to harmful products such as microplastics and toxic constituents (Wright and Kelly 2017). Theoretically, biobased plastic formulations can be considered in their LCA to deliver products with a fit-for-purpose shelf life. For example, plastic water bottles could be designed to degrade under precise conditions with a shelf-life within 2–5 years. Considering that over half of the global biodegradable plastic demand is for packaging materials (Scarfato et al. 2015), such packaging could yield significant benefits in the future.

The degradation bioplastics process seems, therefore, be one of the most important key points for the future that will have to be taken into account into the BBP's LCA approach.

The most detailed guidelines on bioplastic degradation have been developed for industrial composting systems and these define the time required for degradation, the percentage of CO_2 emitted from the bioplastic, and any toxic residues remaining.

In addition to biodegradability conditions, for a competitive bioplastics market, advancements in biotechnology and processing techniques are also paramount to improve performance and reduce their cost, still considered too higher if compared with traditional plastics.

By remaining in a biorefinery approach, Kwan et al. performed a techno-economic study on a biorefinery design for food waste valorization through fungal hydrolysis and microalgae cultivation, which ultimately leads to the production of plasticizer, lactic acid, and animal feed to lower costs. Economic feasibility was only achieved when the production was focused on plasticizer and lactic acid, which are high-value products (Kwan et al. 2015). The production of multiple added-value products (i.e., several chemical building blocks, biomaterials, precursors of bioplastic) from the same biomass resources is a key point to these strategies, as it helps to offset the relatively high costs (CAPEX and OPEX) very often due to their fractioning and purification. Detailed techno-economic and LCA modeling tools are being developed to fast-track biorefinery systems optimization, the development of robust business models, and to derisk scale up. Through such model, it is possible to identify the most valuable and promising production streams, capital and operational costs associated with these, and to plan biorefinery processes able to deliver good economic (e.g., profitability), social (e.g., energy efficiency), and environmental (e.g., greenhouse gas emissions) performances.

13.5 Conclusion

Driving the sustainable industrial development in a new concept of a biorefinery based on organic waste valorization, it will be essential not only to regulate the ability of biobased products growth but improving their full degradation without the release of harmful residual chemicals. The production of BBPs, as well as their degradation or recycling, represents a long-term challenge, which will likely be addressed as technology progresses develop down the cost curve (Karan et al. 2019). For biobased products to compete with the well-established petrochemical products, maintaining consistency while drafting policies is a major gap that needs to be addressed by the policymakers.

The inclusion of biological components in the Circular Economy needs several policies and new legislation that should be formulated in a short time, and their goals should be addressed until to reach more sustainable regional developments (OECD 2018).

The biorefineries can create value from waste but they require high aknowledgment of the process with "re-thinking waste treatment" and a concrete valorization to resources.

Biorefineries can change the perspective of waste and can deviate it from the standard waste management practices, leading to disruption in its management hierarchy. Introducing the biorefinery also necessitates a "re-defining of raw materials" and a modified waste management policies and regulations.

For a national government or regional authorities to consider waste biorefining and to plan adequate investments, there must be a sufficient knowledge of issues that needs to be considered, as the origin of waste and the quantities of waste generated, their recycling and reuse, therefore to the waste management and valorization techniques, as well as their logistic, transport, and storage. In addition, the type of biorefineries to be constructed, their location, type of chemical products to be extracted, and type of pre-processing of feedstock, should also be considered.

Circularity or cascading processes usually enhance the effective use of resources including the valorizing of food wastes. However, accounting for associated emissions, as well as resources and energy consumed, and of the economical sustainability as a whole, is usually more complex.

Furthermore, renewable energy integration to the production of BBPs will be a further challenge to be reached in a short time. Conduct a comprehensive sustainability assessment (LCA) of cascade bioproducts chains should certainly represent a key requisite. Besides along with the BBP's cascade processes, there is a possibility of toxic chemical accumulation, which can hinder recycling or energy exploitation. Therefore, it is mandatory to have a clearly articulated approach where the research needs more time in order to highlight the main weakenesses. The primary target must not just be focused on the maximization of circularity or cascading but also optimise overall outcomes and rewire the economy for equity and ecological sustainability.

The major challenge, especially in developing countries, will be to trust and creating consolidated networking among Governments, Industries, R&D sector, SMEs and citizenship, by sharing the Best Practices about biomass residues exploitation, and food wastes, in the new added-value products.

All these actions have to be planned before setting up biorefinery and management of wastes plans by policymakers and key stakeholders, with a common vision of the future, especially after pandemic period that upset priorities worldwide.

Accomplish this vision requires involving institutions and entities able to deliver key data, to re-define rules, and local waste licensing regulations for transportation, processing, and disposal of food waste quickly. Furthermore, COVID-19 demonstrated how important is to consider the prevention and risks pertaining to health due to transmission of diseases, and potential contaminations. The implication of food and organic waste recycling, disposal or composting, and further valorization through biorefineries, and the public perceptions, and their impacts on local economies must be checked thoroughly. World Health Organization (WHO) confirms Coronaviruses need a live animal or human host to multiply and survive, therefore COVID-19 cannot multiply on the surface of materials, including food packages or biomaterials. Nevertheless, Kampf (Kampf et al. 2020) after having analyzed more than 22 studies related to the several coronaviruses persistence (both deriving from human that veterinary) on inanimate surfaces like metals, paper, ceramics, glass, including plastics, and others, found that human coronaviruses can remain temporarily infectious on these surfaces until to 9 days, depending on the material type. The same authors point out that the coronaviruses may be inactivated by disinfecting the potentially contaminated surfaces. Data about the lifespan of SARS-CoV-2 in different surfaces have been also summarized by Nghiem, (Nghiem et al. 2020), and show that the virus can remain viable for 3 h in aerosols, 4 h in copper, 24 h in cardboard, 2–3 days in stainless steel, 3–4 days in solid faces, and 3 days in plastics and sewage. Despite the several CoronaVirus cannot multiply in different materials, first evidence confirm their survival on organic waste and several materials for a certain timeframe. During the pandemic period, several measures and recommendations for solid food waste handling and management have been developed by international organizations (Santos et al. 2021). The World Health Organization (WHO) has provided guidance on how to safely manage fecal waste and wastewater, and to manage both healthcare and household waste generated by people in quarantine (WHO 2020). Likewise, other organizations have conducted debates on virtual platforms (ISWA 2020a, b), and developed guidelines aiming at raising awareness and encouraging local actions related to safe solid waste management (SWM) to protect the environment and public health, including the solid waste workers (CDC 2020; European Commission 2020; SWANA 2020). Such initiatives are of utmost importance, since waste traceability and management infected by SARS CoV-2, along with water and sewage treatment, are still unsolved issues, especially in developing countries where most of the organic food wastes are manually collected and recycled. In a future outlook of the circular economy, the current circumstances force certainly, to reassess habits and approaches to manage organic and food waste in an approach called "reboot" able to ensure that all procedures used for food waste valorizing to new biobased products follow regulations to "contamination-proof", with particular attention to the biomass residues originated from the urban and zootechnical sector.

More efforts are required for a real implementation of circular economy and radical changes are necessary to rethink the links between use of resources sustainably and economic prosperity, bolstered by huge investments in financial, technical, and social innovations to achieve the real circularity. Therefore, all solutions require contributions from all stakeholders (producers, transformers, and consumers) and emphasis should be on bringing out participatory initiatives among citizens globally (Venkata Mohan et al. 2019; Dilkes-Hoffman et al. 2019). Tackling global gaps through "holistic vision", covering the entire value chain, can spur growth, create jobs and innovation, making a transition towards reduced GHG's while giving people a cleaner and safer environment. Thus, implementing the strategies and plans with concrete actions, sustainable utilization of waste, developing proper infrastructure can make, in practice, the biorefineries an essential tool to achieve the vision of circular bioeconomy.

Bibliography

Adiletta G, Brachi P, Riianova E, Crescitelli A, Miccio M, Kostryukova N (2019) A simplified biorefinery concept for the valorization of sugar beet pulp: ecofriendly isolation of pectin as a step preceding torrefaction. Waste Biomass Valorizion 2019:11

Alvarenga R, Dewulf J, De Meester S, Wathelet A, Villers J, Thommeret R, Hruska Z (2013a) Life cycle assessment of bioethanol-based PVC. Part 2: consequential approach. BioFpR J 7(4). https://doi.org/10.1002/bbb.1398

Alvarenga R, Dewulf J, De Meester S, Wathelet A, Villers J, Thommeret R, Hruska Z (2013b) Life cycle assessment of bioethanol-based PVC. Part 1: attributional approach. BioFpR J 7(4). https://doi.org/10.1002/bbb.1405

ASTM 6868 (2019) Standard specification for labeling of end items that incorporate plastics and polymers as coatings or additives with paper and other substrates designed to be aerobically composted in municipal or industrial facilities. https://compass.astm.org/EDIT/html_annot.cgi?D6868+19

Bátori V, Åkesson D, Zamani A, Taherzadeh MJ, Horváth IS (2018) Anaerobic degradation of bioplastics: a review. Waste Manag 80. https://doi.org/10.1016/j.wasman.2018.09.040

Belboom S, Léonard A (2016) Does biobased polymer achieve better environmental impacts than fossil polymer? comparison of fossil HDPE and biobased HDPE produced from sugar beet and wheat. Biomass Bioenergy 85. https://doi.org/10.1016/j.biombioe.2015.12.014

Beltrán-Ramírez F, Orona-Tamayo D, Cornejo-Corona I, Luz Nicacio González-Cervantes J, de Jesús Esparza-Claudio J, Quintana-Rodríguez E (2019) Agro-industrial waste revalorization: the growing biorefinery. In: Biomass for bioenergy – recent trends and future challenges. https://doi.org/10.5772/intechopen.83569

Bioways Project (2020). http://www.bioways.eu/. EU Commission

Brooks AL, Wang S, Jambeck JR (2018) The Chinese import ban and its impact on global plastic waste trade. Sci Adv 4(6). https://doi.org/10.1126/sciadv.aat013

Buxton L (2019) Chemical watch – global risk and regulation news. https://chemicalwatch.com/83599/echa-proposes-74-chemicals-for-member-state-evaluation-in-2020-2022?pa=true

Caldeoron OR, Arantes V (2019) A review of commercial scale hihg-value productsthat can be produced alongside cellulosic ethanol. Biotechnol Biofuels. https://doi.org/10.1186/s13068-019-1529-1

Carus M (2017) Biobased economy and climate change – important links, pitfalls, and opportunities. Ind Biotechnol 13(2). https://doi.org/10.1089/ind.2017.29073.mca

Carus M, Dammer L (2013) Food or non-food: which agricultural feedstocks are best for industrial uses? Ind Biotechnol 9(4)

CDC (2020) Centers for Disease Control and Prevention, 2020 CDC (Centers for Disease Control and Prevention). What Waste Collectors and Recyclers Need to Know about COVID-19. https://www.cdc.gov/coronavirus/2019-ncov/community/organizations/waste-collection-recycling-workers.html (accessed Dec. 2020)

Cherubini F, Strømman AH (2011) Life cycle assessment of bioenergy systems: state of the art and future challenges. Bioresour Technol J 102(2):437. https://doi.org/10.1016/j.biortech.2010.08.010

CIEL (2017) Fossils, plastics, & petrochemical feedstocks. Fueling plastic. Center for International Environment Law, Washington, DC

Clark JH, Deswarte F (2008) The biorefinery concept-an integrated approach. In: Introduction to chemicals from biomass. Wiley, New York

Clark JH, Deswarte F (2015) Introduction to chemicals from biomass. In: Sustainable chemistry & green chemistry, 2nd edn. Wiley, New York. ISBN: 978-1-118-714478. https://doi.org/10.1002/9781118714478

Council G.B (2018) Bioeconomy Policy (Part III). Office of the Bioeconomy Council, Berlin, (Germany). Update report of national strategies around the world. Bioeconomy Council, Berlin

Cristóbal J, Caldeira C, Corrado S, Sala S (2018) Techno-economic and profitability analysis of food waste biorefineries at European level. Bioresour Technol J 259:244

Dahuya S, Kumar AN, Sravan JS, Chatterjee S, Sarkar O, Mohan VS (2018) Food waste biorefinery: sustainable strategy for circular bioeconomy. Bioresour Technol J 248(Pt A):2

Daioglou V, Wicke B, Faaij APC, van Vuuren DP (2014) Competing uses of biomass for energy and chemical: impplications for long term global CO_2 mitigation potential. Global Change Biol 7. https://doi.org/10.1111/gcbb.12228

Dammer L, Carus M, Raschka A, Scholz L (2013) Market developments of and opportunities for biobased products and chemicals, Reference N° 52202. Nova Institute, Agentshap

Davis SC, Kauneckis D, Kruse NA, Miller KE, Zimmer M, Dabelko GD (2016) Closing the loop: integrative systems management of waste in food, energy, and water system. J Environ Stud Sci 6(1):11

de Jong E, Higson A, Walsh P, Wellish M (2013) Bio-based chemicals – value added products from biorefineries. Task 42 biorefinery, IEA Bioenergy, 2013. www.ieabioenergy.com/wp-content/uploads/2013/10/Task-42-Biobased-Chemicals-value-added-products-from-biorefineries.pdf

Dilkes-Hoffman L, Ashworth P, Laycock B, Pratt S, Lant P (2019) Public attitudes towards bioplastics–knowledge, perception and end-of-lifemanagement. J Resources Conservation Recycling 151

Edge M, Hayes M, Mohammadian M, Allen NS, Jewitt TS, Brems K, Jones K (1991) Aspects of poly (ethylene terephthalate) degradation for archival life and environmental degradation. J Polymer Degradation Stability 32(2):131

Eerhart AJJ (2012) Replacing fossil based PET with biobased PEF, process analysis, energy and GHG balance. J Energy Environment Sci 4:6407

ENABLING Project (2019) HORIZON 2020 Funding Programme, EU Commission. https://www.enabling-project.com/

European Commission (2012a) Bioeconomy strategy "innovation and sustainable growth: a bioeconomy for Europe" and its action plan. http://ec.europa.eu/information_society/newsroom/image/document/2018/review_of_2012_eu_bes_2E89B85F-950B-9C84

European Commission (2012b) Innovating for sustainable growth: a bioeconomy for Europe Brussels. COM, Belgium

European Commission (2020) Waste management in the context of coronavirus crisis. https://ec.europa.eu/info/sites/info/files/waste_management_guidance_dg-env.pdf (accessed Dec. 2020)

FAO (2019) http://www.fao.org/energy/bioeconomy/en/

Fennelly T. & Associates (2015) Advancing green chemistry: barriers to adoption & ways to accelerate green chemistry in supply chains. http://greenchemistryandcommerce.org/assets/media/images/Publications/Advancing-Green-Chemistry-Report-15-June-2015.pdf

GC3. (2015) Green Chemistry and Commerce Council, agenda to mainstream green chemistry. http://www.greenchemistryandcommerce.org/documents/An_Agenda_to_Mainstream_Green_Chemistry.pdf

Gerssen-Gondelach SJ, Saygin D, Wicke B, Patel MK, Faaij APC (2014) Competing uses of biomass: assessment and comparison of the performance of bio-based heat, power, fuels and materials. Renew Sust Energ Rev 40:964

Geyer R, Jambeck JR, Lavender LK (2017) Production, use, and fate of all plastics ever made. J Sci Adv 3(7):e1700782. https://doi.org/10.1126/sciadv.1700782

González-García S, Gullón EP, Gullón EB (2019) Bio-compounds production from agri-food wastes under a biorefinery approach: exploring environmental and social sustainability. In: Environmental footprints and eco-design of products and processes. https://doi.org/10.1007/978-981-13-2408-6_2. ISBN 978-981-13-2408-6, © Springer Nature Singapore Pte. Ltd

Grabowski A, Selke SEM, Auras R, Patel MK, Narayan R (2015) Life cycle inventory data quality issues for bioplastics feedstocks. Int J Life Cycle Assess 20:584

Hassan SS, Williams GA, Jaiswal AK (2019) Moving towards the second generation of lignocellulosic biorefineries in the EU: drivers, challenges, and opportunities. Renew Sustain Energy Rev 101:590. https://doi.org/10.1016/j.rser.2018.11.041

Hottle TA, Bilec MM, Landis AE (2013) Sustainability assessments of bio-based polymers. J Polym Degrad Stabil 98:1898

IRENA (2013) Production of bio-ethylene-tecnology brief. IEA-ETSAP, IRENA, Abu Dhabi

ISO 17088 (2012) ISO 17088:2012. Specifications for compostable plastics. https://www.iso.org/obp/ui/#iso:std:iso:17088:ed-2:v1:en

ISWA (2020a) Webinar ISWA COVID-19 Roundtable Webinar. https://youtu.be/v3OfBWNDnb (accessed Dec. 2020)

ISWA (2020b) Webinar health and safety of solid waste management workers and employers. https://youtu.be/N4n7H_1J-Tc (accessed Dec. 2020)

Kampf G, Todt D, Pfaender S, STainman E (2020) Persistence of coronaviruses on inanimate surfaces and their inactivation with biocidal agents. Hospital Infection J 80. https://doi.org/10.1016/j.jhin.2020.01.022

Karan H, Funk C, Grabert M, Oey M, Hankamer B (2019) Green bioplastics as part of a circular bioeconomy. Trends Planc Sci Rev 24(3):237–249. https://doi.org/10.1016/j.tplants.2018.11.010

Kwan TH, Pleissner D, Lau KY, Venus J, Pommeret A, Sze C, Lin K (2015) Techno-economic analysis of a food waste valorization process via microalgae cultivation and co-production of plasticizer, lactic acid and animal feed from algal biomass and food waste. Bioresour Technol J 198. https://doi.org/10.1016/j.biortech.2015.09.003

Lebreton L, Slat B, Ferrari F, Sainte-Rose B, Aitken J, Marthouse R, Hajbane S, Cunsolo S, Schwarz A, Levivier A, Noble K, Debeljak P, Maral H, Schoeneich-Argent R, Brambini R,

Reisser J (2018) Evidence that the great Pacific garbage patch is rapidly accumulating plastic. Open Access, Sci Rep 8. htpps://doi:10.1038/s41598-018-22939-w
Liptow C, Tillman A-M (2012) A comparative life cycle assessment study of poly-ethylene based on sugarcane and crude oil. J Ind Ecol 16(3):1. https://doi.org/10.1111/j.1530-9290.2011.00405.x
Matzenberger J, Kranzl L, Tromborg E, Junginger M, Daioglou V, Sheng GC, Keramidas K (2015) Future perspectives of international bioenergy trade. Renew Sustain Energy Rev 43:926
Nghiem L.D, Morgan B, Donner E, Short D. M, (2020) The COVID-19 pandemic: Considerations for the waste and wastewater services sector. Case Studies in Chem and Envir Eng 1. https://doi.org/10.1016/j.cscee.2020.100006
OECD (2009) The bioeconomy to 2030: designing a policy agenda. 9789264056886-en. OECD Publishing, Paris
OECD (2018) Realising the circular bioeconomy. OECD science, technology and industry policy papers, no° 60. http://www.wcbef.com/assets/Uploads/Publications/Realising-the-Circular-Bioeconomy.pdf
Ojeda T (2013) Polymers and the environment. Pol Sci. https://doi.org/10.5772/51057
Patel M, Crank M, Dornburg V, Hermann B, Roes L, Huesing B, Overbeek L, Terragni F, Recchia E (2007) Medium and long-term opportunities and risks of the biotechnological production of bulk chemicals from renewable resources. Report n° 20963467, Euroean BREW Project. http://www.chem.uu.nl/brew/. Department of Science, Technology and Society, Copernicus Institute, Utrecht University, Utrecht
Rorat A, Vandenbulcke F (2019) Earthworms converting domestic and food industry wastes into biofertilizer. J Ind Munic Sludge – Emerg Concerns Scope Resour Recov. https://doi.org/10.1016/B978-0-12-815907-1.00005-2
Rutkowska M, Heimowska A, Krasowska K, Janik HZ (2002) Biodegradability of polyethylene starch blends in sea water. Pol J Environ Stud 11(3):1
Santos C, Penteado G, Soares de Castro MA (2021) Covid-19 effects on municipal solid waste management: What can effectively be done in the Brazilian scenario?. Res Recycling 164. https://doi.org/10.1016/j.resconrec.2020.105152
Scarfato P, Di Maio L, Incarnato L (2015) Recent advances and migration issues in biodegradable polymers from renewable sources for food packaging. Appl Polym Sci. https://doi.org/10.1002/app.42597
Scarlat N, Dallemand J-F, Monforti-Ferrario F, Nita V (2015) The role of biomass and bioenergy in a future bioeconomy. Policies and facts. Environ Dev 15:3
Schieb P-A, Lescieux-Katir H, Thénot M, Clément-Larosière B (2015) An original Business model: the integrated biorefinery. In: Biorefinery 2030 – future prospects for the bioeconomy. ISBN 978-3-662-47374-0. Springer, Berlin/Heidelberg. https://doi.org/10.1007/978-3-662-47374-0
Schipfer F, Kranzl L, Leclere D, Leduc S, Forsell N, Valin H (2017) Advanced biomaterials scenarios for the EU28 up to 2050 and their respective biomass demand. Biomass Bioenergy J 96:19
Stephens E, Ross IL, Mussgnug JH, Wagner LD, Borowitzka MA, Posten C, Kruse O, Hankamer B (2010) Future prospects of microalgal biofuel production system. Trends Plant Sci 15(10):554
SWANA (2020) Coronavirus Resources. https://swana.org/resources/corona-virus-resources. (accessed Dec. 2020)
Tahir M, Amin NS (2013) Recycling of carbon dioxide to renewable fuels by photocatalysis: prospects and challenges. Renew Sustain Energy Rev 25:560. https://doi.org/10.1016/j.rser.2013.05.027
Tickner JA, Becker M (2016) Mainstreaming green chemistry: the need for metrics. Curr Opin Green Sustain Chem 1:1. https://doi.org/10.1016/j.cogsc.2016.07.002
UNI EN 13432 (2002) Imballaggi – Requisiti per imballaggi recuperabili mediante compostaggio e biodegradazione – Schema di prova e criteri di valutazione per l'accettazione finale degli imballaggi. http://store.uni.com/catalogo/index.php/uni-en-13432-2002
UNI EN 14995 (2007) Materie plastiche – Valutazione della compostabilità – Schema di prova e specificazioni. http://store.uni.com/catalogo/uni-en-14995-2007

Vaidya N (2019) Toxicity focus is essential for green chemistry adoption and sustainable product development. J Sustain Chem Pharm 11:12

Van der Harst E, Potting J (2013) A critical comparision of ten disposable cup LCAs. Environ Impact Assess Rev 43:86. https://doi.org/10.1016/j.eiar.2013.06.006

Van Sebille E (2015) The oceans' accumulating plastic garbage. Phys Today 68(2). https://doi.org/10.1063/PT.3.2697

Van Uyvanck PP, Hallmark B, Haire G, Marshall PJ, Dennis JD (2014) Impact of the biomass on industry: using ethylene derived from bioethanol within the polyester value chain. CS Sustain Chem Eng 2(5). https://doi.org/10.1021/sc5000804. Copyright © 2014 American Chemical Society

Venkata Mohan S, Dahiya S, Amulya K, Katakojwala R, Vanithaa TK (2019) Can circular bioeconomy be fueled by waste biorefineries — a closer look. Bioresour Technol Rep 7:10027

White House (2012) National bioeconomy blueprint. Ind Biotechnol 8(3):94

WHO (2020) Water, sanitation, hygiene, and waste management for the COVID-19 virus. https://www.who.int/publications/i/item/water-sanitation-hygiene-and-waste-management-for-the-covid-19-virus-interim-guidance (accessed Dec. 2020)

Wright SL, Kelly FJ (2017) Plastic and human health: a micro issue? Environ Sci Technol 51(12). https://doi.org/10.1021/acs.est.7b00423. Copyright © 2017 American Chemical Society

Correction to: Biobased Products from Food Sector Waste: Bioplastics, Biocomposites, and Biocascading

Teresa Cecchi and Carla De Carolis

Correction to:
T. Cecchi, C. De Carolis, *Biobased products from food sector waste*, https://doi.org/10.1007/978-3-030-63436-0

In the original version of this book, the corresponding author names were inadvertently missed out to be included in the chapter opening page of respective chapters in the e-book. The chapter author names have been updated now in this revised version of the e-book.

The updated online versions of this book can be found at
https://doi.org/10.1007/978-3-030-63436-0

T. Cecchi, C. De Carolis, *Biobased products from food sector waste*,
https://doi.org/10.1007/978-3-030-63436-0_14

Index

T. Cecchi, C. De Carolis, *Biobased products from food sector waste*,
https://doi.org/10.1007/978-3-030-63436-0